Short-Term Load Forecasting by Artificial Intelligent Technologies

Short-Term Load Forecasting by Artificial Intelligent Technologies

Special Issue Editors

Wei-Chiang Hong
Ming-Wei Li
Guo-Feng Fan

MDPI • Basel • Beijing • Wuhan • Barcelona • Belgrade

MDPI

Special Issue Editors

Wei-Chiang Hong
Jiangsu Normal University
China

Ming-Wei Li
Harbin Engineering University
China

Guo-Feng Fan
Pingdingshan University
China

Editorial Office
MDPI
St. Alban-Anlage 66
4052 Basel, Switzerland

This is a reprint of articles from the Special Issue published online in the open access journal *Energies* (ISSN 1996-1073) from 2018 to 2019 (available at: https://www.mdpi.com/journal/energies/special_issues/Short_Term_Load_Forecasting)

For citation purposes, cite each article independently as indicated on the article page online and as indicated below:

LastName, A.A.; LastName, B.B.; LastName, C.C. Article Title. *Journal Name* **Year**, *Article Number, Page Range.*

ISBN 978-3-03897-582-3 (Pbk)
ISBN 978-3-03897-583-0 (PDF)

Contents

About the Special Issue Editors

Wei-Chiang Hong's research interests mainly include computational intelligence (neural networks, evolutionary computation) and the application of forecasting technology (ARIMA, Support vector regression, and Chaos theory).

In May 2012, one of his papers was named the "Top Cited Article 2007–2011" of Applied Mathematical Modelling, Elsevier Publisher. In Aug. 2014, he was nominated for the award "Outstanding Professor Award", by Far Eastern Y. Z. Hsu Science and Technology Memorial Foundation (Taiwan). In Nov. 2014, he was nominated for the "Taiwan Inaugural Scopus Young Researcher Award–Computer Science", by Elsevier Publisher, in the Presidents' Forum of Southeast and South Asia and Taiwan Universities. In Jun. 2015, he was named as one of the "Top 10 Best Reviewers" of Applied Energy in 2014. In Aug. 2017, he was named as one of the "Best Reviewers" of Applied Energy in 2016.

Ming-Wei Li received his Ph.D. degree of engineering from Dalian University of Technology, China, in 2013. Since September 2017, he is an associate professor in the College of Shipbuilding Engineering of Harbin Engineering University. His research interests are Intelligent Forecasting Methods, Hybrid Evolutionary Algorithm, Intelligent Ocean and Water Conservancy Engineering, Key Technologies of Marine Renewable Energy.

Guo-Feng Fan received his Ph.D. degree in Engineering from the Research Center of Metallurgical Energy Conservation and Emission Reduction, Ministry of Education, Kunming University of Science and Technology, Kunming, in 2013. His research interests are ferrous metallurgy, energy forecasting, optimization, system identification. In Jan 2018, his paper was a "Top Cited Article" by Nuerocomputing, Elsevier Publisher. In Oct 2018, he won the title of Henan academic and technical leader.

Preface to "Short-Term Load Forecasting by Artificial Intelligent Technologies"

In the last few decades, short-term load forecasting (STLF) has been one of the most important research issues for achieving higher efficiency and reliability in power system operation, to facilitate the minimization of its operation cost by providing accurate input to day-ahead scheduling, contingency analysis, load flow analysis, planning, and maintenance of power system. There are lots of forecasting models proposed for STLF, including traditional statistical models (such as ARIMA, SARIMA, ARMAX, multi-variate regression, Kalman filter, exponential smoothing, and so on) and artificial-intelligence-based models (such as artificial neural networks (ANNs), knowledge-based expert systems, fuzzy theory and fuzzy inference systems, evolutionary computation models, support vector regression, and so on). Recently, due to the great development of evolutionary algorithms (EA), meta-heuristic algorithms (MTA), and novel computing concepts (e.g., quantum computing concepts, chaotic mapping functions, and cloud mapping process, and so on), many advanced hybridizations with those artificial-intelligence-based models are also proposed to achieve satisfactory forecasting accuracy levels. In addition, combining some superior mechanisms with an existing model could empower that model to solve problems it could not deal with before; for example, the seasonal mechanism from ARIMA model is a good component to be combined with any forecasting models to help them to deal with seasonal problems.

This book contains articles from the Special Issue titled "Short-Term Load Forecasting by Artificial Intelligent Technologies", which aims to attract researchers with an interest in the research areas described above. As Fan et al. [1] highlighted, the research trends of forecasting models in the energy sector in recent decades could be divided into three kinds of hybrid or combined models: (1) hybridizing or combining the artificial intelligent approaches with each other; (2) hybridizing or combining with traditional statistical approaches; and (3) hybridizing or combining with the novel evolutionary (or meta-heuristic) algorithms. Thus, the Special Issue, in methodological applications, was also based on these three categories, i.e., hybridizing or combining any advanced/novel techniques in energy forecasting. The hybrid forecasting models should have superior capabilities over the traditional forecasting approaches, and be able to overcome some inherent drawbacks, and, eventually, to achieve significant improvements in forecasting accuracy.

The 22 articles in this compendium all display a broad range of cutting-edge topics of the hybrid advanced technologies in STLF fields. The preface authors believe that the applications of hybrid technologies will play a more important role in STLF accuracy improvements, such as hybrid different evolutionary algorithms/models to overcome some critical shortcoming of a single evolutionary algorithm/model or to directly improve the shortcomings by theoretical innovative arrangements.

Based on these collected articles, an interesting (future research area) issue is how to guide researchers to employ proper hybrid technology for different datasets. This is because for any analysis models (including classification models, forecasting models, and so on), the most important problem is how to catch the data pattern, and to apply the learned patterns or rules to achieve satisfactory performance, i.e., the key success factor is how to successfully look for data patterns. However, each model excels in catching different specific data patterns. For example, exponential smoothing and ARIMA models focus on strict increasing (or decreasing) time series data, i.e., linear pattern, though they have a seasonal modification mechanism to analyze seasonal (cyclic) change; due to artificial learning function to adjust the suitable training rules, the ANN model excels only if the historical data pattern has been learned, there is a lack of systematic explanation on how the accurate forecasting results are obtained; support vector regression (SVR) model could acquire superior performance only with the proper parameters determination search algorithms. Therefore, it is essential to construct an inference system to collect the characteristic rules to determine the data pattern category.

Secondly, it should assign an appropriate approach to implement forecasting for (1) ARIMA or exponential smoothing approaches, the only option is to adjust their differential or seasonal parameters; (2) ANN or SVR models, the forthcoming problem is how to determine the best parameter combination (e.g., numbers of hidden layer, units of each layer, learning rate; or hyper-parameters) to acquire superior forecasting performance. Particularly, for the focus of this discussion, in order to determine the best parameter combination, a series of evolutionary algorithms should be employed to test which data pattern is most familiar. Based on experimental findings, those evolutionary algorithms themselves also have merits and drawbacks, for example, GA and IA are excellent for regular trend data patterns (real number) [2,3], SA excelled for fluctuation or noise data patterns (real number) [4], TA is good for regular cyclic data patterns (real number) [5], and ACO is good for integer number searching [6].

It is possible to build an intelligent support system to improve the efficiency of hybrid evolutionary algorithms/models or to improve them by theoretical innovative arrangements (chaotization and cloud theory) in all forecasting/prediction/classification applications. Firstly, filter the original data by the database with a well-defined characteristic set of rules for the data pattern, such as linear, logarithmic, inverse, quadratic, cubic, compound, power, growth, exponential, etc., to recognize the appropriate data pattern (fluctuation, regular, or noise). The recognition decision rules should include two principles: (1) The change rate of two continuous data; and (2) the decreasing or increasing trend of the change rate, i.e., the behavior of the approached curve. Secondly, select adequate improvement tools (hybrid evolutionary algorithms, hybrid seasonal mechanism, chaotization of decision variables, cloud theory, and any combination of all tolls) to avoid being trapped in a local optimum, improvement tools could be employed into these optimization problems to obtain an improved, satisfied solution.

This discussion of the work by the author of this preface highlights work in an emerging area of hybrid advanced techniques that has come to the forefront over the past decade. These collected articles in this text span a great deal more of cutting edge areas that are truly interdisciplinary in nature.

References

1. Fan, G.F.; Peng, L.L.; Hong, W.C. Short term load forecasting based on phase space reconstruction algorithm and bi-square kernel regression model. Applied Energy 2018, 224, 13–33.
2. Hong, W.C. Application of seasonal SVR with chaotic immune algorithm in traffic flow forecasting. Neural Computing and Applications 2012, 21, 583–593.
3. Hong, W.C.; Dong, Y.; Zhang, W.Y.; Chen, L.Y.; Panigrahi, B.K. Cyclic electric load forecasting by seasonal SVR with chaotic genetic algorithm. International Journal of Electrical Power & Energy Systems 2013, 44, 604–614.
4. Geng, J.; Huang, M.L.; Li, M.W.; Hong, W.C. Hybridization of seasonal chaotic cloud simulated annealing algorithm in a SVR-based load forecasting model. Neurocomputing 2015, 151, 1362–1373.
5. Hong, W.C.; Pai, P.F.; Yang, S.L.; Theng, R. Highway traffic forecasting by support vector regression model with tabu search algorithms. in Proc. the IEEE International Joint Conference on Neural Networks, 2006, pp. 1617–1621.

6. Hong, W.C.; Dong, Y.; Zheng, F.; Lai, C.Y. Forecasting urban traffic flow by SVR with continuous ACO. Applied Mathematical Modelling 2011, 35, 1282–1291.

Wei-Chiang Hong, Ming-Wei Li, Guo-Feng Fan

Special Issue Editors

energies

MDPI

Article

Hybridizing Chaotic and Quantum Mechanisms and Fruit Fly Optimization Algorithm with Least Squares Support Vector Regression Model in Electric Load Forecasting

Ming-Wei Li [1], Jing Geng [1], Wei-Chiang Hong [2,*] and Yang Zhang [1]

[1] College of shipbuilding engineering, Harbin Engineering University, Harbin 150001, Heilongjiang, China; limingwei@hrbeu.edu.cn (M.-W.L.); gengjing@hrbeu.edu.cn (J.G.); zhangyang@hrbeu.edu.cn (Y.Z.)
[2] School of Education Intelligent Technology, Jiangsu Normal University/No. 101, Shanghai Rd., Tongshan District, Xuzhou 221116, Jiangsu, China
* Correspondence: samuelsonhong@gmail.com; Tel.: +86-516-83500307

Received: 13 August 2018; Accepted: 22 August 2018; Published: 24 August 2018

Abstract: Compared with a large power grid, a microgrid electric load (MEL) has the characteristics of strong nonlinearity, multiple factors, and large fluctuation, which lead to it being difficult to receive more accurate forecasting performances. To solve the abovementioned characteristics of a MEL time series, the least squares support vector machine (LS-SVR) hybridizing with meta-heuristic algorithms is applied to simulate the nonlinear system of a MEL time series. As it is known that the fruit fly optimization algorithm (FOA) has several embedded drawbacks that lead to problems, this paper applies a quantum computing mechanism (QCM) to empower each fruit fly to possess quantum behavior during the searching processes, i.e., a QFOA algorithm. Eventually, the cat chaotic mapping function is introduced into the QFOA algorithm, namely CQFOA, to implement the chaotic global perturbation strategy to help fruit flies to escape from the local optima while the population's diversity is poor. Finally, a new MEL forecasting method, namely the LS-SVR-CQFOA model, is established by hybridizing the LS-SVR model with CQFOA. The experimental results illustrate that, in three datasets, the proposed LS-SVR-CQFOA model is superior to other alternative models, including BPNN (back-propagation neural networks), LS-SVR-CQPSO (LS-SVR with chaotic quantum particle swarm optimization algorithm), LS-SVR-CQTS (LS-SVR with chaotic quantum tabu search algorithm), LS-SVR-CQGA (LS-SVR with chaotic quantum genetic algorithm), LS-SVR-CQBA (LS-SVR with chaotic quantum bat algorithm), LS-SVR-FOA, and LS-SVR-QFOA models, in terms of forecasting accuracy indexes. In addition, it passes the significance test at a 97.5% confidence level.

Keywords: least squares support vector regression (LS-SVR); chaos theory; quantum computing mechanism (QCM); fruit fly optimization algorithm (FOA); microgrid electric load forecasting (MEL)

---

1. Introduction

1.1. Motivation

MEL forecasting is the basis of microgrid operation scheduling and energy management. It is an important prerequisite for the intelligent management of distributed energy. The forecasting performance would directly affect the microgrid system's energy trading, power supply planning, and power supply quality. However, the MEL forecasting accuracy is not only influenced by the mathematical model, but also by the associated historical dataset. In addition, compared with the large power grid, microgrid electric load (MEL) has the characteristics of strong nonlinearity, multiple factors, and large fluctuation, which lead to it being difficult to achieve more accurate forecasting performances. Along with the development of artificial intelligent technologies, load forecasting methods have been continuously applied to load forecasting. Furthermore, the hybridization or combination of the intelligent algorithms also provides new models to improve the load forecasting performances. These hybrid or combined models either employ a novel intelligent algorithm or framework to improve the embedded drawbacks or apply the advantages of two of the above models to achieve more satisfactory results. The models apply a wide range of load forecasting approaches and are mainly divided into two categories, traditional forecasting models and intelligent forecasting models.

1.2. Relevant Literature Reviews

Conventional load forecasting models include exponential smoothing models [1], time series models [2], and regression analysis models [3]. An exponential smoothing model is a curve fitting method that defines different coefficients for the historical load data. It can be understood that a series with the forecasted load time has a large influence on the future load, while a series with the long time from the forecasted load has a small influence on the future load [1]. The time series model is applied to load forecasting, which is characterized by a fast forecasting speed and can reflect the continuity of load forecasting, but requires the stability of the time series. The disadvantage is that it cannot reflect the impact of external environmental factors on load forecasting [2]. The regression model seeks a causal relationship between the independent variable and the dependent variables according to the historical load change law, determining the regression equation, and the model parameters. The disadvantage of this model is that there are too many factors affecting the forecasting accuracy. It is not only affected by the parameters of the model itself, but also by the quality of the data. When the external influence factors are too many or the relevant influent factor data are difficult to analyze, the regression forecasting model will result in huge errors [3].

Intelligent forecasting models include the wavelet analysis method [4,5], grey forecasting theory [6,7], the neural network model [8,9], and the support vector regression (SVR) model [10]. In load forecasting, the wavelet analysis method is combined with external factors to establish a suitable load forecasting model by decomposing the load data into sequences on different scales [4,5]. The advantages of the grey model are easy to implement and there are fewer influencing factors employed. However, the disadvantage is that the processed data sequence has more grayscale, which results in large forecasting error [6,7]. Therefore, when this model is applied to load forecasting, only a few recent data points would be accurately forecasted; more distant data could only be reflected as trend values and planned values [7]. Due to the superior nonlinear performances, many models based on artificial neural networks (ANNs) have been applied to improve the load forecasting accuracy [8,9]. To achieve more accurate forecasting performance, these models and other new or novel forecasting approaches have been hybridized or combined [9]. For example, an adaptive network-based fuzzy inference system is combined with an RBF neural network [11], the Monte Carlo algorithm is combined with the Bayesian neural network [12], fuzzy behavior is hybridized with a neural network (WFNN) [13], a knowledge-based feedback tuning fuzzy system is hybridized with a multi-layer perceptron artificial neural network (MLPANN) [14], and so on. However, these ANNs-based models suffer from some serious problems, such as trapping into local optimum easily, it being time-consuming to achieve a functional approximation, and the difficulty of selecting the structural parameters of a network [15,16], which limits its application in load forecasting to a large extent.

The SVR model is based on statistical learning theory, as proposed by Vapnik [17]. It has a solid mathematical foundation, a better generalization ability, a relatively faster convergence rate, and can find global optimal solutions [18]. Because the basic theory of the SVR model is perfect and the model is also easy to establish, it has attracted extensive attention from scholars in the load forecasting fields. In recent years, some scholars have applied the SVR model to the research of load forecasting [18] and achieved superior results. One study [19] proposes the EMD-PSO-GA-SVR model to improve the forecasting accuracy, by hybridizing the empirical mode decomposition (EMD) with two particle swarm optimization (PSO) and the genetic algorithm (GA). In addition, a modified version of the SVR model, namely the LS-SVR model, only considers equality constraints instead of inequalities [20,21]. Focused on the advantages of the LS-SVR model to deal with such problems, this paper tries to simulate the nonlinear system of the MEL time series to receive the forecasting values and improve the forecasting accuracy. However, the disadvantages of the SVR-based models in load forecasting are that when the sample size of the load is large, the time of system learning and training is highly time-consuming, and the determination of parameters mainly depends on the experience of the researchers. This has a certain degree of influence on the accuracy in load forecasting. Therefore, exploring more suitable parameter determination methods has always been an effective way to improve the forecasting accuracy of the SVR-based models. To determine more appropriate parameters of the SVR-based models, Hong and his colleagues have conducted research using different evolutionary algorithms hybridized with an SVR model [22–24]. In the meantime, Hong and his successors have also applied different chaotic mapping functions (including the logistic function [22,23] and the cat mapping function [10]) to diversify the population during modeling processes, and the cloud theory to make sure the temperature continuously decreases during the annealing process, eventually determining the most appropriate parameters to receive more satisfactory forecasting accuracy [10].

The fruit fly optimization algorithm (FOA) is a new swarm intelligent optimization algorithm proposed in 2011, it searches for global optimization based on fruit fly foraging behavior [25,26]. The algorithm has only four control parameters [27]. Compared with other algorithms, FOA has the advantages of being easy to program and having fewer parameters, less computation, and high accuracy [28,29]. FOA belongs to the domain of evolutionary computation; it realizes the optimization of complex problems by simulating fruit flies to search for food sources by using olfaction and vision. It has been successfully applied to the predictive control fields [30,31]. However, similar to those swarm intelligent optimization algorithms with iterative searching mechanisms, the standard FOA also has drawbacks such as a premature convergent tendency, a slow convergent rate in the later searching stage, and poor local search performance [32].

Quantum computing has become one of the leading branches of science in the modern era due to its powerful computing ability. This not only prompted us to study new quantum algorithms, but also inspired us to re-examine some traditional optimization algorithms from the quantum computing mechanism. The quantum computing mechanism (QCM) makes full use of the superposition and coherence of quantum states. Compared with other evolutionary algorithms, the QCM uses a novel encoding method—quantum bit encoding. Through the encoding of qubits, an individual can characterize any linear superposition state, whereas traditional encoding methods can only represent one specific one. As a result, with QCM it is easier to maintain population diversity than with other traditional evolutionary algorithms. Nowadays, it has become a hot topic of research that QCM is able to hybridize with evolutionary algorithms to receive more satisfactory searching results. The literature [33] introduced QCM into genetic algorithms and proposed quantum derived genetic algorithm (QIGA). From the point of view of algorithmic mechanism, it is very similar to the isolated niches genetic algorithm. Han and Kim [34] proposed a genetic quantum algorithm (GQA) based on QCM. Compared with traditional evolutionary algorithms, its greatest advantage is its better ability to maintain population diversity. Han and Kim [35] further introduced the population migration mechanism based on ure [34], and renamed the algorithm a quantum evolution algorithm (QEA). Huang [36], Lee and Lin [37,38], and Li et al. [39] hybridized the particle swam optimization (PSO) algorithm, Tabu search (TS) algorithm, genetic algorithm (GA), and bat algorithm (BA) with the QCM and the cat mapping function, and proposed the CQPSO, CQTS, CQGA, and CQBA algorithms, which were employed to select the appropriate parameters of an SVR model. The results of the application indicate that the improved algorithms obtain more appropriate parameters, and higher forecasting accuracy is achieved. The above applications also reveal that the improved algorithm, by hybridizing with QCM, could effectively avoid local optimal position and premature convergence.

1.3. Contributions

Considering the inherent drawback of the FOA, i.e., suffering from premature convergence, this paper tries to hybridize the FOA with QCM and the cat chaotic mapping function to solve the premature problem of FOA. Eventually, determine more appropriate parameters of an LS-SVR model. The major contributions are as follows:

(1) QCM is employed to empower the search ability of each fruit fly during the searching processes of QFOA. The cat chaotic mapping function is introduced into QFOA and implements the chaotic global perturbation strategy to help a fruit fly escape from the local optima when the population's diversity is poor.

(2) We propose a novel hybrid optimization algorithm, namely CQFOA, to be hybridized with an LS-SVR model, namely the LS-SVR-CQFOA model, to conduct the MEL forecasting. Other similar alternative hybrid algorithms (hybridizing chaotic mapping function, QCM, and evolutionary algorithms) in existing papers, such as the CQPSO algorithm used by Huang [36], the CQTS and CQGA algorithms used by Lee and Lin [37,38], and the CQBA algorithm used by Li et al. [39], are selected as alternative models to test the superiority of the LS-SVR-CQFOA model in terms of forecasting accuracy.

(3) The forecasting results illustrate that, in three datasets, the proposed LS-SVR-CQFOA model is superior to other alternative models in terms of forecasting accuracy indexes; in addition, it passes the significance test at a 97.5% confidence level.

1.4. The Organization of This Paper

The rest of this paper is organized as follows. The modeling details of an LS-SVR model, the proposed CQFOA, and the proposed LS-SVR-CQFOA model are introduced in Section 2. Section 3 presents a numerical example and a comparison of the proposed LS-SVR-CQFOA model with other alternative models. Some insight discussions are provided in Section 4. Finally, the conclusions are given in Section 5.

2. Materials and Methods

2.1. Least Squares Support Vector Regression (LS-SVR)

The SVR model is an algorithm based on pattern recognition of statistical learning theory. It is a novel machine learning approach proposed by Vapnik in the mid-1990s [17]. The LS-SVR model was put forward by Suykens [20]. It is an improvement and an extension of the standard SVR model, which replaces the inequality constraints of an SVR model with equality constraint [21]. The LS-SVR model converts quadratic programming problem into linear programming solving, reduces the computational complexity, and improves the convergent speed. It can solve the load forecasting problems due to its characteristics of nonlinearity, high dimension, and local minima.

2.1.1. Principle of the Standard SVR Model

Set a dataset as $\{(x_i, y_i)\}_{i=1}^{N}$, $x_i \in R^n$ is the input vector of n-dimensional system, $y_i \in R$ is the output (not a single real value, but a n-dimensional vector) of system. The basic idea of the SVR model can be summarized as follows: n-dimensional input samples are mapped from the original space to the high-dimensional feature space F by nonlinear transformation $\varphi(\cdot)$, and the optimal linear regression function is constructed in this space, as shown in Equation (1) [17]:

$$f(x) = w^T \varphi(x) + b, \tag{1}$$

where $f(x)$ represents the forecasting values; the weight, w, and the coefficient, b, would be determined during the SVR modeling processes.

The standard SVR model takes the ε insensitive loss function as an estimation problem for risk minimization, thus the optimization objective can be expressed as in Equation (2) [17]:

$$\min \tfrac{1}{2} w^T w + c \sum_{i=1}^{N} (\xi_i + \xi_i^*)$$

$$s.t. \begin{cases} y_i - w^T \varphi(x_i) - b \le \varepsilon + \xi_i \\ w^T(x_i) + b - y_i \le \varepsilon + \xi_i^* \\ \xi_i \ge 0, \ \xi_i^* \ge 0 \\ i = 1, \cdots, N \end{cases} \tag{2}$$

where c is the balance factor, usually set to 1, and ξ_i and ξ_i^* are the error of introducing the training set, which can represent the extent to which the sample point exceeds the fitting precision ε.

Equation (2) could be solved according to quadratic programming processes; the solution of the weight, w, in Equation (2) is calculated as in Equation (3) [17]:

$$w^* = \sum_{i=1}^{N} (\alpha_i - \alpha_i^*) \varphi(x), \tag{3}$$

where α_i and α_i^* are Lagrange multipliers.

The SVR function is eventually constructed as in Equation (4) [17]:

$$y(x) = \sum_{i=1}^{N} (\alpha_i - \alpha_i^*) \Psi(x_i, x) + b, \tag{4}$$

where $\Psi(x_i, x)$, the so-called kernel function, is introduced to replace the nonlinear mapping function, $\varphi(\cdot)$, as shown in Equation (5) [15]:

$$\Psi(x_i, x_j) = \varphi(x_i)^T \varphi(x_j). \tag{5}$$

2.1.2. Principle of the LS-SVR Model

The LS-SVR model is an extension of the standard SVR model. It selects the binomial of error ξ_t as the loss function; then the optimization problem can be described as in Equation (6) [20]:

$$\min \tfrac{1}{2} w^T w + \tfrac{1}{2} \gamma \sum_{i=1}^{N} \xi_i^2$$

$$s.t. \ y_i = w^T \varphi(x_i) + b + \xi_i, \ i = 1, 2, \cdots, N \tag{6}$$

where the bigger the positive real number γ is, the smaller the regression error of the model is.

The LS-SVR model defines the loss function different from the standard SVR model, and changes its inequality constraint into an equality constraint so that w can be obtained in the dual space. After obtaining parameters α and b by quadratic programming processes, the LS-SVR model is described as in Equation (7) [20]:

$$y(x) = \sum_{i=1}^{N} \alpha_i \Psi(x_i, x) + b. \tag{7}$$

It can be seen that an LS-SVR model contains two parameters, the regularization parameter γ and the radial basis kernel function, σ^2. The forecasting performance of an LS-SVR model is related to the selection of γ and σ^2. The role of γ is to balance the confidence range and experience risk of learning machines. If γ is too large, the goal is only to minimize the experience risk. On the contrary, when the value of γ is too small, the penalty for the experience error will be small, thus increasing the value of experience risk σ controls the width of the Gaussian kernel function and the distribution range of the training data. The smaller σ is, the greater the structural risk there is, which leads to overfitting. Therefore, the parameter selection of an LS-SVR model has always been the key to improve the forecasting accuracy.

2.2. Chaotic Quantum Fruit Fly Algorithm (CQFOA)

FOA is a population intelligent evolutionary algorithm that simulates the foraging behavior of fruit flies [26]. Fruit flies are superior to other species in smell and vision. In the process of foraging, firstly, fruit flies rely on smell to find the food source. Secondly, they visually locate the specific location of food and the current position of other fruit flies, and then fly to the location of food through population interaction. At present, FOA has been applied to the forecasting of traffic accidents, export trade, and other fields [40].

2.2.1. Fruit Fly Optimization Algorithm (FOA)

According to the characteristics of fruit flies searching for food, FOA includes the following main steps.

Step 1. Initialize randomly the fruit flies' location (X_0 and Y_0) of population.

Step 2. Give individual fruit flies the random direction and distance for searching for food by smell, as in Equations (8) and (9) [26]:

$$X_i = X_0 + \text{Random Value} \tag{8}$$

$$Y_i = Y_0 + \text{Random Value.} \tag{9}$$

Step 3. Due to the location of food being unknown, firstly, the distance from the origin (*Dist*) is estimated as in Equation (10) [25], then the determination value of taste concentration (*S*) is calculated as in Equation (11) [25], i.e., the value is the inverse of the distance.

$$Dist_i = \sqrt{X_i^2 + Y_i^2} \tag{10}$$

$$S_i = 1/Dist_i \tag{11}$$

Step 4. The determination value of taste concentration (*S*) is substituted into the determination function of taste concentration (or Fitness function) to determine the individual position of the fruit fly (*Smell_i*), as shown in Equation (12) [26]:

$$Smell_i = \text{Function}(S_i). \tag{12}$$

Step 5. Find the *Drosophila* species (*Best index* and *Best Smell* values) with the highest odor concentrations in this population, as in Equation (13) [26]:

$$\max(Smell_i) \rightarrow (Best\_Smell_i) \text{ and } (Best\_index). \tag{13}$$

Step 6. The optimal flavor concentration value (*Optimal_Smell*) is retained along with the x and y coordinates (with *Best_index*) as in Equations (14)–(16) [25], then the *Drosophila* population uses vision to fly to this position.

$$Optimal\_Smell = Best\_Smell_{i=\text{current}} \tag{14}$$

$$X_0 = X_{Best\_index} \tag{15}$$

$$Y_0 = Y_{Best\_index} \tag{16}$$

Step 7. Enter the iterative optimization, repeat **Steps 2** to **5** and judge whether the flavor concentration is better than that of the previous iteration; if so, go back to **Step 6**.

The FOA algorithm is highly adaptable, so it can efficiently search without calculating partial derivatives of the target function. It overcomes the disadvantage of trapping into local optima easily. However, as a swarm intelligence optimization algorithm, FOA still tends to fall into a local optimal solution, due to the declining diversity in the late evolutionary population.

It is noticed that there are some significant differences between the FOA and PSO algorithms. For FOA, the taste concentration (S) is used to determine the individual position of each fruit fly, and the highest odor concentration in this population is retained along with the x and y coordinates; eventually, the *Drosophila* population uses vision to fly to this position. Therefore, it is based on the taste concentration to control the searching direction to find out the optimal solution. For the PSO algorithm, the inertia weight controls the impact of the previous velocity of the particle on its current one by using two positive constants called acceleration coefficients and two independent uniformly distributed random variables. Therefore, it is based on the inertia weight to control the velocity to find out the optimal solution.

Thus, aiming to deal with the inherent drawback of FOA, i.e., suffering from premature convergence or trapping into local optima easily, this paper tries to use the QCM to empower each fruit fly to possess quantum behavior (namely QFOA) during the modeling processes. At the same time, the cat mapping function is introduced into QFOA (namely CQFOA) to implement the chaotic global perturbation strategy to help a fruit fly escape from the local optima when the population's diversity is poor. Eventually, the proposed CQFOA is employed to determine the appropriate parameters of an LS-SVR model and increase the forecasting accuracy.

2.2.2. Quantum Computing Mechanism for FOA

(1) Quantization of Fruit Flies

In the quantum computing process, a sequence consisting of quantum bits is replaced by a traditional sequence. The quantum fruit fly is a linear combination of state $|0\rangle$ and state $|1\rangle$, which can be expressed as in Equation (17) [34,35]:

$$|\varphi\rangle = \alpha|0\rangle + \beta|1\rangle, \tag{17}$$

where α^2 and β^2 are the probability of states, $|0\rangle$ and $|1\rangle$, respectively, satisfying $\alpha^2 + \beta^2 = 1$, and (α, β) are qubits composed of quantum bits.

A quantum sequence, i.e., a feasible solution, can be expressed as an arrangement of l qubits, as shown in Equation (18) [34,35]:

$$q_i = \left\{ \begin{matrix} \alpha_1 & \alpha_2 & \cdots & \alpha_l \\ \beta_1 & \beta_2 & \cdots & \beta_l \end{matrix} \right\}, \tag{18}$$

where the initial values of α_j and β_j are all set as $1/\sqrt{2}$ to meet the equity principle, $\alpha_j^2 + \beta_j^2 = 1$ ($j = 1, 2, \ldots, l$), which is updated through the quantum revolving door during the iteration.

Conversion between quantum sequence and binary sequence is the key to convert FOA to QFOA. Randomly generate a random number of $[0,1]$, $rand_j$, if $rand_j \geq \alpha_j^2$, the corresponding binary quantum bit value is 1, otherwise, 0, as shown in Equation (19):

$$x_j = \begin{cases} 1 & rand_j \geq \alpha_j^2 \\ 0 & else \end{cases}. \tag{19}$$

Using the above method, the quantum sequence, q, can be transformed into a binary sequence, x; then the optimal parameter problem of an LS-SVR model can be determined using QFOA.

(2) Quantum Fruit Fly Position Update Strategy

In the QFOA process, the position of quantum fruit flies represented by a quantum sequence is updated to find more feasible solutions and the best parameters. This paper uses quantum rotation to update the position of quantum fruit flies. The quantum position of individual i (there are in total N quantum fruit flies) can be extended from Equation (18) and is expressed as in Equation (20):

$$q_i = \left\{ \begin{array}{cccc} \alpha_{i1} & \alpha_{i2} & \cdots & \alpha_{il} \\ \beta_{i1} & \beta_{i2} & \cdots & \beta_{il} \end{array} \right\}, \tag{20}$$

where $\alpha_{ij}^2 + \beta_{ij}^2 = 1$; $i = 1, 2, \ldots, N$; $j = 1, 2, \ldots, l$; and $0 \le \alpha_{ij} \le 1, 0 \le \beta_{ij} \le 1$.

Quantum rotation is a quantum revolving door determined by the quantum rotation angle, which updates the quantum sequence and conducts a random search around the position of quantum fruit flies to explore the local optimal solution. The θ_{ij}^g is the jth quantum rotation angle of the population iterated to the ith fruit fly of generation, g, and the quantum bit q_{ij}^g (due to the nonnegative position constraint of q_{ij}^g, the absolute function, abs() is used to take the absolute value of each element in the calculation result) is updated according to the quantum revolving gate $U\left(\theta_{ij}^g\right)$, as shown in Equations (21) and (22) [34,35]:

$$q_{ij}^{g+1} = \text{abs}\left(U\left(\theta_{ij}^{g+1}\right) \times q_{ij}^g\right) \tag{21}$$

$$U\left(\theta_{ij}^g\right) = \left[\begin{array}{cc} \cos\theta_{ij}^g & -\sin\theta_{ij}^g \\ \sin\theta_{ij}^g & \cos\theta_{ij}^g \end{array} \right]. \tag{22}$$

In special cases, when the quantum rotation angle, θ_{ij}^{g+1}, is equal to 0, the quantum bit, q_{ij}^{g+1}, uses quantum non-gate $\overline{N}$ to update with some small probability, as indicated in Equation (23) [35]:

$$q_{ij}^{t+1} = \overline{N} \times q_{ij}^t = \left[\begin{array}{cc} 0 & 1 \\ 1 & 0 \end{array} \right] \times q_{ij}^t. \tag{23}$$

2.2.3. Chaotic Quantum Global Perturbation

For a bionic evolutionary algorithm, it is a general phenomenon that the population's diversity would be poor, along with the increased iterations. This phenomenon would also lead to being trapped into local optima during modeling processes. As mentioned, the chaotic mapping function can be employed to maintain the population's diversity to avoid trapping into local optima. Many studies have applied chaotic theory to improve the performances of these bionic evolutionary algorithms, such as the artificial bee colony (ABC) algorithm [41], and the particle swarm optimization (PSO) algorithm [42]. The authors have also employed the cat chaotic mapping function to improve the genetic algorithm (GA) [43], the PSO algorithm [44], and the bat algorithm [39], the results of which demonstrate that the searching quality of GA, PSO, ABC, and BA algorithms could be improved by employing chaotic mapping functions. Hence, the cat chaotic mapping function is once again used as the global chaotic perturbation strategy (GCPS) in this paper, and is hybridized with QFOA, namely CQFOA, which hybridizes GCPS with the QFOA while suffering from the problem of being trapped into local optima during the iterative modeling processes.

The two-dimensional cat mapping function is shown as in Equation (24) [39]:

$$\left\{ \begin{array}{l} y^{t+1} = frac\left(y^t + z^t\right) \\ z^{t+1} = frac\left(y^t + 2z^t\right) \end{array} \right., \tag{24}$$

where *frac* function is used to calculate the fractional parts of a real number, y, by subtracting an approached integer.

The global chaotic perturbation strategy (GCPS) is illustrated as follows.

(1) **Generate 2popsize chaotic disturbance fruit flies.** For each *Fruit fly*^{i} (I = 1, 2, ... , 2popsize), Equation (24) is applied to generate d random numbers, z_j, $j = 1, 2, \ldots, d$. Then, the qubit (with quantum state, $|0\rangle$) amplitude, $\cos\theta_j^i$, of *Fruit fly*^{i} is shown in Equation (25):

$$\cos\theta_j^i = y_j = 2z_j - 1. \tag{25}$$

(2) **Select 0.5 popsize better chaotic disturbance fruit flies.** Compute the fitness value of each *Fruit fly* from 2 popsize chaotic disturbance fruit flies, and arrange these fruit flies to be a sequence based on the order of fitness values. Then, select the fruit flies with 0.5 popsize ranking ahead in the fitness values; as a result, the 0.5 popsize better chaotic disturbance fruit flies are obtained.

(3) **Determine 0.5 popsize current fruit flies with better fitness.** Compute the fitness value of each *Fruit fly* from current QFOA, and arrange these fruit flies to be a sequence based on the order of fitness values. Then, select the fruit flies with 0.5 popsize ranking ahead in the fitness values.

(4) **Form the new CQFOA population.** Mix the 0.5 popsize better chaotic disturbance fruit flies with 0.5 popsize current fruit flies with better fitness from current QFOA, and form a new population that contains new 1popsize fruit flies, and name it the new CQFOA population.

(5) **Complete global chaotic perturbation.** After obtaining the new population of CQFOA, take it as the new population of QFOA and continue to execute the QFOA process.

2.2.4. Implementation Steps of CQFOA

The steps of the proposed CQFOA for parameter optimization of an LS-SVR model are as follows as shown in Figure 1.

Step 1. Initialization. The population size of quantum *Drosophila* is 1 popsize; the maximum number of iterations is Gen-max; the random search radius is R; and the chaos disturbance control coefficient is N_{GCP}.

Step 2. Random searching. For quantum rotation angle, θ_{ij}, of a random search, according to the quantum rotation angle, fruit fly locations on each dimension are updated, and then, a quantum revolving door is applied to update the quantum sequence, as shown in Equations (26) and (27) [34,35]:

$$\theta_{ij} = \theta(j) + R \times \text{rand}(1) \tag{26}$$

$$q_{ij} = abs\left(\begin{bmatrix} \cos\theta_{ij} & -\sin\theta_{ij} \\ \sin\theta_{ij} & \cos\theta_{ij} \end{bmatrix} \times Q(j)\right), \tag{27}$$

where i is an individual of quantum fruit flies, $i = 1,2,\ldots,1$popsize; j is the position dimension of quantum fruit flies, $j = 1,2,\ldots,l$. As mentioned above, the position of q_{ij} is non-negative constrained, thus, the absolute function, $abs()$ is used to take the absolute value of each element in the calculation result.

Step 3. Calculating fitness. Mapping each Drosophila location, q_i, to the feasible domain of an LS-SVR model parameters to receive the parameters, (γ_i, σ_i). The training data are used to complete the training processes of the $LS-SVR_i$ model and calculate the forecasting value in the training stage corresponding to each set of parameters. Then, the forecasting error is calculated as in Equation (12) of CQFOA by the mean absolute percentage error (MAPE), as shown in Equation (28):

$$\text{MAPE} = \frac{1}{N}\sum_{i=1}^{N}\left|\frac{f_i(x) - \hat{f}_i(x)}{f_i(x)}\right| \times 100\%, \tag{28}$$

where N is the total number of data points; $f_i(x)$ is the actual load value at point i; and $\hat{f}_i(x)$ is the forecasted load value at point i.

Step 4. Choosing the current optimum. Calculate the taste concentration of fruit fly, $Smell_i$, by using Equation (12), and find the best flavor concentration of individual, $Best\_Smell_i$, by Equation (13), as the optimal fitness value.

Step 5. Updating global optimization. Compare whether the contemporary odor concentration, $Best\_Smell_{i=current}$, is better than the global optima, $Best\_Smell_i$. If so, update the global value by

Equation (14), and enable the individual quantum fruit fly to fly to the optimal position with vision, as in Equations (29) and (30), then go to **Step 6**. Otherwise, go to **Step 6** directly.

$$\theta_0 = \theta_{Best\_index} \qquad (29)$$

$$q_0 = q_{Best\_index} \qquad (30)$$

Step 6. Global chaos perturbation judgment. If the distance from the last disturbance is equal to N_{GCP}, go to **Step 7**; otherwise, go to **Step 8**.

Step 7. Global chaos perturbation operations. Based on the current population, conduct the global chaos perturbation algorithm to obtain the new CQFOA population. Then, take the new CQFOA population as the new population of QFOA, and continue to execute the QFOA process.

Step 8. Iterative refinements. Determine whether the current population satisfies the condition of evolutionary termination. If so, stop the optimization process and output the optimal results. Otherwise, repeat **Steps 2 to 8**.

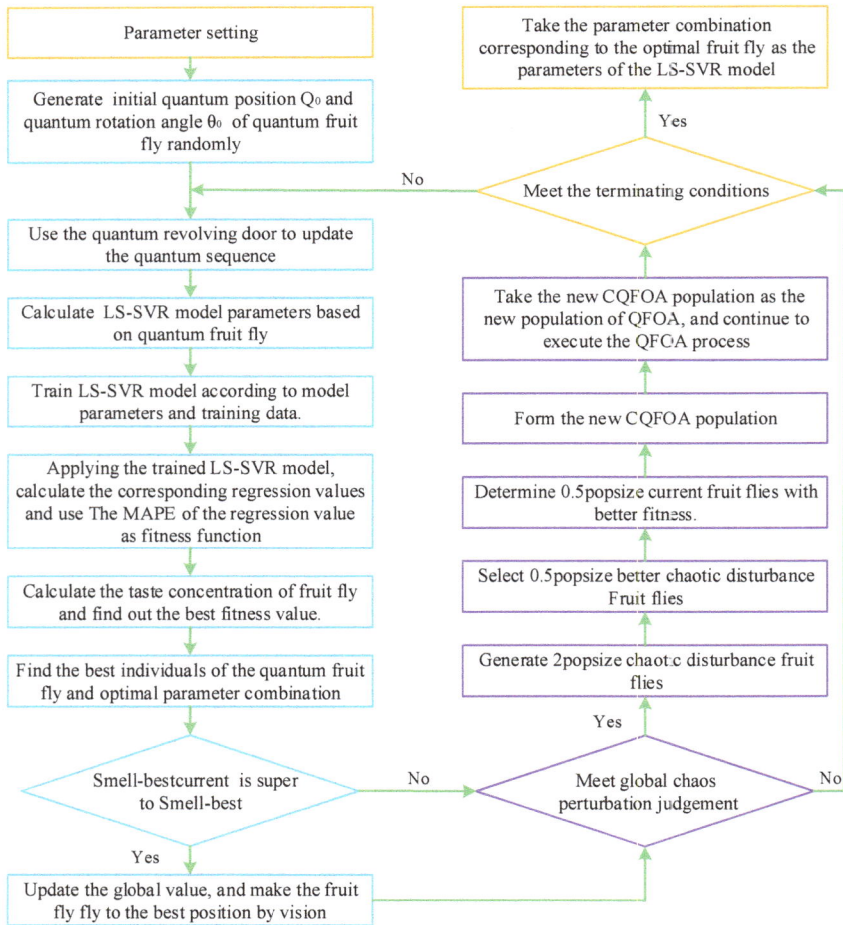

Figure 1. Chaotic quantum FOA algorithm flowchart.

3. Forecasting Results

3.1. Dataset of Experimental Examples

To test the performance of the proposed LS-SVR-CQFOA model, this paper employs the MEL data from an island data acquisition system in 2014 (IDAS 2014) [45] and the data of GEFCom2014-E [46] to carry out a numerical forecast. Taking the whole time of 24 h as the sampling interval, the load data contains 168-hour load values in total, i.e., from 01:00 14 July 2014 to 24:00 20 July 2014 in IDAS 2014 (namely IDAS 2014), and another two load datasets with the same 168-hour load values, i.e., from 01:00 1 January 2014 to 24:00 7 January 2014 (namely GEFCom2014 (Jan.)) and from 01:00 1 July 2014 to 24:00 7 July 2014 (namely GEFCom2014 (July)) in GEFCom2014-E, respectively.

The preciseness and integrity of historical data directly impact the forecasting accuracy. The data of the historical load are collected and obtained by electrical equipment. To some extent, the data transmission and measurement will lead to some "bad data" in the data of historical load, which mainly includes missing and abnormal data. If these data are used for modeling, the establishment of load forecasting model and the forecasting will bring adverse effects. Thus, the preprocessing of historical data is essential to load forecasting. In this paper, before the numerical test, the data of the MEL are preprocessed, including: completing the missing data; identifying abnormal data; eliminating and replacing unreasonable data; and normalizing data. When the input of an LS-SVR model is multidimensional with a large data size (e.g., several orders of magnitude), it may lead to problems when using the raw data to implement model training directly. Therefore, it is essential that the sample data are normalized for processing, to keep all the sample data values in a certain interval (this topic limits [0,1]), ensuring that all of the data have the same order of magnitude.

The normalization of load data is converted according to Equation (31), where $i = 1, 2, \ldots, N$ (N is the number of samples); x_i and y_i represent the values of before and after the normalization of sample data, respectively; and $\min(x_i)$ and $\max(x_i)$ represent the minimal and maximal values of sample data, respectively.

$$y_i = \frac{x_i - \min(x_i)}{\max(x_i) - \min(x_i)} \tag{31}$$

After the end of the forecasting, it is necessary to use the inverse normalization equation to calculate the actual load value, as shown in Equation (32):

$$x_i = (\max(x_i) - \min(x_i))y_i + \min(x_i). \tag{32}$$

The normalized data of the values in IDAS 2014, GEFCom2014 (Jan.) and GEFCom2014 (July) are collected and shown in Tables 1–3, respectively.

During the modeling processes, the load data are divided into three parts: the training set with the former 120 h, the validation set with the middle 24 h, and the testing set with the latter 24 h. Then, the rolling-based modeling procedure, proposed by Hong [18,47], is applied to assist CQFOA to look for appropriate parameters, (γ, σ), of an LS-SVR model during the training stage. Repeat this modeling procedure until all forecasting loads are received. The training error and the validation error can be calculated simultaneously. The adjusted parameters, (γ, σ), would be selected as the most suitable parameters only with both the smallest validation and testing errors. The testing dataset is never used during the training and validation stages; it will only be used to calculate the forecasting accuracy. Eventually, the 24 h's load data are forecasted by the proposed LS-SVR-CQFOA model.

Table 1. Normalization values of load data for IDAS 2014.

Time	14 July	15 July	16 July	17 July	18 July	19 July	20 July
01:00	0.1617	0.1245	0.1526	0.2246	0.1870	0.3354	0.3669
02:00	0.0742	0.0000	0.0826	0.1590	0.1386	0.1924	0.1878
03:00	0.0000	0.0109	0.0000	0.0395	0.0381	0.1022	0.0919
04:00	0.0071	0.1278	0.0937	0.0000	0.0000	0.0000	0.0000
05:00	0.0531	0.1944	0.1419	0.1106	0.1218	0.1570	0.1770
06:00	0.0786	0.0611	0.0920	0.1428	0.1728	0.2558	0.2497
07:00	0.2636	0.1786	0.2724	0.3096	0.3788	0.4038	0.3943
08:00	0.3709	0.4417	0.3464	0.3586	0.4361	0.5129	0.4692
09:00	0.6872	0.5894	0.6549	0.7426	0.7970	0.6051	0.5829
10:00	0.9520	0.8746	0.9028	0.9055	0.9842	0.7632	0.7530
11:00	1.0000	0.9342	0.9650	0.9683	1.0000	0.8130	0.8332
12:00	0.9632	0.9730	0.9087	0.9217	0.9450	0.8935	0.8803
13:00	0.8552	1.0000	0.8135	0.8256	0.8821	0.8077	0.8122
14:00	0.8288	0.9152	0.9257	0.7377	0.8370	0.7185	0.7410
15:00	0.8224	0.8104	0.7663	0.7468	0.7961	0.6037	0.6882
16:00	0.8655	0.9448	0.8542	0.8099	0.8420	0.7347	0.7567
17:00	0.8552	0.7966	0.8340	0.8104	0.8323	0.7593	0.8439
18:00	0.9440	0.8809	0.9155	0.8976	0.9567	0.9286	0.9539
19:00	0.9574	0.8677	1.0000	0.9779	0.9694	0.9734	0.9741
20:00	0.9746	0.9693	0.9657	1.0000	0.9808	1.0000	1.0000
21:00	0.9372	0.8784	0.9236	0.9419	0.9546	0.9575	0.9664
22:00	0.8704	0.7697	0.7977	0.7889	0.8417	0.8634	0.8824
23:00	0.6328	0.5519	0.7193	0.6425	0.6655	0.5858	0.6035
24:00	0.3127	0.2114	0.2794	0.2559	0.3357	0.1080	0.0975

Table 2. Normalization values of load data for GEFCom2014 (Jan.).

Time	1 January	2 January	3 January	4 January	5 January	6 January	7 January
01:00	0.1769	0.0568	0.1127	0.1314	0.1648	0.0769	0.0532
02:00	0.0877	0.0206	0.0338	0.0480	0.0765	0.0222	0.0123
03:00	0.0234	0.0000	0.0000	0.0000	0.0087	0.0000	0.0000
04:00	0.0000	0.0084	0.0035	0.0044	0.0063	0.0076	0.0140
05:00	0.0175	0.0746	0.0634	0.0497	0.0268	0.0565	0.0862
06:00	0.0863	0.2155	0.2134	0.1368	0.0938	0.2122	0.2569
07:00	0.1835	0.4382	0.4345	0.3082	0.2090	0.4740	0.5389
08:00	0.2763	0.5802	0.5894	0.4813	0.3517	0.6277	0.6503
09:00	0.4028	0.6453	0.6972	0.6705	0.5039	0.6849	0.6581
10:00	0.5212	0.7110	0.7683	0.7860	0.6136	0.7300	0.6693
11:00	0.5819	0.7455	0.8106	0.8073	0.6333	0.7446	0.6861
12:00	0.6016	0.7751	0.8042	0.7726	0.6080	0.7573	0.6900
13:00	0.6089	0.7684	0.7592	0.6936	0.5623	0.7300	0.6788
14:00	0.5789	0.7712	0.7176	0.5950	0.5221	0.7078	0.6754
15:00	0.5563	0.7634	0.6887	0.5400	0.4937	0.6842	0.6676
16:00	0.5768	0.7556	0.6852	0.5560	0.5560	0.7109	0.6928
17:00	0.8165	0.8836	0.8479	0.7913	0.8060	0.8558	0.8411
18:00	1.0000	1.0000	1.0000	1.0000	1.0000	1.0000	1.0000
19:00	0.9810	0.9605	0.9845	0.9423	0.9416	0.9778	0.9955
20:00	0.8984	0.8686	0.8859	0.8188	0.8036	0.8920	0.9379
21:00	0.7807	0.7723	0.7908	0.7087	0.6672	0.7903	0.8489
22:00	0.5885	0.6114	0.6289	0.4982	0.4219	0.6112	0.6933
23:00	0.3596	0.4399	0.4303	0.2860	0.1774	0.4180	0.4980
24:00	0.1923	0.2957	0.2542	0.0719	0.0000	0.2764	0.3553

Table 3. Normalization values of load data for GEFCom2014 (July).

Time	1 July	2 July	3 July	4 July	5 July	6 July	7 July
01:00	0.1562	0.1612	0.1583	0.2747	0.2636	0.1699	0.1063
02:00	0.0728	0.0882	0.0763	0.1302	0.1266	0.0857	0.0394
03:00	0.0238	0.0348	0.0232	0.0456	0.0554	0.0302	0.0054
04:00	0.0000	0.0000	0.0000	0.0000	0.0063	0.0000	0.0000
05:00	0.0222	0.0186	0.0181	0.0190	0.0000	0.0021	0.0302
06:00	0.0945	0.0957	0.1040	0.0589	0.0554	0.0154	0.1187
07:00	0.2811	0.2781	0.3143	0.2091	0.1872	0.0955	0.2972
08:00	0.4692	0.4736	0.5172	0.4316	0.4153	0.2521	0.4903
09:00	0.6244	0.6212	0.6637	0.6873	0.7008	0.4459	0.6424
10:00	0.7396	0.7516	0.7733	0.8878	0.9017	0.6131	0.7476
11:00	0.8306	0.8479	0.8722	0.9734	0.9561	0.7163	0.8425
12:00	0.8979	0.9209	0.9389	1.0000	0.9561	0.7570	0.9051
13:00	0.9378	0.9673	0.9678	0.9876	0.9111	0.7809	0.9434
14:00	0.9737	1.0000	0.9938	0.9287	0.8515	0.7928	0.9865
15:00	0.9879	0.9829	1.0000	0.8546	0.8243	0.8111	0.9995
16:00	0.9970	0.9290	0.9881	0.8032	0.8462	0.8574	1.0000
17:00	1.0000	0.8564	0.9423	0.8004	0.9195	0.9199	0.9962
18:00	0.9960	0.8101	0.9005	0.8279	0.9937	0.9853	0.9833
19:00	0.9687	0.7567	0.8672	0.8203	1.0000	1.0000	0.9579
20:00	0.9176	0.6907	0.7756	0.7386	0.9435	0.9579	0.9213
21:00	0.9044	0.6489	0.7377	0.6787	0.9362	0.9417	0.8975
22:00	0.8291	0.5461	0.6354	0.5428	0.8692	0.8687	0.7875
23:00	0.6138	0.3572	0.4262	0.3279	0.6883	0.6426	0.5701
24:00	0.4095	0.1678	0.2272	0.0913	0.4341	0.4213	0.3927

3.2. Forecasting Accuracy Indexes and Performance Tests

3.2.1. Forecasting Accuracy Index

This study uses the *MAPE* (mentioned in Equation (28)), the root mean square error (RMSE), and the mean absolute error (MAE) as forecasting accuracy indexes. The RMSE and MAE are defined as in Equations (33) and (34), respectively:

$$\text{RMSE} = \sqrt{\frac{\sum_{i=1}^{N} \left(f_i(x) - \hat{f}_i(x) \right)^2}{N}} \tag{33}$$

$$\text{MAE} = \frac{1}{N} \sum_{i=1}^{N} \left| f_i(x) - \hat{f}_i(x) \right|, \tag{34}$$

where N is the total number of data points; $f_i(x)$ is the actual value at point i; and $\hat{f}_i(x)$ is the forecasting value at point i.

3.2.2. Forecasting Performance Improvement Tests

To demonstrate the significant forecasting performances of the proposed model, Diebold and Mariano [48] and Derrac et al. [49] suggest that, for a small data size (24-h load forecasting) test, a Wilcoxon signed-rank test [50] is suitable. Thus, we decided to apply the Wilcoxon signed-rank test. For the same data size, a Wilcoxon test detects the significance of the difference (i.e., the forecasting errors from two forecasting models) in the central tendency. Therefore, let d_i be the absolute forecasting errors from any two models on ith forecasting value: R^+ be the sum of ranks that $d_i > 0$; R^- the sum of ranks that $d_i < 0$. If $d_i = 0$, then, remove this comparison and decrease the sample size. The statistics of Wilcoxon test, W, is calculated as in Equation (37):

$$W = \min\{R^+, R^-\}. \tag{35}$$

If W is smaller than or equal to the critical value, based on the Wilcoxon distribution under n degrees of freedom, then the null hypothesis (i.e., equal performance from the two compared forecasting models) could not be accepted, i.e., the proposed model achieves significance.

3.3. The Forecasting Results of the LS-SVR-CQFOA Model

3.3.1. Parameter Setting of the CQFOA Algorithm

The parameters of the proposed CQFOA algorithm for the numerical example are set as follows: the population size, *popsize*, is set to 200; the maximal iteration, *gen-max*, is set to 1000; and the control coefficient of chaotic disturbance, N_{GCP}, is set to 15. These two parameters of the LS-SVR model are set as, $\gamma \in [0, 1000]$, and $\sigma \in [0, 500]$, respectively. The iterative time of each algorithm is set as the same to ensure the reliability of the forecasting results.

3.3.2. Results and Analysis

Considering the CQPSO, CQTS, and CQGA algorithms have been used to determine the parameters of an SVR-based load forecasting model in [36–39], those existing algorithms are also hybridized with an LS-SVR model to provide forecasting values to compare with the proposed model here. These alternative models include LS-SVR-FOA, LS-SVR-QFOA, LS-SVR-CQPSO (LS-SVR hybridized with chaotic quantum particle swarm optimization algorithm [36]), LS-SVR-CQTS (LS-SVR hybridized with chaotic quantum Tabu search algorithm [37]), LS-SVR-CQGA (LS-SVR hybridized with chaotic quantum genetic algorithm [38]), and LS-SVR-CQBA (LS-SVR hybridized with chaotic quantum bat algorithm [39]), in order to compare the forecasting performance of LS-SVR-based models comprehensively, this article also selects BPNN method as a contrast model. The parameters of an LS-SVR model are selected by the CQPSO, CQTS, CQGA, CQBA, FOA, QFOA, and CQFOA algorithms, respectively. The details of the suitable parameters of all models for the IDAS 2014, the GEFCom2014 (Jan.) and the GEFCom2014 (July) data are shown in Tables 4–6, respectively.

Table 4. LS-SVR parameters, MAPE, and computing times of CQFOA and other algorithms for IDAS 2014.

Optimization Algorithms	LS-SVR Parameters		MAPE of Validation (%)	Computing Times (s)
	γ	σ		
LS-SVR-CQPSO [36]	685	125	1.17	129
LS-SVR-CQTS [37]	357	118	1.13	113
LS-SVR-CQGA [38]	623	137	1.11	152
LS-SVR-CQBA [39]	469	116	1.07	227
LS-SVR-FOA	581	109	1.29	87
LS-SVR-QFOA	638	124	1.32	202
LS-SVR-CQFOA,	734	104	1.02	136

Table 5. Parameters combination of LS-SVR determined by CQFOA and other algorithms for GEFCom2014 (Jan.).

Optimization Algorithms	Parameters		MAPE of Validation (%)	Computation Times (s)
	γ	σ		
LS-SVR-CQPSO [36]	574	87	0.98	134
LS-SVR-CQTS [37]	426	68	1.02	109
LS-SVR-CQGA [38]	653	98	0.95	155
LS-SVR-CQBA [39]	501	82	0.9	231
LS-SVR-FOA	482	94	1.54	82
LS-SVR-QFOA	387	79	1.13	205
LS-SVR-CQFOA,	688	88	0.86	132

Table 6. Parameters combination of LS-SVR determined by CQFOA and other algorithms for GEFCom2014 (July).

Optimization Algorithms	Parameters		MAPE of Validation (%)	Computation Times (s)
	γ	σ		
LS-SVR-CQPSO [36]	375	92	0.96	139
LS-SVR-CQTS [37]	543	59	1.04	107
LS-SVR-CQGA [38]	684	62	0.98	159
LS-SVR-CQBA [39]	498	90	0.95	239
LS-SVR-FOA	413	48	1.51	79
LS-SVR-QFOA	384	83	1.07	212
LS-SVR-CQFOA,	482	79	0.79	147

Based on the same training settings, another representative model, the back-propagation neural network (BPNN) is compared with the proposed model. The forecasting results of these models mentioned above and the actual values for IDAS 2014, GEFCom2014 (Jan.) and GEFCom2014 (July) are given in Figures 2–4, respectively. This indicates that the proposed LS-SVR-CQFOA model achieves a better performance than the other alternative models, i.e., closer to the actual load values.

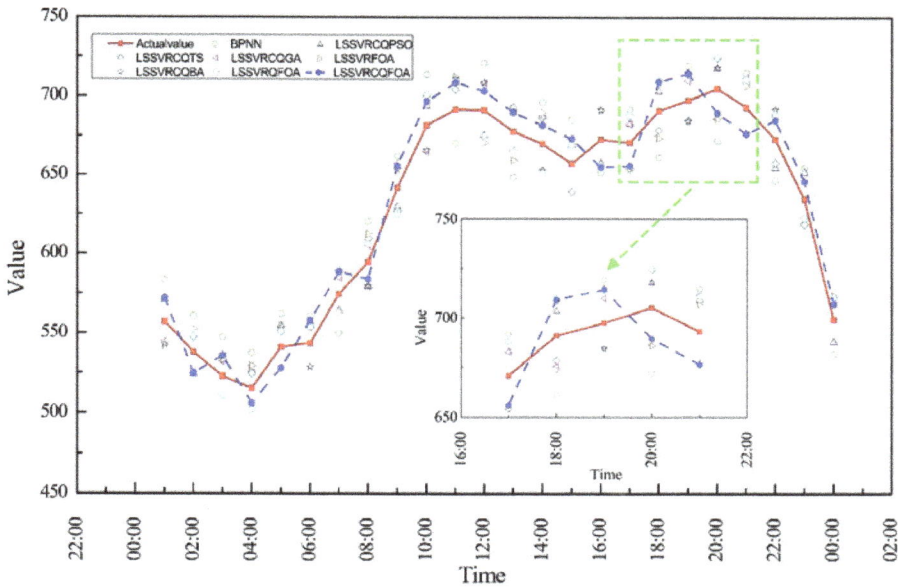

Figure 2. Forecasting values of LS-SVR-CQFOA and other models for IDAS 2014.

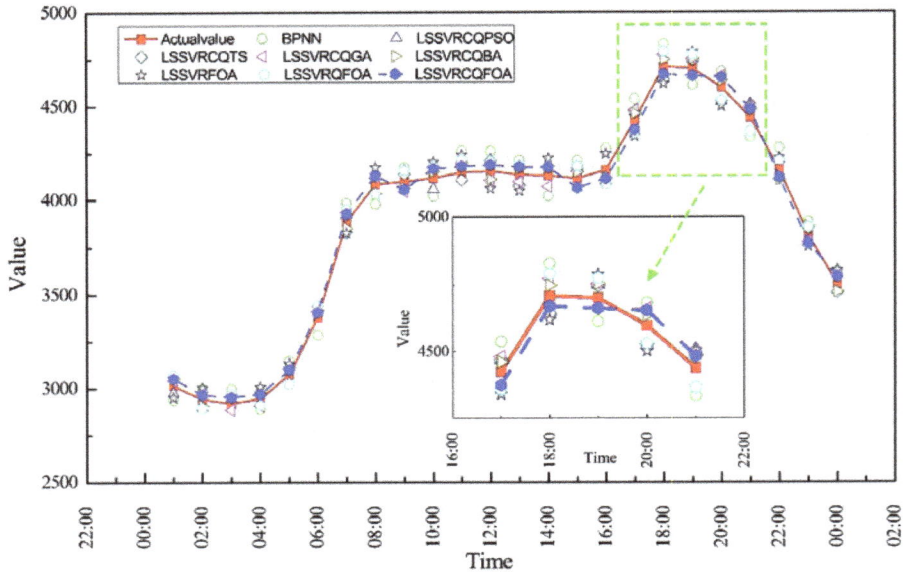

Figure 3. Forecasting values of LS-SVR-CQFOA and other models for GEFCom2014 (Jan.).

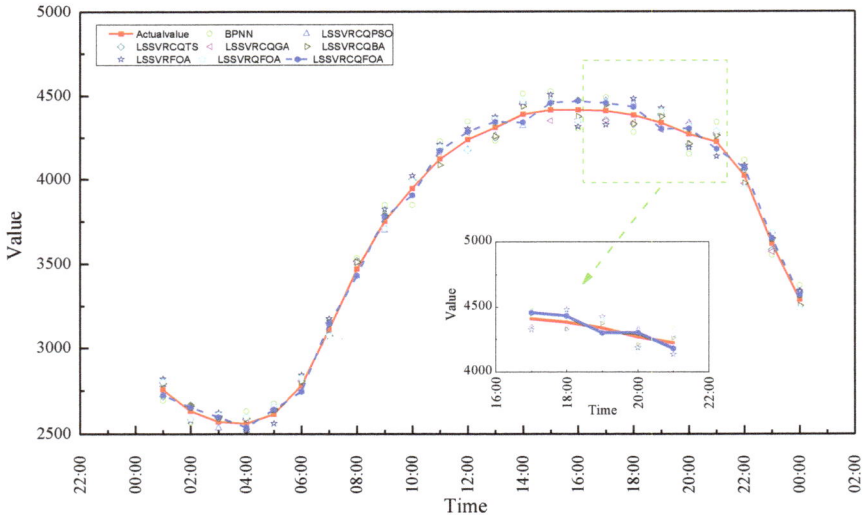

Figure 4. Forecasting values of LS-SVR-CQFOA and other models for GEFCom2014 (July).

Tables 7–9 indicate the evaluation results from different forecasting accuracy indexes for IDAS 2014, GEFCom2014 (Jan.) and GEFCom2014 (July), respectively. For Table 7, the proposed LS-SVR-CQFOA model achieves smaller values for all employed accuracy indexes than the seven other models: RMSE (14.10), MAPE (2.21%), and MAE (13.88), respectively. For Table 8, similarly, the proposed LS-SVR-CQFOA model also achieves smaller values for all employed accuracy indexes compared to the seven other models: RMSE (40.62), MAPE (1.02%), and MAE (39.76), respectively. Similarly in Table 9, the proposed LS-SVR-CQFOA model also achieves smaller values for all employed accuracy indexes than the other seven models: RMSE (38.70), MAPE (1.01%), and MAE (37.48), respectively. The details of the analysis results are as follows.

Table 7. Forecasting indexes of LS-SVR-CQFOA and other models for IDAS 2014.

Compared Models	RMSE	MAPE (%)	MAE
BPNN	24.89	3.92	24.55
LS-SVR-CQPSO [36]	14.40	2.27	14.21
LS-SVR-CQTS [37]	14.50	2.26	14.24
LS-SVR-CQGA [38]	14.41	2.24	14.13
LS-SVR-CQBA [39]	14.45	2.25	14.18
LS-SVR-FOA	15.90	2.48	15.62
LS-SVR-QFOA	15.03	2.32	14.69
LS-SVR-CQFOA	14.10	2.21	13.88

Table 8. Forecasting indexes of LS-SVR-CQFOA and other models for GEFCom2014 (Jan.).

Compared Models	RMSE	MAPE (%)	MAE
BPNN	92.30	2.34	90.74
LS-SVR-CQPSO [36]	51.46	1.31	50.69
LS-SVR-CQTS [37]	50.85	1.27	49.70
LS-SVR-CQGA [38]	46.36	1.16	45.31
LS-SVR-CQBA [39]	42.76	1.07	41.80
LS-SVR-FOA	75.55	1.89	73.88
LS-SVR-QFOA	59.74	1.47	57.96
LS-SVR-CQFOA	40.62	1.02	39.76

Table 9. Forecasting indexes of LS-SVR-CQFOA and other models for GEFCom2014 (July).

Compared Models	RMSE	MAPE (%)	MAE
BPNN	88.24	2.31	85.51
LS-SVR-CQPSO [36]	51.03	1.33	49.35
LS-SVR-CQTS [37]	45.73	1.22	44.68
LS-SVR-CQGA [38]	46.18	1.19	44.46
LS-SVR-CQBA [39]	40.75	1.09	39.85
LS-SVR-FOA	72.00	1.88	69.69
LS-SVR-QFOA	56.33	1.49	54.81
LS-SVR-CQFOA	38.70	1.01	37.48

Finally, to test the significance in terms of forecasting accuracy improvements from the proposed LS-SVR-CQFOA model, the Wilcoxon signed-rank test is conducted under two significant levels, $\alpha = 0.025$ and $\alpha = 0.05$, by one-tail test. The test results for the IDAS 2014, the GEFCom2014 (Jan.), and the GEFCom2014 (July) datasets are described in Tables 10–12, respectively. In these three tables, the results demonstrate that the proposed LS-SVR-CQFOA model achieved significantly better forecasting performance than the other alternative models. For example, in the IDAS 2014 dataset, for LS-SVR-CQFOA vs. LS-SVR-CQPSO, the statistic of Wilcoxon test, W = 72, is smaller than the critical statistics, W** = 81 (under $\alpha = 0.025$) and W* = 91 (under $\alpha = 0.05$), thus we could conclude that the proposed LS-SVR-CQFOA model is significantly outperform the LS-SVR-CQPSO model. In addition, the *p*-value = 0.022 is also smaller than the critical $\alpha = 0.025$ and $\alpha = 0.05$, which support the conclusion.

Table 10. Results of Wilcoxon signed-rank test for IDAS 2014.

Compared Models	Wilcoxon Signed-Rank Test		
	$T_{0.025} = 81$	$T_{0.05} = 91$	*p*-Value
LS-SVR-CQFOA vs. BPNN	0 [T]	0 [T]	0.000 **
LS-SVR-CQFOA vs. LS-SVR-CQPSO	72 [T]	72 [T]	0.022 **
LS-SVR-CQFOA vs. LS-SVR-CQTS	64 [T]	64 [T]	0.017 **
LS-SVR-CQFOA vs. LS-SVR-CQGA	67 [T]	67 [T]	0.018 **
LS-SVR-CQFOA vs. LS-SVR-CQBA	60 [T]	60 [T]	0.012 **
LS-SVR-CQFOA vs. LS-SVR-FOA	50 [T]	50 [T]	0.009 **
LS-SVR-CQFOA vs. LS-SVR-QFOA	68 [T]	68 [T]	0.019 **

[T] Denotes that the LS-SVR-CQGA model significantly outperforms the other models. ** implies the *p*-value is lower than $\alpha = 0.025$; * implies the *p*-value is lower than $\alpha = 0.05$.

Table 11. Results of Wilcoxon signed-rank test for GEFCom2014 (Jan.).

Compared Models	Wilcoxon Signed-Rank Test		
	$T_{0.025} = 81$	$T_{0.05} = 91$	*p*-Value
LS-SVR-CQFOA vs. BPNN	0 [T]	0 [T]	0.000 **
LS-SVR-CQFOA vs. LS-SVR-CQPSO	74 [T]	74 [T]	0.023 **
LS-SVR-CQFOA vs. LS-SVR-CQTS	75 [T]	75 [T]	0.024 **
LS-SVR-CQFOA vs. LS-SVR-CQGA	78 [T]	78 [T]	0.026 **
LS-SVR-CQFOA vs. LS-SVR-CQBA	80 [T]	80 [T]	0.027 **
LS-SVR-CQFOA vs. LS-SVR-FOA	65 [T]	65 [T]	0.018 **
LS-SVR-CQFOA vs. LS-SVR-QFOA	72 [T]	72 [T]	0.022 **

[T] Denotes that the LS-SVR-CQGA model significantly outperforms the other models. ** implies the *p*-value is lower than $\alpha = 0.025$; * implies the *p*-value is lower than $\alpha = 0.05$.

Table 12. Results of Wilcoxon signed-rank test for GEFCom2014 (July).

Compared Models	Wilcoxon Signed-Rank Test		
	$T_{0.025} = 81$	$T_{0.05} = 91$	*p*-Value
LS-SVR-CQFOA vs. BPNN	0 [T]	0 [T]	0.000 **
LS-SVR-CQFOA vs. LS-SVR-CQPSO	73 [T]	73 [T]	0.023 **
LS-SVR-CQFOA vs. LS-SVR-CQTS	76 [T]	76 [T]	0.024 **
LS-SVR-CQFOA vs. LS-SVR-CQGA	77 [T]	77 [T]	0.026 **
LS-SVR-CQFOA vs. LS-SVR-CQBA	79 [T]	79 [T]	0.027 **
LS-SVR-CQFOA vs. LS-SVR-FOA	65 [T]	65 [T]	0.018 **
LS-SVR-CQFOA vs. LS-SVR-QFOA	71 [T]	71 [T]	0.022 **

[T] Denotes that the LS-SVR-CQGA model significantly outperforms the other models. ** implies the *p*-value is lower than $\alpha = 0.025$; * implies the *p*-value is lower than $\alpha = 0.05$.

4. Discussion

Taking the IDAS 2014 dataset as an example, firstly, the forecasting results of these LS-SVR-based models are all closer to the actual load values than the BPNN model. This shows that LS-SVR-based models can simulate nonlinear systems of microgrid load more accurately than the BPNN model, due to its advantages in dealing with nonlinear problems.

Secondly, in Table 4, the selected FOA and QFOA algorithms could achieve the best solution, $(\gamma, \sigma) = (581, 109)$ and $(\gamma, \sigma) = (638, 124)$, with forecasting error, (RMSE = 15.93, MAPE = 2.48%, MAE = 15.63) and (RMSE = 14.87, MAPE = 2.32%, MAE = 14.61), respectively. However, the solution can be further improved by the proposed CQFOA algorithm to $(\gamma, \sigma) = (734, 104)$ with more accurate forecasting performance, (RMSE = 14.10, MAPE = 2.21%, MAE = 13.88). Similar results could also be learned in the GEFCom2014 (Jan.) and the GEFCom2014 (July) from Tables 5 and 6, respectively. This illustrates that the proposed approach is feasible, i.e., hybridizing the FOA with QCM and chaotic mapping function to determine more appropriate parameters of an LS-SVR model to improve the forecasting accuracy.

Comparing the LS-SVR-QFOA model with the LS-SVR-FOA model, the forecasting accuracy of the LS-SVR-QFOA model is superior to that of the LS-SVR-FOA model. This demonstrates that the QCM empowers the fruit fly to have quantum behaviors, i.e., the QFOA find more appropriate parameters of an LS-SVR model, which improves the forecasting accuracy of the LS-SVR-FOA model in which the FOA is hybridized with an LS-SVR model. For example, in Table 4, the usage of the QCM in FOA changes the forecasting performances (RMSE = 15.93, MAPE = 2.48%, MAE = 15.63) of the LS-SVR-FOA model to the much better performance (RMSE = 14.87, MAPE = 2.32%, MAE = 14.61) of the LS-SVR-QFOA model. Similar results are demonstrated in the GEFCom2014 (Jan.) and the GEFCom2014 (July) from Tables 5 and 6, respectively.

For forecasting performance comparison between the LS-SVR-CQFOA and LS-SVR-QFOA models, the values of RMSE, MAPE, and MAE for the LS-SVR-CQFOA model are smaller than those of the LS-SVR-QFOA model. This reveals that the introduction of cat chaotic mapping function into QFOA plays a positive role in searching appropriate parameters when the population of QFOA algorithm is trapped into local optima. Then, the CQFOA finds more appropriate parameters. As a result, as shown in Table 4, employing CQFOA to select the parameters for an LS-SVR model markedly improves the performance (RMSE = 14.87, MAPE = 2.32%, MAE = 14.61) of the LS-SVR-QFOA model to the much better one (RMSE = 14.10, MAPE = 2.21%, MAE = 13.88) of the LS-SVR-CQFOA model. Similar results are illustrated in the GEFCom2014 (Jan.) and the GEFCom2014 (July) from Tables 5 and 6, respectively.

Comparing the time-consuming problem during the parameter searching processes in all the IDAS 2014, the GEFCom2014 (Jan.), and the GEFCom2014 (July) datasets, the proposed CQFOA is less than that of the CQGA and CQBA algorithms, but more than that of the CQPSO and CQTS algorithms. However, considering the time requirements of the actual application, the increase in time compared with CQPSO (more than 7 s) and CQTS (more than 23 s) is acceptable.

Finally, some limitations should be noticed. This paper only employs an existing dataset to establish the proposed model; thus, for different seasons, months, weeks, and dates, the electricity load patterns should be changed season by season, month by month, and week by week. For real-world applications, this paper should be a good beginning to guide planners and decision-makers to establish electricity load forecasting models overlapping the seasons, months, and weeks to achieve more comprehensive results. Thus, our planned future research direction is to explore the feasibility of hybridizing more powerful novel optimization frameworks (e.g., chaotic mapping functions, quantum computing mechanism, and hourly, daily, weekly, monthly adjusted mechanism) and novel meta-heuristic algorithms with an LS-SVR model to overcome the drawbacks of evolutionary algorithms to achieve excellent forecasting accuracy.

5. Conclusions

This paper proposes a novel hybrid forecasting model by hybridizing an LS-SVR model with the QCM, the cat chaotic mapping function, and the FOA. The forecasting results show that the proposed model achieves better performance than the alternative forecasting models, by hybridizing chaotic mapping function, QCM, and other evolutionary algorithms with an LS-SVR-based model. Employing the cat chaotic mapping function to enrich the diversity of searching scope and enhance the ergodicity of the population could successfully avoid trapping into local optima, and, also proves that applying QCM to overcome the limitations of the fruit fly's searching behaviors empowers the fruit fly to undertake quantum searching behaviors, thereby achieving more satisfactory results for MEL forecasting. The global chaotic perturbation strategy based on the cat mapping function is employed to jump out of local minima while the population of QFOA suffers from premature convergence, and also helps to improve the forecasting performance.

Author Contributions: M.-W.L. and W.-C.H. conceived and designed the experiments; G.J. and Z.Y. performed the experiments; M.-W.L. and W.-C.H. analyzed the data and wrote the paper.

Energies **2018**, *11*, 2226

Funding: Funding: This research was funded by the National Natural Science Foundation of China (51509056); the Heilongjiang Province Natural Science Fund (E2017028); the Fundamental Research Funds for the Central Universities (HEUCFG201813); the Open Fund of the State Key Laboratory of Coastal and Offshore Engineering (LP1610); Heilongjiang Sanjiang Project Administration Scientific Research and Experiments (SGZL/KY-08); and the Jiangsu Distinguished Professor Project (no. 9213618401), Jiangsu Normal University, Jiangsu Provincial Department of Education, China.

Acknowledgments: Ming-Wei Li, Jing Geng, and Yang Zhang acknowledge the support from the project grants: the National Natural Science Foundation of China (51509056); the Heilongjiang Province Natural Science Fund (E2017028); the Fundamental Research Funds for the Central Universities (HEUCFG201813); the Open Fund of the State Key Laboratory of Coastal and Offshore Engineering (LP1610); and Heilongjiang Sanjiang Project Administration Scientific Research and Experiments (SGZL/KY-08). Wei-Chiang Hong acknowledges the support from the Jiangsu Distinguished Professor Project (no. 9213618401) of Jiangsu Normal University, Jiangsu Provincial Department of Education, China.

Conflicts of Interest: The authors declare no conflict of interest.

References

1. Maçaira, P.M.; Souza, R.C.; Oliveira, F.L.C. Modelling and forecasting the residential electricity consumption in Brazil with pegels exponential smoothing techniques. *Procedia Comput. Sci.* **2015**, *55*, 328–335. [CrossRef]
2. Pappas, S.S.; Ekonomou, L.; Karampelas, P.; Karamousantas, D.C.; Katsikas, S.K.; Chatzarakis, G.E.; Skafidas, P.D. Electricity demand load forecasting of the Hellenic power system using an ARMA model. *Electr. Power Syst. Res.* **2010**, *80*, 256–264. [CrossRef]
3. Dudek, G. Pattern-based local linear regression models for short-term load forecasting. *Electr. Power Syst. Res.* **2016**, *130*, 139–147. [CrossRef]
4. Chen, Y.; Luh, P.B.; Guan, C.; Zhao, Y.; Michel, L.D.; Coolbeth, M.A.; Friedland, P.B.; Rourke, S.J. Short-term load forecasting: Similar day-based wavelet neural networks. *IEEE Trans. Power Syst.* **2010**, *25*, 322–330. [CrossRef]
5. Li, S.; Wang, P.; Goel, L. Short-term load forecasting by wavelet transform and evolutionary extreme learning machine. *Electr. Power Syst. Res.* **2015**, *122*, 96–103. [CrossRef]
6. Fan, G.F.; Wang, A.; Hong, W.C. Combining grey model and self-adapting intelligent grey model with genetic algorithm and annual share changes in natural gas demand forecasting. *Energies* **2018**, *11*, 1625. [CrossRef]
7. Ma, X.; Liu, Z. Application of a novel time-delayed polynomial grey model to predict the natural gas consumption in China. *J. Comput. Appl. Math.* **2017**, *324*, 17–24. [CrossRef]
8. Lou, C.W.; Dong, M.C. A novel random fuzzy neural networks for tackling uncertainties of electric load forecasting. *Int. J. Electr. Power Energy Syst.* **2015**, *73*, 34–44. [CrossRef]
9. Ertugrul, Ö.F. Forecasting electricity load by a novel recurrent extreme learning machines approach. *Int. J. Electr. Power Energy Syst.* **2016**, *78*, 429–435. [CrossRef]
10. Geng, J.; Huang, M.L.; Li, M.W.; Hong, W.C. Hybridization of seasonal chaotic cloud simulated annealing algorithm in a SVR-based load forecasting model. *Neurocomputing* **2015**, *151*, 1362–1373. [CrossRef]
11. Hooshmand, R.A.; Amooshahi, H.; Parastegari, M. A hybrid intelligent algorithm Based short-term load forecasting approach. *Int. J. Electr. Power Energy Syst.* **2013**, *45*, 313–324. [CrossRef]
12. Niu, D.X.; Shi, H.; Wu, D.D. Short-term load forecasting using Bayesian neural networks learned by hybrid Monte Carlo algorithm. *Appl. Soft Comput.* **2012**, *12*, 1822–1827. [CrossRef]
13. Hanmandlu, M.; Chauhan, B.K. Load forecasting using hybrid models. *IEEE Trans. Power Syst.* **2011**, *26*, 20–29. [CrossRef]
14. Mahmoud, T.S.; Habibi, D.; Hassan, M.Y.; Bass, O. Modelling self-optimised short term load forecasting for medium voltage loads using tunning fuzzy systems and artificial neural networks. *Energy Convers. Manag.* **2015**, *106*, 1396–1408. [CrossRef]
15. Suykens, J.A.K.; Vandewalle, J.; De Moor, B. Optimal control by least squares support vector machines. *Neural Netw.* **2001**, *14*, 23–35. [CrossRef]
16. Sankar, R.; Sapankevych, N.I. Time series prediction using support vector machines: A survey. *IEEE Comput. Intell. Mag.* **2009**, *4*, 24–38.
17. Vapnik, V.N. *The Nature of Statistical Learning Theory*; Springer: New York, NY, USA, 1995.

18. Hong, W.C. Electric load forecasting by seasonal recurrent LS-SVR (support vector regression) with chaotic artificial bee colony algorithm. *Energy* **2011**, *36*, 5568–5578. [CrossRef]

19. Fan, G.F.; Peng, L.L.; Zhao, X.; Hong, W.C. Applications of hybrid EMD with PSO and GA for an SVR-based load forecasting model. *Energies* **2017**, *10*, 1713. [CrossRef]

20. Suykens, J.A.K.; Vanddewalle, J. Least squares support vector machines classifiers. *Neural Netw. Lett.* **1999**, *19*, 293–300. [CrossRef]

21. Wang, J.; Hu, J. A robust combination approach for short-term wind speed forecasting and analysis—Combination of the ARIMA (Autoregressive Integrated Moving Average), ELM (Extreme Learning Machine), SVM (Support Vector Machine) and LSSVM (Least Square SVM) forecasts using a GPR (Gaussian Process Regression) model. *Energy* **2015**, *93*, 41–56.

22. Hong, W.C.; Dong, Y.; Zhang, W.; Chen, L.Y.; Panigrahi, B.K. Cyclic electric load forecasting by seasonal LS-SVR with chaotic genetic algorithm. *Int. J. Electr. Power Energy Syst.* **2013**, *44*, 604–614. [CrossRef]

23. Ju, F.Y.; Hong, W.C. Application of seasonal SVR with chaotic gravitational search algorithm in electricity forecasting. *Appl. Math. Model.* **2013**, *37*, 9643–9651. [CrossRef]

24. Fan, G.; Peng, L.L.; Hong, W.C.; Sun, F. Electric load forecasting by the SVR model with differential empirical mode decomposition and auto regression. *Neurocomputing* **2016**, *173*, 958–970. [CrossRef]

25. Pan, W.T. *Fruit Fly Optimization Algorithm*; Tsanghai Publishing: Taipei, Taiwan, China, 2011.

26. Pan, W.T. A new fruit fly optimization algorithm: Taking the financial distress model as an example. *Knowl.-Based Syst.* **2012**, *26*, 69–74. [CrossRef]

27. Mitić, M.; Vuković, N.; Petrović, M.; Miljković, Z. Chaotic fruit fly optimization algorithm. *Knowl.-Based Syst.* **2015**, *89*, 446–458. [CrossRef]

28. Wu, L.; Liu, Q.; Tian, X.; Zhang, J.; Xiao, W. A new improved fruit fly optimization algorithm IAFOA and its application to solve engineering optimization problems. *Knowl.-Based Syst.* **2018**, *144*, 153–173. [CrossRef]

29. Han, X.; Liu, Q.; Wang, H.; Wang, L. Novel fruit fly optimization algorithm with trend search and co-evolution. *Knowl.-Based Syst.* **2018**, *141*, 1–17. [CrossRef]

30. Zhang, X.; Lu, X.; Jia, S.; Li, X. A novel phase angle-encoded fruit fly optimization algorithm with mutation adaptation mechanism applied to UAV path planning. *Appl. Soft Comput.* **2018**, *70*, 371–388. [CrossRef]

31. Han, S.Z.; Pan, W.T.; Zhou, Y.Y.; Liu, Z.L. Construct the prediction model for China agricultural output value based on the optimization neural network of fruit fly optimization algorithm. *Future Gener. Comput. Syst.* **2018**, *86*, 663–669. [CrossRef]

32. Yang, X.S.; Gandomi, A.H. Bat algorithm: A novel approach for global engineering optimization. *Eng. Comput.* **2012**, *29*, 464–483. [CrossRef]

33. Narayanan, A.; Moore, M. Quantum-inspired genetic algorithms. In Proceedings of the IEEE International Conference on Evolutionary Computation, Nagoya, Japan, 20–22 May 1996; pp. 61–66.

34. Han, K.H.; Kim, J.H. Genetic quantum algorithm and its application to combinatorial optimization problem. In Proceedings of the 2000 Congress on Evolutionary Computation, La Jolla, CA, USA, 16–19 July 2000; pp. 1354–1360.

35. Han, K.H.; Kim, J.H. Quantum-inspired evolutionary algorithm for a class of combinatorial optimization. *IEEE Trans. Evol. Comput.* **2002**, *6*, 580–593. [CrossRef]

36. Huang, M.L. Hybridization of chaotic quantum particle swarm optimization with SVR in electric demand forecasting. *Energies* **2016**, *9*, 426. [CrossRef]

37. Lee, C.W.; Lin, B.Y. Application of hybrid quantum tabu search with support vector regression for load forecasting. *Energies* **2016**, *9*, 873. [CrossRef]

38. Lee, C.W.; Lin, B.Y. Applications of the chaotic quantum genetic algorithm with support vector regression in load forecasting. *Energies* **2017**, *10*, 1832. [CrossRef]

39. Li, M.W.; Geng, J.; Wang, S.; Hong, W.C. Hybrid chaotic quantum bat algorithm with SVR in electric load forecasting. *Energies* **2017**, *10*, 2180. [CrossRef]

40. Shi, D.Y.; Lu, L.J. A judge model of the impact of lane closure incident on individual vehicles on freeways based on RFID technology and FOA-GRNN method. *J. Wuhan Univ. Technol.* **2012**, *34*, 63–68.

41. Yuan, X.; Wang, P.; Yuan, Y.; Huang, Y.; Zhang, X. A new quantum inspired chaotic artificial bee colony algorithm for optimal power flow problem. *Energy Convers. Manag.* **2015**, *100*, 1–9. [CrossRef]

42. Peng, A.N. Particle swarm optimization algorithm based on chaotic theory and adaptive inertia weight. *J. Nanoelectron. Optoelectron.* **2017**, *12*, 404–408. [CrossRef]

43. Li, M.W.; Geng, J.; Hong, W.C.; Chen, Z.Y. A novel approach based on the Gauss-vLS-SVR with a new hybrid evolutionary algorithm and input vector decision method for port throughput forecasting. *Neural Comput. Appl.* **2017**, *28*, S621–S640. [CrossRef]

44. Li, M.W.; Hong, W.C.; Geng, J.; Wang, J. Berth and quay crane coordinated scheduling using chaos cloud particle swarm optimization algorithm. *Neural Comput. Appl.* **2017**, *28*, 3163–3182. [CrossRef]

45. Xiong, Y. Study on Short-Term Micro-Grid Load Forecasting Based on IGA-PSO RBF Neural Network. Master's Thesis, South China University of Technology, Guangzhou, China, 2016.

46. Hong, T.; Pinson, P.; Fan, S.; Zareipour, H.; Troccoli, A.; Hyndman, R.J. Probabilistic energy forecasting: Global Energy Forecasting Competition 2014 and beyond. *Int. J. Forecast.* **2016**, *32*, 896–913. [CrossRef]

47. Hong, W.C. Application of seasonal SVR with chaotic immune algorithm in traffic flow forecasting. *Neural Comput. Appl.* **2012**, *21*, 583–593. [CrossRef]

48. Diebold, F.X.; Mariano, R.S. Comparing predictive accuracy. *J. Bus. Econ. Stat.* **1995**, *13*, 134–144.

49. Derrac, J.; García, S.; Molina, D.; Herrera, F. A practical tutorial on the use of nonparametric statistical tests as a methodology for comparing evolutionary and swarm intelligence algorithms. *Swarm Evol. Comput.* **2011**, *1*, 3–18. [CrossRef]

50. Wilcoxon, F. Individual comparisons by ranking methods. *Biom. Bull.* **1945**, *1*, 80–83. [CrossRef]

energies

MDPI

Article

A Hybrid Seasonal Mechanism with a Chaotic Cuckoo Search Algorithm with a Support Vector Regression Model for Electric Load Forecasting

Yongquan Dong, Zichen Zhang and Wei-Chiang Hong *

School of Computer Science and Technology (School of Education Intelligent Technology), Jiangsu Normal University/101, Shanghai Rd., Tongshan District, Xuzhou 221116, Jiangsu, China; tomdyq@jsnu.edu.cn (Y.D.); zzcpkzzw@126.com (Z.Z.)
* Correspondence: samuelsonhong@gmail.com; Tel.: +86-516-8350-0307

Received: 24 March 2018; Accepted: 18 April 2018; Published: 20 April 2018

Abstract: Providing accurate electric load forecasting results plays a crucial role in daily energy management of the power supply system. Due to superior forecasting performance, the hybridizing support vector regression (SVR) model with evolutionary algorithms has received attention and deserves to continue being explored widely. The cuckoo search (CS) algorithm has the potential to contribute more satisfactory electric load forecasting results. However, the original CS algorithm suffers from its inherent drawbacks, such as parameters that require accurate setting, loss of population diversity, and easy trapping in local optima (i.e., premature convergence). Therefore, proposing some critical improvement mechanisms and employing an improved CS algorithm to determine suitable parameter combinations for an SVR model is essential. This paper proposes the SVR with chaotic cuckoo search (SVRCCS) model based on using a tent chaotic mapping function to enrich the cuckoo search space and diversify the population to avoid trapping in local optima. In addition, to deal with the cyclic nature of electric loads, a seasonal mechanism is combined with the SVRCCS model, namely giving a seasonal SVR with chaotic cuckoo search (SSVRCCS) model, to produce more accurate forecasting performances. The numerical results, tested by using the datasets from the National Electricity Market (NEM, Queensland, Australia) and the New York Independent System Operator (NYISO, NY, USA), show that the proposed SSVRCCS model outperforms other alternative models.

Keywords: support vector regression; tent chaotic mapping function; cuckoo search algorithm; seasonal mechanism; load forecasting

---

1. Introduction

Accurate electric load forecasting is important to facilitate the decision-making process for power unit commitment, economic load dispatch, power system operation and security, contingency scheduling, and so on [1,2]. As indicated in existing papers, a 1% electric load forecasting error increase would lead to a £10 million additional operational cost [3], on the contrary, decreasing forecasting errors by 1% would produce appreciable operation benefits [2]. Therefore, looking for more accurate forecasting models or applying novel intelligent algorithms to achieve satisfactory load forecasting results, to optimize the decisions of electricity supplies and load plans, to improve the efficiency of the power system operations, eventually, reduces the system risks to within a controllable range. However, due to lots of factors, such as energy policy, urban population, socio-economical activities, weather conditions, holidays, and so on [4], the electric load data display seasonality, non-linearity, and a chaotic nature, which complicates electric load forecasting work [5].

Lots of electric load forecasting models have been proposed to continue improving forecasting performances. These forecasting models can be of two types, the first one is based on the statistical methodology, and the other one involves applications of artificial intelligence technology. The statistical models, which include the ARIMA models [6,7], regression models [8,9], exponential smoothing models [10], Kalman filtering models [11,12], Bayesian estimation models [13,14], and so on use historical data to find out the linear relationships among time periods. However, due to their theoretical definitions, these statistical models can only deal well with linear relationships among electric loads and the other factors mentioned above. Therefore, these models could only produce unsatisfactory forecasting performances [15].

Due to its superior nonlinear processing capability, artificial intelligence technology methods such as artificial neural networks (ANNs) [16,17], expert system models [18,19], and fuzzy inference systems [20,21] have been widely applied to improve the performance of electric load forecasting. To overcome the inherent shortcomings of these artificial intelligent models, hybrid models (hybridizing two artificial intelligent models with each other) and combined models (combining two models with each other) have been the research hotspots recently. For example, hybridized or combined with each other models [22] and with evolutionary algorithms [23]. However, these artificial intelligence models (including hybrid or combined models) also have shortcomings themselves, such as being time consuming, difficult to determine structural parameters, and trapping into local minima. Readers may refer to [24] for more discussions regarding load forecasting.

With outstanding nonlinear processing capability, composed of high dimensional mapping ability and kernel computing technology, the support vector regression (SVR) model [25–27] has already produced superior abundant application results in many fields. The application experience demonstrates that an SVR model with well-computed parameters by any evolutionary algorithm could provide significant satisfactory forecasting performance, and overcome the shortcomings of evolutionary algorithms to compute appropriate parameters. For applications in electric load forecasting, Hong and his successors [28,29] have used two types of chaotic mapping functions (i.e., logistic function and cat mapping function) to keep the diversity of population during the search process to avoid trapping into local optima, to significantly improve the forecasting accuracy level.

The cuckoo search (CS) algorithm [30] is a novel meta-heuristic optimization algorithm inspired by the brood reproductive strategy of cuckoo birds via an interesting brood parasitic mechanism, i.e., mimicking the pattern and color of the host's eggs, throwing the eggs out or not, or building a new nest, etc. In [31], the authors demonstrate that, by applying various test functions, it is superior to other algorithms, such as genetic algorithm (GA), differential evolution (DE), simulated annealing (SA) algorithm, and particle swarm optimization (PSO) algorithm in searching for a global optimum. Nowadays, the CS algorithm is widely applied in engineering applications, such as unit maintenance scheduling [32], data clustering optimization [33], medical image recognition [34], manufacturing engineering optimization [35], and software cost estimation [36], etc. However, as mentioned in [37], the original CS algorithm has some inherent limitations, such as its initialization settings of the host nest location, Lévy flight parameter, and boundary handling problem. In addition, because it is a population-based optimization algorithm, the original CS algorithm also suffers from slow convergence rate in the later searching period, homogeneous searching behaviors (low diversity of population), and a premature convergence tendency [33,38,39].

Due to its easy implementation and ability to enrich the cuckoo search space and diversify the population to avoid trapping into local optima, this paper would like to apply a chaotic mapping function to overcome the core shortcomings of the original CS algorithm, to produce more accurate electric load forecasting results. Thus, a tent chaotic mapping function, demonstrating a range of dynamical behavior ranging from predictable to chaos, is hybridized with a CS algorithm to determine three parameters of an SVR model. A new electric load forecasting model, obtained by hybridizing a tent chaotic mapping function and CS algorithm with an SVR model, namely the SVR with chaotic cuckoo search (SVRCCS) model, is thus proposed. In the meanwhile, as mentioned in existing papers [5,28,29], electric load data, particularly short term load data, illustrate an obvious cyclic tendency, thus, the seasonal mechanism proposed in the authors' previous papers [5,28,29] would be further improved and combined with the SVRCCS model. Finally, the proposed seasonal SVR with CCS, namely the SVR with chaotic cuckoo search (SSVRCCS) model, is employed to improve the forecasting accuracy level by sufficiently capturing the non-linear and cyclic tendency of electric load changes. Furthermore, the forecasting results of the SSVRCCS model are used to compare them with other alternative models, such as the SARIMA, GRNN, SVRCCS, and SVRCS models, to test the forecasting accuracy improvements achieved. The principal contribution of this paper is in continuing to hybridize the SVR model with a tent chaotic computing mechanism, CS algorithm, and eventually, combine a seasonal mechanism, to widely explore the electric load forecasting model to produce higher accuracy performances.

The remainder of this article is organized as follows: the basic formulation of an SVR model, the proposed CCS algorithm, seasonal mechanism, and the modeling details of the proposed SSVRCCS model are described in Section 2. A numerical example and forecasting accuracy comparisons among the proposed model and other alternative models are presented in Section 3. Finally, conclusions are given in Section 4.

2. The Proposed SVR with Chaotic Cuckoo Search (SSVRCCS) Model

2.1. Support Vector Regression (SVR) Model

The modeling details of an SVR model are presented briefly as follows. The training data set, $\{(\mathbf{x}_i, y_i)\}_{i=1}^{N}$, is mapped into a high dimensional feature space by a non-linear mapping function, $\varphi(\mathbf{x})$. Then, in the high dimensional feature space, the SVR function, f, is theoretically used to formulate the nonlinear relationships between the input training data ($\mathbf{x}_i$) and the output data (y_i). This can be shown as Equation (1):

$$f(\mathbf{x}) = \mathbf{w}^T \varphi(\mathbf{x}) + b \qquad (1)$$

where $f(\mathbf{x})$ represents the forecasted values; the weight, $\mathbf{w}$, and the coefficient, b, are computed along with minimizing the empirical risk, as shown in Equation (2):

$$R(f) = C\frac{1}{N}\sum_{i=1}^{N}\Theta_{\varepsilon}(y_i, f(\mathbf{x}_i)) + \frac{1}{2}\mathbf{w}^T\mathbf{w} \qquad (2)$$

$$\Theta_{\varepsilon}(\mathbf{y}, f(\mathbf{x})) = \begin{cases} 0, & if |f(\mathbf{x}) - \mathbf{y}| \leq \varepsilon \\ |f(\mathbf{x}) - \mathbf{y}| - \varepsilon, & otherwise \end{cases} \qquad (3)$$

where $\Theta_{\varepsilon}(\mathbf{y}, f(\mathbf{x}))$ is so-called ε-insensitive loss function, as shown in Equation (3). It is used to determine the optimal hyperplane to separate the training data into two subsets with maximal distance, i.e., minimizing the training errors between these two separated training data subsets and $\Theta_{\varepsilon}(\mathbf{y}, f(\mathbf{x}))$, respectively. C is a parameter to penalize the training errors. The second term, $\frac{1}{2}\mathbf{w}^T\mathbf{w}$, is then used to represent the maximal distance between mentioned two separated data subsets, meanwhile, it also determines the steepness and the flatness of $f(\mathbf{x})$.

Then, the SVR modeling problem could be demonstrated as minimizing the total training errors. It is a quadratic programming problem with two slack variables, ξ and ξ^*, to measure the distance between the training data values and the edge values of ε-tube. Training errors under ε are denoted as ξ^*, whereas training errors above ε are denoted as ξ, as shown in Equation (4):

$$\operatorname*{Min}_{w, \xi, \xi^*} R(w, \xi, \xi^*) = \frac{1}{2}\|w\|^2 + C\sum_{i=1}^{N}(\xi_i + \xi_i^*) \qquad (4)$$

with the constraints:

$$y_i - f(\mathbf{x}_i) \leq \varepsilon + \xi_i^*,$$
$$-y_i - f(\mathbf{x}_i) \leq \varepsilon + \xi_i,$$
$$\xi_i^* \geq 0$$
$$\xi_i \geq 0$$
$$i = 1, 2, \ldots, N$$

The solution of Equation (4) is optimized by using Lagrange multipliers, β_i^*, and β_i, the weight vector, $\mathbf{w}$, in Equation (1) is computed as Equation (5):

$$\mathbf{w}^* = \sum_{i=1}^{N} (\beta_i^* - \beta_i) \varphi(\mathbf{x}_i) \tag{5}$$

Eventually, the SVR forecasting function is calculated as Equation (6):

$$f(\mathbf{x}) = \sum_{i=1}^{N} (\beta_i^* - \beta_i) K(\mathbf{x}_i, \mathbf{x}_j) + b \tag{6}$$

where $K(\mathbf{x}_i, \mathbf{x}_j)$ is the so-called kernel function, and its value could be computed by the inner product of $\varphi(\mathbf{x}_i)$ and $\varphi(\mathbf{x}_j)$, i.e., $K(\mathbf{x}_i, \mathbf{x}_j) = \varphi(\mathbf{x}_i) \times \varphi(\mathbf{x}_j)$. The are several kinds of kernel function, such as Gaussian function (Equation (7)) and the polynomial kernel function. Due to its superior ability to map nonlinear data into high dimensional space, a Gaussian function is used in this paper:

$$K(\mathbf{x}_i, \mathbf{x}_j) = \exp\left(-\frac{\|\mathbf{x}_i - \mathbf{x}_j\|^2}{2\sigma^2}\right) \tag{7}$$

Therefore, determining the three parameters, σ, C, and ε of an SVR model would play the critical role to achieve more accurate forecasting performances [5,28,29]. The parameter ε decides the number of support vectors. If ε is large enough, it implies few support vectors with low forecasting accuracy; if ε has a value that is too small, it would increase the forecasting accuracy but be too complex to adopt. Parameter C, as mentioned, penalizes the training errors. If C is large enough, it would increase the forecasting accuracy but suffer from being difficult to adopt; if C has a too small value, the model would suffer from large training errors. Parameter σ represents the relationships among data and the correlations among support vectors. If σ is large enough, the correlations among support vectors are strong and we can obtain accurate forecasting results, but if the value of σ is small, the correlations among support vectors are weak, and adoption is difficult.

However, structural methods to determine the SVR parameters are lacking. Hong and his colleagues have pointed out the advanced exploration way by hybridizing chaotic mapping functions with evolutionary algorithms to overcome the embedded premature convergence problem, to select suitable parameter combination, to achieve highly accurate forecasting performances. To continue this valuable exploration, the chaotic cuckoo search algorithm, the CCS algorithm, is proposed to be hybridized with an SVR model to determine an appropriate parameter combination.

2.2. Chaotic Cuckoo Search (CCS) Algorithm

2.2.1. Tent Chaotic Mapping Function

The chaotic mapping function is an optimization technique to map the original data series to show sensitive dependence on the initial conditions and infinite different periodic responses (chaotic ergodicity), to maintain the diversity of population in the whole optimization procedures, to enrich the search behavior, and to avoid premature convergence. The most popular chaotic mapping function is the logistic function, however, based on the analysis on the chaotic characteristics of the different mapping functions, a tent chaotic mapping function [39] demonstrates a range of dynamical behavior ranging from predictable to chaos, i.e., with good ergodic uniformity [40]. This paper thus applies the tent chaotic mapping function to be hybridized with the CS algorithm to determine the three parameters of an SVR model.

The tent chaotic mapping function is shown as Equation (8):

$$x_{n+1} = \begin{cases} 2x_n & x \in [0, 0.5] \\ 2(1 - x_n) & x \in (0.5, 1] \end{cases} \tag{8}$$

where x_n is the iterative value of the variable x in the nth step, and n is the number of iteration steps.

2.2.2. Cuckoo Search (CS) Algorithm

The CS algorithm is a novel meta-heuristic optimization algorithm, inspired by cuckoo birds' obligate brood parasitic behavior of laying their eggs in the nests of other host birds. Meanwhile, by applying Lévy flight behaviors, the search speed is much faster than that of the normal random walk. Therefore, cuckoo birds can reduce the number of iterations and thus speed up the local search efficiency. For CS algorithm implementation, each egg in a nest represents a potential solution. The cuckoo birds could choose, by Lévy flight behaviors, recently-spawned nests to lay their eggs in the host nests to ensure their eggs could hatch first due to the natural phenomenon that cuckoo eggs usually hatch before the host birds' eggs. It takes times for the host birds to discover that the eggs in their nests do not belong to them, based on the probability, p_a. When these "stranger" eggs are discovered, they either throw out those eggs or abandon the whole nest to build a new nest in a new location. The cuckoo birds would continuously lay new eggs (solutions), and they would choose the nest, by Lévy flight behaviors, around the current best solutions.

The CS algorithm contains three famous idealized rules [31]: (1) each cuckoo lays one egg at a time in a randomly selected host; (2) high-quality eggs and their host nests would survive to the next generation; (3) the number of available host nests is fixed, and the host bird detects the "stranger" egg with a probability $p_a \in [0, 1]$. In this case, the host bird can either throw away the egg or abandon the nest, and build a completely new nest. The last rule can be approximated by a fraction (p_a) of the n host nests that are replaced by new nests (with new random solutions). The value of p_a is often set as 0.25 [37].

The CS algorithm could maintain the balance between two kinds of search (random walks), the local search and the global search, by a switching parameter, p_a. The switching parameter p_a determines the cuckoo birds to abandon a fraction of the worst nests and build new ones for discovering new and more promising regions in the search space. These two random walks are defined by Equations (9) and (10), respectively:

$$x_i^{t+1} = x_i^t + \alpha s \otimes H(p_a - \delta) \otimes \left(x_j^t - x_k^t \right) s \tag{9}$$

$$x_i^{t+1} = x_i^t + \alpha \mathcal{L}(s, \lambda) \tag{10}$$

where x_j^t and x_k^t are current positions randomly selected; α is the positive Lévy flight step size scaling factor; s is the step size; $H(\cdot)$ is the Heavy-side function; δ is a random number from uniform distribution; $\otimes$ represents the entry-wise product of two vectors; $\mathcal{L}(s, \lambda)$ is the Lévy distribution and is used to define the step size of random walk, it is defined as Equation (11):

$$\mathcal{L}(s, \lambda) = \frac{\lambda \Gamma(\lambda) \sin(\pi \lambda / 2)}{\pi} \frac{1}{s^{1+\lambda}} \tag{11}$$

where λ is the standard deviation of step size; the gamma function, $\Gamma(\lambda)$, is defined as $\Gamma(\lambda) = \int_0^\infty t^{\lambda-1} e^{-t} dt$, and represents an extension of factorial function, if λ is a positive integer, then, $\Gamma(\lambda) = (\lambda - 1)!$. Lévy flight distribution enables a series of straight jumps chosen from any flight movements, it is also capable to find out the global optimum, i.e., it could ensure that the system will not be trapped in a local optimum [41].

2.2.3. Implementation Steps of CCS Algorithm

The procedure of the hybrid CCS algorithm with an SVR model is illustrated as followings. The relevant flowchart is shown in Figure 1.

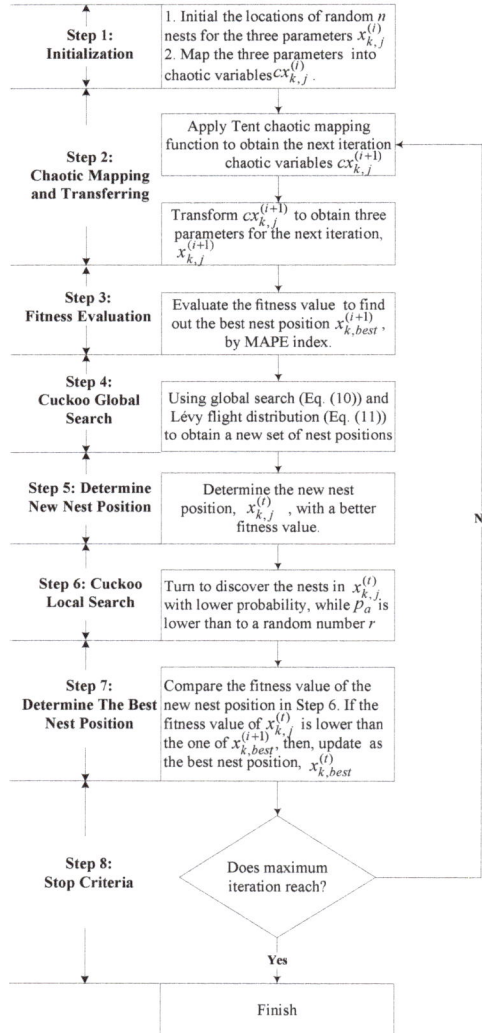

Figure 1. Chaotic cuckoo search algorithm flowchart.

Step 1: Initialization.

The locations of random n nests for the three parameters of an SVR model as $x_{k,j}^{(i)} = \left[x_{k,1}^{(i)}, x_{k,2}^{(i)}, \ldots, x_{k,n}^{(i)} \right]^T$, $k = C, \sigma, \varepsilon$; i represents the iteration number; j represents the number of nests. Let $i = 0$, and normalize the parameters as chaotic variables, $cx_{k,j}^{(i)}$, within the interval $[0, 1]$ by Equation (12):

$$cx_{k,j}^{(i)} = \frac{x_{k,j}^{(i)} - Min_k}{Max_k - Min_k} \tag{12}$$

where Min_k and Max_k are the minima and the maxima of the three parameters, respectively.

Step 2: Chaotic Mapping and Transferring.

Apply the tent chaotic mapping function, defined as Equation (8), to obtain the next iteration of chaotic variables, $cx_{k,j}^{(i+1)}$, as shown in Equation (13):

$$cx_{k,j}^{(i+1)} = \begin{cases} 2cx_{k,j}^{(i)} & cx_{k,j}^{(i)} \in [0, 0.5] \\ 2\left(1 - cx_{k,j}^{(i)}\right) & cx_{k,j}^{(i)} \in (0.5, 1] \end{cases} \tag{13}$$

Then, transform $cx_{k,j}^{(i+1)}$ to obtain three parameters for the next iteration, $x_{k,j}^{(i+1)}$, by the following Equation (14):

$$x_{k,j}^{(i+1)} = Min_k + cx_{k,j}^{(i+1)}(Max_k - Min_k) \tag{14}$$

Step 3: Fitness Evaluation.

Evaluate the fitness value with $x_{k,j}^{(i+1)}$ for all nests to find out the best nest position, $x_{k,best}^{(i+1)}$, in terms of smaller forecasting accuracy index value. In this paper, the forecasting error is calculated as the fitness value by the mean absolute percentage error (MAPE), as shown in Equation (15):

$$MAPE = \frac{1}{N} \sum_{i=1}^{N} \left| \frac{a_i - f_i}{a_i} \right| \times 100\% \tag{15}$$

where N is the total number of data; a_i is the actual electric load value at point i; f_i is the forecasted electric load value at point i.

Step 4: Cuckoo Global Search.

Implement a cuckoo global search, i.e., Equation (10), by using the best nest position, $x_{k,best}^{(i+1)}$, and update other nest positions by Lévy flight distribution (Equation (11)) to obtain a new set of nest positions, then, compute the fitness value.

Step 5: Determine New Nest Position.

Compare the fitness value of the new nest positions with the fitness value of the previous iteration, and update the nest position with a better one. Then determine the new nest position as $x_{k,j}^{(t)} = \left[x_{k,1}^{(t)}, x_{k,2}^{(t)}, \ldots, x_{k,n}^{(t)} \right]^T$.

Step 6: Cuckoo Local Search.

If p_a is lower than to a random number r, then turn to discover the nests in $x_{k,j}^{(t)}$ with lower probability instead of the higher one. Then, compute the fitting value of the new nests and continue updating the nest position $x_{k,j}^{(t)}$ with smaller MAPE value by comparing it with the previous fitness value.

Step 7: Determine The Best Nest Position.

Compare the fitness value of the new nest position, $x_{k,j}^{(t)}$, in Step 6, with the fitness value of the best nest position, $x_{k,best}^{(i+1)}$. If the fitness value of $x_{k,j}^{(t)}$ is lower than the one of $x_{k,best}^{(i+1)}$, then, update $x_{k,j}^{(t)}$ as the best nest position, $x_{k,best}^{(t)}$.

Step 8: Stop Criteria.

If the number of search iterations are greater than a given maximum search iterations, then, the best nest position, $x_{k,best}^{(t)}$, among the current population is determined as parameters (C, σ, ε) of an SVR model; otherwise, go back to Step 2 and continue searching the next iteration.

2.3. Seasonal Mechanism

As indicated in existing papers [5,28,29] the short term electric load data often display cyclic tendencies due to the cyclic nature of economic activities (production, transportation, operation, etc.) or the seasonal climate in Nature (air conditioners and heaters in summer and winter, respectively). It is useful to increase the forecasting accuracy by calculating these seasonal effects (or seasonal indexes) to adjust the seasonal biases. Several researchers have proposed seasonal adjustment approaches to determine the seasonal effects, such as Koc and Altinay [42], Goh and Law [43], and Wang et al. [44], who all apply regression models to decompose the seasonal component. Martens et al. [45] apply a flexible Fourier transform to estimate the daily variation of the stock exchange, and compute a seasonal estimator. Deo et al. [46] composed two Fourier transforms in a cyclic period to further identify the seasonal estimator. Comparing these seasonal adjustment models, Deo's model extends Martens's model for application to general cycle-length data, particularly for hour-based or other shorter cycle-length data. Considering that this paper deals with half-hour based short term electric load data, this paper would like to employ the seasonal mechanism proposed by Hong and his colleagues in [5,28,29]. That is, firstly apply the ARIMA model to identify the seasonal length of the target time series data set; secondly, calculate these seasonal indexes to adjust cyclic effects to receive more satisfied forecasting performances, as shown in Equation (16):

$$Seasonratio_q = \ln\left(\frac{a_q}{f_q}\right)^2 = 2\left(\ln a_q - \ln f_q\right) \tag{16}$$

where $q = j, l + j, 2l + j, \ldots, (m - 1)l + j$ with m seasonal (cyclic) periods and l seasonal length in each period. Thirdly, the seasonal index (SI) for each seasonal point j in each period is calculated as Equation (17):

$$SI_j = \exp\left(\frac{1}{m}\sum_{q=j}^{(m-1)l+j} Seasonratio_q\right)/2 \tag{17}$$

where $j = 1,2, \ldots l$. The seasonal mechanism is demonstrated in Figure 2.

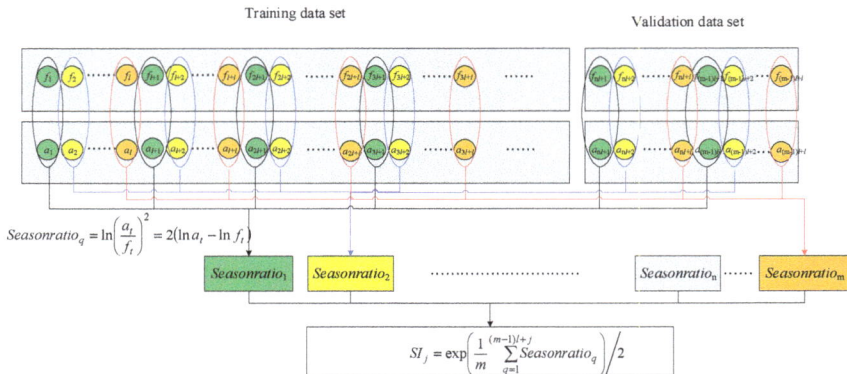

Figure 2. Seasonal mechanism.

3. Numerical Examples of the Proposed SSVRCCS Model

3.1. Data Set of Numerical Examples

To demonstrate the superiorities of the tent chaotic mapping function and seasonal mechanism of the proposed SSVRCCS model, this paper uses the half-hour electric load data from the Queensland regional market of the National Electricity Market (NEM, Queensland, Australia) [47], named Example 1, and the New York Independent System Operator (NYISO, New York, NY, USA) [48], named Example 2. The employed electric load data contains a total of 768 half-hour electric load values in Example 1, i.e., from 00:30 01 October 2017 to 00:00 17 October 2017. Based on Schalkoff's [49] recommendation that the ratio of validation data set to training data set should be approximately one to four, therefore, the electric load data set is divided into three sub-sets. The training set has 432 half-hour electric load values (i.e., from 00:30 01 October 2017 to 00:00 09 October 2017). The validation set contains 144 half-hour electric load values (i.e., from 00:30 09 October 2017 to 00:00 13 October 2017). The testing set has 192 half-hour electric load values (i.e., from 00:30 13 October 2017 to 00:00 17 October 2017). Similarly, in Example 2, the used electric load data also contains a total of 768 hourly electric load values, i.e., from 00:00 01 January 2018 to 23:00 1 February 2018. The electric load data set is also divided into three sub-sets. The training set has 432 hourly electric load values (i.e., from 00:00 01 January 2018 2017 to 23:00 18 January 2018). The validation set has 144 hourly electric load values (i.e., from 00:00 19 January 2018 to 23:00 24 January 2018). The testing set has 192 hourly electric load values (i.e., from 00:00 25 January 2018 to 23:00 1 February 2018). To be based on the same comparison conditions, all compared models thus have the same data division sets.

During the modeling processes, in the training stage, the rolling-based procedure, proposed by Hong [28], is also applied to assist CCS algorithm to implement well searching for an appropriate parameter combination (σ, C, ε) of an SVR model. Specifically, the CCS algorithm minimizes the empirical risk, as shown in Equation (4), to obtain the potential parameter combination by employing the first n electric load data in the training set; then, it receives the first forecasted electric load by the SVR model with these potential parameter combination, i.e., the $(n + 1)$th forecasting electric load. For the second round, the next n electric load data, from 2nd to $(n + 1)$th electric load values, are then used by the SVR model to obtain new potential parameter combination, then, similarly, the $(n + 2)$th forecasting electric load is receive. This procedure would never be stopped till the totally 432 forecasting electric load are computed. The training error and the validation error are also calculated in each iteration.

Only with the smallest validation and testing errors, a potential parameter combination could be finalized as the determined parameter combination of an SVR model. Then, the never used testing data set would be employed to demonstrate the forecasting performances, i.e., eventually, the 192 half-hour/hourly electric load would be forecasted by the proposed SSVRCCS model.

3.2. The SVR with Chaotic Cuckoo Search (SSVRCCS) Electric Load Forecasting Model

3.2.1. Embedded Parameter Settings of the CCS Algorithm

The embedded parameters of CCS algorithm for modeling are set as follows: the number of host nests is set to be 50; the maximum number of iterations is set as 500; the initial probability parameter p_a is set as 0.25. During the parameter optimizing process of an SVR model, the searching feasible ranges of the three parameters are set as following, $\sigma \in [0.01, 5]$, $\varepsilon \in [0.01, 1]$, and $C \in [0.01, 60,000]$. In addition, considering that the iteration time would affect the performance of each model, the given optimization time for each model with an evolutionary algorithm is set at the same inasmuch as possible.

3.2.2. Forecasting Accuracy Indexes

Three forecasting accuracy evaluation indexes are used to compare the forecasting performances for each model: (1) the MAPE mentioned in Equation (5); (2) the root mean square error (RMSE); and (3) the mean absolute error (MAE). The latter two indexes could be calculated by Equations (18) and (19), respectively:

$$\text{RMSE} = \sqrt{\frac{\sum_{i=1}^{N}(a_i - f_i)^2}{N}} \cdot s \tag{18}$$

$$\text{MAE} = \frac{1}{N}\sum_{i=1}^{N}|a_i - f_i| \tag{19}$$

where N is the total number of data; a_i is the actual electric load value at point i; f_i is the forecasted electric load value at point i.

3.2.3. Forecasting Accuracy Significance Tests

To demonstrate the significant superiority of the proposed SSVRCCS model in terms of forecasting accuracy, some famous statistical tests are implemented. Based on Diebold and Mariano's [50] and Derrac et al. [51] research suggestions, the Wilcoxon signed-rank test [52] and Friedman test [53] are simultaneously applied in this paper.

The Wilcoxon signed-rank test is used to compare the significant differences in terms of central tendency between two data set with the same size. Let d_i represent the i-th pair difference of the i-th forecasting errors from any two forecasting models, the differences are ranked according to their absolute values. Let r^+ represent the sum of ranks that the first model larger than the second one; r^- represent the sum of ranks that the second model larger than the first one. In case of $d_j = 0$, then, exclude the j-th pair and reduce sample size. The statistic W of the Wilcoxon signed-rank test is shown as Equation (20):

$$W = \min\{r^+, r^-\} \tag{20}$$

If W meets the criterion of the Wilcoxon distribution under N degrees of freedom, then, the null hypothesis of equal performance of these two compared models cannot be accepted. It also implies that the proposed model is significantly superior to the other model. Of course, if the comparison size is larger than the critical size, the sampling distribution of W would approximate to the normal distribution instead of Wilcoxon distribution, and the associated p-value would also be provided.

On the other hand, due to the non-parametric statistical test in the ANOVA analysis procedure, the Friedman test is devoted to compare the significant differences among two or more models. The statistic F of the Friedman test is shown as Equation (21):

$$F = \frac{12N}{k(k+1)}\left[\sum_{j=1}^{k}R_j^2 - \frac{k(k+1)^2}{4}\right] \tag{21}$$

where N is the total number of forecasting results; k is the number of compared models; R_j is the average rank sum obtained in each forecasting value for each compared model as shown in Equation (22),

$$R_j = \frac{1}{N}\sum_{i=1}^{N}r_i^j \tag{22}$$

where r_i^j is the rank sum from 1 (the smallest forecasting error) to k (the worst forecasting error) for ith forecasting result, for jth compared model.

Similarly, if the associated p-value of F meets the criterion of not acceptance, the null hypothesis, equal performance among all compared models, could also not be held.

3.2.4. Forecasting Results and Analysis for Example 1

To compare the improved forecasting performance of the tent chaotic mapping function, a SVR with the original CS algorithm (without the tent chaotic mapping function), namely the SVRCS model, will also be taken into comparison. Therefore, according to the rolling-based procedure mentioned above, by using the training data set from Example 1 (mentioned in Section 3.1) to conduct the training work, and the parameters for SVRCS and SVRCCS models are eventually determined. These trained models are further used to forecast the electric load. Then, the forecasting results and the suitable parameters of SVRCS and SVRCCS models are listed in Table 1. It is clearly indicated that the proposed SVRCCS model has achieved smaller forecasting performances in terms of the forecasting accuracy indexes, MAPE, RMSE, and MAE.

Table 1. Three parameters of SVRCS and SVR with chaotic cuckoo search (SVRCCS) models for Example 1.

Evolutionary Algorithms	Parameters			MAPE of Testing (%)	RMSE of Testing	MAE of Testing
	σ	C	ε			
SVRCS	1.4744	17,877.54	0.3231	2.63	217.19	151.72
SVRCCS	0.5254	5,885.65	0.7358	1.51	126.92	87.94

As shown in Figure 3, the employed electric load data demonstrates seasonal/cyclic changing tendency in Example 1. In addition, the data recording frequency is on a half-hour basis, therefore, to comprehensively reveal the electric load changing tendency, the seasonal length is set as 48. Therefore, there are 48 seasonal indexes for the proposed SVRCCS and SVRCS models. The seasonal indexes for each half-hour are computed based on the 576 forecasting values of the SVRCCS and SVRCS models in the training (432 forecasting values) and validation (144 forecasting values) processes. The 48 seasonal indexes for the SVRCCS and SVRCS models are listed in Table 2, respectively.

Table 2. The 48 seasonal indexes for SVRCCS and SVRCS models for Example 1.

Time Points	Seasonal Index (SI)		Time Points	Seasonal Index (SI)		Time Points	Seasonal Index (SI)		Time Points	Seasonal Index (SI)	
	SVRCCS	SVRCS		SVRCCS	SVRCS		SVRCCS	SVRCS		SVRCCS	SVRCS
00:00	0.9615	0.9201	06:00	1.0360	1.0536	12:00	1.0025	1.0076	18:00	1.0071	1.0176
00:30	0.9881	0.9241	06:30	1.0518	1.0729	12:30	0.9960	1.0032	18:30	1.0034	1.0109
01:00	0.9893	0.9401	07:00	1.0671	1.0924	13:00	0.9935	0.9992	19:00	0.9694	0.9767
01:30	0.9922	0.9729	07:30	1.0394	1.0810	13:30	0.9975	1.0022	19:30	0.9913	0.9875
02:00	0.9919	0.9955	08:00	1.0088	1.0575	14:00	1.0026	1.0083	20:00	0.9820	0.9812
02:30	0.9948	0.9980	08:30	1.0076	1.0322	14:30	1.0015	1.0088	20:30	0.9789	0.9700
03:00	0.9950	0.9998	09:00	1.0004	1.0148	15:00	1.0000	1.0070	21:00	0.9830	0.9641
03:30	0.9915	0.9961	09:30	0.9903	0.9982	15:30	1.0022	1.0089	21:30	0.9780	0.9547
04:00	1.0082	1.0129	10:00	1.0031	1.0067	16:00	1.0033	1.0115	22:00	0.9906	0.9622
04:30	1.0075	1.0176	10:30	0.9912	0.9981	16:30	1.0097	1.0173	22:30	0.9932	0.9778
05:00	1.0124	1.0245	11:00	0.9928	0.9973	17:00	1.0098	1.0188	23:00	0.9659	0.9645
05:30	1.0139	1.0253	11:30	0.9967	1.0025	17:30	1.0053	1.0164	23:00	0.9601	0.9348

The forecasting comparison curves of six models, including the SARIMA$_{(9,1,8) \times (4,1,4)}$, GRNN ($\sigma = 0.04$), SSVRCCS, SSVRCS, SVRCCS, and SVRCS models mentioned above and actual values are shown in Figure 4. It illustrates that the proposed SSVRCCS model is closer to the actual electric load values than other compared models. To further illustrate the tendency capturing capability of the proposed SSVRCCS model during the electric peak loads, Figures 5–8 are enlargements from four peaks in Figure 4 to clearly demonstrate how closer the SSVRCCS model matches to the actual electric load values than other alternative models. For example, for each peak, the red real line (SSVRCCS model) always follows closely with the black real line (actual electric load), whether climbing up the peak or climbing down the hill.

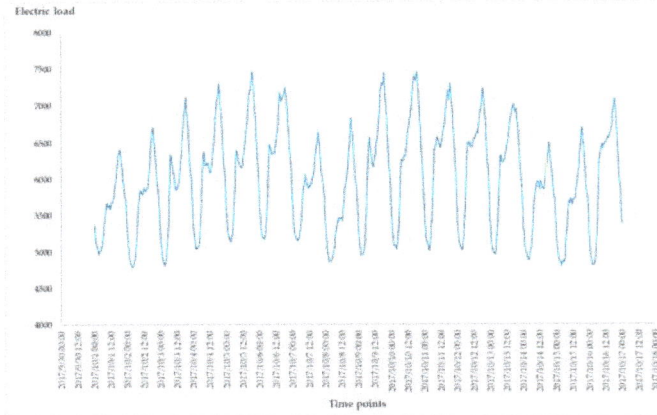

Figure 3. The seasonal tendency of actual half-hour electric load in Example 1.

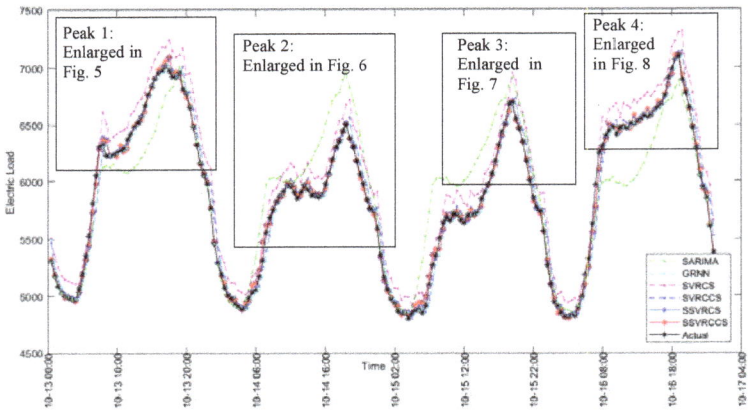

Figure 4. Forecasting values of SSVRCCS model and other alternative models for Example 1.

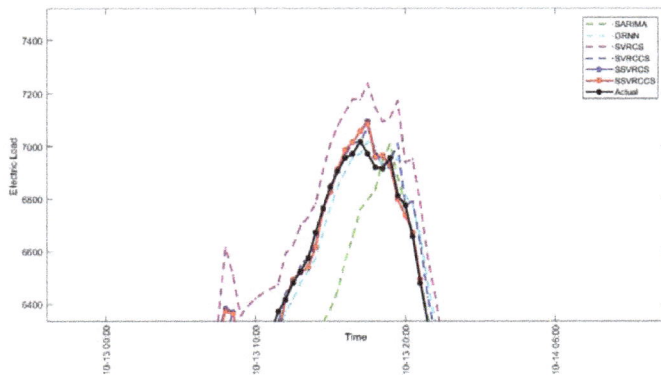

Figure 5. The enlargement comparison of Peak 1 from the compared models for Example 1.

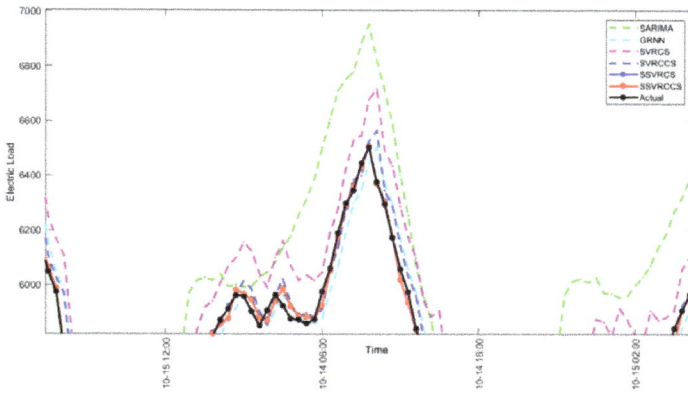

Figure 6. The enlargement comparison of Peak 2 from the compared models for Example 1.

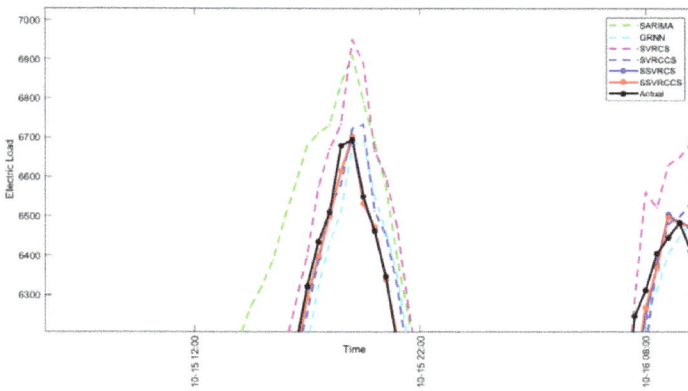

Figure 7. The enlargement comparison of Peak 3 from the compared models for Example 1.

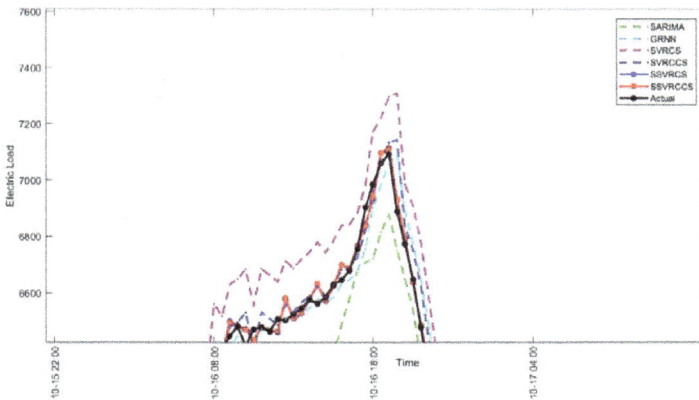

Figure 8. The enlargement comparison of Peak 4 from the compared models for Example 1.

Table 3 illustrates the forecasting accuracy indexes for the proposed SSVRCCS model and other alternative compared models. It is clearly to see that the MAPE, RMSE, and MAE of the proposed SSVRCCS model are 0.70%, 56.90, and 40.79, respectively, which are superior to the other five alternative models. It also implies that the proposed SSVRCCS model contributes great improvements in terms of load forecasting accuracy.

Table 3. Forecasting accuracy indexes of the compared models for Example 1.

Forecasting Accuracy Indexes	$SARIMA_{(9,1,8)\times(4,1,4)}$	$GRNN(\sigma = 0.04)$	SSVRCCS	SSVRCS	SVRCCS	SVRCS
MAPE (%)	3.62	1.53	0.70	0.99	1.51	2.63
RMSE	280.05	114.30	56.90	80.42	126.92	217.19
MAE	217.67	88.63	40.79	57.69	87.94	151.72

Finally, to ensure the significant contribution in terms of forecasting accuracy improvement for the proposed SSVRCCS model, the Wilcoxon signed-rank test and the Friedman test are conducted. Where Wilcoxon signed-rank test is implemented under two significance levels, $\alpha = 0.025$ and $\alpha = 0.05$, by two-tail test; the Friedman test is then implemented under only one significance level, $\alpha = 0.05$. The test results in Table 4 show that the proposed SSVRCCS model almost reaches a significance level in terms of forecasting performance than other alternative compared models.

Table 4. Results of Wilcoxon signed-rank test and Friedman test for Example 1.

Compared Models	Wilcoxon Signed-Rank Test				Friedman Test
	$\alpha = 0.025$; $W = 9264$	p-Value	$\alpha = 0.05$; $W = 9264$	p-Value	$\alpha = 0.05$;
SSVRCCS vs. $SARIMA_{(9,1,8)\times(4,1,4)}$	842 [a]	0.00000 **	842 [a]	0.00000 **	
SSVRCCS vs. $GRNN(\sigma = 0.04)$	3025 [a]	0.00000 **	3025 [a]	0.00000 **	$H_0 : e_1 = e_2 = e_3 = e_4 = e_5 = e_6$
SSVRCCS vs. SSVRCS	2159 [a]	0.00000 **	2159 [a]	0.00000 **	$F = 23.49107$
SSVRCCS vs. SVRCCS	3539 [a]	0.00000 **	3539 [a]	0.00000 **	$p = 0.000272$ (Reject H_0)
SSVRCCS vs. SVRCS	4288 [a]	0.00000 **	4288 [a]	0.00000 **	

[a] Denotes that the SSVRCCS model significantly outperforms the other alternative compared models; * represents that the test indicates not to accept the null hypothesis under $\alpha = 0.05$. ** represents that the test indicates not to accept the null hypothesis under $\alpha = 0.025$.

3.2.5. Forecasting Results and Analysis for Example 2

Similar to Example 1, SVRCS and SVRCCS models are also trained based on the rolling-based procedure by using the training data set from Example 2 (mentioned in Section 3.1). The forecasting results and the suitable parameters of SVRCS and SVRCCS models are shown in Table 5. It is also obviously that the proposed SVRCCS model has achieved a smaller forecasting performance in terms of forecasting accuracy indexes, MAPE, RMSE, and MAE.

Table 5. Three parameters of SVRCS and SVRCCS models for Example 2.

Evolutionary Algorithms	Parameters			MAPE of Testing (%)	RMSE of Testing	MAE of Testing
	σ	C	ε			
SVRCS	0.6628	36,844.57	0.2785	3.42	886.67	631.40
SVRCCS	0.3952	42,418.21	0.7546	2.30	515.10	426.42

Figure 9 also demonstrates the seasonal/cyclic changing tendency from the used electric load data in Example 2. Based on the hourly recording frequency, to completely address the changing tendency of the employed data, the seasonal length is set as 24. Therefore, there are 24 seasonal indexes for the proposed SVRCCS and SVRCS models. The seasonal indexes for each hour are computed based on the 576 forecasting values of the SVRCCS and SVRCS models in the training (432 forecasting values) and validation (144 forecasting values) processes. The 24 seasonal indexes for the SVRCCS and SVRCS models are listed in Table 6, respectively.

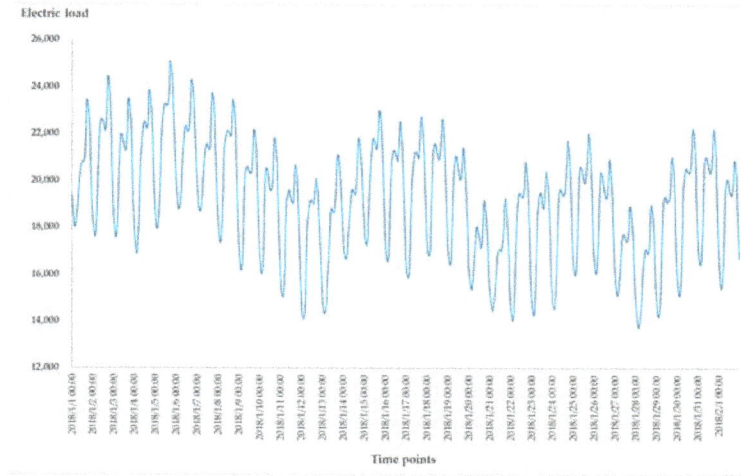

Figure 9. The seasonal tendency of actual hourly electric load in Example 2.

Table 6. The 24 seasonal indexes for SVRCCS and SVRCS models for Example 2.

Time Points	Seasonal Index (SI)		Time Points	Seasonal Index (SI)		Time Points	Seasonal Index (SI)		Time Points	Seasonal Index (SI)	
	SVRCCS	SVRCS		SVRCCS	SVRCS		SVRCCS	SVRCS		SVRCCS	SVRCS
00:00	0.9718	0.9317	06:00	1.0545	1.1043	12:00	0.9848	0.9911	18:00	0.9753	1.0242
01:00	0.9848	0.9670	07:00	1.0383	1.1133	13:00	0.9896	0.9959	19:00	0.9707	0.9743
02:00	0.9894	0.9960	08:00	0.9854	1.0833	14:00	0.9898	0.9960	20:00	0.9711	0.9754
03:00	0.9937	1.0001	09:00	0.9913	1.0259	15:00	0.9994	1.0058	21:00	0.9610	0.9674
04:00	1.0076	1.0140	10:00	0.9860	0.9951	16:00	1.0144	1.0208	22:00	0.9519	0.9435
05:00	1.0343	1.0407	11:00	0.9841	0.9903	17:00	1.0252	1.0441	23:00	0.9567	0.9245

The forecasting comparison curves of six models in Example 2, including $SARIMA_{(9,1,10)\times(4,1,4)}$, GRNN ($\sigma = 0.07$), SSVRCCS, SSVRCS, SVRCCS, and SVRCS models and actual values are shown as in Figure 10. It indicates that the proposed SSVRCCS model is closer to the actual electric load values than the other compared models. Similarly, the enlarged figures, Figures 11–14, from eight peaks in Figure 10 are provided to demonstrate the tendency capturing capability of the proposed SSVRCCS model and how closer the SSVRCCS model matches the actual electric load values than other alternative models. It is clear that for each peak, the red real line (SSVRCCS model) always follows closely with the black real line (actual electric load), whether climbing up the peak or climbing down the hill.

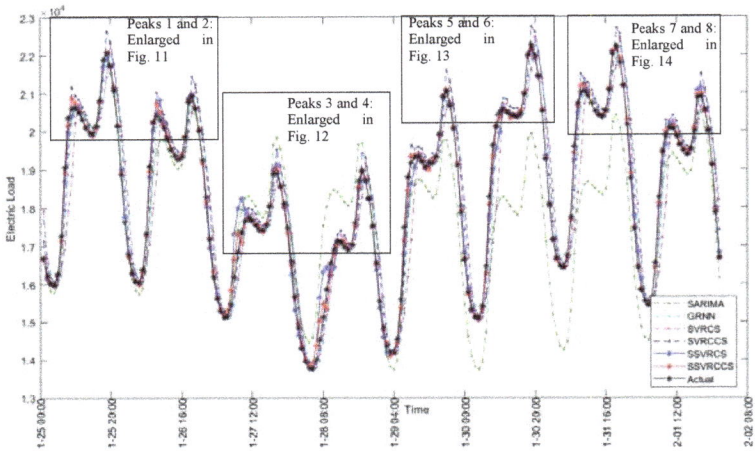

Figure 10. Forecasting values of SVR with chaotic cuckoo search (SSVRCCS) model and other alternative models for Example 2.

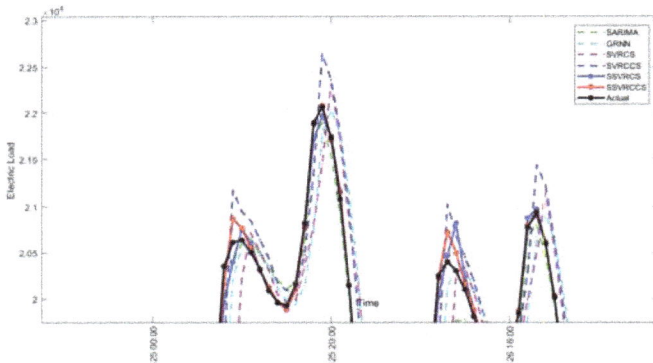

Figure 11. The enlargement comparison of Peaks 1 and 2 from the compared models for Example 2.

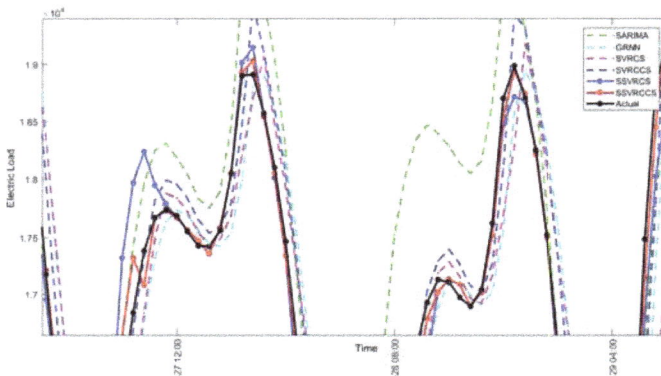

Figure 12. The enlargement comparison of Peaks 3 and 4 from the compared models for Example 2.

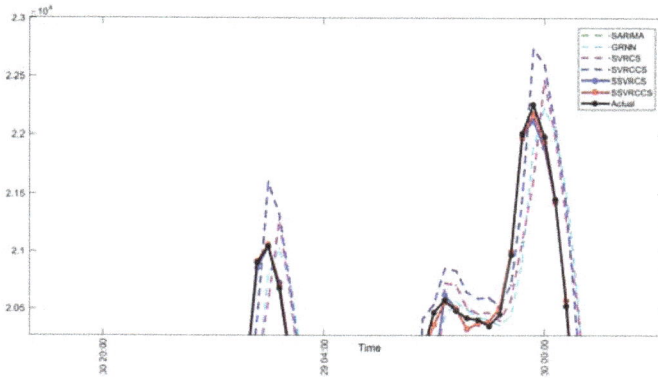

Figure 13. The enlargement comparison of Peaks 5 and 6 from the compared models for Example 2.

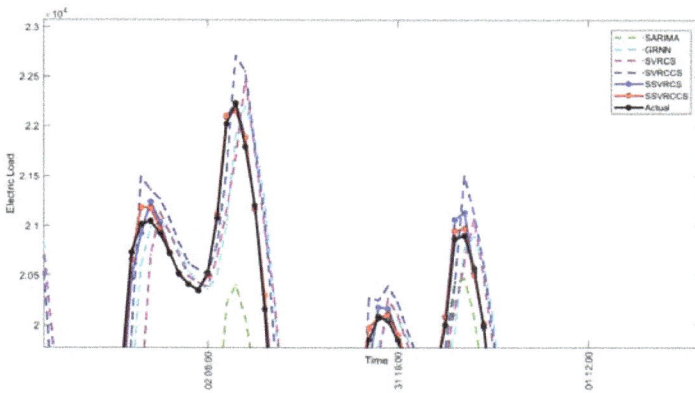

Figure 14. The enlargement comparison of Peaks 7 and 8 from the compared models for Example 2.

For comparison with other alternative models, Table 7 demonstrates the forecasting accuracy indexes for each compared model. Obviously, the proposed SSVRCCS model almost achieves the smallest index values in terms of the MAPE (0.46%), RMSE (126.10), and MAE (80.85), respectively. It is superior to the other five compared models. Once again, it indicates that the proposed SSVRCCS model could produce more accurate forecasting performances.

Table 7. Forecasting accuracy indexes of compared models for Example 2.

Forecasting Accuracy Indexes	$SARIMA_{(9,1,10) \times (4,1,4)}$	$GRNN(\alpha = 0.07)$	SSVRCCS	SSVRCS	SVRCCS	SVRCS
MAPE (%)	5.16	3.19	0.46	0.86	2.30	3.42
RMSE	1233.09	753.97	126.10	262.02	515.10	886.67
MAE	956.14	577.48	80.85	152.02	426.42	631.40

Finally, two statistical tests are also conducted to ensure the significant contribution in terms of forecasting accuracy improvement for the proposed SSVRCCS model. The test results are illustrated in Table 8 that the proposed SSVRCCS model almost reaches significance level in terms of forecasting performance than other alternative compared models.

Table 8. Results of Wilcoxon signed-rank test and Friedman test for Example 2.

Compared Models	Wilcoxon Signed-Rank Test				Friedman Test
	$\alpha = 0.025$; $W = 9264$	*p*-Value	$\alpha = 0.05$; $W = 9264$	*p*-Value	$\alpha = 0.05$;
SSVRCCS vs. $SARIMA_{(9,1,10)\times(4,1,4)}$	152 [a]	0.00000 **	152 [a]	0.00000 **	$H_0 : e_1 = e_2 = e_3 = e_4 = e_5 = e_6$
SSVRCCS vs. $GRNN(\sigma = 0.07)$	396 [a]	0.00000 **	396 [a]	0.00000 **	$F = 149.8006$
SSVRCCS vs. SSVRCS	482 [a]	0.00000 **	482 [a]	0.00000 **	$p = 0.0000$ (Reject H_0)
SSVRCCS vs. SVRCCS	745 [a]	0.00000 **	745 [a]	0.00000 **	
SSVRCCS vs. SVRCS	5207 [a]	0.00000 **	5207 [a]	0.00000 **	

[a] Denotes that the SSVRCCS model significantly outperforms the other alternative compared models; * represents that the test indicates not to accept the null hypothesis under $\alpha = 0.05$. ** represents that the test indicates not to accept the null hypothesis under $\alpha = 0.025$.

3.2.6. Discussions

To learn about the effects of the tent chaotic mapping function in both Examples 1 and 2, comparing the forecasting performances (the values of MAPE, RMSE, and MAE in Tables 3 and 7) between SVRCS and SVRCCS models, the forecasting accuracy of SVRCCS model is superior to that of SVRCS model. It reveals that the CCS algorithm could determine more appropriate parameter combinations for an SVR model by introducing the tent chaotic mapping function to enrich the cuckoo search space and the diversity of the population when the CS algorithm is going to be trapped in the local optima. In Example 1, as shown in Table 1, the parameter searching of an SVR model by CCS algorithm could be moved to a much better solution, (σ, C, ε) = (0.5254, 5885.65, 0.7358) with forecasting accuracy, (MAPE, RMSE, MAE) = (1.51%, 126.92, 87.94) from the local solution, (σ, C, ε) = (1.4744, 17877.54, 0.3231) with forecasting accuracy, (MAPE, RMSE, MAE) = (2.63%, 217.19, 151.72). It almost improves 1.12% (=2.63% − 1.51%) forecasting accuracy in terms of MAPE by employing Tent chaotic mapping function. The same in Example 2, as shown in Table 5, the CCS algorithm also helps to improve the result by 1.12% (=3.42% − 2.30%). These two examples both reveal the great contributions from the tent chaotic mapping function. In future research, it would be worth applying another chaotic mapping function to help to avoid trapping into local optima.

Furthermore, the seasonal mechanism can successfully help to deal with the seasonal/cyclic tendency changes of the electric load data to improve the forecasting accuracy, by determining seasonal length and calculating associate seasonal indexes (per half-hour for Example 1, and per hour for Example 2) from training and validation stages for each seasonal point. In this paper, authors hybridize the seasonal mechanism with SVRCS and SVRCCS models, namely SSVRCS and SSVRCCS models, respectively, by using their associate seasonal indexes, as shown in Tables 2 and 6, respectively. Based on these seasonal indexes, the forecasting results (in terms of MAPE) of the SVRCS and SVRCCS models for Example 1 are further revised from 2.63% and 1.51%, respectively, to achieve more acceptable forecasting accuracy, 0.99% and 0.70%, respectively. They almost improve 1.64% (=2.63% − 0.99%) and 0.81% (=1.51% − 0.70%) forecasting accuracy by applying seasonal mechanism. The same in Example 2, as shown in Table 7, the seasonal mechanism also improves 2.56% (=3.42% − 0.86%) and 1.84% (=2.30% − 0.46%) for SVRCS and SVRCCS models, respectively. In the meanwhile, based on Wilcoxon signed-rank test and Friedman test, as shown in Tables 4 and 8 for Examples 1 and 2, respectively, the SSVRCCS models also achieve statistical significance among other alternative models. Based on above discussions, this seasonal mechanism is also a considerable contribution, and it is worth the time cost to deal with the seasonal/cyclic information during modeling processes.

Therefore, it could be remarked that by hybridizing novel intelligent technologies, such as chaotic mapping functions, advanced searching mechanism, seasonal mechanism, and so on, to overcome some inherent drawbacks of the existing evolutionary algorithms could significantly improve forecasting accuracy. This kind of research paradigm also inspires some interesting future research.

4. Conclusions

This paper proposes a novel SVR-based hybrid electric load forecasting model, by hybridizing the seasonal mechanism, the tent chaotic mapping function, and the CS algorithm with an SVR model, namely the SSVRCCS model. The experimental results indicate that the proposed SSVRCCS model significantly outperforms other alternative compared forecasting models. This paper continues to overcome some inherent shortcomings of the CS algorithm, by actions such as enriching the search space and the diversity of the population by using the tent chaotic mapping function to avoid premature convergence problems and applying seasonal mechanism to provide useful adjustments caused from seasonal/cyclic effects of the employed data set. Eventually, the proposed SSVRCCS model achieves significant accurate forecasting performances.

This paper concludes some important findings. Firstly, by applying appropriate chaotic mapping functions it could help empower the search variables to possess ergodicity characteristics, to enrich the searching space, then, determine well appropriate parameter combinations of an SVR model, to eventually improve the forecasting accuracy. Therefore, any novel hybridizations of existed evolutionary algorithms with other optimization methods or mechanisms which could consider those actions mentioned above during modeling process are all deserving to take a trial to achieve more interesting results. Secondly, only hybridizing different single evolutionary algorithm with an SVR model could contribute minor forecasting accuracy improvements. It is more worthwhile to hybridize different novel intelligent technologies with single evolutionary algorithms to achieve more high forecasting accurate levels. This could be an interesting future research tendency in the SVR-based electric load forecasting field.

Acknowledgments: Yongquan Dong thanks the support from the project grants: National Natural Science Foundation of China (No. 61100167), Natural Science Foundation of Jiangsu Province, China (No. BK2011204), and Qing Lan Project, the National Training Program of Innovation and Entrepreneurship for Undergraduates (No. 201710320058); Zichen Zhang thanks the support from the project grant: Postgraduate Research & Practice Innovation Program of Jiangsu Province (No. 2017YXJ214); Wei-Chiang Hong thanks the support from Jiangsu Distinguished Professor Project by Jiangsu Provincial Department of Education.

Author Contributions: Yongquan Dong and Wei-Chiang Hong conceived, designed the experiments, and wrote the paper; Zichen Zhang collected the data, performed and analyzed the experiments.

Conflicts of Interest: The authors declare no conflict of interest.

References

1. Wan, C.; Zhao, J.; Member, S.; Song, Y. Photovoltaic and solar power forecasting for smart grid energy management. *CSEE J. Power Energy Syst.* **2015**, *1*, 38–46. [CrossRef]
2. Xiao, L.; Wang, J.; Hou, R.; Wu, J. A combined model based on data pre-analysis and weight coefficients optimization for electrical load forecasting. *Energy* **2015**, *82*, 524–549. [CrossRef]
3. Bunn, D.W.; Farmer, E.D. Comparative models for electrical load forecasting. *Int. J. Forecast.* **1986**, *2*, 241–242.
4. Fan, G.; Peng, L.-L.; Hong, W.-C.; Sun, F. Electric load forecasting by the SVR model with differential empirical mode decomposition and auto regression. *Neurocomputing* **2016**, *173*, 958–970. [CrossRef]
5. Ju, F.-Y.; Hong, W.-C. Application of seasonal SVR with chaotic gravitational search algorithm in electricity forecasting. *Appl. Math. Model.* **2013**, *37*, 9643–9651. [CrossRef]
6. Hussain, A.; Rahman, M.; Memon, J.A. Forecasting electricity consumption in Pakistan: The way forward. *Energy Policy* **2016**, *90*, 73–80. [CrossRef]
7. Pappas, S.S.; Ekonomou, L.; Karampelas, P.; Karamousantas, D.C.; Katsikas, S.K.; Chatzarakis, G.E.; Skafidas, P.D. Electricity demand load forecasting of the Hellenic power system using an ARMA model. *Electr. Power Syst. Res.* **2010**, *80*, 256–264. [CrossRef]
8. Vu, D.H.; Muttaqi, K.M.; Agalgaonkar, A.P. A variance inflation factor and backward elimination based robust regression model for forecasting monthly electricity demand using climatic variables. *Appl. Energy* **2015**, *140*, 385–394. [CrossRef]
9. Dudek, G. Pattern-based local linear regression models for short-term load forecasting. *Electr. Power Syst. Res.* **2016**, *130*, 139–147. [CrossRef]

10. Maçaira, P.M.; Souza, R.C.; Oliveira, F.L.C. Modelling and forecasting the residential electricity consumption in Brazil with pegels exponential smoothing techniques. *Procedia Comput. Sci.* **2015**, *55*, 328–335. [CrossRef]

11. Al-Hamadi, H.M.; Soliman, S.A. Short-term electric load forecasting based on Kalman filtering algorithm with moving window weather and load model. *Electr. Power Syst. Res.* **2004**, *68*, 47–59. [CrossRef]

12. Zhang, M.; Bao, H.; Yan, L.; Cao, J.; Du, J. Research on processing of short-term historical data of daily load based on Kalman filter. *Power Syst. Technol.* **2003**, *9*, 39–42.

13. Hippert, H.S.; Taylor, J.W. An evaluation of Bayesian techniques for controlling model complexity and selecting inputs in a neural network for short-term load forecasting. *Neural Netw.* **2010**, *23*, 386–395. [CrossRef] [PubMed]

14. Zhang, W.; Yang, J. Forecasting natural gas consumption in China by Bayesian model averaging. *Energy Rep.* **2015**, *1*, 216–220. [CrossRef]

15. Kelo, S.; Dudul, S. A wavelet Elman neural network for short-term electrical load prediction under the influence of temperature. *Int. J. Electr. Power Energy Syst.* **2012**, *43*, 1063–1071. [CrossRef]

16. Li, H.Z.; Guo, S.; Li, C.J.; Sun, J.Q. A hybrid annual power load forecasting model based on generalized regression neural network with fruit fly optimization algorithm. *Knowl.-Based Syst.* **2013**, *37*, 378–387. [CrossRef]

17. Ertugrul, Ö.F. Forecasting electricity load by a novel recurrent extreme learning machines approach. *Int. J. Electr. Power Energy Syst.* **2016**, *78*, 429–435. [CrossRef]

18. Bennett, C.J.; Stewart, R.A.; Lu, J.W. Forecasting low voltage distribution network demand profiles using a pattern recognition based expert system. *Energy* **2014**, *67*, 200–212. [CrossRef]

19. Lahouar, A.; Slama, J.B.H. Day-ahead load forecast using random forest and expert input selection. *Energy Convers. Manag.* **2015**, *103*, 1040–1051. [CrossRef]

20. Akdemir, B.; Çetinkaya, N. Long-term load forecasting based on adaptive neural fuzzy inference system using real energy data. *Energy Procedia* **2012**, *14*, 794–799. [CrossRef]

21. Chaturvedi, D.K.; Sinha, A.P.; Malik, O.P. Short term load forecast using fuzzy logic and wavelet transform integrated generalized neural network. *Int. J. Electr. Power Energy Syst.* **2015**, *67*, 230–237. [CrossRef]

22. Lou, C.W.; Dong, M.C. A novel random fuzzy neural networks for tackling uncertainties of electric load forecasting. *Int. J. Electr. Power Energy Syst.* **2015**, *73*, 34–44. [CrossRef]

23. Bahrami, S.; Hooshmand, R.-A.; Parastegari, M. Short term electric load forecasting by wavelet transform and grey model improved by PSO (particle swarm optimization) algorithm. *Energy* **2014**, *72*, 434–442. [CrossRef]

24. Hahn, H.; Meyer-Nieberg, S.; Pickl, S. Electric load forecasting methods: Tools for decision making. *Eur. J. Oper. Res.* **2009**, *199*, 902–907. [CrossRef]

25. Vapnik, V.; Golowich, S.; Smola, A. Support vector machine for function approximation, regression estimation, and signal processing. *Adv. Neural Inf. Process. Syst.* **1996**, *9*, 281–287.

26. Zhang, X.; Ding, S.; Xue, Y. An improved multiple birth support vector machine for pattern classification. *Neurocomputing* **2017**, *225*, 119–128. [CrossRef]

27. Hua, X.; Ding, S. Weighted least squares projection twin support vector machines with local information. *Neurocomputing* **2015**, *160*, 228–237. [CrossRef]

28. Hong, W.C. Electric load forecasting by seasonal recurrent SVR (support vector regression) with chaotic artificial bee colony algorithm. *Energy* **2011**, *36*, 5568–5578. [CrossRef]

29. Hong, W.-C.; Dong, Y.; Zhang, W.; Chen, L.-Y.; Panigrahi, B.K. Cyclic electric load forecasting by seasonal SVR with chaotic genetic algorithm. *Int. J. Electr. Power Energy Syst.* **2013**, *44*, 604–614. [CrossRef]

30. Gandomi, A.H.; Yang, X.S.; Alavi, A.H. Cuckoo search algorithm: A metaheuristic approach to solve structural optimization problems. *Eng. Comput.* **2013**, *29*, 17–35. [CrossRef]

31. Yang, X.S.; Deb, S. Cuckoo search via Lévy flights. In Proceedings of the World Congress on Nature and Biologically Inspired Computing (NaBic), Coimbatore, India, 9–11 December 2009; IEEE Publications: Coimbatore, India, 2009; pp. 210–214.

32. Lakshminarayanan, S.; Kaur, D. Optimal maintenance scheduling of generator units using discrete integer cuckoo search optimization algorithm. *Swarm Evolut. Comput.* **2018**. [CrossRef]

33. Boushaki, S.I.; Kamel, N.; Bendjeghaba, O. A new quantum chaotic cuckoo search algorithm for data clustering. *Expert Syst. Appl.* **2018**, *96*, 358–372. [CrossRef]

34. Daniel, E.; Anitha, J.; Gnanaraj, J. Optimum laplacian wavelet mask based medical image using hybrid cuckoo search – grey wolf optimization algorithm. *Knowl.-Based Syst.* **2017**, *131*, 58–69. [CrossRef]

35. Dao, T.-P.; Huang, S.-C.; Thang, P.T. Hybrid Taguchi-cuckoo search algorithm for optimization of a compliant focus positioning platform. *Appl. Soft Comput.* **2017**, *57*, 526–538. [CrossRef]
36. Puspaningrum, A.; Sarno, R. A hybrid cuckoo optimization and harmony search algorithm for software cost estimation. *Procedia Comput. Sci.* **2017**, *124*, 461–469. [CrossRef]
37. Huang, L.; Ding, S.; Yu, S.; Wang, J.; Lu, K. Chaos-enhanced Cuckoo search optimization algorithms for global optimization. *Appl. Math. Model.* **2016**, *40*, 3860–3875. [CrossRef]
38. Li, X.; Yin, M. A particle swarm inspired cuckoo search algorithm for real parameter optimization. *Soft Comput.* **2016**, *20*, 1389–1413. [CrossRef]
39. Sheng, Y.; Pan, H.; Xia, L.; Cai, Y.; Sun, X. Hybrid chaos particle swarm optimization algorithm and application in benzene-toluene flash vaporization. *J. Zhejiang Univ. Technol.* **2010**, *38*, 319–322.
40. Li, M.; Hong, W.-C.; Kang, H. Urban traffic flow forecasting using Gauss-SVR with cat mapping, cloud model and PSO hybrid algorithm. *Neurocomputing* **2013**, *99*, 230–240. [CrossRef]
41. Yang, X.S.; Deb, S. Cuckoo search: Recent advances and applications. *Neural Comput. Appl.* **2014**, *24*, 169–174. [CrossRef]
42. Koc, E.; Altinay, G. An analysis of seasonality in monthly per person tourist spending in Turkish inbound tourism from a market segmentation perspective. *Tour. Manag.* **2007**, *28*, 227–237. [CrossRef]
43. Goh, C.; Law, R. Modeling and forecasting tourism demand for arrivals with stochastic nonstationary seasonality and intervention. *Tour. Manag.* **2002**, *23*, 499–510. [CrossRef]
44. Wang, J.; Zhu, W.; Zhang, W.; Sun, D. A trend fixed on firstly and seasonal adjustment model combined with the ε-SVR for short-term forecasting of electricity demand. *Energy Policy* **2009**, *37*, 4901–4909. [CrossRef]
45. Martens, K.; Chang, Y.C.; Taylor, S. A comparison of seasonal adjustment methods when forecasting intraday volatility. *J. Financ. Res.* **2002**, *25*, 283–299. [CrossRef]
46. Deo, R.; Hurvich, C.; Lu, Y. Forecasting realized volatility using a long-memory stochastic volatility model: Estimation, prediction and seasonal adjustment. *J. Econom.* **2006**, *131*, 29–58. [CrossRef]
47. The Electricity Demand Data of National Electricity Market. Available online: https://www.aemo.com.au/Electricity/National-Electricity-Market-NEM/Data-dashboard#aggregated-data (accessed on 2 March 2018).
48. The Electricity Demand Data of the New York Independent System Operator (NYISO). Available online: http://www.nyiso.com/public/markets_operations/market_data/load_data/index.jsp (accessed on 2 April 2018).
49. Schalkoff, R.J. *Artificial Neural Networks*; McGraw-Hill: New York, USA, 1997.
50. Diebold, F.X.; Mariano, R.S. Comparing predictive accuracy. *J. Bus. Ecosn. Stat.* **1995**, *13*, 134–144.
51. Derrac, J.; García, S.; Molina, D.; Herrera, F. A practical tutorial on the use of nonparametric statistical tests as a methodology for comparing evolutionary and swarm intelligence algorithms. *Swarm Evolut. Comput.* **2011**, *1*, 3–18. [CrossRef]
52. Wilcoxon, F. Individual comparisons by ranking methods. *Biom. Bull.* **1945**, *1*, 80–83. [CrossRef]
53. Friedman, M. A comparison of alternative tests of significance for the problem of m rankings. *Ann. Math. Stat.* **1940**, *11*, 86–92. [CrossRef]

energies

MDPI

Article

Short-Term Load Forecasting in Smart Grids: An Intelligent Modular Approach

Ashfaq Ahmad [1,*], Nadeem Javaid [2], Abdul Mateen [2], Muhammad Awais [2] and Zahoor Ali Khan [3]

[1]   School of Electrical Engineering and Computing, The University of Newcastle, Callaghan 2308, Australia
[2]   Department of Computer Science, COMSATS University Islamabad, Islamabad 44000, Pakistan; nadeemjavaid@comsats.edu.pk (N.J.); ammateen49@gmail.com (A.M.); amawais@hotmail.com (M.A.)
[3]   Computer Information Science, Higher Colleges of Technology, Fujairah 4114, UAE; zkhan1@hct.ac.ae
*   Correspondence: ashfaqahmad@ieee.org; Tel.: +61-416-618-613

Received: 11 November 2018; Accepted: 1 January 2019; Published: 4 January 2019

Abstract: Daily operations and planning in a smart grid require a day-ahead load forecasting of its customers. The accuracy of day-ahead load-forecasting models has a significant impact on many decisions such as scheduling of fuel purchases, system security assessment, economic scheduling of generating capacity, and planning for energy transactions. However, day-ahead load forecasting is a challenging task due to its dependence on external factors such as meteorological and exogenous variables. Furthermore, the existing day-ahead load-forecasting models enhance forecast accuracy by paying the cost of increased execution time. Aiming at improving the forecast accuracy while not paying the increased executions time cost, a hybrid artificial neural network-based day-ahead load-forecasting model for smart grids is proposed in this paper. The proposed forecasting model comprises three modules: (i) a pre-processing module; (ii) a forecast module; and (iii) an optimization module. In the first module, correlated lagged load data along with influential meteorological and exogenous variables are fed as inputs to a feature selection technique which removes irrelevant and/or redundant samples from the inputs. In the second module, a sigmoid function (activation) and a multivariate auto regressive algorithm (training) in the artificial neural network are used. The third module uses a heuristics-based optimization technique to minimize the forecast error. In the third module, our modified version of an enhanced differential evolution algorithm is used. The proposed method is validated via simulations where it is tested on the datasets of DAYTOWN (Ohio, USA) and EKPC (Kentucky, USA). In comparison to two existing day-ahead load-forecasting models, results show improved performance of the proposed model in terms of accuracy, execution time, and scalability.

Keywords: artificial neural network; load prediction; smart grid; heuristic optimization; energy trade; accuracy

1. Introduction

An existing/traditional grid system needs renovation to bridge the ever-increasing gap between demand and supply and also to meet essential challenges such as grid reliability, grid robustness, customer electricity cost minimization, etc. [1]. In this regard, recent integration of advanced communication technologies and infrastructures into traditional grids have led to the formation of so called smart grids (SGs) [2]. The national national institute of standards and technology (NIST) [3] conceptual diagram of smart grid (SG) is shown in Figure 1. This conceptual diagram can be used as a reference model for standardization works in seven SG domains: generation, transmission, distribution, end users, markets, operations, and service providers. Each domain involves one or more SG actors (e.g., devices, systems, programs, etc.) to make decisions for realizing an application based on exchange of information. Further details on each domain, its involved actors, and respective applications can be found in [3]. One of the advantages of this integration is customer engagement, which plays a key role in the economies of energy trade. In other words, the old concept of uni-directional energy flow is replaced by the new and smart concept of bi-directional energy flow—transformation from traditional consumer to a smart prosumer [4].

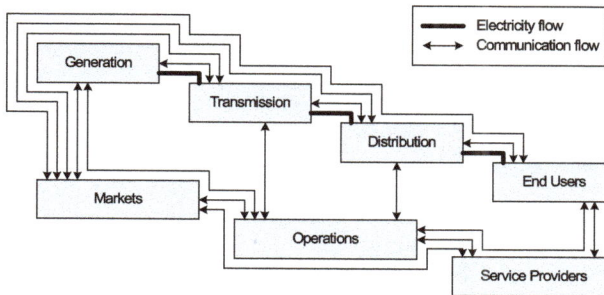

Figure 1. Conceptual diagram of SG.

The resulting/new grid, integrated with advanced metering infrastructure, faces many challenges such as [5]: (i) designing new techniques to meet the load while not increasing the generation capacity; and (ii) devising new ways/policies to ensure customer engagement with utility. When installing new technologies, utilities aim for a maximum possible return on an investment. However, this maximization would require that the daily operations of an SG utility (such as strategic decisions to bridge the gap between demand and supply, and fuel resource planning) are properly conveyed. All these decisions are highly influenced by load forecast strategy(ies) [6]. Accurate load forecast means that both utility and prosumer can maximize their electricity price savings due to spot price establishment—one of the major reasons that utilities show growing interest towards SG implementation. The concerned utility forecasts the future price/load signal which is based on the past activities of users' energy consumption patterns. In response to the forecast price/load signal, the users adjust their energy consumption schedules subject to minimization of electricity cost and/or their comfort level [7]. In reference [8], Hippert et al. classify load forecast based on time to be predicted (Figure 2): short-term, medium-term and long-term. Short-term load forecasting is further categorized into two types: (i) very short-term; and (ii) short-term forecasting. The first one has a prediction duration from seconds/minutes to hours and model applications in flow control. The second one has prediction horizon from hours to weeks and model applications in adjusting generation and demand, therefore, used to launch offers to the electrical market. The short-term forecasting models are vital in day-to-day operations, evaluation of net interchange, unit commitment and scheduling functions, and system security analysis. In medium term forecasting, the prediction horizon is typically between months. These models are used by utilities for fuel scheduling, maintenance planning, and hydro reservoir management. In long-term forecasting, the prediction horizon is for years. Utilities use these types of models for planning capacity of the grid and maintenance scheduling. Since accurate load forecast is needed by utilities to properly plan the ongoing grid operations for efficient management of their resources, this paper aims at an accurate load-forecasting model. However, the scope of this paper is limited to short-term load forecasting with a day-ahead prediction horizon only. In the literature, two types of day-ahead load forecasting (DALF) models have been presented: linear and non-linear [9]. Also, [10] has highlighted the relative limitation(s) of linear models as compared to non-linear models. In reference [9], the non-linear models are investigated in five classes: (i) support vector machine-based models; (ii) Markov chain-based models; (iii) artificial neural network (ANN)-based models; (iv) fuzzy ANN-based models; and (v) stochastic distribution-based models. The support vector machine-based models [11–13] achieve relatively moderate accuracy, but at the cost of high execution time (slow convergence rate) due to high complexity. Whereas, the Markov chain-based models [14–16] have low execution time, but at the cost of reduced forecast accuracy. Furthermore, the stochastic distribution-based models [17–20] need improvement in terms of both accuracy and execution time. The fuzzy ANN-based models [21–26] achieve moderate accuracy, but at the cost of high execution time. Finally, hybrid ANN-based models improve the accuracy of ANN-based models to an extent, but at the cost of high execution time. Among the hybrid ANN-based models, reference [27] selects features via MI technique and ANN-based prediction to forecast the day-ahead load (DAL) of SGs. To improve the accuracy of [27], the authors in [28] add a heuristic optimization-based technique with [27]. Similarly, another hybrid strategy is presented in [29] subject to DALF of SGs. However, reference [27,29] achieve relatively high forecast accuracy while taking high time to execute the algorithm. Furthermore, the forecast error of the existing works [28,29] significantly increases due to meteorological variables (such as dew point temperature, dry bulb temperature, etc.), and exogenous variables (such as cultural and social events, human impact, etc.). Thus, we aim at improving the forecast accuracy of DALF models without increasing their execution time, and in the presence of meteorological and exogenous variables.

In our proposed work, a hybrid ANN-based DALF model for SGs is presented which is a multi-model forecasting ANN with a supervised architecture and MARA for training. The proposed model follows a modular structure (it has three functional modules): a pre-processor, a forecaster, and an optimizer. Given the correlated lagged load data along with influential meteorological and exogenous variables as inputs, the first module removes two types of features from it: (i) redundant; and (ii) irrelevant. Given the selected features, the second module employs ANN to predict future values of load. The AN is activated by sigmoid function and the ANN is trained by MARA. We further minimize the forecast/prediction error by using an optimization module in which a a heuristics-based optimization technique is implemented. The proposed DALF strategy for SGs is validated via simulations which show that our proposed strategy forecasts the future load of SGs with approximately 98.76% accuracy. To sum up, this paper has the following contributions/advantages:

- The proposed model takes into account external DALF influencing factors such as meteorological and exogenous variables.
- Due to better accuracy and less execution time, we have used MARA for training which none of the existing forecast models has used for training.
- To improve the forecast accuracy and minimize the execution of the forecast model, we have performed local training which none of the existing forecast models has used.
- We have used our modified version of the EDE in the error minimization module. The existing Bi-level strategy [28] has used EDE algorithm in the error minimization module.
- We have tested our proposed model on the datasets of two USA grids: DAYTOWN and EKPC. For evaluation and validation purposes, we have compared our proposed model with two existing forecast models (bi-level forecast and MI+ANN forecast) and provided extensive simulation results.

Please note that this work is continuation of our previous work in [30,31], where in both [30,31] we have not considered exogenous and meteorological variables. The rest of the paper is organized as follows. Section 2 discusses recent/relevant DALF works, Section 3 briefly describes the newly proposed ANN and modified evolutionary algorithm-based DALF model for SGs, simulation results are discussed in Section 4, and Section 5 states the concluding points drawn from this work along with future work.

2. Related Work

For the sake of better understanding, the existing techniques are discussed in two classes (linear and non-linear) according to the type of model used [9]. The model to be used is totally the choice of researcher due to specific design considerations.

2.1. Linear Models

Linear models give continuous response which is a function or linear combination of one or more prediction variables. These models depend on the synthesis of all features of a problem that is more or less solved by complex equations. Examples of these models include spectral decomposition-based models, ordinary least square-based models, ARMA, etc. Since the prediction of demand is complex due non-linearities, the linear forecast models predict with high relative errors due to their inability to map the complex relationship between input and output. Thus, development of linear models is highly challenging. Furthermore, Hagan et al. [10] highlighted the relative limitation(s) of linear models as compared to non-linear models. Therefore, this research work is focused towards the discussion of non-linear models only.

2.2. Non-Linear Models

When the observational data is modeled by non-linear combination of one or more prediction variables, the model is said to be non-linear. To describe the relation between residual and periodical components, Bunn and Farmer [32] realize/conclude the ability of non-linear models to overcome the limitation(s) of linear models. In reference [9], the non-linear models are further categorized into five classes: (i) support vector machine-based models; (ii) Markov chain-based models; (iii) ANN-based models; (iv) fuzzy neural network-based models; and (v) stochastic distribution-based models. These models are discussed as follows.

(i) *Support vector machine-based models:* In reference [11], Niu et al. propose support vector machine and ant colony optimization-based load-forecasting technique for an SG. The authors use ant colony optimization technique for preprocessing of the input data. In this paper, system mining technique is used for feature selection. The selected features are fed into the forecaster which is a support vector machine-based predictor. Another important work has been presented by Li et al. in [12]. This varied version of the authors is least squares-based support vector machine. Similarly, reference [13] models the cyclic nature of demand by support vector machine-based linear regression. In conclusion, the support vector machine-based works are better in terms of accuracy; however, development of these models is highly challenging due to high complexity.

(ii) *Markov chain-based models:* Subject to robustness of DALF forecast strategy, authors in [14] propose a Markov chain-based strategy. This stochastic strategy aims to tackle load-time series fluctuations associated with energy consumption of users in a heterogeneous environment. The Markov chains are used to predict the future duty cycles of appliances. The technique is robust due to their memoryless nature (predicted pattern only depends on the current states; past states are not considered). In reference [15], Markov chain Monte Carlo method is used to model the switching pattern of household appliances. In simulations, they consider 100 households for one weak. However, this model limited in scope as it applies to situations in the Netherlands only. Another work in [16] proposes explicit duration hidden Markov model along with differential observation-based model to predict individual load of appliances. The authors collect the aggregated power signals by ordinary smart meters. The memoryless nature of Markov chains not only makes the DALF strategy robust but also relatively less complex in comparison to the aforementioned techniques. However, the memory less nature of Markov chains also has a drawback; less accuracy.

(iii) *ANN-based models:* ANNs learn from experience/training to predict future values while being fed with relevant input information. The advantages of these networks include but are not limited to self-organization, adaptive learning, fault tolerance, ease of integration with existing network/technology, and real time operation. The abilities to generalize and to capture non-linearity in complex environments make ANNs very attractive in problems of load forecasting. There are two basic architectures of ANN; *feed forward* and *feedback*. The former one carries information from input to output via hidden layer in forward direction only, i.e., the information of each layer is independent from that of the others. Feed forward ANNs are widely used for pattern recognition and forecasting problems. The later one carries information in both directions, forward and feedback, such that the information of each layer is dependent on that of the others. Feedback ANNs are appropriate for complex and time varying problems [33–35]. On the other hand, the learning modes of ANNs fall under three categories: *supervised* [36], *unsupervised* [37], and *re-enforced* [38]. In the first category, the ANN attempts to minimize minimum square error (MSE) for known target vector (i.e., the input/output vectors are specified). For a given input/output, error is calculated between output and the target values. This error is used to update the weights and biases of the ANN to minimize the MSE to a certain threshold. In the second category, the ANN does not need explicit target data. The system adjusts its output based on self-learning from different input patterns. In the third category, the connections between ANs are reinforced every time these are activated. Since this research work is limited in scope to supervised learning only, we discuss some of these latest/relevant works as follows.

In reference [27], authors present a hybrid technique subject to short-term price forecasting of SGs. This hybrid technique comprises two steps; feature selection and prediction. In the first step, a mutual information-based technique is implemented to remove redundancy and irrelevancy from the input load-time series. In the second step, ANN along with evolutionary algorithm is used to forecast the time series of the future load. In this process, the authors assume sigmoid activation function for artificial neurons (ANs), and Levenburg-Marquardt algorithm for training. In addition, the authors fine-tune some adjustable parameters during the first and second steps via an iterative search procedure which is part of their work. Subject to forecast accuracy, this technique is efficient as it embeds various techniques; however, the cost paid is high execution time. In reference [28], the authors investigate stochastic characteristics of SG's load. More importantly, the authors present a bi-level DALF technique for SGs. In the first/lower level, ANN and evolutionary algorithm are implemented to forecast the future load/price curve. In the second/upper level, an EDE algorithm is implemented to further minimize the prediction errors. Effectiveness of this work is reflected via MATLAB simulations which demonstrate that the proposed strategy performs DALF in SGs with a reasonable accuracy by paying the cost of high execution time. The hybrid methodology in [39] completes the DALF task in four steps: (i) data selection; (ii) transformation; (iii) forecast; and (iv) error correction. In step one, some well-known techniques of data selection are used to minimize the high dimensionality curse of input load-time series characteristics. Step two deals wavelet transformation of the selected characteristics of input load-time series to enable redundancy and irrelevancy filter implementation. Followed by step three, which uses ANN and a training algorithm subject to DALF in SGs. More importantly, they choose sigmoid activation function for ANs due non-linear capturability. Finally, error correcting functions are used in step four to improve the proposed DALF methodology in terms of accuracy. In simulations, this methodology is tested against practical household load which demonstrates that this methodology is very good for improving the accuracy by paying the cost of high complexity. Another novel strategy is presented in [40] to predict the occurrence of price spikes in SGs. The proposed strategy uses wavelet transformation for input feature selection. An ANN is then used to predict future price spikes based on the training of the selected inputs.

(iv) *Fuzzy neural network-based models:* Doveh et al. [21] present fuzzy ANN-based model for load forecasting. In their work, the input variables are heterogeneous. They also model the seasonal effect via a fuzzy indicator. In reference [22], the authors present a self-adaptive load-forecasting model for SGs. To correlate demand profile information and the operational conditions, a knowledge-based feedback fuzzy system is proposed. For optimization of error, a multilayered perceptron ANN structure is used where training is done via back propagation method. Some other hybrid strategies such as [23,24] focus on fuzzy ANN as well. Wang [23] presents electric demand forecasting model using fuzzy ANN model, whereas, Che et al. [24] present an adaptive fuzzy combination model. Che et al. iteratively combine different subgroups while calculating fuzzy functions for all the subgroups. A few more works combining fuzzy ANN with other schemes are presented in [25,26]. Subject to fuzzy neural network controller design for improving prediction accuracy, membership functions to express the inference rules by linguistic terms need proper definitions. As fuzzy systems lack such formal definitions, optimization of these functions is thus a potential research area. However, the integration of optimization technique further complicates the overall methodology.

(v) *Stochastic distribution-based models:* The model in [17] predicts the power usage time series by using a probability-based approach. The model also configures household appliances between holidays and working days. A major assumption in this work is the gaussian distribution-based on-off cycles of household appliances, number of appliances, and power consumption pattern of appliances. In this work, not only a wide range of appliances is considered but also high flexibility degree of appliances is considered. However, absence of closed form solution makes the gaussian-based forecast strategy very complex. Moreover, these assumptions cannot be always true, thus, accuracy of the predicted load-time series is highly questionable. An improvement over [17] is presented in [18]. This research work uses $\frac{1}{2}$ regulizer to overcome the computational complexity of gaussian distribution-based DALF strategy in [17]. Moreover, the proposed DALF strategy can capture heteroscedasticity of load in a more efficient way as compared [17]. Simulations are conducted to prove that the proposed DALF strategy performs better than the existing one. To sum up, we conclude that [18] has overcome the complexity of [17] to some extent; however, the basic assumptions (gaussian distribution-based on-off cycles of household appliances, number of appliances, and power consumption pattern of appliances) still hold the bases and thus make the proposal highly questionable in terms of accuracy. A semi-parametric additive forecast model is presented in [19]. This work is based on point forecast and calculates the prediction intervals via a modified bootstrap algorithm. Similarly, another semi-parametric generalized additive load forecast model is presented in [20]. In terms of forecast horizon, the generalized additive forecast model is better than the non-generalized one due to its dual forecast capability; short-term and middle term. However, both the forecast models are not sufficient in terms of accuracy when compared to the ANN-based models. The overall classification hierarchy of forecast techniques is shown in Figure 2, and their summary is given in Table 1.

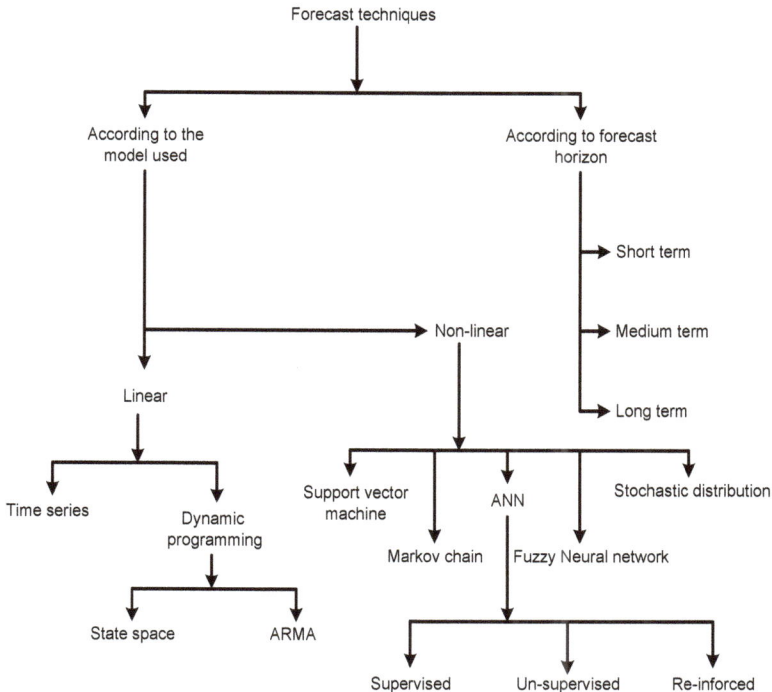

Figure 2. Classification of existing forecast techniques.

Table 1. Performance analyses of the selected forecast classes.

Forecast Class	Accuracy	Execution Time	Convergence Rate	Remarks
Support vector machine-based models [11–13]	Moderate	High	Slow	These models achieve relatively moderate accuracy, however, at the cost of high execution time (slow convergence rate) due to high complexity.
Markov chain-based models [14–16]	Low	Low	Fast	Forecast accuracy of these models needs improvement.
ANN-based models [27,28,39,40]	Low to moderate	Low to high	Fast to slow	Hybrid ANN-based models improve the forecast accuracy of ANN-based models, but at the cost of high execution time (slow convergence rate).
Fuzzy ANN-based models [21–26]	Low to moderate	High	Slow	Execution time (convergence rate) need improvement.
Stochastic distribution-based models [17–20]	Low	High	Slow	Both forecast accuracy, and execution time (convergence rate) need improvement.

3. The Proposed Forecast Strategy

ANNs are widely used as forecasters because these networks can predict the non-linearities of SGs' load with low convergence time. However, sometimes the achieved prediction accuracy is not up to the mark. Thus, leading to the adoption of optimization techniques that can significantly enhance the prediction accuracy of ANNs. However, the cost paid to achieve high accuracy is increased convergence time. Therefore, we aim towards the development of a new DALF strategy using the concept of hybrid integration subject to: (i) improvement of prediction accuracy; and (ii) reduction of convergence time.

Our proposed DALF strategy is implemented in three interconnected modules: (i) a pre-processing module; (ii) a forecast module; and (iii) an optimization module. Given the input data, the pre-processing module removes redundant and irrelevant samples from the input data. Using sigmoid activation function and MARA, the hybrid ANN-based forecast module predicts the DAL of an SG. Finally, the optimization module minimizes prediction errors to improve accuracy of the overall DALF strategy. Block diagram of the proposed model is shown in Figure 3. Detailed description of each module is as follows.

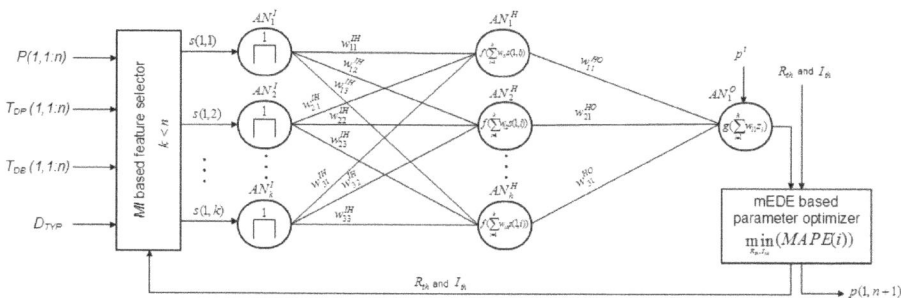

Figure 3. Block diagram of the proposed modular approach for an hour.

3.1. Pre-Processing Module

Since the ANN-based forecaster predicts load of the next day, the input data must be pre-processed subject to removal of redundant and irrelevant samples due to two reasons: (i) redundant features do not provide more information and thus unnecessarily increase the execution time during the training process (will be later discussed in the forecast module); and (ii) irrelevant features do not provide useful information and act as outliers. Detailed description of the pre-processor module is as follows.

As mentioned earlier, the data preparation module receives the input load-time series (historical). Suppose, following is the input load data:

$$P = \begin{bmatrix} p(h_1, d_1) & p(h_2, d_1) & p(h_3, d_1) & \cdots & p(h_m, d_1) \\ p(h_1, d_2) & p(h_2, d_2) & p(h_3, d_2) & \cdots & p(h_m, d_2) \\ p(h_1, d_3) & p(h_2, d_3) & p(h_3, d_3) & \cdots & p(h_m, d_3) \\ p(h_1, d_4) & p(h_2, d_4) & p(h_3, d_4) & \cdots & p(h_m, d_4) \\ p(h_1, d_5) & p(h_2, d_5) & p(h_3, d_5) & \cdots & p(h_m, d_5) \\ \vdots & \vdots & \vdots & \ddots & \vdots \\ p(h_1, d_n) & p(h_2, d_n) & p(h_3, d_n) & \cdots & p(h_m, d_n) \end{bmatrix} \tag{1}$$

where, d_n is the nth day, h_m is the mth hour of the day, and $p(h_m, d_n)$ is power usage value of the of the nth day at the mth hour. Similarly, we have input dew point temperature data in a matrix T_{DP}, input dry bulb temperature data in a matrix T_{DB}, and the input type of day (working day or holiday) data in a matrix D_T. Choosing n is totally dependent on the choice of designer. Greater value of n means that more historical lagged samples are available (fine tuning). This fine tuning however results in greater time during execution of the algorithm. Thus, there is a trade-off between convergence rate and accuracy. Before feeding the forecast/prediction module with P, the values of P are normalized. In this process, a local maximum value $'p_{max}^{c_i}'$ is computed in each column of P:

$$p_{max}^{c_i} = max(p(h_i, d_1), p(h_i, d_2), p(h_i, d_3), \ldots$$
$$, p(h_i, d_n)), \quad \forall\, i \in \{1, 2, 3, \ldots, m\} \tag{2}$$

By local normalization we mean normalization of each P's column by local maxima (one maximum in each column); results are saved in P_{nrm} (range of $P_{nrm} \in [0, \ldots, 1]$). Similarly, the matrices $T_{DP,nrm}$, $T_{DB,nrm}$ and $D_{T,nrm}$ are normalized forms of T_{DP}, T_{DB} and D_T, respectively.

These input matrices P_{nrm}, $T_{DP,nrm}$, $T_{DB,nrm}$ and $D_{T,nrm}$ not only contain irrelevant features but also contain redundant features. To remove these two types of features, we use mutual information technique that is proposed in [27] and later used in [28] as well. According to this technique, the relative amount of mutual information between two quantities; input K and target G, is as follows:

$$MI(K, G) = \sum_i \sum_j p(K_i, G_j) log_2 \left(\frac{p(K_i, G_j)}{p(K_i) p(K_i)} \right) \tag{3}$$

In reference (3), $MI(K, G) = 0$ reflects that the input and target variables and independent, high value of $MI(K, G)$ reflects that there is a strong relation between K and G two and low value of $MI(K, G)$ reflects that there is loose relation between K and G.

By using (3), we calculate $MI(K, G)$ with the help of which two types of samples (redundant plus irrelevant) are discarded from the given input data matrices P_{nrm}, $T_{DP,nrm}$, $T_{DB,nrm}$ and $D_{T,nrm}$. According to [27,28], this MI technique achieves acceptable accuracy while not taking high time for execution.

Remark 1. *The data set used for training is historical, i.e., for tomorrow's load forecast we need measured load values of previous days. Yes! The historical data was time dependent however with respect to the current day these values do not undergo any change. In other words, we deal with previously recorded data which means that the stationary assumption is not violated. Thus, the computation of MI is applicable here.*

Remark 2. *The power consumption/demand of a user is different for days such as holidays or working days. It even shows variation for different hours such as on-peak and off-peak hours. To better explain our choice, let us consider the following example:*

Considering matrix P in Equation (1), let $p(h_1, d_1)$ be the prediction variable. Then there are two possible cases for training:

(a)   *The ANN is trained by all elements of the matrix P except the first row.*
(b)   *The ANN is trained only by the 1st column of the matrix P except $p(h_1, d_1)$.*

The training samples in case (a) lead to greater prediction error due to the presence of outliers. Whereas, the training samples in case (b) lead to smaller prediction error because the outliers are removed.

Remark 3. *To improve accuracy of a forecast/prediction model, the samples used for training must be a-priori made relevant. Also, minimized number of samples will decrease algorithm's execution time. Due to these two reasons, we prefer/chose local training for each hour. In our approach, the historical load values are locally normalized by local maxima. Then the normalized values are binary encoded with respect to local median. This encoding represents two classes of values: high and low. The classes are used for selecting features only, i.e., the mutual information is easily calculated for binary variables. This selection reduces the computational complexity of the mutual information-based feature selection strategy. Once we get rid of redundant and irrelevant samples are removed from the data set, the actual values against the binary encoded values are used* *for training and optimization in the rest of the modules to prevent information loss. Thus, we have used a compromising approach between computational complexity and information loss.*

Remark 4. *Feature selection is done at beginning, and the selected features are then used for training during the operational life of the technique. From simulations, we conclude the following:*

(i)   *If the data set size is small (≤1 month), feature selection has no significant impact on the computational complexity of the overall strategy.*

(ii) *If the data set size is moderate (≥1 month and ≤3 months), feature selection somehow affects the computational complexity of the overall strategy.*

(iii) *If the data set size is large (≥3 months), feature selection has a significant impact on the computational complexity of the overall strategy.*

3.2. Forecast Module

From the works discussed in Section 2, it is concluded that any DALF strategy must ensure non-linear prediction capability. Therefore, we choose ANNs because these can capture the highly volatile characteristics of load-time series with reasonable accuracy.

For DALF, two strategies are used; direct forecasting and iterative forecasting [28]. However, it is discussed in [41] that the first strategy may introduce significant round off errors and the second one introduces large forecast errors. To overcome these imperfections, reference [28] has introduced the idea of cascaded strategy. Thus, our proposed forecast module implements the cascaded strategy. Our forecast module consists of an ANN; 24 consecutive cascaded forecasters such that each one of the 24 forecasters has an output for forecasting an hour's load of the upcoming day. It is worth mentioning that the 24 h′ forecasters/predictors are modeled explicitly instead of a single implicit/complex one. These 24 one hour ahead forecasters allow improvement in terms of accuracy [28]. The cascaded ANN forecast structure is a combination of direct and iterative structures such that load of each hour of the next day is directly predicted and each forecaster yields exactly one output.

In the forecast module, each forecaster is an AN that implements sigmoid function for activation. We have chosen sigmoid activation function because for enabling ANs in terms of capturing the highly volatile (non-linear) SG's time variant load characteristics. To update the weights during training process of the ANN, different algorithms have been used previously. For example, reference [42] include Gradient Descent Back Propagation algorithm. Similarly, references [27,28] suggest Levenberg-Marquardt algorithm as it can train the ANN 1–100 times faster than the Gradient Descent Back Propagation algorithm. We use multivariate auto regressive algorithm (MARA) [43] because it can train the ANN faster than Levenberg-Marquardt algorithm and Gradient Descent Back Propagation algorithm [42]. According to Kolmogrov theorem, if the ANN is provided with proper number of ANs then it can solve a problem by adopting one hidden layer. Thus, we have considered one hidden layer in the cascaded ANN structure of all 24 ANs. From the selected features $S_f(.)$ of the pre-processing module, the forecast module constructs training and validation samples, $S_T = S_f(i, j)$ and $S_V = S_f(1, j)$, respectively (where $i \in [2, m]$ and $j \in [1, n]$). These samples illustrate that the training of ANN by all the candidate inputs except the last/final one. The set of last samples of historical load-time series is used for validation purpose. In fact, the validation set is a part of the training load set constructed from it the training. Thus, the validation set becomes unseen for ANN. To make the validation error as a true representative of the forecast error, validation set needs to be as close to the forecast horizon as possible. While forecasting tomorrow's load we choose one day backward samples due to two reasons: (i) daily periodicity; and (ii) short-run trend [44]. Thus, each of the 24 ANs is trained as per multi-variate MARA using the aforementioned training and validation sets. Further details of the training process to update the weights can be found in [43] and pictorial view of the learning process is shown in Figure 4.

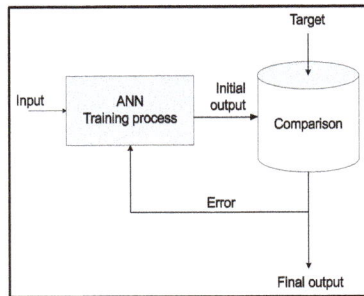

Figure 4. Supervised learning of the ANN.

For a set of finite input-target pairs, once the weights are adaptively adjusted as per MARA [43], the forecast module returns the forecast error signal; mean absolute percentage error' $MAPE(i) = \frac{1}{m}\sum_{j=1}^{m}\frac{|p^a(i,j)-p^f(i,j)|}{p^a(i,j)}$ ', to the optimization module. Where $p^a(i,j)$ is the actual load value and $p^f(i,j)$ is the forecasted load value. Stepwise operations of the proposed forecast module are shown in Figure 5a.

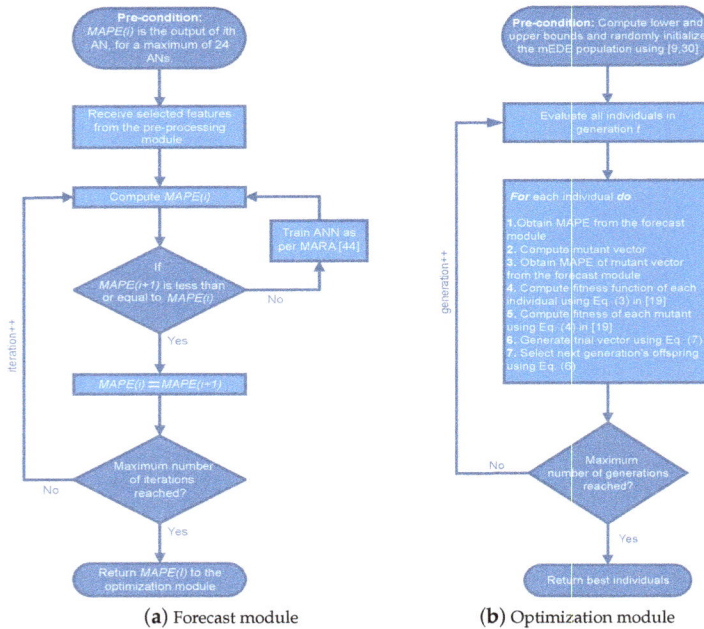

(**a**) Forecast module   (**b**) Optimization module

Figure 5. Flow charts of our modular approach.

3.3. Optimization Module

Based on the nature of the overall forecast strategy, the basic objective of optimization module is to minimize the forecast error, $E_F(.)$,

$$\underset{I_{th},R_{th}}{minimize}\ \mathrm{MAPE}(i) \tag{4}$$

where $i \in [1, m]$, I_{th} and R_{th} represent thresholds for irrelevancy and redundancy, respectively. Optimization module gives I_{th}'s and R_{th}'s optimized values to the MI-based feature selection module which uses these threshold values for feature selection. For this purpose, various choices are available such as linear programming, non-linear programming, quadratic programming, convex optimization, heuristic optimization, etc. However, the first one is not applicable here because the problem is highly non-linear. The non-linear problem can be converted into a linear one; however, the overall process would become very complex. The second one is applicable here and gives accurate results by paying execution time's cost. Similarly, the third and fourth ones suffer from slow convergence time. It is worth mentioning here that optimization does not imply exact reachability to optimum set of solutions, rather, near optimal solution(s) is(are) obtained. To sum up, heuristic optimization techniques are preferred in these situations because these provide near optimal solution(s) in relatively less execution time.

DE is one of the heuristic optimization techniques proposed in [45] and its enhanced version is used for forecast error minimization in [28]. In this paper, we modify the EDE algorithm for the sake accuracy improvement. Thus, in the upcoming paragraphs, detailed discussion is presented.

According to [28], in generation t, the jth trial vector y for ith individual is given as:

$$y_{i,j}^{'t} = \begin{cases} u_{i,j}^t & \text{if } rnd(j) \leq FF_N(U_i^t) \\ x_{i,j}^t & \text{if } rnd(j) > FF_N(U_i^t) \end{cases} \tag{5}$$

where, $x_{i,j}^t$ and $u_{i,j}^t$ are the corresponding parent and mutant vectors, respectively. In (5), $FF_N(.)$ denotes the fitness function $(0 < FF_N(.) < 1)$ and $Rand(j) \in [0,1]$ is a random number complying to uniform distribution. Between X_i^t and Y_i^t, the corresponding offspring of the next generation $X_i^{(t+1)}$ is selected as follows:

$$y_{i,j}^t = \begin{cases} y_{i,j}^{'t} & \text{if MAPE}(y_i^{'t}) \leq E_F(x_i^t) \\ x_{i,j}^t & \text{otherwise} \end{cases} \tag{6}$$

where, MAPE(.) is the objective function. From (5) and (6), it is clear that offspring selection depends on the trial vector which in turn depends on the random number and the fitness function. From this discussion, we conclude that the selected offspring is not the fittest. To make the fittest one, our approach eliminates the chances of offspring selection under the influence of random number, i.e., we modify (5) as follows:

$$y_{i,j}^{'t} = \begin{cases} u_{i,j}^t & \text{if } \frac{X_i^t}{X_{i_{max}}^t} < FF_N(U_i^t) \\ x_{i,j}^t & \text{if } \frac{X_i^t}{X_{i_{max}}^t} \geq FF_N(U_i^t) \end{cases} \tag{7}$$

From (7), it is clear that the trial vector no longer depends on the random number instead its dependence in now totally on the mutant vector which in turn depends on the parent vector. Offspring selection by this method will ensure selection of the fittest ones subject to accuracy improvement. Stepwise operations of the optimization module are shown in Figure 5b.

4. Simulation Results

For evaluation of our proposed model, we conduct simulations. For simulations, we have used MATLAB installed on Intel(R) Core(TM) i3-2370M CPU @ 2.4GHz and 2GB RAM with Windows 7. The proposed MI+ANN+mEDE-based forecast model is compared with two existing DALF models: MI+ANN forecast [27], and bi-level forecast [28]. For simulation purpose, traces of real time data for DAYTOWN and EKPC (the two USA grids) are taken from PJM electricity market. This data is freely available at [46]. We have used January–December 2014 load values for training the ANN, and January–December 2015 data for testing the ANN. Following are the simulation parameters that are used in our experiments (refer to Table 2). Justification of these parameters can be found in [27,28,42,43].

The newly proposed MI+ANN+mEDE model is tested against the two existing models in terms of three performance metrics: (i) accuracy; and (ii) execution time or convergence rate.

- *Accuracy:* $Accuracy(.) = 100 - MAPE(.)$. We have measured this metric in %.
- *Variance:* $Var(i) = \frac{1}{m} \sum_{j=1}^{m} |p^f(i,j) - \overline{p^a(i,j)}|$. Where $\overline{p^a(i,j)}$ is the mean value of $p^a(i,j)$. Monthly variance is calculated by using the same formula while considering the calculated daily variances.
- *Execution time:* During simulations, the time taken by the system to completely execute a given forecast strategy. The strategy for which execution time is small converges more quickly and vice versa. In simulations, we have measured execution time in seconds.

Table 2. Parameters used in simulations.

Parameter	Value
Forecasters	24
Hidden layers	1
Maximum iterations	100
Neurons (in the hidden layer)	5
Bias	0
Initial weights	0.1
Momentum	0
Load data (historical)	1 year
Maximum generations	100

Referring to Figure 6a–f and Tables 3–6, which are graphical/tabular illustrations/representations of the proposed MI+ANN+mEDE-based forecast model versus the two existing DALF models: MI+ANN and bi-level. From Figure 6a,b, it is clear that the proposed MI+ANN+mEDE model effectively predicts/forecasts the future load of the two selected SGs. The ANN-based forecaster captures the non-linearities in the history load-time series. This non-linear prediction capability is not only due to sigmoid activation function but also due to the selected training algorithm; MARA. When we look at the hourly forecast results in Figure 6c,d, the % error of the MI+ANN-based forecast model is 3.8% and 3.81 for DAYTOWN and EKPC, respectively. The % error of the bi-level forecast model is 2.2% and 2.23% for DAYTOWN and EKPC, respectively. The % error of the proposed MI+ANN+mEDE-based forecast model is 1.24% for both DAYTOWN and EKPC, respectively. Similarly, the daily forecast results of the two simulated models for January 2015 are shown in Tables 3 and 5 for the two selected USA grids, respectively. From these results, it is clear that the existing MI+ANN-based forecast model predicts the future load with the highest % error and the highest variance. Also, the monthly forecast results of the three simulated models for January–December 2015 are shown in Tables 4 and 6 for EKPC and DAYTOWN, respectively. From Tables 4 and 6, it is evident that the proposed MI+ANN+mEDE model forecasts the future load with the least prediction error and the least variance as compared to the other two existing models. This result is obvious due to absence of optimization module in MI+ANN-based forecast model. To minimize this forecast error, the bi-level forecast model uses EDE algorithm. Subject to further minimization of the forecast error, we have integrated an mEDE optimization technique. Please note that mEDE is our modified version of existing EDE algorithm for down scaling forecast error. Results show that integration of mEDE algorithm yields fruitful results; the MI+ANN+mEDE-based DALF model is relatively more accurate than the other two existing DALF models. These figures show the positive impact of optimization module on the forecast error minimization between target curve and the forecast curve. It is obvious that the error curve decreases as the number of generations of the mEDE algorithm are increased. As the proposed MI+ANN+mEDE forecast model compares the forecast curve's error (next generation) with the existing one (existing generation) and updates the weights if the forecast curve's error is less than the existing one (survival of the fittest). Thus, as expected, the forecast error is significantly minimized as the forecast strategy is subjected to step ahead generations. However, during simulations, we observed that from 89th to 100th generation, the forecast error does not exhibit significant improvement. Therefore, the proposed and the existing forecast models are not subjected to further generations. There exists a possible trade-off between accuracy of a forecast strategy and its convergence rate (refer to Sections 1–3). This trade-off is shown in Figure 6c–f. From these figures, it is clear that the bi-level forecast model improves the accuracy of MI+ANN forecast model while paying cost in terms of relatively slow convergence rate. On the other hand, the newly proposed MI+ANN+mEDE model modifies the EDE algorithm to further improve the accuracy of the bi-level forecast model. More importantly, the MI+ANN+mEDE model improves the prediction accuracy by not paying surplus cost in terms of execution time. However, the execution time of our proposed forecast model is still greater than the MI+ANN forecast model due to integration of optimization module.

Figure 7 shows the impact of dataset size (number of training data samples) on error performance (see Figure 7a) and execution time (see Figure 7b) of the three selected models. By observing Figure 7a, an improvement of error performance for all the compared STLF models is evident when the number of lagged input samples increase from 30 to 120. This result follows Equation (1), i.e., the ANN is more finely tuned by increasing the value of n (30 to 120) which improves the forecast error performance. However, this improvement is not significant at much higher tuning when the number of training samples are increased from 60 to 120 (stability can be seen in the curves). On the other hand, Figure 7b shows the cost of high execution time paid by the fine tuning to achieve relative improvement in forecast accuracy. This is obvious because training of the ANN takes additional time when the number of training samples are increase. From Figure 7a,b, it is clear that the proposed modular model is more scalable (relatively higher degree of stability can be seen for MI+ANN+mEDE forecast) as compared to the other two models. The reasons for this higher scalability are: usage of selected features for training of the ANN, training the ANN via MARA algorithm with local normalization, and usage mEDE algorithm for error minimization.

Table 7 shows the relationship between MAPE and the number of iterations of the three compared STLF models when tested on DAYTOWN and EKPC datasets. The convergence characteristics (i.e., the number of iterations) indicate that the proposed MI+ANN+mEDE model and the bi-level model converge at an optimal value in almost the same number of iterations. On the other hand, the MI+ANN model takes only 20–23 iterations for converging into an optimal target value. This result is obvious due to the added computational burden in the bi-level and the MI+ANN+mEDE models (i.e., these models use the optimization module) which is not the case in MI+ANN model (i.e., this model does not use the optimization model). In other words, the MI+ANN model achieves its target of the required training, testing, and validation with the least number of iterations. However, this least computational burden is achieved by paying the high cost of forecast accuracy. In this regard, a regression analysis of the network was performed to evaluate confidence interval of the training, testing and validation performance of the compared forecast models, and the results are shown in Table 7. Clearly, the proposed MI+ANN+mEDE model achieves the highest confidence interval (i.e., 98%) as compared to bi-level (i.e., 97%) and MI+ANN (i.e., 96%) models. This means that only 2% of the estimated data is not statistically significant for the network in case of the proposed MI+ANN+mEDE model. As a result, the forecasted load demand of the proposed MI+ANN+mEDE model is rather closer to its actual value as compared to the other two models (see Figure 6a,b).

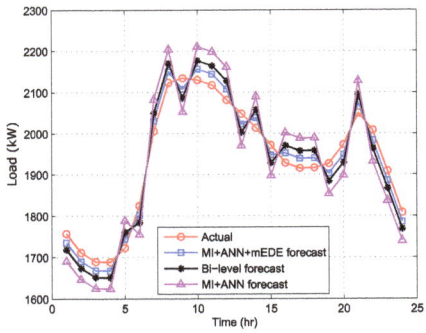

(**a**) DAYTOWN grid: actual vs. forecast load curves

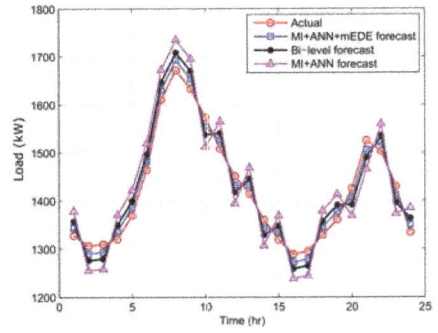

(**b**) EKPC grid: actual vs. forecast load curves

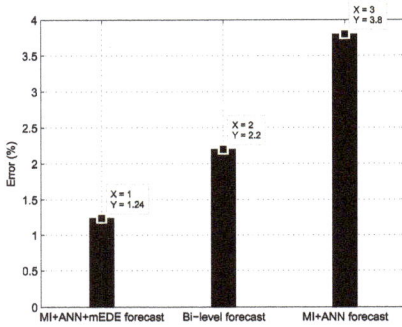

(**c**) DAYTOWN grid: forecast error

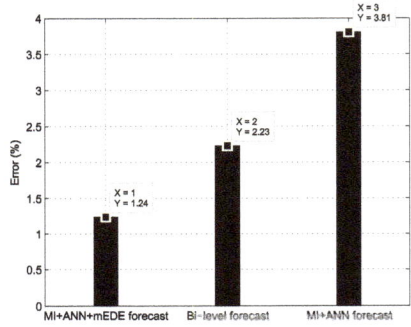

(**d**) EKPC grid: forecast error

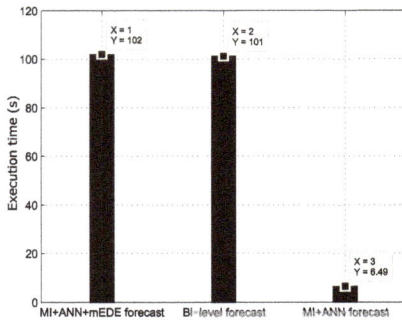

(**e**) DAYTOWN grid: convergence rate or execution time behavior

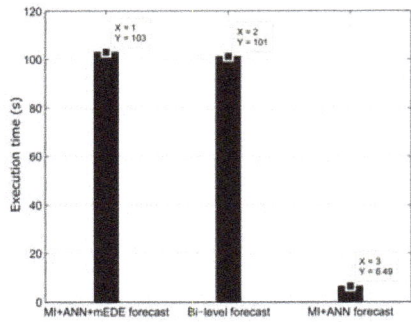

(**f**) EKPC grid: Convergence rate or execution time behavior

Figure 6. Relative performance of the proposed intelligent modular approach tested on historical data of DAYTOWN and EKPC grid: STLF results for 27 January 2015.

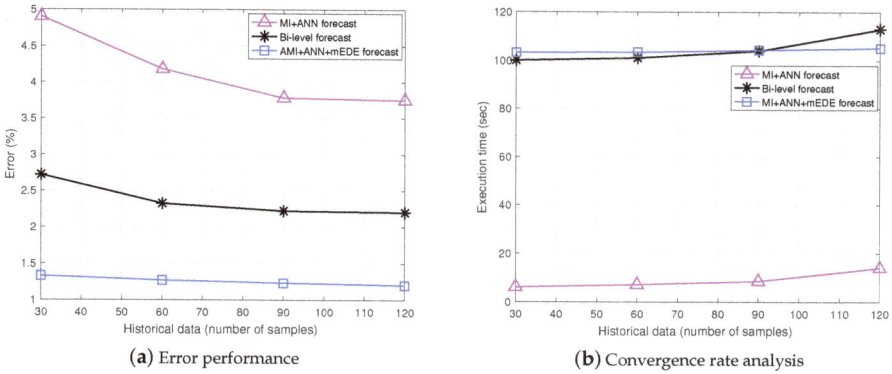

(**a**) Error performance   (**b**) Convergence rate analysis

Figure 7. Relative scalability analysis of the proposed intelligent modular approach.

Table 3. EKPC: Results for January 2015.

| Day | Forecast Model | | | | | |
| | MI+ANN | | Bi-Level | | MI+ANN+mEDE | |
	MAPE	Variance	MAPE	Variance	MAPE	Variance
1	3.99	1.89	2.40	1.50	1.04	1.12
2	3.42	1.78	1.97	1.46	1.32	0.97
3	4.10	2.08	2.61	1.26	1.15	1.09
4	3.67	1.91	2.13	1.41	1.44	0.96
5	3.79	1.70	1.97	1.37	1.16	1.05
6	3.62	1.88	2.43	1.48	1.29	0.97
7	3.93	1.73	2.62	1.39	1.40	1.11
8	3.97	1.94	1.92	1.28	1.19	1.03
9	3.54	2.04	2.18	1.42	1.39	0.90
10	3.46	1.79	2.21	1.36	1.10	1.03
11	4.05	1.72	1.85	1.39	1.25	1.05
12	4.21	1.84	1.97	1.29	1.29	0.90
13	3.89	2.00	1.94	1.33	1.07	1.03
14	3.62	1.75	1.84	1.46	1.36	1.10
15	3.79	1.99	2.11	1.26	1.14	0.93
16	3.47	1.81	2.44	1.38	1.36	1.07
17	4.24	2.10	2.26	1.26	1.20	1.04
18	4.20	1.74	2.61	1.41	1.23	1.08
19	3.86	1.97	2.44	1.46	1.07	0.96
20	3.61	1.80	2.52	1.42	1.18	0.98
21	3.82	1.95	2.29	1.48	1.36	1.12
22	3.77	2.03	2.62	1.45	1.42	0.99
23	4.23	1.86	2.53	1.51	1.34	1.01
24	3.94	1.77	2.38	1.29	1.11	0.92
25	3.44	1.73	2.20	1.47	1.32	1.14
26	3.56	1.94	2.23	1.34	1.10	0.97
27	3.81	1.78	2.29	1.40	1.24	1.11
28	3.39	1.82	1.94	1.29	1.39	1.03
29	4.19	2.05	2.43	1.32	1.08	0.98
30	3.52	1.77	1.98	1.42	1.12	1.06
31	4.01	1.99	1.82	1.42	1.33	0.99
Average	3.81	1.84	2.23	1.38	1.24	1.03

Table 4. EKPC: Results for 2015.

Month	MI+ANN		Bi-Level		MI+ANN+mEDE	
	MAPE	Variance	MAPE	Variance	MAPE	Variance
January	3.81	1.84	2.23	1.38	1.24	1.03
February	3.85	1.75	2.15	1.44	1.20	0.99
March	4.76	1.90	2.26	1.39	1.26	1.05
April	3.84	1.76	2.19	1.41	1.29	1.00
May	3.80	1.71	1.20	1.47	1.23	1.02
June	3.73	1.73	2.16	1.35	1.21	1.01
July	3.72	1.81	2.29	1.40	1.24	1.07
August	3.84	1.70	1.28	1.40	1.25	1.03
September	3.82	2.90	2.22	1.33	1.20	0.99
October	3.82	1.88	2.15	1.36	1.30	1.01
November	4.77	1.75	1.17	1.48	1.22	1.06
December	4.80	1.82	1.27	1.32	1.27	1.02
Average	3.79	1.80	2.13	1.39	1.24	1.01

Table 5. DAYTOWN: Results for January 2015.

Day	MI+ANN		Bi-Level		MI+ANN+mEDE	
	MAPE	Variance	MAPE	Variance	MAPE	Variance
1	3.72	1.70	2.59	1.36	1.20	1.02
2	3.60	1.86	2.38	1.30	1.31	1.10
3	3.54	1.90	2.20	1.51	1.35	0.97
4	3.81	1.88	1.77	1.27	1.25	0.95
5	3.78	1.92	2.57	1.41	1.32	1.07
6	4.07	1.83	2.65	1.33	1.21	0.96
7	3.88	1.79	2.58	1.43	1.35	1.11
8	3.62	1.81	2.25	1.28	1.22	1.01
9	4.30	1.88	2.25	1.50	1.15	0.90
10	3.71	1.93	2.43	1.44	1.27	1.03
11	3.59	1.77	2.27	1.30	1.34	1.12
12	3.82	1.74	2.34	1.37	1.24	0.95
13	3.77	1.84	2.50	1.25	1.29	1.06
14	4.15	1.83	2.64	1.31	1.16	1.13
15	3.69	1.91	1.88	1.40	1.28	0.93
16	3.87	1.89	2.47	1.52	1.30	1.12
17	4.27	2.76	2.60	1.33	1.29	1.10
18	3.64	1.78	2.15	1.42	1.31	1.00
19	4.18	1.84	1.86	1.40	1.21	1.12
20	3.75	1.99	2.31	1.28	1.19	0.99
21	3.58	1.97	2.05	1.39	1.18	1.05
22	3.83	2.72	2.70	1.30	1.32	0.98
23	4.88	1.99	2.60	1.38	1.37	1.09
24	3.73	1.88	2.44	1.29	1.18	1.12
25	4.21	2.01	1.91	1.47	1.33	0.92
26	3.59	1.76	1.79	1.32	1.21	1.04
27	3.80	1.96	2.20	1.37	1.24	1.10
28	3.66	1.89	1.97	1.27	1.22	1.03
29	4.25	1.81	2.33	1.49	1.15	0.98
30	3.51	1.92	1.90	1.24	1.36	1.03
31	4.03	1.95	1.88	1.43	1.20	1.06
Average	3.86	1.92	2.27	1.36	1.25	1.03

Table 6. DAYTOWN: Results for 2015.

| Month | Forecast Model | | | | | |
| | MI+ANN | | Bi-Level | | MI+ANN+mEDE | |
	MAPE	Variance	MAPE	Variance	MAPE	Variance
January	3.86	1.92	3.27	1.36	1.25	1.03
February	3.85	1.71	2.30	1.47	1.20	0.99
March	3.80	1.75	2.20	1.44	1.22	1.05
April	3.71	1.79	2.24	1.38	1.27	1.06
May	3.79	1.87	2.28	1.40	1.22	1.02
June	3.72	1.85	2.13	1.30	1.24	1.07
July	3.76	1.76	2.22	1.36	1.28	0.99
August	3.87	1.76	2.18	1.43	1.26	1.08
September	3.70	2.70	2.29	1.38	1.23	1.02
October	3.77	1.88	2.17	1.36	1.21	1.09
November	3.83	1.83	2.27	1.50	1.27	1.00
December	3.80	1.81	2.25	1.33	1.21	1.01
Average	3.78	1.88	2.31	1.39	1.23	1.03

Table 7. Comparison of training iterations (convergence) and regression analysis (accuracy).

Dataset	Forecast Model	Iterations	Training	Testing	Validation
	MI+ANN	20	0.9626	0.9619	0.9556
DAYTOWN	Bi-Level	94	0.9787	0.9799	0.9776
	MI+ANN+mEDE	95	0.9876	0.9890	0.9872
	MI+ANN	23	0.9622	0.9617	0.9551
EKPC	Bi-Level	95	0.9769	0.9783	0.9766
	MI+ANN+mEDE	96	0.9877	0.9892	0.9878

5. Conclusions and Future Work

In SGs, DALF is an essential task because its accuracy has a direct impact on the planning schedules of utilities that strongly affects the energy trade market. Moreover, high volatility in the history load curves makes DALF in SGs relatively more challenging when compared to load forecast for longer duration. Taking into account DALF influencing factors such as exogenous variables and meteorological variables, we have presented a hybrid ANN-based DALF model for SGs which is a multi-model forecasting ANN with a supervised architecture and MARA for training. The proposed model significantly reduced the execution time and enhanced the forecast accuracy by distinctly carrying local normalization and local training. Moreover, sigmoid activation function and MARA enable the forecast strategy to capture non-linearities in load-time series. Integration of optimization module (based on our proposed modifications) with the forecast strategy also improved the forecast accuracy. Tests are conducted on three USA grids: DAYTOWN, EKPC and FE. Results show that the proposed model achieves relatively better forecast accuracy (98.76%) in comparison to an existing bi-level technique and an MI+ANN technique. Moreover, improvement in forecast accuracy is achieved while not paying the cost of slow convergence rate. Thus, the trade-off between convergence rate and forecast is not created. Finally, from application perspective, the proposed model can be used by utilities to launch better offers in the electricity market. This means that the utilities can save significant amount of money due to better adjustment of their generation and demand schedules simply because of high accuracy of the proposed model.

In future, we are interested in advanced signal processing techniques for feature selection and extraction purposes. Moreover, exploration of particle swarm optimization-based techniques and a complete forecast plus scheduling-based technique is also under consideration.

Author Contributions: Conceptualization, A.A.; Formal analysis, A.M. and M.A.; Investigation, A.A.; Methodology, A.A. and N.J.; Software, A.A.; Supervision, N.J.; Validation, N.J., A.M. and M.A.; Writing—original draft, A.A.; Writing—review & editing, N.J. and Z.A.K.

Funding: This research received no external funding.

Conflicts of Interest: The authors declare no conflicts of interest.

Nomenclature

SG	Smart grid
DAL	Day-ahead load
DALF	Day-ahead load forecast(ing)
AN	Artificial neuron
ANN	Artificial neural network
MARA	Multivariate auto regressive algorithm
ARMA	Auto regressive and moving average
EDE	Enhanced differential evolution algorithm
mEDE	Modified version of EDE algorithm
NIST	National institute of standards and technology
MSE	Minimum square error
P	Historical load data matrix
T_{DP}	Historical dew point temperature data matrix
T_{DB}	Historical boiling point temperature data matrix
D_{TYP}	Historical dew point temperature data matrix
p_{h_m,d_n}	Load value at mth hour of the nth day
$p_{max}^{c_i}$	Local maxima for each column of P
P_{nrm}	Locally normalized P
$T_{DP,nrm}$	Locally normalized T_{DP}
$T_{DB,nrm}$	Locally normalized T_{DB}
$MI(K,G)$	Relative mutual information between input K and target G
$p_r(K,G)$	Joint probability between K and G
$p_r(K)$	Individual probability of K
S_f	Selected features
S_T	Training samples
S_V	Validation samples
MAPE	Mean absolute percentage error
p^a	Actual load
p^f	Forecasted load
I_{th}	Irrelevancy threshold value
R_{th}	Redundancy threshold value
$y'^t_{i,j}$	jth trial vector y' for ith individual in generation t
$x^t_{i,j}$	jth parent vector x for ith individual in generation t
$u^t_{i,j}$	jth mutant vector u for ith individual in generation t
$y^t_{i,j}$	jth offspring vector y for ith individual in generation t
rnd	Random number
$FF_N(.)$	Fitness function
E_F	Forecast error

References

1. Gelazanskas, L.; Gamage, K.A. Demand side management in smart grid: A review and proposals for future direction. *Sustain. Cities Soc.* **2014**, *11*, 22–30. [CrossRef]
2. Yan, Y.; Qian, Y.; Sharif, H.; Tipper, D. A Survey on Smart Grid Communication Infrastructures: Motivations, Requirements and Challenges. *IEEE Commun. Surv. Tutor.* **2013**, *15*, 5–20. [CrossRef]

3. National Institute of Standards and Technology. NIST Framework and Roadmap for Smart Grid Interoperability Standards. Release 1.0. 2010. Available online: http://www.nist.gov/publicaffairs/releases/upload/smartgridinteroperabilityfinal.pdf (accessed on 10 November 2018).

4. Leiva, J.; Palacios, A.; Aguado, J.A. Smart metering trends, implications and necessities: A policy review. *Renew. Sustain. Energy Rev.* **2016**, *55*, 227–233. [CrossRef]

5. *How Does Forecasting Enhance Smart Grid Benefits?* SAS Institute Inc.: Cary, NC, USA, 2015; pp. 1–9.

6. Hernandez, L.; Baladron, C.; Aguiar, J.M.; Carro, B.; Sanchez-Esguevillas, A.J.; Lloret, J.; Massana, J. A survey on electric power demand forecasting: Future trends in smart grids, microgrids and smart buildings. *IEEE Commun. Surv. Tutor.* **2014**, *16*, 1460–1495. [CrossRef]

7. Vardakas, J.S.; Zorba, N.; Verikoukis, C.V. A Survey on Demand Response Programs in Smart Grids: Pricing Methods and Optimization Algorithms. *IEEE Commun. Surv. Tutor.* **2015**, *17*, 152–178. [CrossRef]

8. Hippert, H.S.; Pedreira, C.E.; Souza, C.R. Neural Networks for Short-Term Load Forecasting: A review and Evaluation. *IEEE Trans. Power Syst.* **2001**, *16*, 44–51. [CrossRef]

9. Raza, M.Q.; Khosravi, A. A review on artificial intelligence based load demand forecasting techniques for smart grid and buildings. *Renew. Sustain. Energy Rev.* **2015**, *50*, 1352–1372. [CrossRef]

10. Hagan, M.T.; Behr, S.M. The Time Series Approach to Short Term Load Forecasting. *IEEE Trans. Power Syst.* **1987**, *2*, 785–791. [CrossRef]

11. Niu, D.; Wang, Y.; Wu, D. Power load forecasting using support vector machine and ant colony optimization. *Exp. Syst. Appl.* **2010**, *37*, 2531–2539. [CrossRef]

12. Li, H.; Guo, S.; Zhao, H.; Su, C.; Wang, B. Annual Electric Load Forecasting by a Least Squares Support Vector Machine with a Fruit Fly Optimization Algorithm. *Energies* **2012**, *5*, 4430–4445. [CrossRef]

13. Aung, Z.; Toukhy, M.; Williams, J.R.; S'anchez, A.; Herrero, S. Towards Accurate Electricity Load Forecasting in Smart Grids. In Proceedings of the Fourth International Conference on Advances in Databases, Knowledge, and Data Applications, Athens, Greece, 2–6 June 2012; pp. 51–57.

14. Meidani, H.; Ghanem, R. Multiscale Markov models with random transitions for energy demand management. *Energy Build.* **2013**, *61*, 267–274. [CrossRef]

15. Nijhuis, M.; Gibescu, M.; Cobben, J.F. Bottom-up Markov Chain Monte Carlo approach for scenario based residential load modelling with publicly available data. *Energy Build.* **2016**, *112*, 121–129. [CrossRef]

16. Guo, Z.; Wang, Z.J.; Kashani, A. Home appliance load modeling from aggregated smart meter data. *IEEE Trans. Power Syst.* **2015**, *30*, 254–262. [CrossRef]

17. Gruber, J.K.; Prodanovic, M. Residential energy load profile generation using a probabilistic approach. In Proceedings of the IEEE UKSim-AMSS 6th European Modelling Symposium, Valetta, Malta, 14–16 November 2012; pp. 317–322.

18. Kou, P.; Gao, F. A sparse heteroscedastic model for the probabilistic load forecasting in energy-intensive enterprises. *Electr. Power Energy Syst.* **2014**, *55*, 144–154. [CrossRef]

19. Fan, S.; Hyndman, R.J. Short-Term Load Forecasting Based on a Semi-Parametric Additive Model. *IEEE Trans. Power Syst.* **2012**, *27*, 134–141. [CrossRef]

20. Goude, Y.; Nedellec, R.; Kong, N. Local Short and Middle Term Electricity Load Forecasting with Semi-Parametric Additive Models. *IEEE Trans. Power Syst.* **2014**, *5*, 440–446. [CrossRef]

21. Doveh, E.; Feigin, P.; Greig, D.; Hyams, L. Experience with FNN Models for Medium Term Power Demand Predictions. *IEEE Trans. Power Syst.* **1999**, *14*, 538–546. [CrossRef]

22. Mahmoud, T.S.; Habibi, D.; Hassan, M.Y.; Bass, O. Modelling self-optimised short term load forecasting for medium voltage loads using tunning fuzzy systems and Artificial Neural Networks. *Energy Convers. Manag.* **2015**, *106*, 1396–1408. [CrossRef]

23. Wang, Z.Y. Development Case-based Reasoning System for Shortterm Load Forecasting. In Proceedings of the IEEE Russia Power Engineering Society General Meeting, Montreal, QC, Canada, 18–22 June 2006; pp. 1–6.

24. Che, J.; Wang, J.; Wang, G. An adaptive fuzzy combination model based on self-organizing map and support vector regression for electric load forecasting. *Energy* **2012**, *37*, 657–664. [CrossRef]

25. Nadimi, V.; Azadeh, A.; Pazhoheshfar, P.; Saberi, M. An Adaptive-Network-Based Fuzzy Inference System for Long-Term Electric Consumption Forecasting (2008–2015): A Case Study of the Group of Seven (G7) Industrialized Nations: USA, Canada, Germany, United Kingdom, Japan, France and Italy. In Proceedings of the Fourth UKSim European Symposium on Computer Modeling and Simulation, Pisa, Italy, 17–19 November 2010; pp. 301–305.

26. Lou, C.W.; Dong, M.C. Modeling data uncertainty on electric load forecasting based on Type-2 fuzzy logic set theory. *Eng. Appl. Artif. Intell.* **2012**, *25*, 1567–1576. [CrossRef]

27. Amjaday, N.; Keynia, F. Day-Ahead Price Forecasting of Electricity Markets by Mutual Information Technique and Cascaded Neuro-Evolutionary Algorithm. *IEEE Trans. Power Syst.* **2009**, *24*, 306–318. [CrossRef]

28. Amjady, N.; Keynia, F.; Zareipour, H. Short-Term Load Forecast of Microgrids by a New Bilevel Prediction Strategy. *IEEE Trans. Smart Grid* **2014**, *1*, 286–294. [CrossRef]

29. Liu, N.; Tang, Q.; Zhang, J.; Fan, W.; Liu, J. A Hybrid Forecasting Model with Parameter Optimization for Short-term Load Forecasting of Micro-grids. *Appl. Energy* **2014**, *129*, 336–345. [CrossRef]

30. Ahmad, A.; Javaid, N.; Alrajeh, N.; Khan, Z.A.; Qasim, U.; Khan, A. A modified feature selection and artificial neural network-based day-ahead load forecasting model for a smart grid. *Appl. Sci.* **2015**, *5*, 1756–1772. [CrossRef]

31. Ahmad, A.; Javaid, N.; Guizani, M.; Alrajeh, N.; Khan, Z.A. An accurate and fast converging short-term load forecasting model for industrial applications in a smart grid. *IEEE Trans. Ind. Inform.* **2017**, *13*, 2587–2596. [CrossRef]

32. Bunn, D.W.; Farmer, E.D. *Comparative Models for Electrical Load Forecasting*; Wiley: New York, NY, USA, 1985; pp. 13–30.

33. Ahmad, I.; Abdullah, A.B.; Alghamdi, A.S. Application of artificial neural network in detection of probing attacks. *IEEE Sympos. Ind. Electron. Appl.* **2009**, 57–62.

34. Malki, H.A.; Karayiannis, N.B.; Balasubramanian, M. Short term electric power load forecasting using feedforward neural networks. *Exp. Syst.* **2004**, *21*, 157–167. [CrossRef]

35. Hahn, H.; Meyer-Nieberg, S.; Pickl, S. Electric load forecasting methods: Tools for decision making. *Eur. J. Oper. Res.* **2009**, *199*, 902–907. [CrossRef]

36. Amakali, S. Development of Models for Short-Term Load Forecasting Using Artficial Neural Networks. Master's Thesis, Cape Peninsula University of Technology, Cape Town, South Africa, 2008.

37. Valova, I.; Szer, D.; Gueorguieva, N.; Buer, A. A parallel growing architecture for self-organizing maps with unsupervised learning. *Neurocomputing* **2005**, *68*, 177–195. [CrossRef]

38. Anderson, J.; Silverstein, J.; Ritz, S.; Jones, R. Distinctive features, categorical perception and probability learning: Some applications on a neural model. *Psychol. Rev.* **1977**, *84*, 413–451. [CrossRef]

39. Yang, H.T.; Liao, J.T.; Lin, C.I. A Load Forecasting Method for HEMS Applications. In Proceedings of the 2013 IEEE Grenoble Conference, Grenoble, France, 16–20 June 2013; pp. 1–6.

40. Amjady, N.; Keynia, F. Electricity market price spike analysis by a hybrid data model and feature selection technique. *Electr. Power Syst. Res.* **2010**, *80*, 318–327. [CrossRef]

41. Amjady, N.; Keynia, F. Short-term load forecasting of power systems by combination of wavelet transform and neuro-evolutionary algorithm. *J. Energy* **2009**, *34*, 46–57. [CrossRef]

42. Engelbrecht, A.P. *Computational Intelligence: An Introduction*, 2nd ed.; John Wiley & Sons: New York, NY, USA, 2007.

43. Anderson, C.W.; Stolz, E.A.; Shamsunder, S. Multivariate autoregressive models for classification of spontaneous electroencephalographic signals during mental tasks. *IEEE Trans. Biomed. Eng.* **1998**, *45*, 277–286. [CrossRef] [PubMed]

44. Lasseter, R.H.; Piagi, P. Microgrid: A conceptual solution. In Proceedings of the IEEE International Conference on Power Electronics Specialists, Aachen, Germany, 20–25 June 2004; pp. 4285–4290.

45. Storn, R.; Price, K. Differential evolution—A simple and efficient heuristic for global optimization over continuous spaces. *J. Glob. Optim.* **2009**, *11*, 341–359. [CrossRef]

46. PJM Electricity Market. Available online: www.pjm.com (accessed on 1 February 2015).

energies

MDPI

Article

Deep Learning Based on Multi-Decomposition for Short-Term Load Forecasting

Seon Hyeog Kim[ID]**, Gyul Lee**[ID]**, Gu-Young Kwon**[ID]**, Do-In Kim and Yong-June Shin ***[ID]

Department of Electrical and Electronic Engineering, Yonsei University, Seoul 03722, Korea;
goodguy7@yonsei.ac.kr (S.H.K.); thyecho@yonsei.ac.kr (G.L.); kgy926@yonsei.ac.kr (G.-Y.K.);
penpony109@yonsei.ac.kr (D.-I.K.)
* Correspondence: yongjune@yonsei.ac.kr; Tel.: +82-2-2123-4625

Received: 31 October 2018; Accepted: 3 December 2018; Published: 7 December 2018

Abstract: Load forecasting is a key issue for efficient real-time energy management in smart grids. To control the load using demand side management accurately, load forecasting should be predicted in the short term. With the advent of advanced measuring infrastructure, it is possible to measure energy consumption at sampling rates up to every 5 min and analyze the load profile of small-scale energy groups, such as individual buildings. This paper presents applications of deep learning using feature decomposition for improving the accuracy of load forecasting. The load profile is decomposed into a weekly load profile and then decomposed into intrinsic mode functions by variational mode decomposition to capture periodic features. Then, a long short-term memory network model is trained by three-dimensional input data with three-step regularization. Finally, the prediction results of all intrinsic mode functions are combined with advanced measuring infrastructure measured in the previous steps to determine an aggregated output for load forecasting. The results are validated by applications to real-world data from smart buildings, and the performance of the proposed approach is assessed by comparing the predicted results with those of conventional methods, nonlinear autoregressive networks with exogenous inputs, and long short-term memory network-based feature decomposition.

Keywords: deep learning; empirical mode decomposition (EMD); long short-term memory (LSTM); load forecasting; neural networks; variational mode decomposition (VMD); weekly decomposition

---

1. Introduction

Accurate load forecasting optimizes power loads, reducing costs and stabilizing electric power distribution. Load forecasting accuracy depends on the time series data of non-stationary and non-linearity characteristics. These characteristics are influenced by the prediction time scale and energy consumption scale. Depending on the prediction time scale, load forecasting is classified into four types.

Long-term load forecasting (LTLF) has a time scale of more than a year, medium-term load forecasting (MTLF) a time scale from one week to one year, and short-term load forecasting (STLF) a time scale from one hour to one week. System operators typically estimate demand by referring to load profiles from several hours ago. Ultra-short-term load forecasting (USTLF) is a key issue for smart grids, real-time demand side management (DSM), and energy transactions because energy trading in DSM requires precise load forecasting in the order of minutes, and profit is strongly related to forecast accuracy. Therefore, the USTLF time scale is from several minutes to one hour [1].

Conventional load forecasting methods use statistical models based on inherent characteristics of historical data. Previous STLF studies have proposed auto-regressive integrated moving average (ARIMA), Gaussian processing regression (GPR), support vector regression (SVR), and neural network models [2]. ARIMA is a common method for linear time series-based methods. GPR and SVR provide alternative methods to model time series loads, using external data such as weather data to consider non-linearity and non-stationarity. GPR is a supervised machine learning model based on statistical regression and a kernel function that refines variance and step length [3].

To reduce the nonlinearity of the time series data and to analyze their statistical characteristics, a seasonal analysis combined prediction method is used [4,5]. Recent research activities divide profiles into sub-profiles according to the load patterns of customers based on human factor, contract type, and region. After dividing the profiles into sub-profiles, a clustering algorithm is used for hierarchical classification [6–9].

To improve the accuracy of load forecasting using external data such as temperature, humidity, weather information, and electricity prices, a method has been proposed [10–13]. However, measuring such data is a challenging task for low-level distribution and small-scale loads. Furthermore, data processing and data storage of each piece of the dataset are required because the resolution of time-series data is different. Therefore, recent research trends use the technique of feature selection [14,15] or decomposing the load profile to extract the characteristics of the load using signal processing theory [16–24].

Wavelet decomposition with neural networks [1,16–18] has been employed to increase prediction accuracy. In [1], a wavelet algorithm dealt with the noise of the actual electrical load data, and load forecasting based on artificial neural networks (ANN) was proposed. Empirical mode decomposition (EMD) with machine learning has also been proposed for load forecasting, wind speed, or energy prices [19–22]. However, EMD lacks a mathematical definition and has weaknesses that diverge at end-points when decomposing the signal. To overcome the weakness of EMD, load forecasting studies using variational mode decomposition (VMD) have been proposed [22–25]. Existing regression methods with various decompositions, clustering algorithms, and probabilistic analyses have been investigated, as they can be used to identify load characteristics; however, they increase the dimension of the input [26–28]. Clustering and decomposition methods are applied in the pre-processing stage to improve the accuracy of the load prediction, and current state-of-art load forecasting studies have improved the performance of the prediction model through deep learning [29–33].

A recurrent neural network (RNN) has a memory structure and a hidden layer suitable for processing big data using deep learning techniques. However, an RNN has vanishing gradient problems caused by an increase in the number of layers. Nonlinear autoregressive exogenous (NARX) RNNs offer an orthogonal mechanism for dealing with the vanishing gradient problem by allowing direct connections or delays from the distant past data [34,35]. However, NARX RNNs have a limited impact on vanishing gradients, and the delay structure increases the computation time. Most successful RNN architectures have long short-term memory (LSTM), which uses nearly additive connections between states, to alleviate the vanishing gradient problem [36–41]. In [42], gated recurrent unit (GRU) neural networks with K-mean clustering were proposed. A GRU is a variant LSTM with a simpler structure, but it has similar performance, and convolutional neural networks (CNNs) are also widely used in deep learning for image classification [43,44]. As the load prediction model becomes more sophisticated, shorter prediction time scales [1,5] and lower level feeders of distributions, such as behind-the-meter individual load, business buildings, and household electric usage, are being studied [6,26–28].

This paper proposes a deep learning method whereby features are extracted through multi-decomposition for short-term load forecasting. The scale of the predicted load is a feeder-level business building. Feeder-level load forecasting is more complicated than that of an aggregated load because the statistical characteristics are greatly changed even with a slight change in power consumption. The proposed decomposition method significantly captures intrinsic load pattern components and periodic features.

A load forecasting method based on LSTM with VMD is designed and implemented in this paper. The proposed two-stage decomposition analysis identifies the characteristics of the load profile with

AMI only, i.e., without external data. In addition, the three-step regularization process removes the problem of data processing in deep running and improves LSTM. The proposed method simulates load forecasting within a few minutes (USTLF) to several days (STLF) using real-world building data and shows the advantages that LSTM has over the traditional models.

The rest of the paper is organized as follows. Section 2 introduces the proposed feature extraction method and provides background information. Section 3 presents deep learning. Section 4 introduces the experiments, and Section 5 presents the analysis results using the proposed multi-decomposition. In Section 6, the prediction results with different models are compared, and Section 7 summarizes and concludes the paper.

2. The Proposed Multi-Decomposition for Feature Extraction

2.1. Enhanced AMI for Small-Scale Load and Real Time

Load forecasting aims to determine the future power plan based on a series of given historical datasets. For efficient power planning, a minimum weekly load must be predicted according to the time scale of the task, e.g., demand side management, economic dispatch, and energy scheduling [2]. As the prediction time scale and load scale become smaller, the non-linearity problem must be solved through a more sophisticated prediction method. State-of-the-art AMI with 5-min sampling provides more samples per hour than conventional 15-min AMI. As a result, power consumption measurements that are close to real-time measurements are achieved. However, as the amount of data increases, conventional machine learning causes problems such as overfitting, the vanishing gradient problem, the long-term dependency problem, and increased calculation times.

2.2. Empirical Mode Decomposition

Decomposition methods are widely used to analyze similar signals and extract features. The EMD decomposition method uses extreme signal values, and the VMD method decomposes the signal by reflecting frequency characteristics to compensate for the weaknesses EMD. Both methods were employed to analyze time series data in [22]. The EMD method preprocesses data by recursively detecting local minima and maxima in a signal and estimating lower and upper envelopes by spline interpolation of the extreme values, then removing lower and upper envelope averages. To decompose a signal into a sum of intrinsic mode functions (IMFs), the following two conditions must be satisfied [18–22]:

- In the entire dataset, the number of zero crossings must either be equal to or differ from the number of extrema by no more than one;
- The lower and upper envelope means, defined by interpolating the local signal minima and maxima, respectively, must equal zero.

2.3. Variational Mode Decomposition

The goal of VMD is to decompose a signal into a discrete number of sub-signals (modes) that have specific sparsity properties while reproducing the signal. VMD replaces the most recent definition of IMFs; for example, an EMD mode is defined as a signal whose number of local extrema and zero-crossings differ at most by one or as AM-FMsignals by the corresponding narrow band property [23].

Variational mode decomposition provides an analytical expression that relates AM-FM parameter descriptors to the estimated signal bandwidth, i.e., each mode k is required to be mostly compact around a center pulsation, w_k, that is determined along with the decomposition. This IMF definition complements the weakness of EMD of lacking a mathematical definition. VMD also reduces EMD end-point effects because it decomposes the signal into k discrete IMFs, whereas each IMF is band limited in the spectral domain [23–25].

VMD Algorithm

1. For each mode, v_k, compute the associated analytic signal using the Hilbert transform to obtain the unilateral frequency spectrum;
2. For each mode, v_k, shift the mode frequency spectrum to the baseband (narrow frequency) by mixing it with an exponential tuned to the corresponding estimated center frequency;
3. Estimate the bandwidth using the Gaussian smoothed demodulated signal.

The resulting constrained variational problem is expressed as:

$$\min_{v_k, w_k} \left\{ \sum_{k}^{K} \left\| \partial_t \left[\left(\delta(t) + \frac{j}{\pi t} \right) * v_k(t) \right] e^{-jw_k t} \right\|_2^2 \right\} \tag{1}$$

subject to

$$\sum_{k}^{K} v_k = W_p(t). \tag{2}$$

where $W_p(t)$ is the p weekly load profile with mode v and frequency w, δ is the Dirac distribution, k is the mode index, K is the total number of modes and the decomposition level, and $*$ denotes convolution. Mode v with high order k represents low frequency components. In contrast to that of EMD, the decomposition level of VMD, k, must be pre-determined [22–25].

2.4. Decomposition for Feature Selection

Figure 1 shows the proposed load profile decomposition method. The building load profile has similar weekday patterns, and the load is measured at 5-min intervals by AMI. To classify seasonal patterns, the typical load profile (x_t) of the building is decomposed on a weekly basis for weekly seasonality features (x_t^p). The typical load profile is decomposed into two dimensions. The load variations can be extracted if they are periodic because the VMD decomposes the load profile in terms of the frequency ($x_{k,t}^p$). Thus, all the IMFs exhibit periodic characteristics. As each IMF has a specific frequency, the VMD identifies periods that cannot be identified in the typical load profile and the weekly load profile. As a result, the typical load profile is decomposed into three-dimensional data according to time, weekly seasonality, and IMF-level. The feature decomposition process of the load profile contributes to the load characteristics without external data, such as the calendar information about holidays, temperature, and humidity.

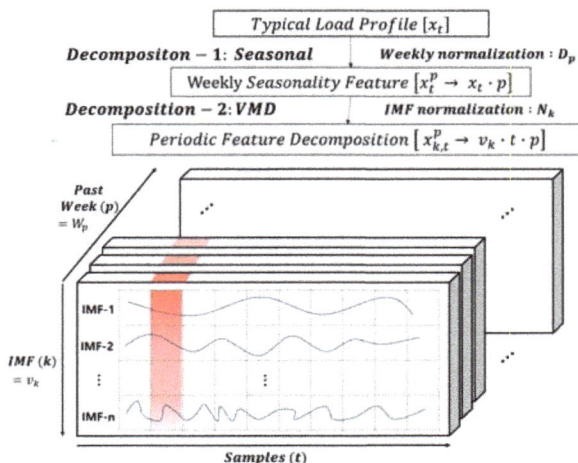

Figure 1. Load profile feature decomposition process.

2.5. Three-Step Regularization Process

The sampling of AMI used in this study is three-times larger than that of conventional AMI, which collects data at 15-min intervals. In addition, as the load profile is decomposed into sub-profiles, the sub-profiles that have detailed frequency characteristics can be learned as the input variables, but the number of input variables increases. As the number of input variables increases, the curse of dimensionality degrades the learning ability because the number of hidden nodes increases. As a result, the number of hidden layers is increased to solve the curse of dimensionality, but this causes the vanishing gradient problem. Moreover, without feature selection, overfitting occurs. The learned hypothesis may fit the training set very well, but it cannot be extended to new samples. In addition, without the normalization process, the covariate shift problem degrades performance. The covariate shift, which refers to the change in the distribution of the input variables present in the training and test data, should be prevented. Therefore, the proposed method includes a three-step regularization to solve each of the above-mentioned problems. First, the delay factor of weekly data is estimated. Although a large amount of data can be beneficial for deep learning, the distant past data can result in overfitting problems and increase the computation time. A similar problem was addressed in [36] to solve the dependence on distant historical data. In [36], a decay factor was used to solve the long-term dependency problem of NARX-RNN.

In this paper, the weekly decay's exponent by a factor is proposed as Equation (3):

$$D_p = 2^{-(p-1)} \tag{3}$$

where p is the number of weeks, which gives high weights to nearby weekly data in time and lower weights to distant weekly past data.

Secondly, the separated IMF signals $(x_{k,t}^p)$ are normalized against the original signal size (x_t^p). This is because these signals correspond to residual noise such as frequencies that are too high or too low to be identified in a certain pattern. The IMF normalization process is performed to identify features that degrade learning. The IMF normalization factor given by Equation (4) and T' is the number of samples of the weekly data.

$$N_k = \frac{\sum_{t=1}^{T'} x_{k,t}^p / T'}{\sum_{t=1}^{T'} x_t^p / T'} \tag{4}$$

Finally, as the number of hidden layers increases, the internal covariance can be shifted. The internal covariate shift causes the distribution of the training set and test set to differ, which can lead to local points. Batch normalization (BN) is used to address internal covariate shift. BN normalizes the output of a previous activation layer by subtracting the batch mean and dividing by the batch standard deviation. The advantages of BN are (1) fast learning, (2) less careful initialization, and (3) a regularization effect. BN is one of the regularization techniques used in the deep learning field [41,42]. The regularization process contributes to the accuracy of the load forecasting and the optimization of the model by applying a high weight to the input data having the most definite period, reducing dependency on the past distant data and avoiding the covariate shift of the data group. The three-step regularization process increases the accuracy of the load forecasting by minimizing problems that can occur when several inputs are learned.

3. Deep Learning

Deep learning is one of the machine learning techniques that proposes to model high-level abstractions in data by using ANN architectures composed of multiple non-linear transformations. Deep learning refers to stacking multiple layers of neural networks and relying on stochastic optimization to perform efficient machine learning tasks. To take advantage of deep learning, three technical constraints must be solved. The three technical constraints are (1) the lack of sufficient data, (2) the lack of computing resources for a large network size, and (3) the lack of an efficient training algorithm. Recently, these constraints were solved by the development of big data applications, the

Internet-of-Things, and high performance smart computing [37–39]. One of the most efficient deep learning processes is RNN.

RNNs are fundamentally different from traditional feed-forward neural networks; RNNs have a tendency to retain information acquired through subsequent time-stamps. This characteristic of RNNs is useful for load forecasting. Even though RNNs have good approximation capabilities, they are not fit to handle long-term dependencies of data. Learning long-range dependencies with RNNs is challenging because of the vanishing gradient problem. The increase in the number of layers and the longer paths to the past cause the vanishing gradient problem because of the back-propagation algorithm, which has the very desirable characteristic of being very flexible, although causes the vanishing gradient problem [30,32–34].

3.1. Long Short-Term Memory Neural Networks

The long-short term memory network has been employed to approach the best performance of state-of-the-art RNNs. The problem of the vanishing gradient is solved by replacing nodes in the RNN with memory cells and a gating mechanism. Figure 2 shows the LSTM block structure. The overall support in a cell is provided by three gates. The memory cell state s_{t-1} interacts with the intermediate output h_{t-1}. The sub-sequent input x_t determines whether to remember or forget the cell state. The forget gate f_t determines the input for the cell state s_{t-1} using the sigmoid function. The input gate i_t, input node g_t, and output node o_t determine the values to be updated by each weight matrix, where σ represents the sigmoid activation function, while ϕ represents the tanh function. The weight matrices in the LSTM network model are determined by the back-propagation algorithm [37–42].

The LSTM has become the state-of-the-art RNN model for a variety of deep learning techniques. Several variants of the LSTM model for recurrent neural networks have been proposed. Variant LSTM models have been proposed to improve performance by solving issues such as computation time and the model complexity of the standard LSTM structure. Among the variants, the GRU maintains performance by simplifying the structure with an update gate that is coupled with an input gate and forget gate. The structure of the GRU is advantageous for forecasting in a large-scale grid to reduce calculation time [42]. In [45], variants of the LSTM architecture were designed and their performances were compared through implementation. The results revealed that none of the variants of LSTM could improve upon the standard LSTM. In other words, a clear winner could not be declared. Therefore, the popular LSTM networks are used in this study [45,46].

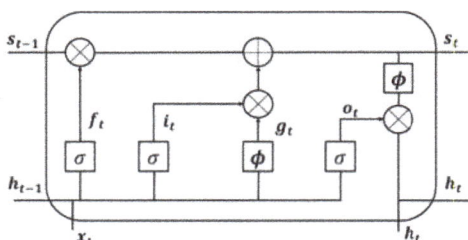

Figure 2. The structure of the LSTM.

3.2. Nonlinear Autoregressive Network with Exogenous Inputs

NARX RNNs and LSTM solve the vanishing gradient problem with different mechanisms. NARX RNNs allow delays from the distant past layer, but this structure increases computation time and has a small effect on long-term dependencies. The LSTM solves the vanishing gradient problem by replacing nodes in the RNN with memory cells and a gating mechanism [36].

4. Experiments

This section describes the process used to obtain time series models for load forecasting. Figure 3 shows the proposed load forecasting model using LSTM with multi-decomposition for feature extraction. We will discuss each step in detail.

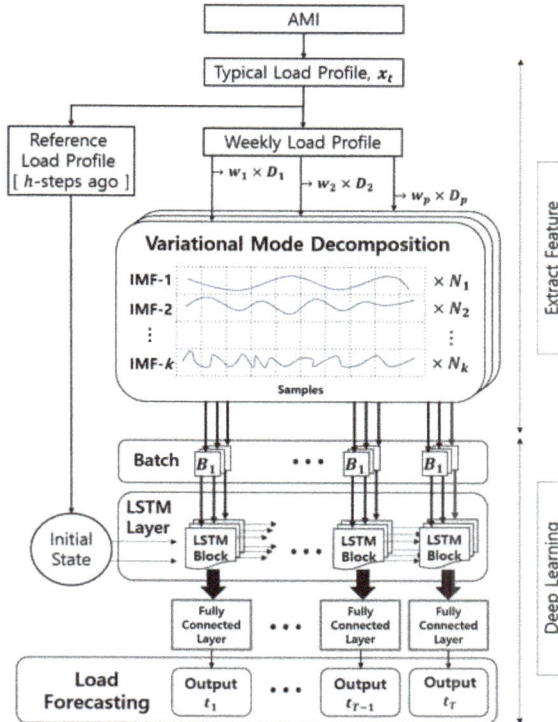

Figure 3. Deep-learning load forecasting based on the multi-decomposition method.

4.1. Prediction of the Time Scale

Reference load profiles reflect the load profiles that are close to real-time load profiles before h steps ago, where h determines the prediction time scales, which depend on the purpose of the load forecasting. STLF techniques can be used for a variety of purposes by enabling smaller scales and faster prediction. USTLF, which predicts the load within a few minutes to one hour, can be used for electricity theft detection or can provide information for emergency power supply [47]. STLF, which predicts the load from one hour to a day, can be used for electricity transactions or economic dispatch of renewable energy resources [2].

4.2. Extract Feature Layer

Through the multi-decomposition method, the features of time-series data are extracted. The number of decomposition levels (K) is 10, which is the value obtained when the decomposition loss rate is 0.1% or less. The weight of the weekly load profile (D_p) considers the trend of load patterns according to seasonal changes. Each IMF decomposed through the VMD has a frequency characteristic and is normalized to make the feature stand out (N_k).

4.3. Long Short-Term Memory Layer

The LSTM can capture long-term dependencies in time-stamps; therefore, it can address the vanishing gradient problems. In the proposed method, the number of hidden layers increases due to the decomposition of input data, but the vanishing gradient problem is solved through the memory cell structure with three-step regularization. In addition, to minimize the covariate shift problem, batch normalization is performed prior to the activation phase of the input. IMFs and reference load profiles are trained at each LSTM layer and have predictive values, all of which are summed to predict the load profile.

4.4. Model Construction

4.4.1. Hyperparameter Tuning and Training Options

The LSTM model has several hyperparameters such as the number of input neurons, hidden layers, input window size, number of epochs, regularization weight, batch size, and learning rate. The window size of input and output parameters depends on the time scale of load forecasting. The input neuron parameter is determined by the dimensions of the input data. The input dimension of the proposed method is 11, which is the sum of the reference profile and 10 IMF signals. We selected the hyperparameters and used ADAM optimization, one of the optimization techniques used in deep learning [30–40].

4.4.2. Training and Testing

The overall AMI dataset of each day is divided into a ratio of 70:15:15 for the purposes of model training, validation, and testing, respectively.

4.4.3. Performance Measures

The root mean squared error (RMSE) is used to compare differences between the predicted value $\hat{y}_t$ and measured value y_t and is computed for T (which is the number of samples of the weekly load profile) different predictions as the square root of the mean of the squares of the deviations:

$$\text{RMSE} = \sqrt{\frac{\sum_{t=1}^{T} (\hat{y}_t - y_t)^2}{T}}. \tag{5}$$

The mean absolute error (MAE) is one of a number of ways of comparing forecasts with their eventual outcomes.

$$\text{MAE} = \frac{1}{T} \sum_{t=1}^{T} |y_t - \hat{y}_t|. \tag{6}$$

The mean absolute percent error (MAPE) is also widely used to evaluate accuracy. Accuracy can be compared via MAPE using percentages when the scale of the loads is different [37–40].

$$\text{MAPE} = \frac{100}{T} \sum_{t=1}^{T} \left| \frac{y_t - \hat{y}_t}{y_t} \right|. \tag{7}$$

5. Load Profile Analysis by Multi-Decomposition Methods

5.1. Weekly Seasonality

This study used real-world load profile data from the R&D business building that utilized enhanced AMI for demand side management. Figure 4 shows the real-world load profile of the business building. The building generates 288 samples per day, 2016 samples per week, and 8640 samples per month. The load profile is measured and stored in data storage.

The electrical load profile of the office building is usually light on weekends compared to weekdays because energy is consumed according to the business schedule. In contrast to those of residential load profile patterns, energy the increase and decrease times of office building load profiles are related to commute time and have similar daily characteristics. Figure 4 shows a typical profile for building electricity load over one week, from which a clear weekly seasonality pattern can be observed. The weekly load pattern is quite similar over four weeks, with a weekly average correlation of 0.93. Therefore, many studies have proposed load forecasting methods using weekly statistical methods or dividing the time series data into holidays, weekends, and weekdays [4,5].

However, the process of dividing the time series data in a database into weekdays, weekends, and holidays is inefficient because the calendar information may not be provided in advance, and each consumer group may have different days off. Moreover, the simple method of dividing the data into weekdays and holidays cannot capture the periodicity of the load profile such as the commute time and periodic power system on/off states. In Figure 4, the fourth week load pattern deviates somewhat from the previous pattern, with significant peak load shift in the afternoon, particularly on Wednesday and Friday (average correlations for Wednesday and Friday are 0.82 & 0.84, respectively). As the patterns deviated greatly on weekends (the weekend average correlation is 0.71), it is difficult to predict accurately energy consumption using daily statistical data alone. Therefore, feature extraction from the load profile is required to capture periodic components caused by commuting time, meal times, thermal control change, elevator system operation, etc.

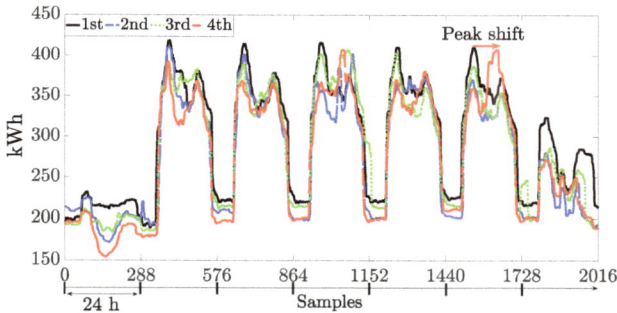

Figure 4. The typical load profile of the business building.

5.2. Comparison of Decomposition Performance

Figures 5 and 6 show the load profiles of Figure 4 decomposed by EMD and VMD, respectively, where each IMF of each load profile covers four weeks. To analyze various frequency components and preserve the signal energy, in EMD, the standard deviation as the stop criterion is determined as 0.1%; hence, the weekly load profiles are decomposed into 10 IMFs.

As EMD decomposes the signal using extrema envelopes (Figure 5), the results are similar to those obtained with a low pass filter. However, VMD is similar to a high pass filter, as it decomposes the load profile from low frequency components. VMD IMFs (VMFs) are band limited; hence, they are similar to harmonic components. Therefore, VMD efficiently identifies periodic characteristics in non-linear and non-stationary signals compared to EMD IMFs (EMFs).

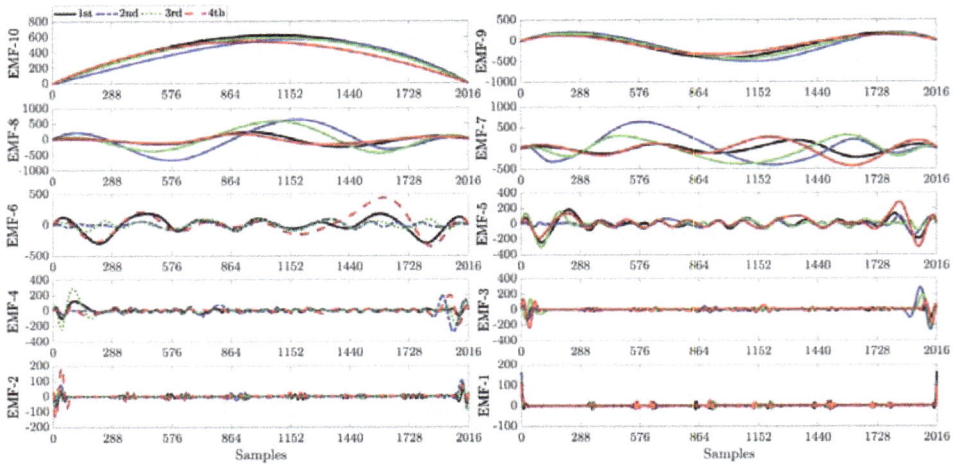

Figure 5. Weekly load profile decomposition using EMD (*k* = 10). EMF, EMD IMF.

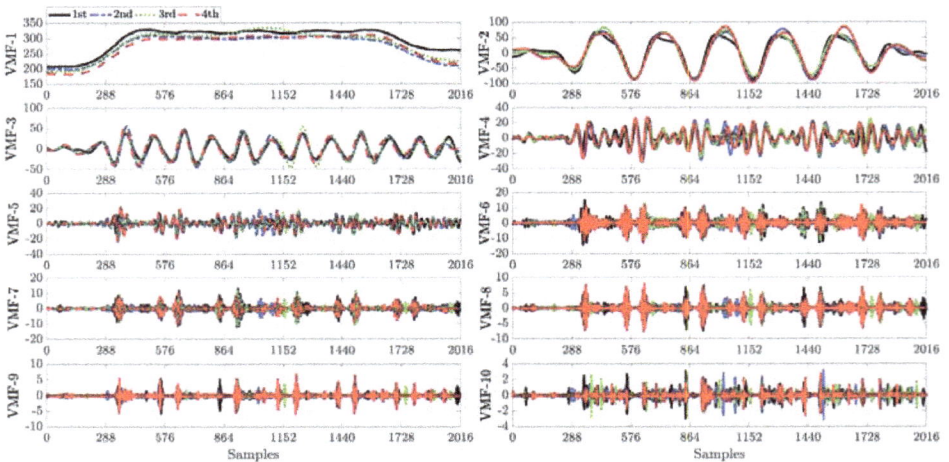

Figure 6. Weekly load profile decomposition using VMD (*k* = 10). VMF, VMD IMF.

The first VMF (VMF-1) is effectively the DC bias (Figure 6), i.e., the average daily load consumption. VMF-2 and VMF-3 show high correlation signal periodicities. Office buildings typically exhibit a commute period, and this appears in VMF-2. This R&D building has two peaks around the commute time, and this pattern appears in VMF-3. On the other hand, EMF-10 and EMF-9 show high correlation trends, whereas the other EMFs show low correlations. High frequency EMFs (EMF-5–EMF-10) also include end-point problems, whereas VMD decomposes the signal into band-limited signals; hence, VMFs have no end-point issues.

Table 1 shows the correlations for each IMF. The VMFs capture similar frequency signals better than the EMFs and decompose high frequency signals well. As VMD is done mathematically, the correlation between VMFs is gradually reduced, whereas EMD IMFs are irregular. Therefore, in the case of high sampling or short prediction time scales, VMD shows better performance than EMD because VMD can reflect the high frequency characteristics of the dataset.

Table 1. Correlation index comparison of EMD and VMD.

Decomposition	IMF									
	1	2	3	4	5	6	7	8	9	10
EMD	0.58	0.42	0.40	0.28	0.46	0.35	0.02	0.43	0.82	0.96
VMD	0.98	0.83	0.80	0.63	0.53	0.26	0.15	0.01	−0.02	−0.02

In addition, VMD can remove the inherent noise. Actual AMI data have noise owing to the interference due to peripheral electronic devices. VMD can improve the accuracy of the load forecasting through the deep learning training and regularization process by reducing the weight of high frequencies that are susceptible to noise, such as VMF-8, VMF-9, and VMF-10, which have low correlation indices of less than 1%. The AMI used in this study has a three-times higher sampling than conventional AMI and can reduce the model uncertainty as more samples are measured. The proposed method reduces the prediction uncertainty by training the decomposed signal with the high sampling AMI.

6. Case Studies

The time series forecasting models were simulated on real-world datasets of business buildings. We conducted the case studies with different prediction models and prediction time scales. The weekly prediction results for one hour ahead load forecasting are shown in Figure 7.

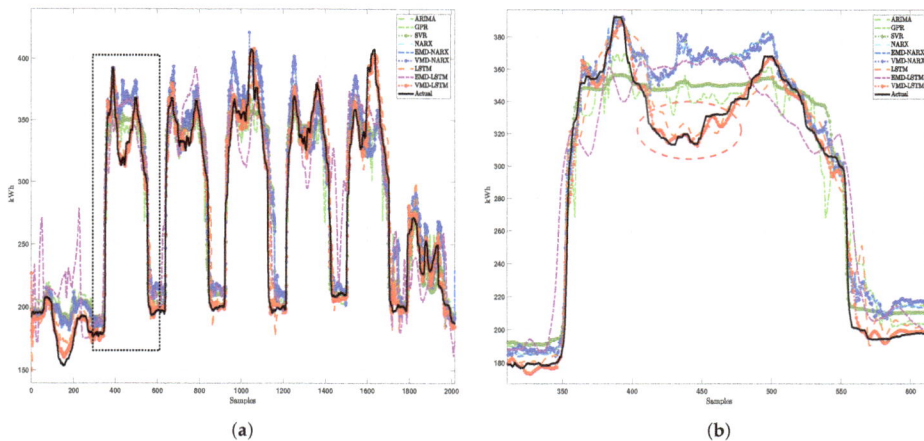

(a)   (b)

Figure 7. Actual and different load forecasting for a week. (a) Weekly load forecasting; (b) Monday load forecasting.

6.1. Comparative Conventional Load Forecasting Models

To validate the efficacy of the proposed VMD-LSTM RNN, eight load forecasting models, including ARIMA, SVR, GPR, NARX, NARX with EMD, NARX with VMD, LSTM, LSTM with EMD, and LSTM with VMD, were compared under the same benchmarks (RMSE, MAE, and MAPE).

The ARIMA model has been used for time-series prediction. However, with the rise of machine learning, the GPR and SVR models are being utilized. To account for seasonality in an ARIMA model, three hyperparameters were used: autoregression, stationarity, and moving average. The GPR model uses statistical hyperparameters, including variance and length, whereas the SVR model depends on kernel parameters, a penalty factor, and insensitive zone thickness. The ARIMA, GPR, and SVR models are trained through cross-validation and ADAM optimization or particle swarm optimization [2,26–29]. To compare the performance of the RNNs, we compared the results of applying two decomposition methods to the NARX and LSTM models The prediction results of all models are shown in Figure 7, and the prediction accuracy by day of the week is shown in Figure 8. Table 2 also summarizes the performance at different time scales.

6.2. Weekly Load Forecasting

Figure 7 illustrates the STLF for building load with one hour ago (12 steps ahead). To check the performance of the proposed method based on VMD and LSTM, the prediction results of different methods were compared. A closer look at the prediction results reveals the Monday load forecasting in Figure 7b. The proposed model showed robust performance under abrupt load increases and decreases in 400 samples and 500 samples, respectively. Conventional models exhibited conservative changes to sudden load changes, and EMD-LSTM exhibited excessive weight changes.

Figure 8 shows the average predictive error of the different methods. The result of load forecasting with one-month AMI data is shown in Figure 8a, and Figure 8b is the prediction result with three months of AMI data. There are distinct load characteristics for each day of the week. EMD-LSTM had large errors with an RMSE of 32.68 kWh, MAE of 28.61 kWh, and MAPE of 12.24% on Sunday in Figure 8a. However, if the size of the dataset is sufficiently large or the prediction time scale is long enough, the initial error can be corrected. When the data are insufficient with a short time scale, the input of the reference load profile (which is measurement data at the maximum observable time before load forecasting) can be a dominant feature of machine learning, which causes a large error. Figure 8 shows that, if the LSTM correctly decomposed periodic features, it had high accuracy even with small amounts of data, but if there was an error in the feature, the prediction error also increased because of the memory cell structure of LSTM.

VMD can reflect more dominant patterns than EMD with distinct periodicity. The performance difference of decomposition between EMD and VMD is shown in Figures 4 and 5. The RNNs using VMD showed performance improvements. However, there was a difference in performance improvement between NARX and LSTM because the vanishing gradient problem was solved differently, where NARX used the delay factor and LSTM had the memory cell structure. As LSTM preserved characteristics of dominant features through the memory cell, LSTM showed higher accuracy than NARX in STLF.

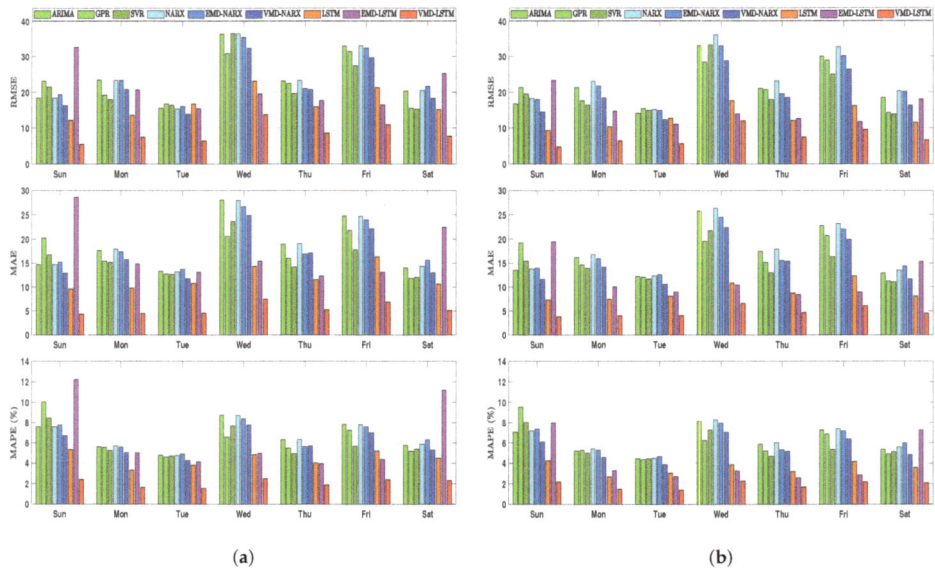

Figure 8. Benchmarks of different models. (**a**) One-month AMI data; (**b**) Three-month AMI data.

The MAPE of VMD-LSTM was around 2%. In the weekly comparison, the least error occurred on Tuesday: RMSE of 6.49 kWh, MAE of 3.98 kWh, and MAPE of 1.48%. This was because the correlation between days of the week was the highest on Tuesday. On the other hand, there was a large error on Wednesday and Friday because the correlation was relatively lower than on other days of the week.

The proposed VMD-LSTM reflecting the mixed periodic pattern of the load profile based on multi-decomposition with deep learning had the lowest error.

6.3. Benchmark for Different Prediction Time Scales

Finally, in this section, we analyze the accuracy of the load forecasting methods for the case study considering different prediction time scales (5 m, 1 h, 3 h, 24 h, 48 h, 72 h). The accuracy results are summarized in Table 2. The best accuracies were obtained for the shortest prediction time scale (5 m) for all models. The proposed model, VMD-LSTM, showed the best accuracy with an MAE of 1.95 kWh, RMSE of 4.28 kWh, and MAPE of 0.71%.

In addition, EMD-LSTM and VMD-LSTM showed better accuracy on the previous day when compared to the 36 steps ahead (3 h) and one day to three days ahead (24 h, 48 h, 72 h) time scales. The 24 h, 48 h, and 72 h cases show that RNN-based models had higher accuracies than ARIMA or GPR, but eventually showed similar errors, and their performances were saturated. This result was obtained because the reference load profile was learned as a dominant input according to the prediction time scale to reflect the power consumption trend, so the 288 steps ahead case and 576 steps ahead case, which had similar patterns, were slightly more accurate than the 36 steps ahead case (3 h).

Table 2. Load forecasting errors of different models.

Prediction Horizon	Index	ARIMA	GPR	SVR	NARX	EMD NARX	VMD NARX	LSTM	EMD LSTM	VMD LSTM
1 step ahead (5 min)	MAE	7.45	6.03	3.43	7.52	7.33	3.25	2.92	5.53	1.95
	RMSE	11.77	10.21	6.89	11.89	11.21	6.62	4.98	8.72	4.28
	MAPE (%)	3.46	2.67	1.96	3.61	3.39	1.84	1.12	2.21	0.71
12 steps ahead (1 h)	MAE	17.28	16.11	14.76	17.71	17.02	15.12	9.01	11.69	4.81
	RMSE	22.12	20.94	20.12	24.12	22.49	19.31	12.87	15.08	7.53
	MAPE (%)	6.20	6.06	5.70	6.35	6.27	5.43	3.54	4.27	1.90
36 steps ahead (3 h)	MAE	57.14	53.96	48.72	58.85	56.54	50.69	30.25	38.52	16.27
	RMSE	64.50	61.35	59.31	70.64	66.80	56.38	38.05	43.66	22.40
	MAPE (%)	20.62	19.91	18.17	21.79	20.97	20.03	11.63	14.26	6.01
288 steps ahead (24 h)	MAE	51.22	48.55	43.25	52.68	51.50	45.12	28.19	32.65	15.60
	RMSE	59.38	58.81	56.88	58.12	57.24	56.85	35.98	37.38	21.80
	MAPE (%)	18.90	17.91	16.13	19.16	19.06	16.71	10.62	11.78	5.75
576 steps ahead (48 h)	MAE	57.24	52.87	46.57	57.48	55.65	48.23	28.60	32.48	15.85
	RMSE	63.28	60.31	59.72	62.51	61.49	57.53	36.24	42.27	22.11
	MAPE (%)	22.08	19.72	17.76	21.49	20.99	17.18	10.92	12.18	5.89
864 steps ahead (72 h)	MAE	60.45	58.55	51.92	59.75	59.01	53.44	29.12	34.37	16.09
	RMSE	68.24	62.42	58.35	67.38	66.26	58.72	36.85	43.52	22.18
	MAPE (%)	24.14	21.76	18.72	22.43	21.19	19.05	11.05	12.86	5.96

7. Conclusions

This paper proposes short-term load forecasting using deep learning based on multi-decomposition. The results of the proposed approach were validated by applications to real-world data from a business building, and the performance of the proposed approach was assessed by comparing the predicted results with those of other models.

To monitor small-scale load and demand side management, an enhanced AMI that provides three-times more sample data points per hour than conventional AMI was used, increasing the accuracy of the load forecasting using deep learning. In this study, to detect the features of the load profile, the load profile was decomposed by a weekly seasonality and variational mode decomposition. Two decomposition methods can identify features such as seasonality, load increase/decrease pattern, and periodicity without any external data, such as temperature.

The three-step regularization process reduced the long-term dependency, overfitting, and covariate shift problem caused by feature decomposition, which increases the data samples and dimensions. The results also reveal the effectiveness of the long short-term memory neural networks based on variational mode decomposition with different prediction time scales. We expect the proposed method to be a key technique for demand side management, electrical power theft detection, energy storage system scheduling, and energy trading platforms in future smart grids.

Author Contributions: S.H.K. developed the main idea and designed the proposed model; he conducted the simulation studies and wrote the paper with the support of G.L. and G.-Y.K. under the supervision of the corresponding author, Y.-J.S. D.-I.K. contributed to the editing of the paper. All authors have read and approved the final manuscript.

Acknowledgments: This work was supported by a National Research Foundation of Korea (NRF) grant funded by the Ministry of Science, ICT & Future Planning (No. NRF-2017R1A2A1A05001022) and the framework of the international cooperation program managed by the NRF of Korea (No. NRF-2017K1A4A3013579). This research was also supported by the Korea Electric Power Corporation (KEPCO) (No. R18XA05).

Conflicts of Interest: The authors declare no conflicts of interest.

Nomenclature

AMI	advanced measuring infrastructure
ANN	artificial neural network
LSTM	long short-term memory
EMD	empirical mode decomposition
VMD	variational mode decomposition
LTLF	long-term load forecasting
MTLF	medium-term load forecasting
STLF	short-term load forecasting
USTLF	ultra-short-term load forecasting
DSM	demand side management
ARIMA	auto-regressive integrated moving average
GPR	Gaussian processing regression
GRU	gated recurrent unit
SVR	support vector regression
RNN	recurrent neural network
NARX	nonlinear autoregressive exogenous
CNN	convolutional neural network
IMFs	intrinsic mode functions
k	the mode index
v_k	k^{th} intrinsic mode
$W_p(t)$	p^{th} weekly load profile
w	frequency of mode
K	the total number of modes
x_t	the typical load profile
x_t^p	the load p^{th} weekly seasonality feature
$x_{k,t}^p$	k^{th} IMF of the load p^{th} weekly seasonality feature
δ	the Dirac distribution
D_p	the weekly decay's exponent factor
N_k	the IMF normalization factor
BN	batch normalization
s_t	the memory cell state of LSTM
f_t	the forget gate of LSTM

i_t	the input gate of LSTM
g_t	the input node of LSTM
o_t	the output gate of LSTM
h_t	the output value of LSTM
RMSE	root mean squared error
MAE	mean absolute error
MAPE	mean absolute percent error
VMF	variational mode function
EMF	empirical mode function

References

1. Ekonomou, L.; Christodoulou, C.A.; Mladenov, V. A short-term load forecasting method using artificial neural networks and wavelet analysis. *Int. J. Power Syst.* **2016**, *1*, 64–68.
2. Mirowski, P.; Chen, S.; Ho, T.K.; Yu, C.-N. Demand forecasting in smart grids. *Bell Syst. Tech. J.* **2014**, *18*, 135–158. [CrossRef]
3. Zhang, X. Short-term load forecasting for electric bus charging stations based on fuzzy clustering and least squares support vector machine optimized by wolf pack algorithm. *Energies* **2018**, *11*, 1449. [CrossRef]
4. Fiot, J.-B.; Dinuzzo, F. Electricity demand forecasting by multi-task learning. *IEEE Trans. Smart Grid* **2018**, *9*, 544–551. [CrossRef]
5. Dahl, M.; Brun, A.; Kirsebom, O.; Andresen, G. Improving short-term heat load forecasts with calendar and holiday data. *Energies* **2018**, *11*, 1678. [CrossRef]
6. Teeraratkul, T.; O'Neill, D.; Lall, S. Shape-based approach to household electric load curve clustering and prediction. *IEEE Trans. Smart Grid* **2018**, *9*, 5196–5206. [CrossRef]
7. Wang, Y.; Zhang, N.; Chen, Q.; Kirschen, D.S.; Li, P.; Xia, Q. Data-driven probabilistic net load forecasting with high penetration of behind-the-meter pv. *IEEE Trans. Power Syst.* **2018**, *33*, 3255–3264. [CrossRef]
8. Haben, S.; Singleton, C.; Grindrod, P. Analysis and clustering of residential customers energy behavioral demand using smart meter data. *IEEE Trans. Smart Grid* **2016**, *7*, 136–144. [CrossRef]
9. Stephen, B.; Tang, X.; Harvey, P.R.; Galloway, S.; Jennett, K.I. Incorporating practice theory in sub-profile models for short term aggregated residential load forecasting. *IEEE Trans. Smart Grid* **2017**, *8*, 1591–1598. [CrossRef]
10. Hayes, B.P.; Gruber, J.K.; Prodanovic, M. Multi-nodal short-term energy forecasting using smart meter data. *IET Gener. Transm. Dis.* **2018**, *12*, 2988–2994. [CrossRef]
11. Xie, J.; Chen, Y.; Hong, T.; Laing, T.D. Relative humidity for load forecasting models. *IEEE Trans. Smart Grid* **2018**, *9*, 191–198. [CrossRef]
12. Xie, J.; Hong, T. Temperature scenario generation for probabilistic load forecasting. *IEEE Trans. Smart Grid* **2018**, *9*, 1680–1687. [CrossRef]
13. Li, P.; Zhang, J.; Li, C.; Zhou, B.; Zhang, Y.; Zhu, M.; Li, N. Dynamic similar sub-series selection method for time series forecasting. *IEEE Access* **2018**, *6*, 32532–32542. [CrossRef]
14. Lin, L.; Xue, L.; Hu, Z.; Huang, N. Modular predictor for day-ahead load forecasting and feature selection for different hours. *Energies* **2018**, *11*, 1899. [CrossRef]
15. Xie, J.; Hong, T. Variable selection methods for probabilistic load forecasting: Empirical evidence from seven states of the united states. *IEEE Trans. Smart Grid* **2018**, *9*, 6039–6046. [CrossRef]
16. Li, B.; Zhang, J.; He, Y.; Wang, Y. Short-term load-forecasting method based on wavelet decomposition with second-order gray neural network model combined with adf test. *IEEE Access* **2017**, *5*, 16324–16331. [CrossRef]
17. Rafiei, M.; Niknam, T.; Aghaei, J.; Shafie-khah, M.; Catalão, J.P.S. Probabilistic load forecasting using an improved wavelet neural network trained by generalized extreme learning machine. *IEEE Trans. Smart Grid* **2018**, *9*, 6961–6971. [CrossRef]
18. Auder, B.; Cugliari, J.; Goude, Y.; Poggi, J.-M. Scalable clustering of individual electrical curves for profiling and bottom-up forecasting. *Energies* **2018**, *11*, 1893. [CrossRef]
19. Qiu, X.; Ren, Y.; Suganthan, P.N.; Amaratunga, G.A.J. Empirical mode decomposition based ensemble deep learning for load demand time series forecasting. *Appl. Soft Comput.* **2017**, *54*, 246–255. [CrossRef]

20. Bedi, J.; Toshniwal, D. Empirical mode decomposition based deep learning for electricity demand forecasting. *IEEE Access* **2018**, *6*, 49144–49156. [CrossRef]

21. Liu, H.; Mi, X.; Li, Y. An experimental investigation of three new hybrid wind speed forecasting models using multi-decomposing strategy and elm algorithm. *Renew. Energy* **2018**, *123*, 694–705. [CrossRef]

22. Lahmiri, S. Comparing variational and empirical mode decomposition in forecasting day-ahead energy prices. *IEEE Syst. J.* **2017**, *11*, 1907–1910. [CrossRef]

23. Dragomiretskiy, K.; Zosso, D. Variational mode decomposition. *IEEE Trans. Signal Process.* **2014**, *62*, 531–544. [CrossRef]

24. Huang, N.; Yuan, C.; Cai, G.; Xing, E. Hybrid short term wind speed forecasting using variational mode decomposition and a weighted regularized extreme learning machine. *Energies* **2016**, *9*, 989. [CrossRef]

25. Lin, Y.; Luo, H.; Wang, D.; Guo, H.; Zhu, K. An ensemble model based on machine learning methods and data preprocessing for short-term electric load forecasting. *Energies* **2017**, *10*, 1186. [CrossRef]

26. Ruiz-Abellón, M.; Gabaldón, A.; Guillamón, A. Load forecasting for a campus university using ensemble methods based on regression trees. *Energies* **2018**, *11*, 2038. [CrossRef]

27. Dong, Y.; Zhang, Z.; Hong, W.-C. A hybrid seasonal mechanism with a chaotic cuckoo search algorithm with a support vector regression model for electric load forecasting. *Energies* **2018**, *11*, 1009. [CrossRef]

28. Li, M.-W.; Geng, J.; Hong, W.-C.; Zhang, Y. Hybridizing chaotic and quantum mechanisms and fruit fly optimization algorithm with least squares support vector regression model in electric load forecasting. *Energies* **2018**, *11*, 2226. [CrossRef]

29. Sheng, H.; Xiao, J.; Cheng, Y.; Ni, Q.; Wang, S. Short-term solar power forecasting based on weighted gaussian process regression. *IEEE Trans. Ind. Electron.* **2018**, *65*, 300–308. [CrossRef]

30. Manic, M.; Amarasinghe, K.; Rodriguez-Andina, J.J.; Rieger, C. Intelligent buildings of the future: Cyberaware, deep learning powered, and human interacting. *IEEE Ind. Electron. Mag.* **2016**, *10*, 32–49. [CrossRef]

31. Li, C.; Ding, Z.; Yi, J.; Lv, Y.; Zhang, G. Deep belief network based hybrid model for building energy consumption prediction. *Energies* **2018**, *11*, 242. [CrossRef]

32. Wang, Y.; Zhang, N.; Tan, Y.; Hong, T.; Kirschen, D.S.; Kang, C. Combining probabilistic load forecasts. *IEEE Trans. Smart Grid* **2018**. Available online: https://arxiv.org/abs/1803.06730 (accessed on 5 November 2018).

33. Wang, J.; Gao, Y.; Chen, X. A novel hybrid interval prediction approach based on modified lower upper bound estimation in combination with multi-objective salp swarm algorithm for short-term load forecasting. *Energies* **2018**, *11*, 1561. [CrossRef]

34. Sun, W.; Zhang, C. A hybrid ba-elm model based on factor analysis and similar-day approach for short-term load forecasting. *Energies* **2018**, *11*, 1282. [CrossRef]

35. Ruiz, L.G.B.; Cuéllar, M.P.; Calvo-Flores, M.D.; Jiménez, M.D.C.P. An application of non-linear autoregressive neural networks to predict energy consumption in public buildings. *Energies* **2016**, *9*, 684. [CrossRef]

36. DiPietro, R.; Rupprecht, C.; Navab, N.; Hager, G.D. Analyzing and exploiting narx recurrent neural networks for long-term dependencies. *arXiv* **2017**, arXiv:1702.07805.

37. Bouktif, S.; Fiaz, A.; Ouni, A.; Serhani, M. Optimal deep learning lstm model for electric load forecasting using feature selection and genetic algorithm: Comparison with machine learning approaches. *Energies* **2018**, *11*, 1636. [CrossRef]

38. Kong, W.; Dong, Z.Y.; Jia, Y.; Hill, D.J.; Xu, Y.; Zhang, Y. Short-term residential load forecasting based on lstm recurrent neural network. *IEEE Trans. Smart Grid* **2018**. [CrossRef]

39. Chen, K.; Chen, K.; Wang, Q.; He, Z.; Hu, J.; He, J. Short-term load forecasting with deep residual networks. *IEEE Trans. Smart Grid* **2018**. Available online: https://arxiv.org/abs/1805.11956 (accessed on 5 November 2018).

40. Shi, H.; Xu, M.; Li, R. Deep learning for household load forecasting—A novel pooling deep rnn. *IEEE Trans. Smart Grid* **2018**, *9*, 5271–5280. [CrossRef]

41. Kuo, P.-H.; Huang, C.-J. A high precision artificial neural networks model for short-term energy load forecasting. *Energies* **2018**, *11*, 213. [CrossRef]

42. Wang, Y.; Liu, M.; Bao, Z.; Zhang, S. Short-term load forecasting with multi-source data using gated recurrent unit neural networks. *Energies* **2018**, *11*, 1138. [CrossRef]

43. Merkel, G.; Povinelli, R.; Brown, R. Short-term load forecasting of natural gas with deep neural network regression. *Energies* **2018**, *11*, 2008. [CrossRef]

Energies **2018**, *11*, 3433

44. Li, Y.; Huang, Y.; Zhang, M. Short-term load forecasting for electric vehicle charging station based on niche immunity lion algorithm and convolutional neural network. *Energies* **2018**, *11*, 1253. [CrossRef]
45. Greff, K.; Srivastava, R.K.; Koutník, J.; Steunebrink, B.R.; Schmidhuber, J. LSTM: A search space odyssey. *IEEE Trans. Neural Netw. Learn. Syst.* **2017**, *28*, 2222–2232. [CrossRef] [PubMed]
46. Chung, J.; Gulcehre, C.; Cho, K.; Bengio, Y. Empirical Evaluation of Gated Recurrent Neural Networks on Sequence Modeling. *arXiv* **2014**, arXiv:1412.3555.
47. Zhan, T.-S.; Chen, S.-J.; Kao, C.-C.; Kuo, C.-L.; Chen, J.-L.; Lin, C.-H. Non-technical loss and power blackout detection under advanced metering infrastructure using a cooperative game based inference mechanism. *IET Gener. Transm. Dis.* **2016**, *10*, 873–882. [CrossRef]

![energies logo] *energies*

MDPI

Article

Impacts of Load Profiles on the Optimization of Power Management of a Green Building Employing Fuel Cells

Fu-Cheng Wang * and **Kuang-Ming Lin**

Department of Mechanical Engineering, National Taiwan University, Taipei 10617, Taiwan;
r07522809@ntu.edu.tw
* Correspondence: fcw@ntu.edu.tw; Tel.: +886-2-3366-2680

Received: 25 October 2018; Accepted: 24 December 2018; Published: 25 December 2018

Abstract: This paper discusses the performance improvement of a green building by optimization procedures and the influences of load characteristics on optimization. The green building is equipped with a self-sustained hybrid power system consisting of solar cells, wind turbines, batteries, proton exchange membrane fuel cell (PEMFC), electrolyzer, and power electronic devices. We develop a simulation model using the Matlab/SimPowerSystem^{TM} and tune the model parameters based on experimental responses, so that we can predict and analyze system responses without conducting extensive experiments. Three performance indexes are then defined to optimize the design of the hybrid system for three typical load profiles: the household, the laboratory, and the office loads. The results indicate that the total system cost was reduced by 38.9%, 40% and 28.6% for the household, laboratory and office loads, respectively, while the system reliability was improved by 4.89%, 24.42% and 5.08%. That is, the component sizes and power management strategies could greatly improve system cost and reliability, while the performance improvement can be greatly influenced by the characteristics of the load profiles. A safety index is applied to evaluate the sustainability of the hybrid power system under extreme weather conditions. We further discuss two methods for improving the system safety: the use of sub-optimal settings or the additional chemical hydride. Adding 20 kg of NaBH$_4$ can provide 63 kWh and increase system safety by 3.33, 2.10, and 2.90 days for the household, laboratory and office loads, respectively. In future, the proposed method can be applied to explore the potential benefits when constructing customized hybrid power systems.

Keywords: hybrid power system; fuel cell; solar; wind; fuel cell; optimization; cost; reliability

---

1. Introduction

Today's energy crises and pollution problems have increased the current interest in fuel cell research. One of the most popular fuel cells is the proton exchange membrane fuel cell (PEMFC), which can transform chemical energy into electrical energy with high energy conversion efficiency by electrochemical reactions. At the anode, the hydrogen molecule ionizes, releasing electrons and H$^+$ protons. At the cathode, oxygen reacts with electrons and H$^+$ protons through the membrane to form water. The electrons pass through an electrical circuit to create current output of the PEMFC. The PEMFC has several advantageous properties, including a low operating temperature and high efficiency. However, it also has very complex electrochemical reactions, so attempts to develop dynamic models for PEMFC systems have become an active research focus. For example, Ceraolo et al. [1] developed a PEMFC model that contained the Nernst equation, the cathodic kinetics equation, and the cathodic gas diffusion equation. Similarly, Gorgun [2] presented a dynamic PEMFC model that included water phenomena, electro-osmotic drag and diffusion, and a voltage ancillary. These models have served as the basis of many advanced control techniques aimed at improving the performance of PEMFC systems. For instance, Woo and Benziger [3] tried to improve PEMFC efficiency using a proportional-integral-derivative (PID) controller to regulate the hydrogen flow rate. Vega-Leal et al. [4] controlled the air and hydrogen flow rates to optimize the PEMFC output power. Park et al. [5] considered load perturbations and applied a sliding mode control to maintain the pressures of hydrogen and oxygen regardless of current changes. Wang et al. [6] designed a robust controller to regulate the air flow rate to ensure that the PEMFC provided a steady output voltage. This idea was further extended to a multi-input multi-output (MIMO) PEMFC model to reduce hydrogen consumption while providing a steady voltage [7]. Reduced-order robust control [8] and robust PID control [9] were also proposed for hardware simplification and industrial applications.

A PEMFC can supply sustainable power as long as the hydrogen supply is continuous; therefore, the PEMFC has been widely applied in transportation [10–19] and stationary power systems [20–29]. A PEMFC can also supply sustainable energy regardless of weather conditions, making it a reliable power source when solar and wind energy are unavailable. However, the price of hydrogen energy is generally high when compared to other green (e.g., solar) energy, so the PEMFC is typically integrated with other energy sources and storage systems to form hybrid power systems. For example, Zervas et al. [30] presented a hybrid system that contained photovoltaics (PV), a PEMFC, and an electrolyzer with metal hydride tanks. Rekioua et al. [31] considered a hybrid photovoltaic-electrolyzer-fuel cell system and discussed its optimization by selection of different topologies. Nizetic et al. [29] proposed a system for household application that used a high-temperature PEMFC to drive a modified heat pump system, with a cost of less than 0.16 euro/kWh.

The role of the PEMFC in hybrid power systems is unique, because it can act as both an energy source and an energy storage system. It serves as an energy source to provide backup power when the load requirement is greater than the energy supply from other energy sources and as an energy storage system to store hydrogen electrolyzed by redundant energy when the energy supply is greater than the consumption [32]. Some hybrid power systems have recently been implemented in practice. For instance, Singh et al. [22] presented a PEMFC/PV hybrid system for stand-alone applications in India. Das et al. [23] introduced the PV/battery/PEMFC and PV/battery systems installed in Malaysia. Al-Sharafi et al. [24] considered six different systems in the Kingdom of Saudi Arabia. Martinez-Lucas et al. [25] demonstrated a system based on wind turbine (WT) and pump storage hydropower on the Canary Island of El Hierro, Spain. Kazem et al. [27] evaluated four different hybrid power systems on Masirah Island, Oman.

Because of the influence of weather conditions and loads, the costs of these hybrid systems can be optimized by changing the system configurations. For example, Ettihir et al. [26] applied the adaptive recursive least square method to find the best efficiency and power operating points. Singh et al. [22] applied a fuzzy logic program to calculate system costs and concluded that the PEMFC and battery are the most significant modules for meeting load demands late at night and in the early morning. Kazem et al. [27] showed that that a PV/WT/battery/diesel hybrid system had the lowest cost for energy production. Cozzolino et al. [28] analyzed the Tunisia and Italy (TUNeIT) Project and showed that this almost self-sustaining renewable power plant, consisting of a WT, PV, battery, PEMFC, and diesel engine, ran at a cost of 0.522 €/kWh. Wang et al. [33] studied a hybrid system that consisted of a WT, PV, battery, and an electrolyzer and concluded that the costs and reliability of hybrid power systems can be greatly improved by adjusting the component sizes. They also showed that power management can help to reduce system costs [32]. The present paper extends these ideas by discussing the impacts of load profiles on the optimization of system costs. We applied three typical load profiles to a hybrid system and discussed the cost and energy distribution. We also evaluated the guaranteed operation durations (called system safety) of hybrid systems and discussed the applications of two methods to extend system safety.

The remainder of this paper is arranged as follows: Section 2 introduces the green building and its hybrid power system. Based on the system characteristics, we build a general hybrid power model consisting of solar cells, WTs, batteries, a PEMFC, hydrogen electrolysis, and chemical hydrogen generation. The model parameters were tuned based on experimental data to allow the prediction of system responses under different operation conditions. Historical irradiation and wind data were applied to estimate the power supplied by the PV and WT, while three typical load profiles were considered to understand their impacts on system optimization. Section 3 defines three performance indexes for evaluating hybrid power systems equipped with different components and management strategies. We applied three typical loads to optimize system design by tuning the component sizes and power management. The results showed that the optimization processes can effectively reduce the energy costs by 38.9%, 40.0%, and 28.6% and greatly improve system reliability by 4.89%, 26.42%, and 5.08% for household, laboratory, and office loads, respectively. The guaranteed sustainable operation periods under extreme weather conditions were also estimated. The results revealed that system sustainability can be improved by the use of a sub-optimal design or chemical hydrides. We also discuss the critical prices of implementing a chemical hydrogen generation system. Conclusions are then drawn in Section 4.

2. System Description and Modelling

The green building, as shown in Figure 1 [34], is located in Miao-Li County in Taiwan. It was constructed by China Engineering Consultants Inc. (CECI) and was equipped with a hybrid power system that consisted of 10 kW PV arrays, 6 kW WTs, 800Ah lead-acid batteries, a 3 kW PEMFC, and a 2.5 kW electrolyzer with a hydrogen production rate of 500 L/h. The building was autonomous and did not connect to the main grid, i.e., its electricity was supplied completely by green energy, such as solar and wind. The energy can be stored for use when the green energy is less than the load demands. These components were originally selected to provide a daily energy supply of about 20 kWh based on the National Aeronautics and Space Administration (NASA) data [34], as illustrated in Table 1. Solar energy was abundant in the summer but poor in the winter, so wind energy was expected to compensate for solar energy in the winter. However, Chen and Wang [32] applied the Vantage Pro2 Plus Stations [35] to measure the real weather data on the building site and found that the wind energy was not sufficient to compensate for the reduced solar energy in the winter. Further analyses of the energy costs also revealed that the wind energy was not economically efficient for this building, as illustrated in Table 2. Therefore, the following component selection principles were suggested to improve system performance [32]:

(1) Energy sources: the use of PV and PEMFC in the green building was suggested, because solar energy was the most economical energy source and the PEMFC could guarantee energy sustainability. The PEMFC can be regarded as an energy source that provides steady energy and as an energy storage system when coupled with a hydrogen electrolyzer. Considering the transportation, storage, and efficiency of energy conversion, the PEMFC with chemical hydrogen generation by $NaBH_4$ [36] was suggested for the system.

(2) Energy storage: the lead-acid battery was suggested because of its greater than 90% efficiency [37]. Though the PEMFC with a hydrogen electrolyzer can also store energy, the conversion efficiency from electricity into hydrogen was only about 60% [33]. Therefore, the total energy storage efficiency was about 36%, because the PEMFC converted hydrogen into electricity with an efficiency of about 60% [38]. Note that the LiFe battery has a higher efficiency (more than 95%) but is much more expensive than a lead-acid battery. Therefore, the lead-acid battery was preferred for the green building.

That is, the selection of multiple energy sources and storages depended on the local conditions and load requirements.

Figure 1. The green building.

Table 1. The daily average weather data on the building site [32,34].

Data Source	Irradiance (W/m^2)		Wind Speed (m/s)	
	Summer	Winter	Summer	Winter
NASA [34]	267	109	4.95	8.70
Measured	239	115	2.42	3.96

Table 2. Energy cost analyses ($/kWh) [32].

Energy Sources	Summer	Winter
Photovoltaic (PV) arrays	0.11	0.23
Wind turbine (WT)	7.76	0.69
Proton exchange membrane fuel cell (PEMFC) with chemical H$_2$ generation	$\geqslant 1.76$	$\geqslant 1.76$

2.1. The Hybrid Power Model

A general hybrid power model, as shown in Figure 2, was developed to evaluate system performance at different operating conditions (e.g., varying the component sizes and power management strategies) [32,33,39]. The model consisted of a PV module, a WT module, a battery module, a PEMFC module, an electrolyzer module, a chemical hydrogen generation module, and a load module. The power management strategies were applied to operate these modules based on battery state-of-charge (SOC). The module parameters were adjusted by the component characteristics and experimental responses to allow prediction and analysis of the system dynamics without the need for extensive experiments [39,40].

First, the 1 kW PV module was developed based on the following equation [32,41]:

$$P_{PV} = 0.69(E - 1.52) \tag{1}$$

where P_{PV} (Watt) and E (Watt per meter square) represent solar power and irradiance, respectively. Second, the WT module was presented as a look-up table, according to the relation between wind power and wind speed [33,42]. Third, the PEMFC acted as a back-up power source to guarantee system sustainability based on the following management strategies (see Figure 3a) [39]:

(1) When the battery SOC dropped to the lower bound, SOC_{low}, the PEMFC was switched on to provide a default current of 20 A at the highest energy efficiency [41].

(2)   When the SOC continuously dropped to $SOC_{low} - 5\%$, the PEMFC current was increased according to the required load until the SOC was raised to $SOC_{low} + 5\%$, where the PEMFC provided a default current of 20 A.

(3)   When the battery SOC reached SOC_{high}, the PEMFC was switched off.

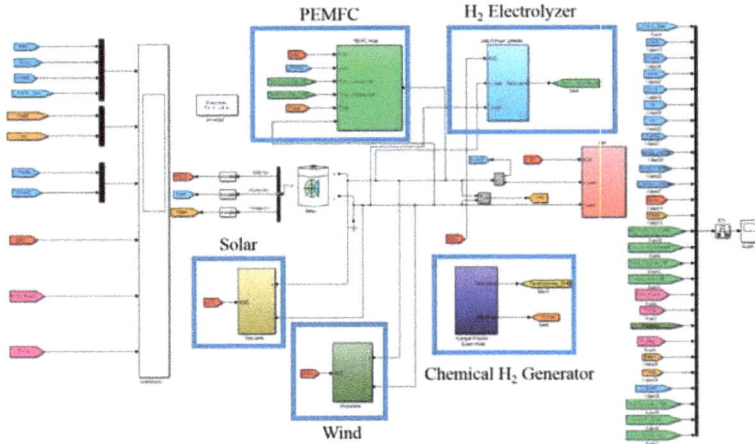

Figure 2. The hybrid power model.

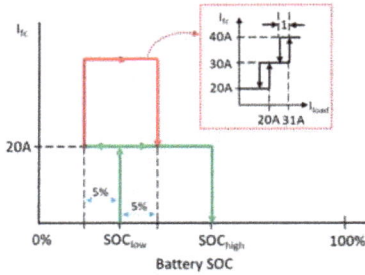

(a)

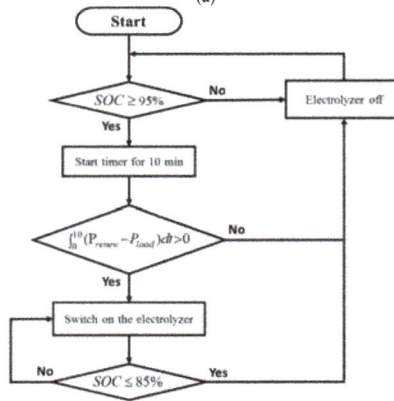

(b)

Figure 3. Power management. (**a**) The proton exchange membrane fuel cell (PEMFC) management [39]. (**b**) The electrolyzer management [33].

Therefore, the power management can be adjusted by tuning SOC_{low} and SOC_{high}. As a last stage, the hydrogen electrolyzer transferred redundant energy to hydrogen storage based on the following strategies (see Figure 3b) [33]:

(1) When the battery SOC was higher than 95%, the extra renewable energy was regarded as redundant.

(2) The electrolyzer module would wait for ten minutes to avoid chattering. If the total redundant energy increased during this period, the electrolyzer was switched on.

(3) When the hydrogen tank was full or the battery SOC dropped to 85%, the electrolyzer was switched off.

Thus, the electrolyzer produced hydrogen when the battery SOC was between 85% and 95%. The electrolyzer module was set to produce hydrogen at a rate of 1.14 L/min by consuming a constant power of 410 W, based on the experimental results [33].

2.2. Inputs Energy and Output Loads

We applied the historical irradiation and wind speed data [32], as shown in Figure 4, to the PV and WT modules, respectively. As shown in Figure 4, solar radiation was abundant in the summer but poor in the winter; therefore, solar energy in the summer can be stored for use in the winter. Conversely, the wind speed was high in the winter but low in the summer, so wind energy was expected to compensate for the lack of solar energy in the winter. However, the compensation effects were not as significant as originally designed because the wind was not sufficiently strong and the energy cost was much higher (see Table 2) when compared to other energy sources. Note that both solar and wind energy were concentrated in the daytime, indicating that this energy should be stored for use at night.

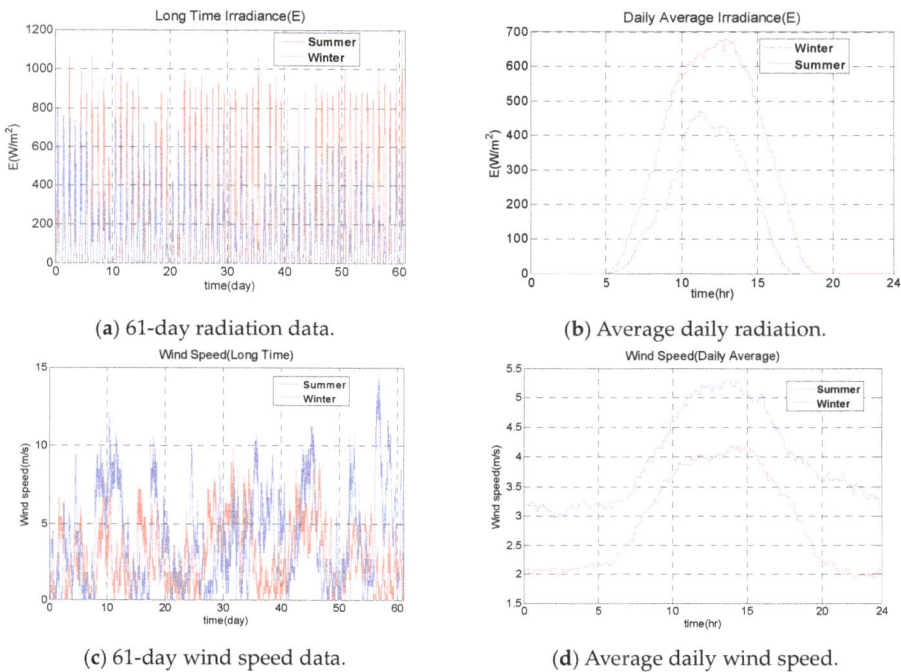

(**a**) 61-day radiation data.

(**b**) Average daily radiation.

(**c**) 61-day wind speed data.

(**d**) Average daily wind speed.

Figure 4. Radiation and wind data.

Three standard load profiles [43,44], as illustrated in Figure 5, were applied to the load module to investigate the impacts of loads on the optimization of the hybrid power system. The 61-day historical data were used for simulation and optimization analyses. Table 3 illustrates the statistical data of these load profiles, where the household had the largest historically peak and the office had the largest daily average peak, while the laboratory load had the greatest energy consumption. Therefore, we used these three typical loads to demonstrate how load characteristics can affect the performance optimization of the hybrid power system.

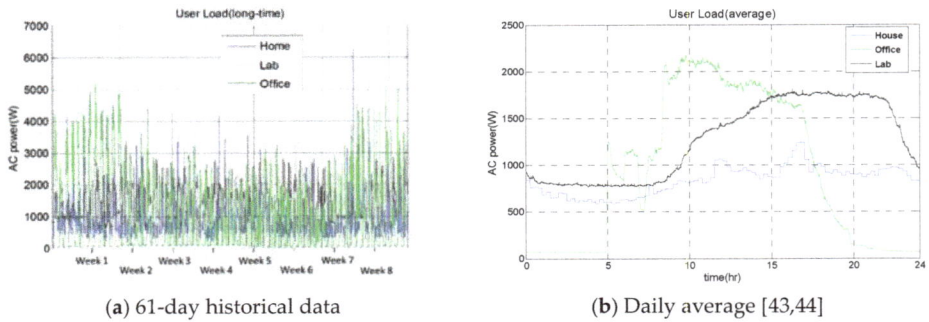

(a) 61-day historical data   (b) Daily average [43,44]

Figure 5. Three standard load profiles.

Table 3. The statistical data of load profiles [39].

	Household	Lab	Office
Historic peak (W)	6220	3395	5333
Daily average peak (W)	1237	1811	2178
Daily average (kWh)	19.96	30.41	22.32

3. Design Optimization of the Hybrid Power System

The hybrid power model was applied to predict system responses under different conditions, such as the use of varying components and loads. We defined three indexes to evaluate the performance of the hybrid power system: cost, reliability, and safety, as described by the following:

(1) System cost: the system cost $J_{(b, s, w)}$ consisted of two parts, J_i and J_o, as follows [39]:

$$J_{(b, s, w)} = J_{i(b, s, w)} + J_{o(b, s, w)} \tag{2}$$

where J_i and J_o indicate the initial and operation costs, respectively. The subscripts b, s, and w represent the numbers of batteries, PV arrays, and WTs in units of 100Ah, 1kW, and 3kW, respectively. The initial cost J_i accounted for the investment in the components, such as the PEMFC, power electric devices, PV arrays, WT, hydrogen electrolyzer, chemical hydrogen generator, and battery set, as follows:

$$J_{i(b, s, w)} = \sum_k J_{i(b, s, w)}^k \tag{3}$$

where k = PEMFC, DC, solar, WT, HE, CHG, and batt, respectively.

The operation cost J_o included the hydrogen consumption and the maintenance of the WT and PV arrays, as in the following:

$$J_{o(b, s, w)} = \sum_l J_{o(b, s, w)}^l \tag{4}$$

where l = NaBH$_4$, WT, and solar, respectively.

We calculated the initial costs $J^k_{i(b, s, w)}$ and the operation costs $J^l_{o(b, s, w)}$ as follows:

$$J^k_{i(b, s, w)} = C_k \cdot n_k \cdot CRF_k \tag{5}$$

$$J^l_{o(b, s, w)} = C_l \cdot n_l \tag{6}$$

in which C and n are the price per unit and the installed units, respectively, for each component k. CRF represented the capital recovery factor that was defined as [32,33,39]:

$$CRF = \frac{ir(1 + ir)^{ny}}{(1 + ir)^{ny} - 1} \tag{7}$$

where ir is the inflation rate, which was set as 1.26% in this paper by referring to the average annual change of consumer price index of Taiwan [39], and ny is the expected life of the components. The price and expected life of the components are illustrated in Table 4 were used to calculate the system costs in the following examples.

Table 4. Component life and price [32,33,39].

Component	Life (year)	Price ($)
Hybrid system	15	N/A
Wind turbine (3 kW)	15	9666
PV arrays (1 kW)	15	1833
Power electronic devices (6 kW)	15	1666
Chemical H_2 generation	15	10,666
NaBH$_4$ (60 g)	N/A	0.33
Electrolyzer (2.5 kW–500 L/h)	N/A	10,666
PEMFC (3 kW)	N/A	6000
Battery (48 V–100 Ah)	N/A	866

(2) System reliability: the reliability of the hybrid system was defined as the loss of power supply (LPSP) as follows [32,33,39]:

$$LPSP = \frac{\int LPS(t)dt}{\int P(t)dt} \tag{8}$$

in which $LPS(t)$ was the shortage (lost) of power supply at time t, while $P(t)$ was the power demand of the load profile at time t. Therefore, $\int LPS(t)dt$ indicated the insufficient energy supply and $\int P(t)dt$ represented the total energy demand for the entire simulation. If the power supply met the load demand at all times, (i.e., $LPS(t) = 0, \forall t$), then the system was completely reliable with $LPSP = 0$.

(3) System safety: system safety was defined as the guaranteed sustainable period of the hybrid power system under extreme weather conditions when no solar or wind energy was available. Suppose the energy stored in the system was E_{store} and the average daily energy consumption was E_{day}; then, the system safety can be defined as follows:

$$Safety = \frac{E_{store}}{E_{day}} \tag{9}$$

For example, average daily energy demand is 19.96, 30.41, and 22.32 kWh for the household, laboratory, and office, respectively (see Table 3). Therefore, if the energy stored in the battery and hydrogen is 60 kWh, the system safety is 3.01, 1.97, and 2.69 days for the laboratory, office, and household, respectively. When considering the efficiency of the battery and inverter both as 90%, then the system safety is 2.70, 1.78, and 2.42 days, respectively.

We applied the three typical loads to investigate their impacts on the optimization of the hybrid power system by tuning the component sizes and power management strategies.

3.1. Household Load

Applying the household load (see Figure 5) to the original system layout $(b, s, w) = (8, 10, 2)$ and management settings of $(SOC_{low}, SOC_{high}) = (40\%, 50\%)$ gave the system's reference plot shown in Figure 6a, where the system cost was estimated as $J = 1.300$ \$/kWh with $LPSP = 4.89\%$ (see Step 1 of Table 5). From Figure 6a, the system cost can be reduced to $J = 1.169$ \$/kWh by adjusting the components as $(b, s, w) = (18, 9, 2)$ but with a possible power cut ($LPSP = 2.61\%$, see Step 2 of Table 5). If the requirement was $LPSP = 0$, then the optimal system cost was $J = 1.189$ \$/kWh, achieved by setting $(b, s, w) = (18, 10, 2)$ (see Step 3 of Table 5). That is, we can reduce the system cost from $J = 1.300$ to 1.189 \$/kWh, while improving the system reliability from $LPSP = 4.89\%$ to 0.

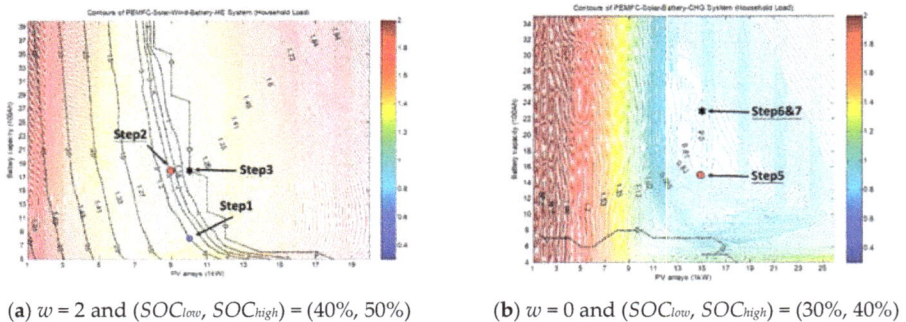

(a) $w = 2$ and $(SOC_{low}, SOC_{high}) = (40\%, 50\%)$ (b) $w = 0$ and $(SOC_{low}, SOC_{high}) = (30\%, 40\%)$

Figure 6. The reference plots for the household load.

Table 5. The optimal design procedure for the house load.

	(b, s, w)	(SOC_{low}, SOC_{high})	$LPSP$ (%)	J (\$/kWh)
Step 1	(8, 10, 2)	(40%, 50%)	4.89%	1.300
Step 2	(18, 9, 2)	(40%, 50%)	2.61%	1.169
Step 3	(18, 10, 2)	(40%, 50%)	0%	1.189
Step 4	(15, 15, 0)	(40%, 50%)	0%	0.822
Step 5	(15, 15, 0)	(30%, 40%)	0%	0.810
Step 6	(23, 15, 0)	(30%, 40%)	0%	0.794
Step 7	(23, 15, 0)	(30%, 40%)	0%	0.794
Optimal	(23, 15, 0)	(30%, 40%)	0%	0.794

Because the cost of wind energy was much higher than the cost of solar energy (see Table 2) and the compensation effects were not significant (see Figure 4), the use of solar and a PEMFC with chemical hydrogen production was viewed as economically efficient for the green building [32]. Therefore, we set $w = 0$ and the resulting optimization showed that the system cost can be significantly reduced to $J = 0.822$ \$/kWh by setting $(b, s, w) = (15, 15, 0)$, as illustrated in Step 4 of Table 5. Furthermore, when we fixed the component settings of $(b, s, w) = (15, 15, 0)$ and tuned the power management strategies $(SOC_{low}, SOC_{high}) = (30\%, 40\%)$, the system cost was further decreased to $J = 0.810$ \$/kWh (see Step 5 of Table 5). Steps 6 and 7 illustrate the iterative tuning of component size and power management, respectively. The results indicated that the system cost converged to $J = 0.794$ \$/kWh with $(b, s, w) = (23, 15, 0)$ and $(SOC_{low}, SOC_{high}) = (30\%, 40\%)$. Compared with the original cost, the cost was reduced by 38.9%, while maintaining complete system reliability. Note that the iterative method can greatly reduce the computation time because the simultaneous optimization of four parameters $(b, s, SOC_{low}, SOC_{high})$ took much longer than iterative optimization, as indicated in [45]. Therefore, the proposed iterative optimization can be applied for a quick estimation of the system behavior. Simultaneous optimization can be considered for potentially better optimization if time permits.

3.2. Laboratory Load

Similarly, the results of applying the laboratory load (see Figure 5) to the hybrid power model are shown in Figure 7 and Table 6. First, the original system layout $(b, s, w) = (8, 10, 2)$ with management settings of $(SOC_{low}, SOC_{high}) = (40\%, 50\%)$ resulted in a system cost of $J = 1.100$ $/kWh and $LPSP = 26.42\%$. Note that the $LSPS$ was much higher than was obtained for the household, because the laboratory load was mainly at night and the stored energy by hydrogen electrolyzation failed to provide sufficient energy. The initial component optimization can reduce the system cost to $J = 0.929$ $/kWh by setting $(b, s, w) = (27, 15, 2)$ but with $LPSP = 2.34\%$ (see Step 2 of Table 6). The sub-optimal settings of $(b, s, w) = (30, 16, 2)$ gave $LPSP = 0$ with $J = 0.944$ $/kWh (see Step 3 of Table 6), i.e., the reliability was improved by 26.42%, while the cost was reduced by 14.18%.

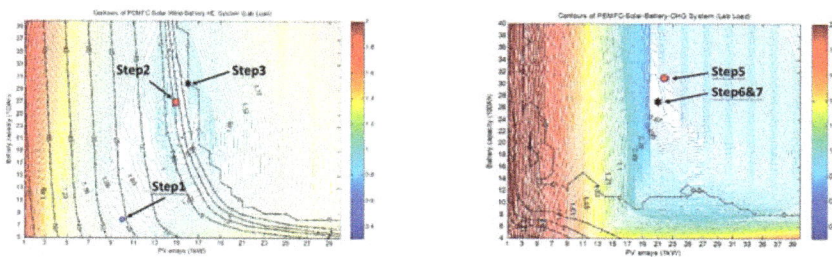

(a) $w = 2$ and $(SOC_{low}, SOC_{high}) = (40\%, 50\%)$ (b) $w = 0$ and $(SOC_{low}, SOC_{high}) = (30\%, 40\%)$

Figure 7. The reference plots for the lab load.

Table 6. The optimal design procedures for the lab load.

	(b, s, w)	(SOC_{low}, SOC_{high})	LPSP (%)	J ($/kWh)
Step 1	(8, 10, 2)	(40%, 50%)	26.42%	1.100
Step 2	(27, 15,2)	(40%, 50%)	2.34%	0.929
Step 3	(30, 16, 2)	(40%, 50%)	0%	0.944
Step 4	(31, 21, 0)	(40%, 50%)	0%	0.684
Step 5	(31, 21, 0)	(30%, 40%)	0%	0.668
Step 6	(27, 21, 0)	(30%, 40%)	0%	0.660
Step 7	(27, 21, 0)	(30%, 40%)	0%	0.660
Optimal	(27, 21, 0)	(30%, 40%)	0%	0.660

Because the WT was not economically efficient for this building, setting w=0 can greatly reduce the system cost to $J = 0.684$ $/kWh with $LPSP = 0$ by $(b, s, w) = (31, 21, 0)$ (see Step 4 of Table 6). The iterative procedures could then further improve the system cost to $J = 0.668$ $/kWh with $LPSP = 0$ by setting the power management as $(SOC_{low}, SOC_{high}) = (30\%, 40\%)$, and the cost finally converged to $J = 0.660$ $/kWh with $LPSP = 0$ by setting $(b, s, w) = (27, 21, 0)$ and $(SOC_{low}, SOC_{high}) = (30\%, 40\%)$. When compared with the original cost, the cost was reduced by 40%, while the system reliability was reduced by 26.42%.

3.3. Office Load

The analyses of the office load (see Figure 5) are shown in Figure 8 and Table 7. First, the original system layout $(b, s, w) = (8, 10, 2)$ with management settings of $(SOC_{low}, SOC_{high}) = (40\%, 50\%)$ gave a system cost of $J = 1.107$ $/kWh and $LPSP = 5.08\%$. Optimizing the settings slightly reduced the system cost to $J = 1.106$ $/kWh with $LPSP = 0$ using $(b, s, w) = (23, 11, 2)$ (see Step 2 of Table 7). Note that the system reliability was better than the house and the laboratory loads at this step, because the office load profile was basically synchronized with the irradiation and wind curves and the solar energy could be used directly to supply the loads. Therefore, we omitted Step 3 that represented the optimization with $w = 2$ and $LPSP = 0$ in Tables 5 and 6.

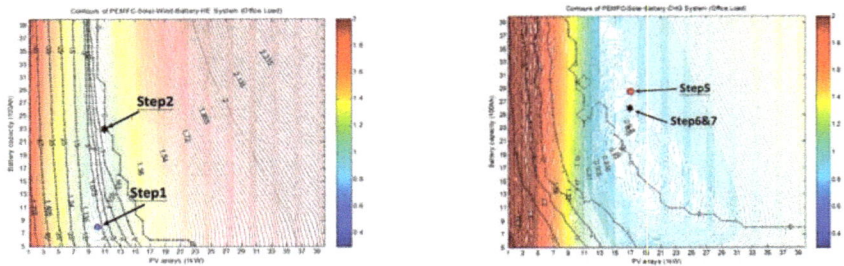

(a) $w = 2$ and $(SOC_{low}, SOC_{high}) = (40\%, 50\%)$ (b) $w = 0$ and $(SOC_{low}, SOC_{high}) = (30\%, 40\%)$

Figure 8. The reference plots for the office load.

Table 7. The optimal design procedures for the office load.

	(b, s, w)	(SOC_{low}, SOC_{high})	$LPSP$ (%)	J ($/kWh)
Step 1	(8, 10, 2)	(40%, 50%)	5.08%	1.107
Step 2	(23, 11, 2)	(40%, 50%)	0%	1.106
Step 3	-	-	-	-
Step 4	(29, 17, 0)	(40%, 50%)	0%	0.818
Step 5	(29, 17, 0)	(30%, 40%)	0%	0.817
Step 6	(26, 17, 0)	(30%, 40%)	0%	0.791
Step 7	(26, 17, 0)	(30%, 40%)	0%	0.791
Optimal	(26, 17, 0)	(30%, 40%)	0%	0.791

Setting $w = 0$ gave a significant cost reduction to $J = 0.818$ \$/kWh with $LPSP = 0$ by setting (b, s, w) = (29, 17, 0) (see Step 4 of Table 7). The iterative procedures then further improved the system cost to $J = 0.817$ \$/kWh with $LPSP = 0$ by adjusting the power management as $(SOC_{low}, SOC_{high}) = (30\%, 40\%)$, and the cost finally converged to $J = 0.791$ \$/kWh with $LPSP = 0$ by setting $(b, s, w) = (26, 17, 0)$ and $(SOC_{low}, SOC_{high}) = (30\%, 40\%)$. When compared with the original cost, the cost was reduced by 28.6% while maintaining complete system reliability.

3.4. Cost and Energy Distributions

The optimal system designs for the three loads, based on the reference plots, are illustrated in Tables 5–7. We further analyzed the cost and energy distributions of these systems, as shown in Table 8. First, the laboratory achieved the lowest unit energy cost because its average daily energy consumption was the largest; therefore, the initial costs were shared. The household load showed an opposite result. Second, the laboratory used the most solar panels and batteries, while the household applied the fewest solar panels and batteries, to produce and store sufficient energy for the load requirements. Third, the optimal battery units for all loads did not differ much (23–27 units); this was not intuitive because the laboratory load was mainly at night, while the office load was mainly in daytime. The reason for this was that the battery life was shortened if only a small amount of the battery energy was used. Therefore, using a large amount of the battery energy increased the initial cost but it also helped to extend the battery life, thereby reducing the battery costs. For instance, for the laboratory load, the battery cost was the lowest even though the laboratory load used the largest amount of battery energy. Because the initial battery SOC was set as 80% in the simulation model, a negative energy supply distribution of battery means the battery SOC is higher than 80% at the end of the simulation, i.e., the battery is charged by the renewable energy so that its final SOC is greater than the initial SOC. Fourth, the costs of the solar panels, battery, and the PEMFC system (including the chemical hydrogen production system, PEMFC, and NaBH₄) are about 40%, 25%, and 20%, respectively, for all loads. That is, the cost distributions are almost the same for all systems after optimization. Finally, solar energy provided nearly 100% of the required load demands because the current high cost of hydrogen requires that the system avoid using the PEMFC unless necessary. The current optimal costs are 0.794, 0.660, and 0.791 for the household, lab, and office loads, respectively. Although the costs cannot compete with the grid power, the system provides a self-sustainable power solution for remote areas and islands without grid power. The energy cost can be greatly reduced when the component prices are reduced with popularity. For example, the analyses in [33] indicated that the critical hydrogen price is about 10 NT$/batch (one batch consumes 60 g of NaBH₄ to produce about 150 L of hydrogen). That is, more hydrogen energy will be used in an optimal hybrid power system if the hydrogen price is less than 1/15 NT$/L.

Table 8. Cost and energy distributions for the optimal systems.

		House	Lab	Office
	Daily average (kWh)	19.96	30.41	22.32
	Optimal cost ($/kWh)	0.794	0.660	0.791
	Optimal sizes (b, s, w)	(23, 15, 0)	(27, 21, 0)	(26, 17, 0)
Cost Distribution (%)	Lead-acid battery	25.34%	23.50%	25.72%
	Power electric devices	10.59%	11.72%	11.41%
	Wind turbine	0	0%	0%
	Solar panels	39.72%	43.91%	40.41%
	Chemical hydrogen production	13.56%	10.71%	12.18%
	PEMFC	7.63%	6.03%	6.85%
	Sodium borohydride (NaBH₄)	3.16%	4.13%	3.43%
Energy Supply Distribution (%)	Wind	0%	0%	0%
	PEMFC	1.27%	1.35%	1.36%
	Solar	100.65%	100.30%	98.50%
	Battery	−1.92%	−1.65%	0.14%

3.5. Safety Analyses

The optimization designs illustrated in Tables 5–7 were based on historical weather data, where the solar and wind energy co-assisted the sustainability of the power system. Because the aim of the hybrid power system is to provide uninterrupted power, we further investigated its ability to perform in extreme weather conditions when no solar or wind energy is available.

We applied the optimal settings in Tables 5–7 to the hybrid power model and recorded the lowest battery SOC during the 61-day simulation to calculate the lowest remaining energy and system safety by Equation (9). The results are illustrated in Figure 9 and Table 9, where the lowest SOC (stored energy) for the household, laboratory, and office loads were 29.99% (11.03 kWh), 26.04% (7.83 kWh), and 27.18% (8.97 kWh), respectively. Therefore, the equivalent sustainable operation periods of the system are 0.49, 0.23, and 0.36 days, respectively, considering the average daily energy consumption shown in Table 3 and assuming a battery efficiency of 90%. If a longer sustainability is required, we can adopt sub-optimal settings. For example, the minimal settings and costs to sustain 1 day or 2 days are labeled in Figure 9. Suppose the safety requirement is 1 day; then, the lowest system costs to guarantee 1 day of operation are 0.8952 USD/kWh, 0.7603 USD/kWh, and 0.8735 USD/kWh, respectively, for the household, laboratory, and office loads. The corresponding component sizes are (b, s, w) = (33, 26, 0), (b, s, w) = (40, 24, 0), and (b, s, w) = (40, 17, 0), respectively.

Another way to extend the guaranteed system sustainability is to use the chemical hydrogen generation system to produce hydrogen for the PEMFC as a means of providing back-up power. Referring to [36], one mole of NaBH₄ can generate four moles of hydrogen, so 20 kg of NaBH₄ can produce 4.16 kg of hydrogen, which would provide 63 kWh of electricity for the system. Therefore, a further sustainability guarantee is possible by stocking more NaBH₄ with the auto-batching system developed in [36], which can produce hydrogen when the system requires energy from the PEMFC. For example, if the system stores 20 kg of NaBH$_4$, the safety indexes for the household, laboratory, and office loads can be extended by 3.33, 2.10, and 2.90 days, respectively, assuming an inverter efficiency of 90%. Installing 40kg of NaBH4 could guarantee 6.17, 3.96, and 5.44 days of operation for the household, laboratory, and office, respectively, in the worst case scenario.

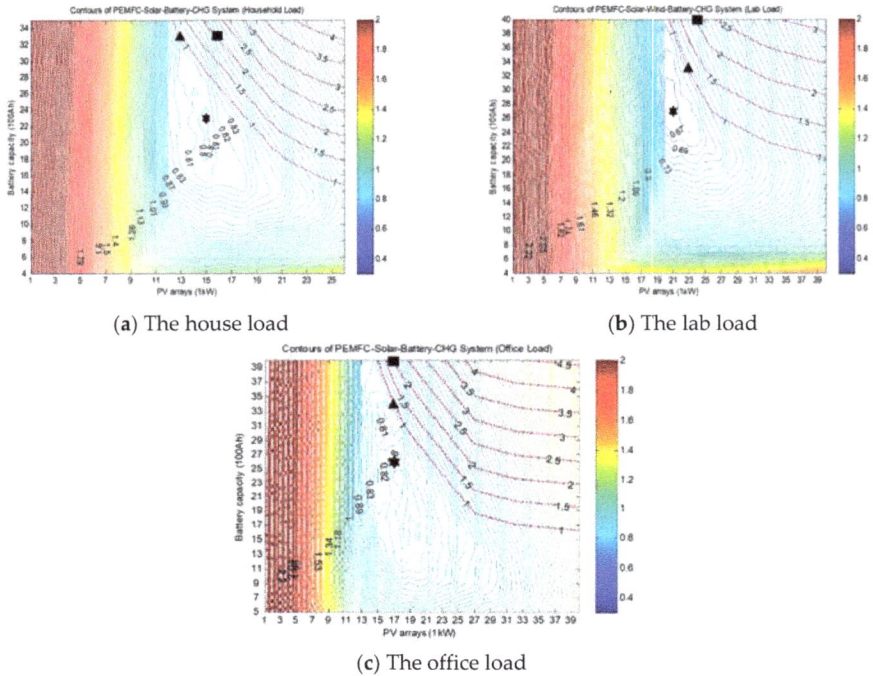

(**a**) The house load

(**b**) The lab load

(**c**) The office load

Figure 9. The reference plots with safety consideration.

Table 9. Safety analyses.

		House	**Lab**	**Office**
Daily average (kWh)		19.96	30.41	22.32
Optimal sizes (b, s, w)		(23, 15, 0)	(27, 21, 0)	(26, 17, 0)
Lowest SOC (%)		29.99	26.04	27.18
Lowest remaining energy (kWh)		11.03	7.83	8.97
Safety (days)		0.49	0.23	0.36
1-day safety requirement	System Sizes (b, s, w)	(33, 13, 0)	(33, 24, 0)	(34, 17, 0)
	System Cost (USD/kWh)	0.8152	0.7062	0.8266
2-days safety requirement	System Sizes (b, s, w)	(33, 16, 0)	(40, 24, 0)	(40, 17, 0)
	System Cost (USD/kWh)	0.8952	0.7603	0.8735

The choice of a sub-optimal design or extra NaBH$_4$ stock would depend on the estimated extreme weather conditions and the price of NaBH$_4$. For instance, if the expected extreme weather happens one day during the 61-day simulation, the total system costs are increased by $25.81, $85.70, and $48.47 for the household, laboratory, and office, respectively, using the sub-optimal settings. Conversely, the required extra NaBH$_4$ to guarantee sustainability under the worst-case conditions are 3.59 kg, 8.26 kg, and 5.04 kg, respectively, assuming an inverter efficiency of 90%. This will increase the system cost by \$35.90, \$82.60, and \$50.40 for the household, laboratory, and office, respectively, if the $NaBH_4$ price is 10 NT\$/kg. Therefore, the second option (using extra $NaBH_4$) will be the better choice if the cost of $NaBH_4$ is less than 7.19, 10.38, and 9.62 NT\$/kg for the household, laboratory, and office, respectively. Note that these analyses are based on the worst-case conditions, where the battery SOC is at the lowest when the extreme weather happens. Hence, in general, the cost should be lower and more benefits are possible by storing extra $NaBH_4$ with the auto batch system [36].

4. Results and Conclusions

This paper has demonstrated the optimization of a green building that was autonomous and did not connect to the main grid. The building can be applied to remote stations and small islands, where no grid power is available. We discussed the impacts of three typical loads on the optimization of a hybrid power system. First, we built a general hybrid power model based on a green building in Taiwan. The model consisted of PV, WT, batteries, PEMFC, electrolyzer, and chemical hydrogen production modules. Second, we evaluated the system performance by applying the household, laboratory, and office load profiles to the model. The results indicated that the combination of PV, battery, PEMFC, and chemical hydrogen production can guarantee system reliability. When compared with the original settings, the total system cost was greatly reduced by 38.9%, 40%, and 28.6% for the household, laboratory, and office loads, respectively, while the system reliability was significantly reduced by 4.89%, 24.42%, and 5.08%, respectively. Third, the cost distribution showed similar results for the three loads: the battery, PV, and PEMFC systems accounted for about 25%, 40%, and 20% of the system costs for all three cases. Note that the current usage of lead-acid battery is a compromise between cost and efficiency. For example, applying the hybrid system with LiFe battery [33], the optimal system costs became 2.237, 1.846, and 1.853 per kWh for the household, lab, and office loads, respectively. That is much higher than the current optimal costs by the lead-acid battery. Fourth, the energy distributions indicated that the PV provided nearly 99% of the required energy, because of the current high price of hydrogen. As shown in [33], hydrogen energy will be compatible when the hydrogen price drops to about one third of the current price. Finally, we evaluated the safety of these systems under extreme weather conditions and proposed two methods for extending system sustainability: using a sub-optimal design or using more $NaBH_4$. The latter method tended to be more flexible and was more able to cope with uncertainties. For example, adding 20 kg of $NaBH_4$ will increase the system safety by 3.33, 2.10, and 2.90 days for the household, laboratory, and office loads, respectively. These findings can be considered when developing customized hybrid power systems in the future.

Author Contributions: Conceptualization, F.-C.W.; Methodology, F.-C.W. and K.-M.L.; Software, K.-M.L.; Validation, F.-C.W., and K.-M.L.; Formal Analysis, F.-C.W. and K.-M.L.; Investigation, F.-C.W. and K.-M.L.; Resources, F.-C.W. and K.-M.L.; Data Curation, F.-C.W. and K.-M.L.; Writing—Original Draft Preparation, F.-C.W. and K.-M.L.; Writing—Review and Editing, F.-C.W.; Visualization, F.-C.W. and K.-M.L.; Supervision, F.-C.W.; Project Administration, F.-C.W.; Funding Acquisition, F.-C.W.

Funding: This research was funded by the Ministry of Science and Technology, R.O.C., in Taiwan under Grands MOST 105-2622-E-002-029 -CC3, MOST 106-2622-E-002-028 -CC3, MOST 106-2221-E-002-165-, MOST 107-2221-E-002-174-, and MOST 107-2221-E-002-174-. This research was also financially supported in part by the Ministry of Science and Technology of Taiwan (MOST 107-2634-F-002-018), National Taiwan University, Center for Artificial Intelligence and Advanced Robotics. The authors would like to thank Ye-Che Yang and I-Ming Fu for helping the simulation. The collaboration and technical support of M-Field™ are much appreciated.

Conflicts of Interest: The authors declare no conflict of interest.

References

1. Ceraolo, M.; Miulli, C.; Pozio, A. Modelling static and dynamic behaviour of proton exchange membrane fuel cells on the basis of electro-chemical description. *J. Power Sources* **2003**, *113*, 131–144. [CrossRef]
2. Gorgun, H. Dynamic modelling of a proton exchange membrane (PEM) electrolyzer. *Int. J. Hydrogen Energy* **2006**, *31*, 29–38. [CrossRef]

3. Woo, C.H.; Benziger, J.B. PEM fuel cell current regulation by fuel feed control. *Chem. Eng. Sci.* **2007**, *62*, 957–968. [CrossRef]
4. Vega-Leal, A.P.; Palomo, F.R.; Barragan, F.; Garcia, C.; Brey, J.J. Design of control systems for portable PEM fuel cells. *J. Power Sources* **2007**, *169*, 194–197. [CrossRef]
5. Park, G.; Gajic, Z. Sliding mode control of a linearized polymer electrolyte membrane fuel cell model. *J. Power Sources* **2012**, *212*, 226–232. [CrossRef]
6. Wang, F.C.; Yang, Y.P.; Huang, C.W.; Chang, H.P.; Chen, H.T. System identification and robust control of a portable proton exchange membrane full-cell system. *J. Power Sources* **2007**, *165*, 704–712. [CrossRef]
7. Wang, F.C.; Chen, H.T.; Yang, Y.P.; Yen, J.Y. Multivariable robust control of a proton exchange membrane fuel cell system. *J. Power Sources* **2008**, *177*, 393–403. [CrossRef]
8. Wang, F.C.; Chen, H.T. Design and implementation of fixed-order robust controllers for a proton exchange membrane fuel cell system. *Int. J. Hydrogen Energy* **2009**, *34*, 2705–2717. [CrossRef]
9. Wang, F.C.; Ko, C.C. Multivariable robust PID control for a PEMFC system. *Int. J. Hydrogen Energy* **2010**, *35*, 10437–10445. [CrossRef]
10. Hwang, J.J.; Wang, D.Y.; Shih, N.C.; Lai, D.Y.; Chen, C.K. Development of fuel-cell-powered electric bicycle. *J. Power Sources* **2004**, *133*, 223–228. [CrossRef]
11. Gao, D.W.; Jin, Z.H.; Lu, Q.C. Energy management strategy based on fuzzy logic for a fuel cell hybrid bus. *J. Power Sources* **2008**, *185*, 311–317. [CrossRef]
12. Hwang, J.J. Review on development and demonstration of hydrogen fuel cell scooters. *Renew. Sustain. Energy Rev.* **2012**, *16*, 3803–3815. [CrossRef]
13. Hsiao, D.R.; Huang, B.W.; Shih, N.C. Development and dynamic characteristics of hybrid fuel cell-powered mini-train system. *Int. J. Hydrogen Energy* **2012**, *37*, 1058–1066. [CrossRef]
14. Wang, F.C.; Chiang, Y.S. 2012, August, Design and Control of a PEMFC Powered Electric Wheelchair. *Int. J. Hydrogen Energy* **2012**, *37*, 11299–11307. [CrossRef]
15. Wang, F.C.; Peng, C.H. The development of an exchangeable PEMFC power module for electric vehicles. *Int. J. Hydrogen Energy* **2014**, *39*, 3855–3867. [CrossRef]
16. Wang, F.C.; Gao, C.Y.; Li, S.C. Impacts of Power Management on a PEMFC Electric Vehicle. *Int. J. Hydrogen Energy* **2014**, *39*, 17336–17346. [CrossRef]
17. Han, J.G.; Charpentier, J.F.; Tang, T.H. An Energy Management System of a Fuel Cell/Battery Hybrid Boat. *Energies* **2014**, *7*, 2799–2820. [CrossRef]
18. Wang, C.; Wang, S.B.; Peng, L.F.; Zhang, J.L.; Shao, Z.G.; Huang, J.; Sun, C.W.; Ouyang, M.G.; He, X.M. Recent Progress on the Key Materials and Components for Proton Exchange Membrane Fuel Cells in Vehicle Applications. *Energies* **2016**, *9*, 603. [CrossRef]
19. Bahrebar, S.; Blaabjerg, F.; Wang, H.; Vafamand, N.; Khooban, M.H.; Rastayesh, S.; Zhou, D. A Novel Type-2 Fuzzy Logic for Improved Risk Analysis of Proton Exchange Membrane Fuel Cells in Marine Power Systems Application. *Energies* **2018**, *11*, 721. [CrossRef]
20. Wang, F.C.; Kuo, P.C.; Chen, H.J. Control design and power management of a stationary PEMFC hybrid power system. *Int. J. Hydrogen Energy* **2013**, *38*, 5845–5856. [CrossRef]
21. Han, Y.; Chen, W.R.; Li, Q. Energy Management Strategy Based on Multiple Operating States for a Photovoltaic/Fuel Cell/Energy Storage DC Microgrid. *Energies* **2017**, *10*, 136. [CrossRef]
22. Singh, A.; Baredar, P.; Gupta, B. Techno-economic feasibility analysis of hydrogen fuel cell and solar photovoltaic hybrid renewable energy system for academic research building. *Energy Convers. Manag.* **2017**, *145*, 398–414. [CrossRef]
23. DAS, H.S.; Tan, C.W.; Yatim, A.H.M.; Lau, K.Y. Feasibility analysis of hybrid photovoltaic/battery/fuel cell energy system for an indigenous residence in East Malaysia. *Renew. Sustain. Energy Rev.* **2017**, *76*, 1332–1347. [CrossRef]
24. AL-Sharafi, A.; Sahin, A.Z.; Ayar, T.; Yilbas, B.S. Techno-economic analysis and optimization of solar and wind energy systems for power generation and hydrogen production in Saudi Arabia. *Renew. Sustain. Energy Rev.* **2017**, *69*, 33–49. [CrossRef]
25. Martinez-Lucas, G.; Sarasua, J.I.; Sanchez-Fernandez, J.A. Frequency Regulation of a Hybrid Wind-Hydro Power Plant in an Isolated Power System. *Energies* **2018**, *11*, 239. [CrossRef]
26. Ettihir, K.; Boulon, L.; Agbossou, K. Optimization-based energy management strategy for a fuel cell/battery hybrid power system. *Appl. Energy* **2016**, *163*, 142–153. [CrossRef]

27. Kazem, H.A.; Al-Badi, H.A.S.; Al Busaidi, A.S.; Chaichan, M.T. Optimum design and evaluation of hybrid solar/wind/diesel power system for Masirah Island. *Environ. Dev. Sustain.* **2017**, *19*, 1761–1778. [CrossRef]
28. Cozzolino, R.; Tribioli, L.; Bella, G. Power management of a hybrid renewable system for artificial islands: A case study. *Energy* **2016**, *106*, 774–789. [CrossRef]
29. Nizetic, S.; Tolj, I.; Papadopoulos, A.M. Hybrid energy fuel cell based system for household applications in a Mediterranean climate. *Energy Convers. Manag.* **2015**, *105*, 1037–1045. [CrossRef]
30. Zervas, P.L.; Sarimveis, H.; Palyvos, J.A.; Markatos, N.C.G. Model-based optimal control of a hybrid power generation system consisting of photovoltaic arrays and fuel cells. *J. Power Sources* **2008**, *181*, 327–338. [CrossRef]
31. Rekioua, D.; Bensmail, S.; Bettar, N. Development of hybrid photovoltaic-fuel cell system for stand-alone application. *Int. J. Hydrogen Energy* **2014**, *39*, 1604–1611. [CrossRef]
32. Chen, P.J.; Wang, F.C. Design Optimization for the Hybrid Power System of a Green Building. *Int. J. Hydrogen Energy* **2018**, *43*, 2381–2393. [CrossRef]
33. Wang, F.C.; Hsiao, Y.S.; Yang, Y.Z. The Optimization of Hybrid Power Systems with Renewable Energy and Hydrogen Generation. *Energies* **2018**, *11*, 1948. [CrossRef]
34. China Engineering Consultants, Inc. Experiments and Demonstration of an Indipendent Grid with Wind, Solar, Hydrogen Energy Compensative Electricity Generation. Available online: http://www. ceci.org.tw/Resources/upload/Cept/Quarterly/ed0887bb-4462-4b24-9a55-5371ef5938aa.pdf (accessed on 23 October 2018).
35. Vantage Pro2 Plus Stations. Davis Instrument. Available online: https://www.davisinstruments.com/product_documents/weather/spec_sheets/6152_62_53_63_SS.pdf (accessed on 26 November 2018).
36. Li, S.C.; Wang, F.C. The Development and Integration of a Sodium Borohydride Hydrogen Generation System. *Int. J. Hydrogen Energy* **2016**, *41*, 3038–3051. [CrossRef]
37. Battery University, Charging Lead-Acid Battery. Available online: http://batteryuniversity.com/learn/article/charging_the_lead_acid_battery (accessed on 23 October 2018).
38. M-Field Eco-Conscious Alternative. Available online: http://www.m-field.com.tw/product_module.php (accessed on 23 October 2018).
39. Wang, F.C.; Chen, H.C. The Development and Optimization of Customized Hybrid Power Systems. *Int. J. Hydrogen Energy* **2016**, *41*, 12261–12272. [CrossRef]
40. Current Transducers. LEM International. Available online: http://www.europowercomponents.com/media/uploads/HTB50...400.pdf?fbclid=IwAR3pYy5N1HBzNW6jCfn5KSFDD5HlOj0ZjVWYbicn8N1cw_sejsGhT7y4f4M (accessed on 26 November 2018).
41. Guo, Y.F.; Chen, H.C.; Wang, F.C. The development of a hybrid PEMFC power system. *Int. J. Hydrogen Energy* **2015**, *40*, 4630–4640. [CrossRef]
42. Digitech Co, Ltd. DGH-3000/PG-3000 Wind Electicity Generation System. Available online: http://140.112.14.7/~{}sic/PaperMaterial/Windturbinedata.pdf (accessed on 23 October 2018).
43. Load Profiles. Available online: http://140.112.14.7/~{}sic/PaperMaterial/Load%20Profile.pdf (accessed on 26 November 2018).
44. Energy in Building and Communities Program. Integration of Micro-Generation and Related Energy Technologies in Buildings. Available online: http://www.iea-ebc.org/Data/publications/EBC_Annex_54_DHW_Electrical_Load_Profile_Survey.pdf (accessed on 26 November 2018).
45. Yang, Y.Z.; Chen, P.J.; Lin, C.S.; Wang, F.C. Iterative Optimization for a Hybrid PEMFC Power System. In Proceedings of the 56th Annual Conference of the Society of Instrument and Control Engineers of Japan (SICE), Kanazawa, Japan, 19–22 September 2017; pp. 1479–1484.

Article

Short-Term Forecasting of Total Energy Consumption for India-A Black Box Based Approach

Habeebur Rahman [1,*], **Iniyan Selvarasan [1]** and **Jahitha Begum A [2]**

[1] Institute for Energy Studies, Anna University, Chennai 600025, India; iniyan777@hotmail.com
[2] Department of Education, Gandhigram Rural Institute, Dindigul 624302, India; jahee_j@yahoo.co.in
* Correspondence: thashabee14@gmail.com; Tel.: +91-9751841499

Received: 31 October 2018; Accepted: 29 November 2018; Published: 9 December 2018

Abstract: Continual energy availability is one of the prime inputs requisite for the persistent growth of any country. This becomes even more important for a country like India, which is one of the rapidly developing economies. Therefore electrical energy's short-term demand forecasting is an essential step in the process of energy planning. The intent of this article is to predict the Total Electricity Consumption (TEC) in industry, agriculture, domestic, commercial, traction railways and other sectors of India for 2030. The methodology includes the familiar black-box approaches for forecasting namely multiple linear regression (MLR), simple regression model (SRM) along with correlation, exponential smoothing, Holt's, Brown's and expert model with the input variables population, GDP and GDP per capita using the software used are IBM SPSS Statistics 20 and Microsoft Excel 1997–2003 Worksheet. The input factors namely GDP, population and GDP per capita were taken into consideration. Analyses were also carried out to find the important variables influencing the energy consumption pattern. Several models such as Brown's model, Holt's model, Expert model and damped trend model were analysed. The TEC for the years 2019, 2024 and 2030 were forecasted to be 1,162,453 MW, 1,442,410 MW and 1,778,358 MW respectively. When compared with Population, GDP per capita, it is concluded that GDP foresees TEC better. The forecasting of total electricity consumption for the year 2030–2031 for India is found to be 1834349 MW. Therefore energy planning of a country relies heavily upon precise proper demand forecasting. Precise forecasting is one of the major challenges to manage in the energy sector of any nation. Moreover forecasts are important for the effective formulation of energy laws and policies in order to conserve the natural resources, protect the ecosystem, promote the nation's economy and protect the health and safety of the society.

Keywords: India; TEC; short-term; forecasting; black box

---

1. Introduction

Energy is the driving force of any nation. Energy security and energy efficiency is the need of the hour. Energy conservation, decentralized energy planning techniques seems to be the solution to meet the energy requirements in almost every sector. The installed capacity out of renewable energy during 2012–2013 was around 12.26% and now later during 2017–2018 it has come to around 18.8% (www.cea.nic.in) [1]. If this trend stays, it is anticipated that the renewable energy sources would come forward to contribute even more in near future, which is a good sign. Renewable energy sector is expanding rapidly and in particular it has already grabbed its attention to be the potential contributor for sustainable energy security. India is one among the mainly swiftly developing countries in the planet. Flourishing industrialization also requires energy to excel, which in turn makes India an energy starving state. At present India depends heavily upon the fossil fuels and also has to expend more, whereas India also has a huge potential for the alternative sources of energy [2]. India is almost certainly urbanizing quicker. With a severe development predicament in the energy sector, energy becomes one of the top focus and an additional major issue in sustainable improvement and also a long-term security [3]. Managing energy consumption and energy resources in parallel has become very important among energy planners and policy framers. Thus an incorporated energy administration approach is vital for the sustainable improvement of India. Models have turned out to be the standard tools in energy planning. For energy modelling, energy forecasting is a basic necessary requirement. This emphasizes the significance of energy forecasting. Demand forecast is similarly a vital job for the effectual function and setting up of systems. Forecasts can be catalogued as long-term, medium and short-term depending upon the time. Long-term prediction, generally keep up a correspondence to several months to even several decades to the front. Overvalued electricity and energy consumption forecast will result in heedless venture in the erection of surplus power amenities and other inventories; whereas undervaluing the consumption might end up with deficient manufacturing, planning and will not be able to bridge the gap with the demand. Short term forecasting always draws attention and it is also paying attention in point forecasts. Density forecasts that is, forecasts that provide approximation of the probability distributions of the likely upcoming values of the consumption be essential for long-term forecast. In the literature a range of forecasting practices has been witnessed in the earlier period, generally presumptuous short and midterm forecasts. To assess the execution of the comparative study the real and forecasted results are figured out and the forecasts based upon the data observed up to 2017 are also worked out. The outcome proves good foreseeing capability of the proposed method at forecasting the country's energy consumption.

The literature is actually huge with the number of competing models and some major contributions among them are read. In support of perspective of single day or with a reduction in it, models employing univariate time series models [4,5] and ANN [6] are quite familiar. Some of the researchers have used forecasting models and techniques. Several forecasting methods were used in energy forecasting leading to different levels of accomplishment. Starting from linear regression, multi-variate regression and so on, several other models have also been used [7–9]. Time series models for various years have been offered with multinomial, linear and also exponential approximation [10].

Mixed Integer Linear Programming model has been evolved for the optimized electricity generation scheme planning for the country to reach a precise CO_2 emission target [11]. Holt's method was used to determine three different circumstances such as business as usual, renewable energy and also regarding how to conserve energy [12]. Long-term dynamic linear programming model was considered to calculate future investments of electricity production technologies of very long-term energy scenarios. Linear Programming (LP) can be implemented to support the choice of renewable energy technology to meet CO_2 emission reduction targets [13]. An estimation of data-driven models was performed by Tardioli et al. at city level [14]. Choi et al. offers an extreme deep learning method to obtain improved building energy consumption forecast [15].

Simple fuzzy models incorporating Artificial Intelligence techniques have been useful to forecast midterm energy and also the peak load [16–18]. A new way of energy demand forecasting at an intra-day ruling using semi parametric regression smoothing which relates for the yearly and weather conditional components is suggested. Dependence upon the residual series is explained by one among the two multiple variables time series models, with the measurement identical to the quantity of intra-day range. The profit of this procedure in the process of forecasting of: (i) Demand for heating steam network of one of the district in Germany; (ii) collective electricity demand in Victoria, a state in Australia. With both studies accounting for meteorological conditions can perk up the predicted value significantly, so do the application of the time series models. A multivariate non-linear regression method for forecasting the mid-term energy power systems in yearly basis by captivating into concern the correlation study of the elected input variables weighing factor and the training epoch that is to be used. A fine forecasting model is framed by [19] for the power system in Greece and for the dissimilar category of low voltage clients. Energy forecasting models in long-term basis are playing key role, provided the concern of GHG discharge and the existing want for evaluating choices for reaching the Kyoto's objective as given by [20] who paid attention on gas supply and also the oil supply projections and also provides helpful insight into the intricacy of forecasting the same and developing an systematic structure that explains the method used by Natural Resources, Canada by setting up oil and gas supply predictions and resolve the model for the same and provide the forecasts of the oil supply and demand and also the natural gas supply and demand for the year 2020. Predicting the energy need for the upcoming markets is among the key policy methods used by the international policy makers. Autoregressive integrated moving average and Seasonal autoregressive integrated moving average procedure is employed to guess Turkey's energy demand in future from the year 2005 to 2020 by [21]. Autoregressive Integrated Moving Average forecasting of the overall prime energy demand was more steadfast over the summing up of the individual predictions. The results are a sign of the average yearly increase rates of entity energy resources and the overall prime energy will diminish except wood and bio remains to hold pessimistic growth rate. A novel method for predicting the rising trend in an optimized univariate discrete grey forecasting method is assumed to predict the sum of energy making as well as utilization and a new Markov model built upon quadratic programming technique is projected for predicting the energy production and consumption trend in China for the year 2015 and 2020. The projected models are able to efficiently imitate and predict the overall quantity and structure of energy production and consumption [22]. To predict energy usage in Jordan using yearly data for 1976–2008, ref. [23] used ANN analyses. Four independent variables viz. Population, exports, GDP and imports are employed to predict the energy utility. The outcome tells that the predictable energy use for Jordan will get to 8349, 9269, 10,189 Ktoe for the years 2015, 2020 and 2025 respectively. The authors perform energy modelling and forecasting of Turkey's existing need for evaluating choices for meeting the socio-economic variables using regression and artificial neural networks. Four dissimilar representations including different variables were used for this purpose. As a result, Model 2 was found to be the appropriate ANN model comprising four independent variables to competently guesstimate Turkey's energy consumption. And the model envisioned healthier than that of the regressive models and also the additional three models from ANN [24]. An inclusive forecasting solution is portrayed by Hyndman et al. [25]. The author reveals and emphasizes the significance to prevent myocardium dysfunction, which is the most general way of death globally. He says 50,000,000 people are vulnerable to cardiac diseases around the world. He collected 744 fragments of ECG signals for one lead, ML II, from 29 patients and he proposed a new model which comprised of longer fragments reveals of ECG signal and the spectral density was estimated using Welch significance to prevent myocardium dysfunction and enables the efficient recognition of heart disorders [26]. Plawiek et al. compares selected approximations of five concentration levels of phenol. The semiconductor gas sensors' outcome formed input vectors for further work. Prior data processing encompassed principal component analysis, data standardization and data normalization in addition to data reduction. Nine systems were made into a single system using fuzzy systems, neural networks and also some hybrid systems. Every system was validated upon the complexity and accuracy. By the combination of the three principal components the input vector was formed. They applied and compared as many as nine CI models for the phenol concentration analysis developed from the metal oxide sensor using signals [27]. The authors propose MARKAL model which takes care of the allocations for various energy sources in India, for the Business As Usual (BAU) scenario and for the case of exploitation of energy. In this scenario, the demand for electrical energy will shoot up every year unto 5000bKwh of the installed capacity with major clients being the domestic, industrial and the service sectors [28]. So as to obtain accurate and enhanced energy consumption for buildings, extreme deep learning approach is given in this article. The model proposed clubs stacked auto encoders with the machine learning to exploit its characteristics. To obtain precise prediction results ELM is used as a predictor. The partial autocorrelation analysis method is adopted to determine the input variables of this deep learning model [29]. In Italy the influence of economic and demographic variables on the yearly electricity consumption was examined with the intention to develop long-term electricity consumption model. Forecasting models were developed using different regression models as gross domestic product and other input variables [30]. Turkey's energy consumption was forecasted based on the demographic and socio-economic variables viz., GDP, population, import, export and employment using regression and ANN [31].

Machine learning is one of the effective methods for pattern recognition in big data. These algorithms find the patterns in the data by nature and help making better predictions and critical decisions in Energy load, peak and price forecasting, image processing, face recognition, motion and object detection, tumour detection, predictive maintenance, natural language processing. Gajowniczek et al. has proposed a data mining technique to find out the electricity peak load for the country by representing the same as a pattern recognition research problem rather than a time series forecasting problem by using ANN. The main innovation is that they detect 96.2% of the peak electricity load accurately up to a day ahead [32]. Singh et al. also presented a data mining model to predict the trend in energy consumption pattern that describe the domestic device usage in connection with hourly, daily, weekly, monthly yearly basis as well as domestic device to domestic device linkages in a house. They proposed unsupervised data clustering and frequent pattern mining analysis on energy time series. Bayesian network prediction was referred for energy usage forecasting. The accuracy of the results outperformed SVM and MLP's accuracy of 81.82%, 85.90% and 89.58% for 25%, 50% and 75% of the size of the data used for training respectively [33].

Thus the literature review of three decades reveals that various technologies and applications were used to predict energy consumption in various sectors which helped us to utilize the proposed approach in computing the energy consumption for India.

The main contribution of this article is that it provides

- A point forecast for the total electricity consumption for the upcoming years up to 2030 is determined which in turn will help the energy planning in a holistic approach for the nation.
- An insight to the policy makers at bridging the gap between the forecasted and the actual data for future.
- The major contribution of the article is that it emphasizes the researchers to get to know the basic statistical models before proceeding to the advanced packages.

The goal of the study is to forecast the short-term TEC of India using the basic and reliable methodologies which seems to be much better than the advanced methods in forecasting the energy consumption of India. The corresponding author has done a forecast of energy consumption of a state in India, Tamil Nadu, in a smaller scale during his post-graduation; which actually is the motivation of the research. Apart from that the authors reviewed many studies pertaining to energy consumption of Turkey, Jordan and China and so forth. which motivated them to undertake the study for India. Dr. Iniyan is the Research supervisor of the corresponding author and is a veteran in this field of energy planning, who has taken up various projects and is also a voracious publisher and is one of the major sources of inspiration. The data used for the analyses is sometimes carried over from the year 1970. Forecasted outcome reveal that it holds good on the historical data taken for the analysis. With the intervention of new methods, there are areas for probable potential enhancement. An added region for progress would be to optimize the forecast further. For India the energy consumption is forecasted for the year 2030 and this shall be done even for years down the lane from then on that is, long time forecasting.

2. Materials and Methods

Data driven models are those which use available prior data to forecast energy behaviour. To perform this, a database is established to train the models, by combining dissimilar techniques for predicting the energy consumption. Among the data driven models the most popular are black-box based approaches which shall be used for energy prediction and forecasting in which regression model, multiple linear regression model, decision trees, ANN, support vector machine and various other optimization techniques shall also be employed. By utilizing the black-box approach the present study is performed with the major objective of predicting the Total Electricity Consumption (TEC) in industry, agriculture, domestic, commercial, traction railways and other sectors of India for the year 2030; using linear programming, multiple linear programming, correlation, exponential smoothing, Holt's, Brown's and expert model using the independent variables viz., population, GDP and GDP/capita. GDP and population are two vital independent variables which are proven to be playing an important role in energy load forecasting in the literature among various countries. Both of the variables are closely, positively and comparatively highly correlated with energy consumption. The predicted value of TEC is compared to that of the actual value of Energy Statistics data of 2017.

This study is also attempted to get back to the basics and reliable methodologies which are sometimes much better than the advanced methods which are basically built upon these methodologies for some of the simple key research problems such as forecasting the energy consumption of Republic India. Nevertheless several other advanced and multifaceted techniques are also in place. The key features are (a) it provides a point forecast for the energy consumption values for the upcoming years with demonstrated errors and (b) the gap between the forecasted and actual data are analysed in close intervals. The data used for the analyses is considered only from the early 1960s and 1970s. With the intervention of new methods, there are areas for probable potential enhancement in near future. An added region for progress would be to optimize the forecast further. For India the energy consumption is forecasted for the year 2030 and this shall be done even for years down the lane from then on that is, long time forecasting. The software used in the analyses is SPSS which stands for 'Statistical Package for the Social Sciences' which was developed by IBM. The various curve estimations and other analyses used for forecasting is carried out by IBM SPSS Statistics 20. The Linear, Compound, Logarithmic and Power curves are also fitted using this tool. Microsoft Excel is also used for various other analyses. The device on which the analysis is carried out configures 2.00 GHz Intel Core2 Duo Processor, with a memory of 4096 MB and a hard drive of 320 GB.

Total Energy Consumption for the period 1960–2013 for sectors such as industry, agriculture, commercial, domestic, traction railways and others were obtained from various energy statistics reports of Ministry of Statistics and Programme Implementation (MOSPI), GoI. The year wise data for the population, GDP per capita and GDP were also obtained from various other sources and other different departments of the GoI [34].

After the independence of the country, that is, in 1947 the TEC is observed to be 4182 GwH. At that time the domestic sector consumed around 10% of the overall and industrial sector consumed almost 71%, these two were the major players till the end of the 3rd plan. During the 1968–1969 that is, during the 3rd Annual plan the domestic sector experienced a dip compared to that of the agriculture sector and the Industrial and agriculture sector were found to be the major consumers till the end of the 9th plan that is, 2001–2002. During that point of time industrial sector consumed around 43% and the agriculture sector engulfed 21.8% which is just short of domestic sector 21.27%. Again from the end of the 19th plan till the end of the 12th plan that is, from 2001–2002 to 2016–2017 the domestic sector again started consuming more compared to the agriculture sector which was found to be 24.11% compared to agriculture's 18.01%. Nevertheless the Industries have always been the major consumer since independence till date even though a decreasing trend is noticed overall. The traction sector in India has always followed a decreasing trend apart from a few periods which has also shown only a feeble increase. The Miscellaneous sector (others) has increased lately to 6.45% since independence compared to its 5.24% even though it is not a key consumer. This is an overview of the sector wise total energy consumption for India since 1947 as demonstrated in Figure 1.

Sl. No.	During financial year ending with	Domestic	% to Total	Commercial	% to Total	Industrial	% to Total	Traction	% to Total	Agriculture	% to Total	Misc.	% to Total	Total
1	1947	423	10.11	178	4.26	2960	70.78	277	6.62	125	2.99	219	5.24	4182
2	1950	525	9.36	309	5.51	4057	72.32	308	5.49	162	2.89	249	4.44	5610
3	1955-56 (End of the 1st Plan)	934	9.20	546	5.38	7514	74.03	405	3.99	316	3.11	435	4.29	10150
4	1960-61 (End of the 2nd Plan)	1492	8.88	848	5.05	12547	74.67	454	2.70	833	4.96	630	3.75	16804
5	1965-66(End of the 3rd Plan)	2355	7.73	1650	5.42	22596	74.19	1057	3.47	1892	6.21	905	2.97	30455
6	1968-69(End of the 3 Annual Plans)	3184	7.69	2126	5.14	29931	72.31	1247	3.01	3465	8.37	1439	3.48	41392
7	1973-74(End of the 4th Plan)	4645	8.36	2988	5.38	37791	68.02	1531	2.76	6310	11.36	2292	4.13	55557
8	1978-79(End of the 5th Plan)	7576	9.02	4330	5.15	54440	64.81	2186	2.60	12028	14.32	3445	4.10	84005
9	1979-80(End of the 2 Annual Plans)	8402	9.85	4657	5.46	53206	62.35	2301	2.70	13452	15.76	3316	3.89	85334
10	1984-86 (End of the 6th Plan)	15506	12.45	6937	5.57	73520	59.02	2880	2.31	20961	16.83	4765	3.83	124569
11	1989-90 (End of the 7th Plan)	29577	15.16	9548	4.89	100373	51.45	4070	2.09	44056	22.58	7474	3.83	195098
12	1991-92(End of the 2 Annual Plans)	35854	15.51	12032	5.20	110844	47.94	4520	1.96	58557	25.33	9394	4.06	231201
13	1996-97(End of the 8th Plan)	55267	17.53	17519	5.56	139253	44.17	6594	2.09	84019	26.65	12642	4.01	315294
14	2001-02(End of the 9th Plan)	79694	21.27	24139	6.44	159507	42.57	8106	2.16	81673	21.80	21651	5.75	374670
15	2006-07(End of 10th Plan)	111002	21.12	40220	7.65	241216	45.89	10800	2.05	99023	18.84	23411	4.45	525672
16	2011-12 (End of 11th Plan)	171104	21.79	65381	8.33	352291	44.87	14206	1.81	140960	17.95	41252	5.25	785194
17	2016-17 (End of 12th Plan)	255826	24.11	89825	8.46	440206	41.48	15683	1.48	191151	18.01	68493	6.45	1061183
18	2017-18 @	273550	24.20	96141	8.51	468825	41.48	14356	1.27	204293	18.08	73079	6.47	1130244

Figure 1. Category wise TEC growth in India (1947–2018).

The graphical representation of the Domestic, Commercial, Industrial, Agricultural and Traction sectors for the period 2000–2018 has been shown in Figure 2. The value for the year 2018 is an estimated value compared to all the other values. The industrial sector has seen a steep increase since the start of the century which usually will be the case for any country. And all the other sectors have shown a gradual increase. The agriculture and the domestic categories have shown some fluctuations whereas the other sectors have shown a steady increase.

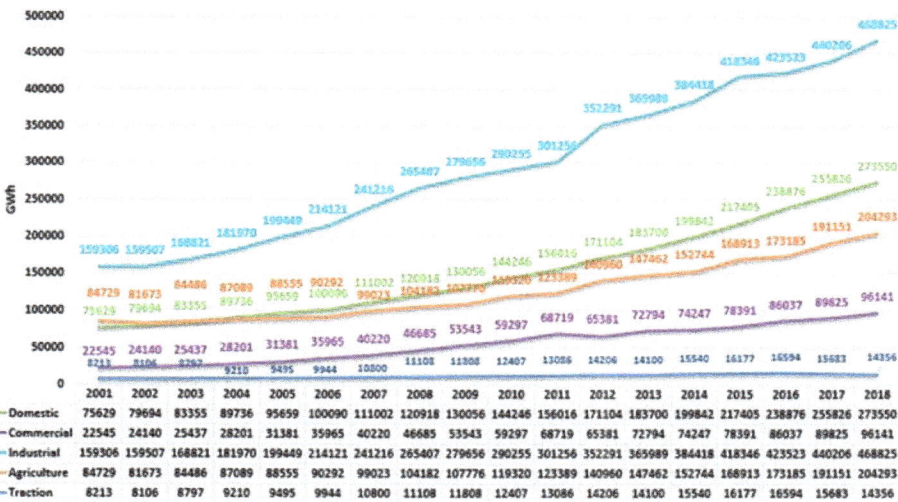

Figure 2. Sector wise Energy Consumption since the Century (2000–2018).

By the end of India's 12th plan the split among all the sectors is shown above in Figure 3. The domestic and the agriculture which were going hand to hand, few decades earlier were found to show a visible contrast of over 6% between them. The industrial sector's consumption over the years has decreased considerably even though it happened to be the vital consumer of all the sectors. And the same trend is expected to continue over the years which might transform India from being an Agriculture based economy.

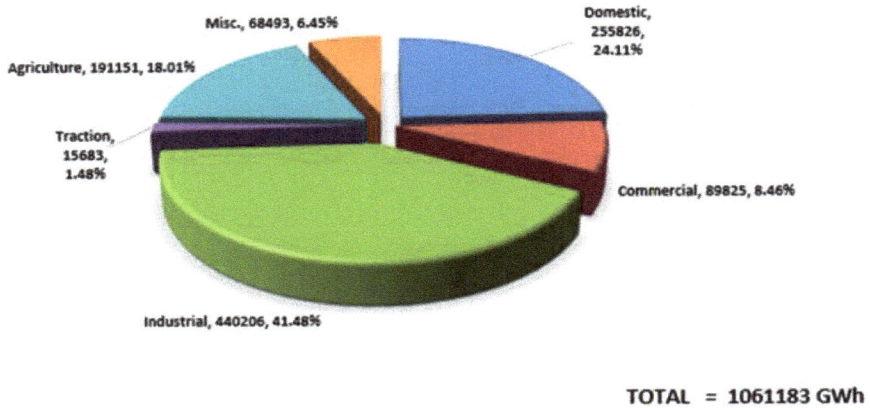

TOTAL = 1061183 GWh

Figure 3. India's Electricity Consumption by the end of 12th Plan (31 March 2017).

The category wise electricity consumption of India compared to other developed and developing countries are analysedin Figures 4 and 5, along with the International Energy Agency's report for the year 2015. India falls short of China and the US in terms of GWh in almost all the sectors except the agriculture sector. When compared to the European Union, India consumes electricity more in the agriculture sector and the others sector which is quite evident, since India is a tropical country and is agriculture based. India consumes 173,185 GWh of electricity which is higher than China's 103,983 GWh which comes next. In the commercial sector US is the major electricity consumer, it tops the list with 1,359,480 GWh. US consumes 1,401,616 GWh in the residential sector which is almost the consumption of the Chinese republic's 756,521 GWh and European Union's 795,406 GWh combined, in spite of world's highly populous nations such as China and India. In the transport category China's 179,638 GWh electricity consumption stands out way ahead of the Russian federation's 82,120 GWh, which is the second largest consumer. China is also the top consumer in the industrial sector in terms of electricity which is 32,121,168 GWh which is more than 26% of the whole world.

Energy Statistics brings out energy indicators meant for the practice of policy framers and for wide-ranging coverage. Indicators participate in a critical job by transforming the data to useful information for the plan makers and also aid in the process of making decisions. List of indicators identification depends upon various factors such as lucidity, technical validity, strength, sensitivity and the degree to which they are gelled to each other. No single factor can determine everything since each indicator needs different set of data. GDP is the country's broadest quantitative gauge of total economic activity. In specific GDP tells us the financial value of all the goods and services manufactured within the country's borders over a time span [34]. The data in the study has been gathered from the respective ministries of the Government of India (GoI). Energy intensity's value has dipped over the latest ten years which might be ascribed to the quicker increase of GDP than the energy need.

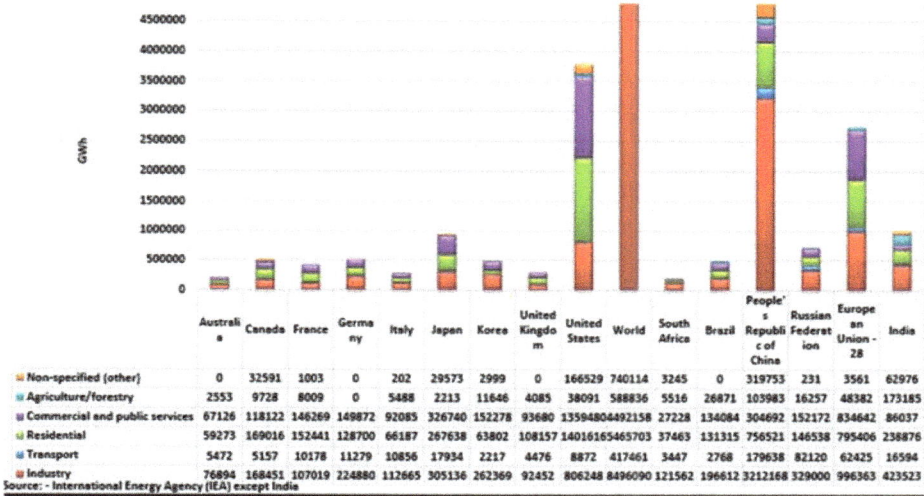

	Australia	Canada	France	Germany	Italy	Japan	Korea	United Kingdom	United States	World	South Africa	Brazil	People's Republic of China	Russian Federation	European Union - 28	India
Non-specified (other)	0	32591	1003	0	202	29573	2999	0	166529	740114	3245	0	319753	231	3561	62976
Agriculture/forestry	2553	9728	8009	0	5488	2213	11646	4085	38091	588836	5516	26871	103983	16257	48382	173185
Commercial and public services	67126	118122	146269	149872	92085	326740	152278	93680	1359480	4492158	27228	134084	304692	152172	834642	86037
Residential	59273	169016	152441	128700	66187	267638	63802	108157	1401616	5465703	37463	131315	756521	146538	795406	238876
Transport	5472	5157	10178	11279	10856	17934	2217	4476	8872	417461	3447	2768	179638	82120	62425	16594
Industry	76894	168451	107019	224880	112665	305136	262369	92452	806248	8496090	121562	196612	3212168	329000	996363	423523

Source: - International Energy Agency (IEA) except India

Figure 4. Category-wise Electricity Consumption across various countries (2015).

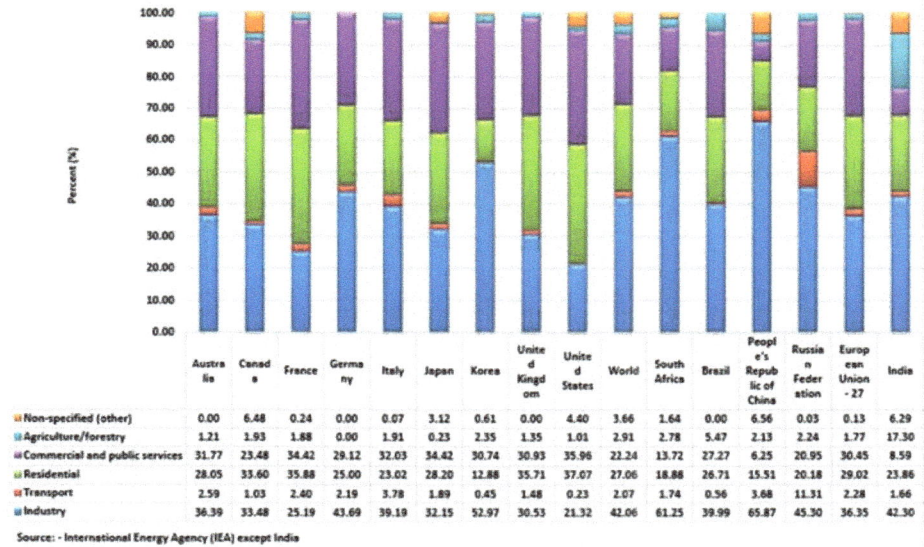

	Australia	Canada	France	Germany	Italy	Japan	Korea	United Kingdom	United States	World	South Africa	Brazil	People's Republic of China	Russian Federation	European Union -27	India
Non-specified (other)	0.00	6.48	0.24	0.00	0.07	3.12	0.61	0.00	4.40	3.66	1.64	0.00	6.56	0.03	0.13	6.29
Agriculture/forestry	1.21	1.93	1.88	0.00	1.91	0.23	2.35	1.35	1.01	2.91	2.78	5.47	2.13	2.24	1.77	17.30
Commercial and public services	31.77	23.48	34.42	29.12	32.03	34.42	30.74	30.93	35.96	21.24	13.72	27.27	6.25	20.95	30.45	8.59
Residential	28.05	33.60	35.88	25.00	23.02	28.20	12.88	35.71	37.07	27.06	18.88	26.71	15.51	20.18	29.02	23.86
Transport	2.59	1.03	2.40	2.19	3.78	1.89	0.45	1.48	0.23	2.07	1.74	0.56	3.68	11.31	2.28	1.66
Industry	36.39	33.48	25.19	43.69	39.19	32.15	52.97	30.53	21.32	42.06	61.25	39.99	65.87	45.30	36.35	42.30

Source: - International Energy Agency (IEA) except India

Figure 5. Category-wise % Shares in Electricity Consumption across various countries (2015).

3. Results

3.1. Linear Regression

In order to predict the influencing variable on Total Electricity Consumption (TEC) for India, Linear Regression is used initially. GDP, Population and GDP per Capita are taken as the input variables which are used one by one in linear regression to predict TEC.

The distance between the actual value and the mean is calculated and also the distance between the estimated line and its mean is calculated in the regression line. The comparison between the two values that is, the difference between the actual and mean and the difference between the estimated and the mean gives the R^2 value.

$$1 - R^2 = \{SSR/SST\}$$

SSR refers to the residual sum of squares and SST refers to the total sum of squares.

The standard error of the estimate is the distance between the estimated and the actual value. The constant value is actually the 'y' intercept of the line. The independent value is the slope of the regression line; since the line is linear the slope is also constant. The significance is the actual 'p' values.

$$\text{Standard Error } (\sqrt{n}) = \sigma$$

where σ refers to the standard deviation, n refers to the sample size.

3.1.1. Population

As depicted in Table 1, the constant value is −591,193.3447. The independent value that is, the slope of the regression line is 959,469.219; since the line is linear the slope is also constant. The regression equation usually frames a prediction and the precision of the prediction is calculated by means of the standard error. It also measures the scatter or dispersion of the observed values around the regression line.

$Y = 959,469.219X − 591,193.347$ is the regression equation.

Table 1. Summary of the model with Population as the variable.

Ind. Variable	R Square	Std. Error	Constant	Slope	Significance
Population	0.845	89,127.342	−591,193.347	959,469.219	0.000

3.1.2. GDP

As illustrated in Table 2, the constant value is 53,096.385. The independent value that is, the slope of the regression line is 417,965.826; since the line is linear the slope is also constant.

$Y = 417,965.826X − 53,096.385$ is the regression equation.

Table 2. Summary of the model with GDP as the variable.

Ind. Variable	R Square	Std. Error	Constant	Slope	Significance
GDP	0.957	46,784.201	53,096.385	417,965.826	0.000

3.1.3. GDP per Capita

The constant value is −2457.344. The independent value that is, the slope of the regression line is 959,469.511; since the line is linear the slope is also constant as in Table 3.

$Y = 546.511X − 2457.344$ is the regression equation.

Table 3. Summary of the model with GDP per capita as the variable.

Ind. Variable	R Square	Std. Error	Constant	Slope	Significance
GDP/Capita	0.951	50,234.297	−2457.344	546.511	0.000

When we forecast Total Electricity Consumption (TEC) using three variables, the GDP plays an important role and it predicts better the Total Electricity Consumption than the GDP per Capita and the population. The R^2 value for GDP and TEC is 0.957 whereas between GDP per capita and TEC it is only 0.951. When compared with Population and TEC it is even as lower as 0.845. Hence it is concluded that GDP foresees TEC better. The lowest std. error, 46,784.201 of all the three is also with the GDP.

3.2. Multiple Linear Regression

To forecast the TEC of electricity for 2030, multiple linear regression method is used now, taking in account the yearly GDP per capita, GDP and historical population data, as in the case of Turkey. During most of the situations, multiple independent variables might be used to predict the significance of a dependent variable for which we use multiple regression. In multiple regression, GDP and Population are taken simultaneously as the predicting variables. Multiple variable regression analysis establishes a relationship between a dependent variable (in this work Total Energy Consumption (TEC)) and two or even more than two independent variables that is, the predictors, population and GDPutilized an application technique for yearly consumption forecasting algorithm on the smart new intelligent electronic devices using multiple regression method which is put into practice in addition to recursive least square.

TEC = (332,023.240) Population + (302,638.253) GDP − 185,039.015is the regression equation from Table 4.

Table 4. Summary of the model with both Population and GDP as the variable.

Ind. Variable	R Square	Std. Error	Constant	Slope	Significance
Population	0.986	27,442.309	−185,039.015	332,023.240	0.000
GDP	0.986	27,442.309	−185,039.015	302,638.253	0.000

With one independent variable of population the R^2 is 0.845 and with that of GDP it is 0.957, whereas with two independent variables GDP and population combined, in multiple linear regression the R^2 increases to 0.986. The standard error of 46,784.201 with one variable, GDP drops to 27,442.309 with two variables. Lower is better. The GDP's standard error is almost less than half of the population's error. So GDP is again the better predictor in terms of Linear Multiple Regression.

3.3. Correlation Analysis

Almost all the independent variables exhibit a higher degree of correlation against the dependent variables, the analysis of correlation from Table 5 illustrates that there is positive high correlation between population and TEC. The Pearson correlation coefficient is found to be 0.919. From Figures 6–8, the analysis of correlation between GDP and TEC proves that there is a very high positive correlation. The Pearson Correlation Coefficient is found to be 0.978Whereas the correlation between GDP per capita and TEC demonstrates that there is a positive comparatively low correlation between GDP and TEC. The Coefficient is found to be 0.975.

Table 5. Correlation Matrix.

Variables	TEC	Outcome	Direction
Population	0.919	High correlation	Positive
GDP	0.978	Very high correlation	Positive
GDP/Capita	0.975	High correlation	Positive

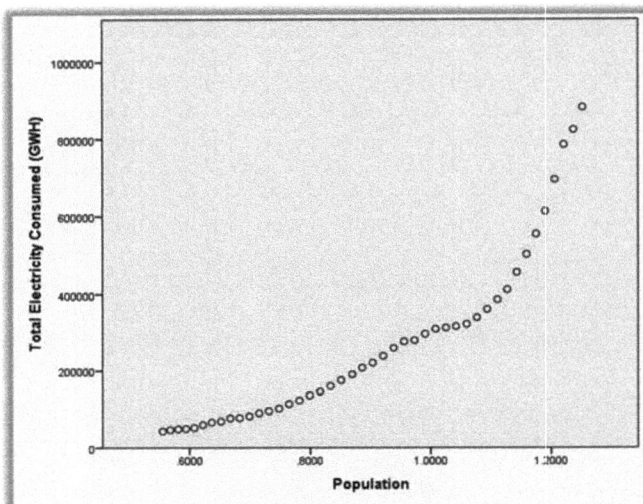

Figure 6. Plot between Population and TEC.

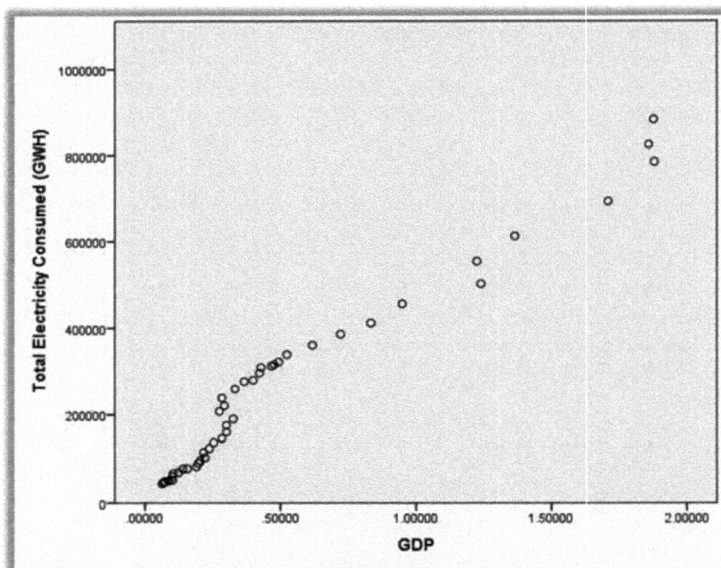

Figure 7. Plot between GDP and TEC.

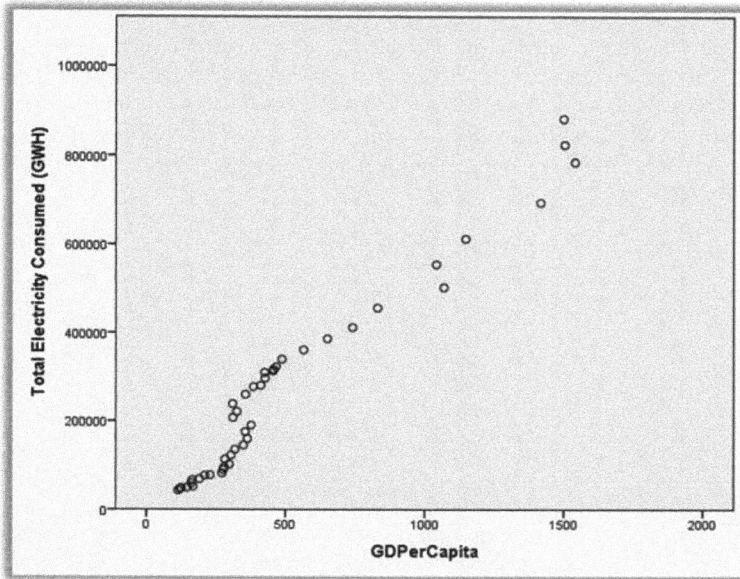

Figure 8. Plot between GDP per Capita and TEC.

3.4. Simple Exponential Smoothing

If the time series vary about base level, simple exponential smoothing might be bring into play to find good estimates or upcoming value of the same series. To depict this phenomenon, let A_t the smoothed average of a time series. Subsequent to observing x_t, A_t is the anticipate for the value of the time series during any upcoming period.

- A_t = smoothed average at the end of the epoch
- $t = ft$,
- k = for the forecast period $(t + k)$ at the end of the epoch t.

Choose α so that it minimizes the MAD.
The key in equation in simple exponential smoothing is that

$$A_t = \alpha x_t + (1 - \alpha) A_{t-1} \tag{1}$$

In the Equation (1), α will be the smoothing constant that suit $0 < \alpha > 1$. To start the forecasting process, we have got to set a value for A_0 before surveying x_1. Typically, we let A_0 be the experiential value for the period right away prior the first period. As among moving-average forecasts, we let $f_{t,k}$ be the estimate for x_{t+k} ready at the final period t. Then

$$A_t = f_{t,k} \tag{2}$$

Pretentious that we attempt to forecast one period ahead, the error for forecasting x_t is

$$E_t = x_t - f_{t-1,1} = x_t - A_{t-1} \tag{3}$$

The smoothing constant value considered for the analysis is $\alpha = 0.3, 0.4$ and 0.5.
The TEC for 2015 was found to be 746,882 MW when $\alpha = 0.3$ and 793,765 MW when $\alpha = 0.4$ and 823,941.3 MW when $\alpha = 0.5$.

3.4.1. Time Series Modeler—Expert Model (Sector Wise)

In the expert time series modeler it automatically assigns the model best suited based on the system's expertise. For the industrial, agricultural and domestic sectors it has assigned Brown's model and it is found to be the appropriate model. For the commercial, traction and others' sector, the expert model has assigned ARIMA (0,1,0) model; ARIMA (2,1,0) model and ARIMA (0,1,0) respectively, automatically as the appropriate models as in Table 6. The respective degrees of freedom and other parameters are shown in Table 7.

Table 6. Summary of Expert model.

Model ID	Model Type
Industry	Brown
Agriculture	Brown
Domestic	Brown
Commercial	ARIMA (0,1,0)
Traction Railways	ARIMA (2,1,0)
Others	ARIMA (0,1,0)

Table 7. Summary of the model.

Model	Statistics	Ljung-Box			No. of Outliers
	Stationary R^2	Statistics	DF	Sig.	
Industry	0.281	3.802	17	1.00	0
Agriculture	0.080	49.040	17	0.000	0
Domestic	0.432	6.242	17	0.991	0
Commercial	1.102×10^{-15}	10.125	18	0.928	0
Traction/Railways	0.331	15.057	17	591	0
Others	5.310×10^{-16}	20.114	18	0.326	0

3.4.2. Holt's Model-Exponential Smoothing with Trend

Several models such as Brown's model, Holt's model, Expert model and damped trend model were analysed. And the analysis of the Holt's model is shown in the Table 8.

Table 8. Summary of Holt's model.

Model	Statistics	Ljung-Box			No. of Outliers
	Stationary R^2	Statistics	DF	Sig.	
Industry	0.291	2.371	16	1.00	0
Agriculture	0.103	43.582	16	0.000	0
Domestic	0.447	6.536	16	0.981	0
Commercial	0.422	3.726	16	0.999	0
Traction/Railways	0.394	35.017	16	0.004	0
Others	0.434	19.379	16	0.250	0
TEC	0.069	14.250	16	0.580	0

3.4.3. Time Series Modeler (Exponential Smoothing-Brown)

The analysis of the Brown model is shown in the Table 9.

Table 9. Summary of Brown model.

Model	Statistics	Ljung-Box			No. of Outliers
	Stationary R^2	Statistics	DF	Sig.	
Industry	0.281	3.802	17	1.00	0
Agriculture	0.80	49.040	17	0.000	0
Domestic	0.432	6.242	17	0.991	0
Commercial	0.421	3.999	17	0.999	0
Traction/Railways	0.393	33.210	17	0.011	0
Others	0.411	23.559	17	0.132	0
TEC	0.067	14.569	17	0.627	0

3.4.4. Time series modeler (Exponential Smoothing—Damped Trend)

The TEC for the years 2019, 2024 and 2030 were forecasted to be 1,162,453 MW, 1,442,410 MW and 1,778,358 MW respectively.

$$\text{RMSE} = \sqrt{\{\Sigma \, (Y_{actual} - Y_{forecast})/N\}}$$

The Expert model selects different models on its own for different variables and produces the above mentioned forecast by means of a low root mean square error value, RMSE of 10,734.649 and a R^2 value of 0.997 which is comparatively high. And the analysis of the Damped trend model is shown in the Table 10.

Table 10. Summary of Damped trend model.

Model	Statistics	Ljung-Box			No. of Outliers
	Stationary R^2	Statistics	DF	Sig.	
Industry	0.392	2.365	15	1.00	0.392
Agriculture	0.337	44.861	15	0.000	0.337
Domestic	0.570	6.543	15	0.969	0
Commercial	0.467	3.720	15	0.999	0
Traction/Railways	0.114	34.625	15	0.003	0
Others	0.062	19.223	15	0.204	0
TEC	0.760	14.245	15	0.507	0

The forecasted values are shown in Table 11 for the above mentioned years.

Table 11. Summary of the results and Time line forecasted values for 2030.

Sector	Model	2019	2024	2030
Industry	Brown	538,089	686,457	864,498
Agriculture	Brown	208,891	258,836	318,770
Domestic	Brown	259,381	321,054	395,062
Commercial	ARIMA	115,130	172,213	279,201
Traction/Railways	ARIMA	20,554	26,906	465,523
Others	ARIMA	68,072	100,343	37,404
TEC	Brown	1,162,453	1,442,410	1,778,358

3.5. Moving Average

The three years' four years' and five years' moving average for the time period of 1974–2014 is computed here and shown in Figure 9. The values were found to be 830,696, 796,620 and 759,825 MW respectively for the year 2014.

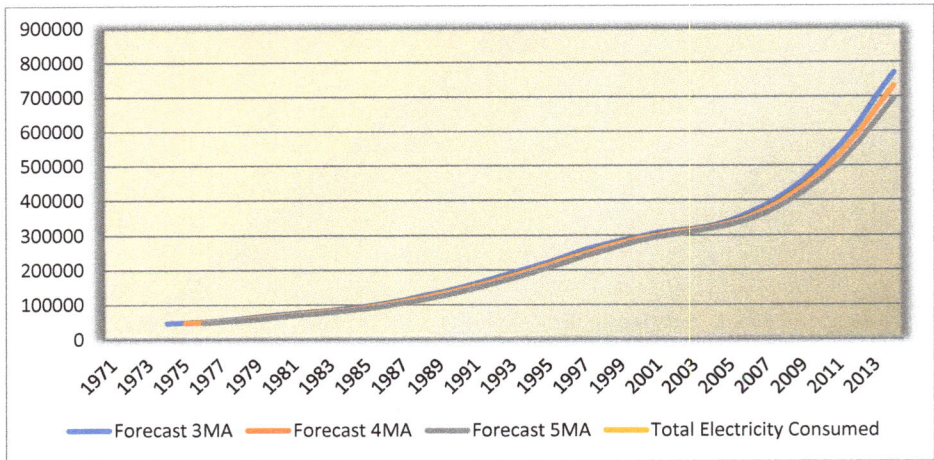

Figure 9. Chart for moving average method (1971–2014).

Moving average is one method which is very much suitable for short-term load forecasting, STLF. The forecast of the fourth year is the average of the first three years and so on.

3.6. Weighted Moving Average

The three years' and four years' weighted moving average for the time period 1971–2015 is calculated here. The values were found to be 786,587.1 and 765,421.5 MW respectively for the year 2015 in Figure 10.

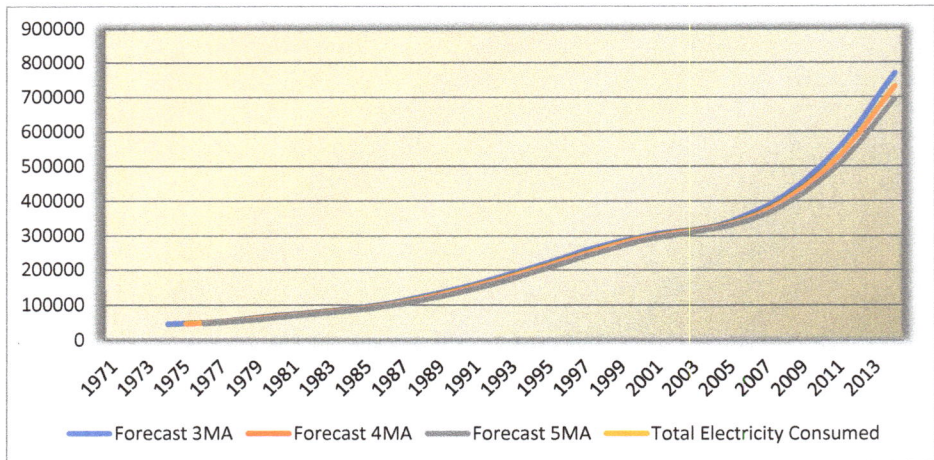

Figure 10. Chart for Weighted moving average method (1971–2015).

For the three years weighted moving average the α value for the previous three years were 0.5, 0.3 and 0.2 respectively. The higher α value is allotted to the immediate month since it influences the outcome more than that of the previous values. For the four years weighted moving average the α value for the previous four years assigned were 0.4, 0.3, 0.2 and 0.1 respectively. It is made sure that the α values add up to 1.

3.7. Curve Estimation

3.7.1. Linear Model

This Table 12 presents the regression coefficients and it is to be made a note of that, the correlation will be of negative value when the slope is negative. The following linear regression equation is determined by these coefficients.

$$y = -31{,}615{,}881.66 + 16{,}010.77x$$

Table 12. Summary of Linear model.

Equation	Summary of the Model				Parameter Estimates
	R^2	df1	df2	Sig.	Constant
Linear	0.844	1	42	0.000	−31,615,881.66

Series 1 in the Figure 11 is the actual TEC and the series 2 is the linear forecast. The forecasted value for the linear curve fitting model for 2030 is 885,981.44 MW.

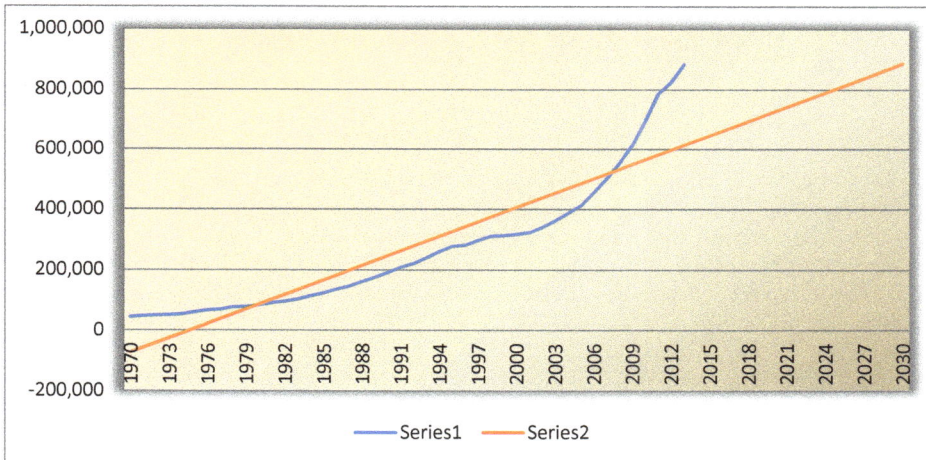

Figure 11. Chart for Linear method (1971–2030).

3.7.2. Compound/Exponential Model

The Table 13 represents the regression coefficients and it is to be taken into account that, the correlation will be in the negative side when the slope is of negative value. The following regression equation is made out of these coefficients.

$$y = 41{,}116.428e^{0.07x}$$

Table 13. Summary of the model.

Equation	Summary of the Model				Parameter Estimates
	R^2	df1	df2	Sig.	Constant
Comp./Exp.	0.991	1	42	0.000	41,116.428

The forecasted value for the compound curve fitting model for 2030 is 2,741,903.862 MW is plotted in Figure 12.

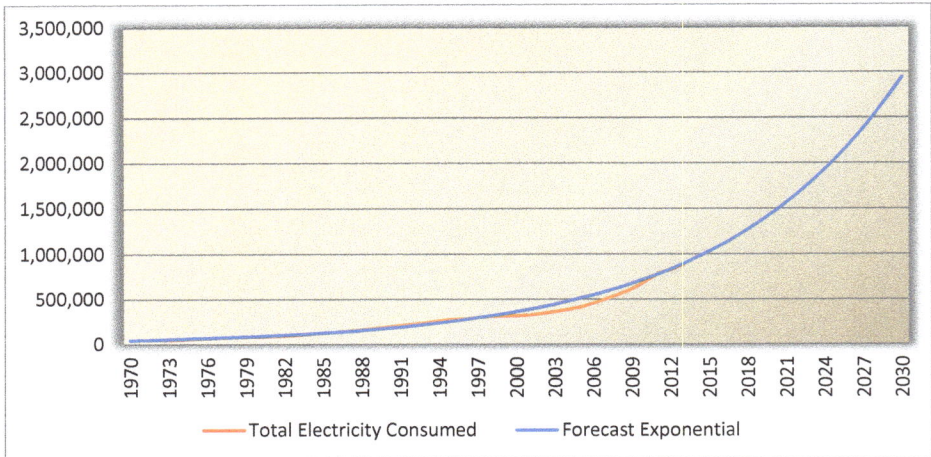

Figure 12. Chart for Exponential method (1971–2030).

Similarly the regression equations for several other models shall also be interpreted.

4. Discussion

The measure of the adequacy of the fit is determined by the sample correlation (r) between the true value and responses got out of the fit. The sample correlation's square is worked out readily out of the statistical package in the ANOVA and is termed the coefficient of determination (R^2). The coefficient of determination is computed directly by estimating Pearson's correlation 'r' between the predicted and the actual data. The coefficients of determination are generally expressed in terms of percentage. The value of R^2 lies in between 0% and 100%. The nearer the value to the upper bound; the healthier will be the fit [35].

LEAP and Holt's exponential smoothing method were also employed to estimate the electricity energy demand for 2030 in Maharashtra, India in that study. ANN, multiple regression approaches and ANOVA were used. It is evident from the analysis of variance in this article that the regression method is able to forecast the cutting forces with a higher accuracy [36] which supports the present study. An optimal renewable energy model, OREM for India was evolved for the year 2020–2021 to meet the increasing energy requirements [37]. An optimization model for various end-uses was formulated by determining the optimum allocation of renewable energy for 2020–2021, by considering the energy requirement of the commercial sector. This study revealed that the social acceptance of bio resources increased by 3% and solar PV utilizationdecreased by 65% [38].Various energy demand forecasting models were reviewed by [39] and found that traditional methods viz., time series regression, econometric analysis are extensively made use for demand side management whereas the TEC is calculated for 2030 in this paper. Regression analysis, linear model analysis and R^2 correlation value was built by [40] for a curved vane demister which supports the using linear model analysis of the current study. The utilization of black box approach to forecast the TEC for India is supported by vast literature among which an optimal renewable energy model for India for 2020–2021 was presented by distributing renewable energy effectively to help the policy framers in marketing the renewable energy resources and to determine the optimized allotment of various non-conventional energy resources for various end-uses. In this study linear as well as multiple regression analysis proves GDP is again the better predictor in terms of Multiple linear regression. Therefore, a sensible energy forecast is needed for the policy framers while taking decision for the future. Thus, the policy framers need to take this boost in energy usage in mind. It is also recommended that the other energy forecasting techniques shall also be used to testify the outcome and also energy prediction shall be recurrently done as the circumstances are dynamic.

Some of the state of art work in the same research area is discussed here

- According to National Energy Map for India: Technology Vision 2030, India's electricity consumption will become fourfold from about 1.1 trillion units to 4 trillion units by2030.Brookings Institution India Centre, in 2013, estimated that the shoot up in global energy consumption is attributed mainly due to India and China [41].
- Asia-Pacific territory lonely contributes to 79% of the hike in international liquids use, which rises from 1281.7 Million tons of oil equivalent in 2010 to 1859.3 Million tons of oil equivalent in 2030. The per capita energy utilization in 2030 for India is expected to rise from 19.58 million Btu to 29.84 million Btu [42].
- The former Coal and power minister of India, Mr. Piyush Goyal stated in May, 2016 that a possible 10% jump is expected in the annual electricity growth for the next 15 or 16 years [43].
- Sugandha Chauhan (2017) studied electricity demand and reported that it will increase from 1115 BU in 2015–2016 to 1692 BU in 2022, 2509 BU in 2027 and 3175 BU in 2030 reflecting the higher end of the demand for electricity [44].
- Iniyan et al. 2000. proposed a model that allocates the renewable energy distribution pattern for the year 2020–2021 for India [45].

5. Conclusions

This work presents the analysis of available data and the predicted one regarding what will be the Total Electricity Consumption (TEC) of India for the year 2030 using various black box based approaches. The forecasting of total electricity consumption for the year 2030–2031 for India is found to be 1,834,349MW while doing so the forecast for 2017 was compared with the actual data given by Energy statistics, GOI which sits close to the forecasted data. And the expert model is forecasted to be the best fit that suits the prediction since the R^2 value is 0.997 which is comparatively high. Obtained results show that this model is of a high precision. The advantages of the model are that it can be computed easily with simple statistical software and available in almost every recent statistical package. Accessibility is not an obstacle and the analysis shall be performed with a device of minimal configuration. The time taken for running the model is very minimal which is a mere 00:00:00.06 s (processor time). The disadvantage of the model is that it selects the best suitable model on its own. The limitation of the work is that we could not apply the popular methodologies of black box approaches such as Decision Trees, ANN, SVM. There are several other variables such as imports, exports, villages electrified, pump sets energized and so forth, which has a futuristic scope for further extensive studies. Energy forecasting can be taken up to the next level, for example, for Asia-Pacific territory. As the need for energy consumption is constantly increasing in manifolds, it is assumed that the findings and forecasts given in this article would be of use to the policy makers and energy strategists to evolve future scenarios for the Indian electricity consumption which should focus greatly in further increasing the overall share of renewable energy resources compared to the conventional sources of the installed capacity as well as in the consumption pattern. The future research may be done considering more input variables such as the quantum of CO_2 emission, GNP per capita, consumer price index, power consumption per capita, wholesale price index, imports, gross domestic savings, exports and so forth. Other methodologies such as computational intelligence forecasts, beyond point forecasts, combined forecasts may also be applied in short term load forecasting of the electrical energy demand.

Author Contributions: Conceptualization, H.R. and I.S.; Methodology, H.R.; Software, H.R.; Validation, H.R.; Formal Analysis, H.R. and I.S.; Investigation, I.S. and J.B.A.; Resources, H.R. and J.B.A.; Data Curation, H.R.; Writing-Original Draft Preparation, H.R.; Writing-Review & Editing, I.S. and J.B.A.; Supervision, I.S. H.R., IniyanSelvarasanand J.B.A. have added inputs in developing strategies for total energy consumption forecasting and the collection of the data set was taken care by H.R. All the authors were involved in drafting and revising the manuscript.

Funding: This research received no external funding.

Acknowledgments: The corresponding author first thanks the Al Mighty. He next expresses his gratitude to his parents who help him greatly to carry out his research work. He whole heartedly thanks his research supervisor for his invaluable guidance. The authors thank K. Padmanathan, who made available some of the data. He also thanks the authorities and faculty in Department of Mechanical engineering, Anna University, Chennai.

Conflicts of Interest: The authors declare no conflict of interest.

References

1. Central Electricity Authority. Available online: www.cea.nic.in (accessed on 15 October 2018).
2. Kalyani, K.A.; Pandey, K.K. Waste to Energy Status in India: A Short Review. *Renew. Sustain. Energy Rev.* **2014**, *31*, 113–120. [CrossRef]
3. Alagh, Y.K. The Food, Water and Energy, Inter Linkages for Sustainable Development in India. *South Asian Surv.* **2010**, *17*, 159–178. [CrossRef]
4. Taylor, J.W. Triple seasonal methods for short-term electricity demand forecasting. *Eur. J. Oper. Res.* **2010**, *204*, 139–152. [CrossRef]
5. Taylor, J.W.; McSharry, P.E. Short-Term Load Forecasting Methods: An Evaluation Based on European Data. *IEEE Trans. Power Syst.* **2008**, *22*, 2213–2219. [CrossRef]
6. Park, D.C.; El-Sharkawi, M.A.; Marks, R.J.; Atlas, L.E.; Damborg, M.J. Electric Load Forecasting Using an Artificial Neural Network. *IEEE Trans. Power Syst.* **1991**, *6*, 442–449. [CrossRef]
7. Mohamed, Z.; Bodger, P. Forecasting electricity consumption in New Zealand using economic and demographic variables. *Energy* **2005**, *30*, 1833–1843. [CrossRef]
8. Haida, T.; Muto, S. Regression based peak load using a transformation technique. *IEEE Trans. Power Syst.* **1994**, *9*, 1788–1794. [CrossRef]
9. Mirasgedis, S.; Safaridis, Y.; Georgopoulou, E.; Lalas, D.P.; Moschovits, M.; Karagiannis, F.; Papakonstantinou, D. Models for mid-term electricity demand forecasting incorporating weather influences. *Energy* **2006**, *31*, 208–227. [CrossRef]
10. Da, X.; Jiangyan, Y.; Jilai, Y. The physical series algorithm of mid-long term load forecasting of power systems. *Electr. Power Syst. Res.* **2000**, *53*, 31–37. [CrossRef]
11. Muis, Z.A.; Hashim, H.; Manan, Z.A.; Taha, F.M.; Douglas, P.L. Optimal planning of renewable energy—Integrated electricity generation schemes with CO_2 reduction target. *Renew. Energy* **2010**, *35*, 2562–2570. [CrossRef]
12. Kale, R.V.; Pohekar, S.D. Electricity demand and supply scenarios for Maharashtra (India) for 2030: An application of long range energy alternatives planning. *Energy Policy* **2014**, *72*, 1–13. [CrossRef]
13. Messner, S.; Golodnikov, A.; Gritsevskii, A. A Stochastic version of the dynamic linear programming model, MESSAGE III. *Energy* **1996**, *21*, 775–784. [CrossRef]
14. Tardioli, G.; Kerrigan, R.; Oates, M.; O'Donnell, J.; Finn, D. Data driven approaches for prediction of building energy consumption at urban level. *Energy Procedia* **2015**, *78*, 3378–3383. [CrossRef]
15. Choi, M.S.; Xiang, L.; Lee, S.J.; Kim, T.W. An Innovative Application Method of Monthly Load Forecasting for Smart IEDs. *J. Electr. Eng. Technol.* **2013**, *8*, 984–990. [CrossRef]
16. Yalcinoz, T.; Eminoglu, U. Short term and medium term power distribution load forecasting by neural networks. *Energy Convers. Manag.* **2005**, *46*, 1393–1405. [CrossRef]
17. Chen, G.J.; Li, K.K.; Chung, T.S.; Sun, H.B.; Tang, G.Q. Application of an innovative combined forecasting method in power system load forecasting. *Electr. Power Syst. Res.* **2001**, *59*, 131–137. [CrossRef]
18. Mestekemper, T.; Kauermann, G.; Smith, M.S. A comparison of periodic autoregressive and dynamic factor models in intraday energy demand forecasting. *Int. J. Forecast.* **2013**, *29*, 1–12. [CrossRef]

19. Tsekouras, G.J.; Dialynas, E.N.; Hatziargyriou, N.D.; Kavatza, S. A non-linear multivariable regression model for midterm energy forecasting of power systems. *Electr. Power Syst. Res.* **2007**, *77*, 1560–1568. [CrossRef]

20. Persaud, J.; Kumar, U. An eclectic approach in energy forecasting: A case of Natural Resources Canada's oil and gas outlook. *Energy Policy* **2001**, *29*, 303–313. [CrossRef]

21. Edigera, V.S.; Akarb, S. ARIMA forecasting of primary energy demand by fuel in Turkey. *Energy Policy* **2007**, *35*, 1701–1708. [CrossRef]

22. Xie, N.-M.; Yuan, C.-Q.; Yang, Y.-J. Forecasting China's energy demand and self-sufficiency rate by grey forecasting model and Markov model. *Elect. Power Energy Syst.* **2015**, *66*, 1–8. [CrossRef]

23. Bassam, M.; Al-Foul, A. Forecasting Energy Demand in Jordan Using Artificial Neural Networks. *Top. Middle East. Afr. Econ.* **2012**, *14*, 473.

24. Kankal, M.; Akpınar, A.; Komurcu, M.I.; Ozsahin, T.S. Modeling and forecasting of Turkey's energy consumption using socio-economic and demographic variables. *Appl. Energy* **2011**, *88*, 1927–1939. [CrossRef]

25. Hyndman, R.J.; Fan, S. Density Forecasting for Long-Term-Peak Electricity Demand. *IEEE Trans. Power Syst.* **2010**, *25*, 1142–1153. [CrossRef]

26. Pławiak, P. Novel Genetic Ensembles of Classifiers Applied to Myocardium Dysfunction Recognition Based on ECG Signals. *Swarm Evol.Comput.* **2018**, *39*, 192–208. [CrossRef]

27. Pławiak, P.; Rzecki, K. Approximation of Phenol Concentration using Computational Intelligence Methods Based on Signals from the Metal Oxide Sensor Array. *IEEE Sens. J.* **2015**, *15*, 1770–1783.

28. Mallah, S.; Bansal, N.K. Allocation of energy resources for power generation in India: Business as usual and energy efficiency. *Energy Policy* **2010**, *38*, 1059–1066. [CrossRef]

29. Li, C.; Ding, Z.; Zhao, D.; Yi, J.; Zhang, G. Building Energy Consumption Prediction: An Extreme Deep Learning Approach. *Energies* **2017**, *10*, 1–20. [CrossRef]

30. Bianco, V.; Manca, O.; Nardini, S. Electricity consumption forecasting in Italy using linear regression models. *Energy* **2009**, *34*, 1413–1421. [CrossRef]

31. Erdogdu, E. Electricity demand analysis using co-integration and ARIMA modeling: A case study of Turkey. *Energy Policy* **2007**, *35*, 1129–1146. [CrossRef]

32. Gajowniczek, K.; Nafkha, R.; Ząbkowski, T. Electricity peak demand classification with artificial neural networks. *Ann.Comput. Sci. Inf. Syst.* **2017**, *11*, 307–315.

33. Singh, S.; Yassine, A. Big Data Mining of Energy Time Series for Behavioral Analytics and Energy Consumption Forecasting. *Energies* **2018**, *11*, 452. [CrossRef]

34. *Energy Statistics 2018*; Central Statistics Office, Ministry of Statistics and Programme Implementation, Government of India: New Delhi, India, 2018.

35. Robert Mason, L.; Richard Gunst, F.; James Hess, L. *Statistical Design and Analysis of Experiments*; John Wiley & Sons Publication: New York, NY, USA, 2003.

36. Hanief, M.; Wani, M.F.; Charoo, M.S. Modeling and prediction of cutting forces during the turning of red brass (C23000) using ANN and regression analysis. *Eng. Sci. Technol. Int. J.* **2017**, *20*, 1220–1226. [CrossRef]

37. Iniyan, S.; Suganthi, L.; Jagadeesan, T.R.; Samuel, A.A. Reliability based socio economic optimal renewable energy model for India. *Renew. Energy* **2000**, *19*, 291–297. [CrossRef]

38. Suganthi, L.; Williams, A. Renewable energy in India—A modelling study for 2020–2021. *Energy Policy* **2000**, *28*, 1095–1109. [CrossRef]

39. Suganthi, L.; Samuel, A.A. Energy models for demand forecasting—A review. *Renew. Sus. Energy Rev.* **2012**, *16*, 1223–1240. [CrossRef]

40. Venkatesan, G.; Kulasekharan, N.; Muthukumar, V.; Iniyan, S. Regression analysis of a curved vane demister with Taguchi based optimization. *Desalination* **2015**, *370*, 33–43. [CrossRef]

41. TERI. *National Energy Map for India: Technology Vision 2030: Summary for Policy-Makers*; The Energy and Resources Institute TERI & Office of the Principal Scientific Adviser, Government of India: New Delhi, India, 2015.

42. Gokarn, S.; Sajjanhar, A.; Sandhu, R.; Dubey, S. *Energy 2030*; Brookings Institution India Center: New Delhi, India, 2013.

43. The Economic Times. Available online: https://economictimes.indiatimes.com/industry/energy/power/indias-electricity-consumption-to-touch-4-trillion-units-by-2030/articleshow/52221341.cms (accessed on 21 November 2018).

44. Chauhan, S.; Shekhar, S.; D'Souza, S.; Gopal, I. *Transitions in Indian Electricity Sector 2017–2030*; The Energy and Resources Institute TERI: New Delhi, India, 2017.

45. Iniyan, S.; Sumathy, K. An optimal renewable energy model for various end-uses. *Energy* **2000**, *25*, 563–575. [CrossRef]

energies

MDPI

Article

Hybrid Short-Term Load Forecasting Scheme Using Random Forest and Multilayer Perceptron [†]

Jihoon Moon [1], Yongsung Kim [2], Minjae Son [1] and Eenjun Hwang [1,*]

[1] School of Electrical Engineering, Korea University, 145 Anam-ro, Seongbuk-gu, Seoul 02841, Korea; johnny89@korea.ac.kr (J.M.); smj5668@korea.ac.kr (M.S.)

[2] Software Policy & Research Institute (SPRi), 22, Daewangpangyo-ro 712 beon-gil, Bundang-gu, Seongnam-si, Gyeonggi-do 13488, Korea; kys1001@spri.kr

* Correspondence: ehwang04@korea.ac.kr; Tel.: +82-2-3290-3256

[†] This paper is an extended version of our paper published in Proceedings of the 2018 IEEE International Conference on Big Data and Smart Computing (BigComp), Shanghai, China, 15–18 January 2018.

Received: 31 October 2018; Accepted: 21 November 2018; Published: 25 November 2018

Abstract: A stable power supply is very important in the management of power infrastructure. One of the critical tasks in accomplishing this is to predict power consumption accurately, which usually requires considering diverse factors, including environmental, social, and spatial-temporal factors. Depending on the prediction scope, building type can also be an important factor since the same types of buildings show similar power consumption patterns. A university campus usually consists of several building types, including a laboratory, administrative office, lecture room, and dormitory. Depending on the temporal and external conditions, they tend to show a wide variation in the electrical load pattern. This paper proposes a hybrid short-term load forecast model for an educational building complex by using random forest and multilayer perceptron. To construct this model, we collect electrical load data of six years from a university campus and split them into training, validation, and test sets. For the training set, we classify the data using a decision tree with input parameters including date, day of the week, holiday, and academic year. In addition, we consider various configurations for random forest and multilayer perceptron and evaluate their prediction performance using the validation set to determine the optimal configuration. Then, we construct a hybrid short-term load forecast model by combining the two models and predict the daily electrical load for the test set. Through various experiments, we show that our hybrid forecast model performs better than other popular single forecast models.

Keywords: hybrid forecast model; electrical load forecasting; time series analysis; random forest; multilayer perceptron

---

1. Introduction

Recently, the smart grid has been gaining much attention as a feasible solution to the current global energy shortage problem [1]. Since it has many benefits, including those related to reliability, economics, efficiency, environment, and safety, diverse issues and challenges to implementing such a smart grid have been extensively surveyed and proposed [2]. A smart grid [1,2] is the next-generation power grid that merges information and communication technology (ICT) with the existing electrical grid to advance electrical power efficiency to the fullest by exchanging information between energy suppliers and consumers in real-time [3]. This enables the energy supplier to perform efficient energy management for renewable generation sources (solar radiation, wind, etc.) by accurately forecasting power consumption [4]. Therefore, for a more efficient operation, the smart grid requires precise electrical load forecasting in both the short-term and medium-term [5,6]. Short-term load forecasting (STLF) aims to prepare for losses caused by energy failure and overloading by maintaining an active power consumption reserve margin [5]. It includes daily electrical load, highest or peak electrical load, and very short-term load forecasting (VSTLF). Generally, STLF is used to regulate the energy system from 1 h to one week [7]. Accordingly, daily load forecasting is used in the energy planning for the next one day to one week [8,9].

A higher-education building complex, such as a university campus, is composed of a building cluster with a high electrical load, and hence, has been a large electric power distribution consumer in Korea [10–12]. In terms of operational cost management, forecasting can help in determining where, if any, savings can be made, as well as uncovering system inefficiencies [13]. In terms of scheduling, forecasting can be helpful for improving the operational efficiency, especially in an energy storage system (ESS) or renewable energy.

Forecasting the electrical load of a university campus is difficult due to its irregular power consumption patterns. Such patterns are determined by diverse factors, such as the academic schedule, social events, and natural condition. Even on the same campus, the electrical load patterns among buildings differ, depending on the usage or purpose. For instance, typical engineering and science buildings show a high power consumption, while dormitory buildings show a low power consumption. Thus, to accurately forecast the electrical load of university campus, we also need to consider the building type and power usage patterns.

By considering power consumption patterns and various external factors together, many machine learning algorithms have shown a reasonable performance in short-term load forecasting [3,4,6,8,14–17]. However, even machine learning algorithms with a higher performance have difficulty in making accurate predictions at all times, because each algorithm adopts a different weighting method [18]. Thus, we can see that there will always be randomness or inherent uncertainty in every prediction [19]. For instance, most university buildings in Korea show various electrical load patterns which differ, depending on the academic calendar. Furthermore, Korea celebrates several holidays, such as Buddha's birthdays and Korean Thanksgiving days, called Chuseok, during the semester, which are counted on the lunar calendar. Since the campus usually remains closed on the holidays, the power consumption of the campus becomes very low. In such cases, it is difficult for a single excellent algorithm to make accurate predictions for all patterns. However, other algorithms can make accurate predictions in areas where the previous algorithm has been unable to do so. For this purpose, a good approach is to apply two or more algorithms to construct a hybrid probabilistic forecasting model [14]. Many recent studies have addressed a hybrid approach for STLF. Abdoos et al. [20] proposed a hybrid intelligent method for the short-term load forecasting of Iran's power system using wavelet transform (WT), Gram-Schmidt (GS), and support vector machine (SVM). Dong et al. [21] proposed a hybrid data-driven model to predict the daily total load based on an ensemble artificial neural network. In a similar way, Lee and Hong [22] proposed a hybrid model for forecasting the monthly power load several months ahead based on a dynamic and fuzzy time series model. Recently, probabilistic forecasting has arisen as an active topic and it could provide quantitative uncertainty information, which can be useful to manage its randomness in the power system operation [23]. Xiao et al. [18] proposed no negative constraint theory (NNCT) and artificial intelligence-based combination models to predict future wind speed series of the Chengde region. Jurado et al. [24] proposed hybrid methodologies for electrical load forecasting in buildings with different profiles based on entropy-based feature selection with AI methodologies. Feng et al. [25] developed an ensemble model to produce both deterministic and probabilistic wind forecasts that consists of multiple single machine learning algorithms in the first layer and blending algorithms in the second layer. In our previous study [12], we built a daily electrical load forecast model based on random forest. In this study, to improve the forecasting performance of that model, we first classify the electrical load data by pattern similarity using a decision tree. Then, we construct a hybrid model based on random forest and multilayer perceptron by considering similar time series patterns.

The rest of this paper is organized as follows. In Section 2, we introduce several previous studies on the machine learning-based short-term load forecasting model. In Section 3, we present all the steps for constructing our hybrid electrical load forecasting model in detail. In Section 4, we describe several metrics for performance a comparison of load forecasting models. In Section 5, we describe how to evaluate the performance of our model via several experiments and show some of the results. Lastly, in Section 6, we briefly discuss the conclusion.

2. Related Work

So far, many researchers have attempted to construct STLF using various machine learning algorithms. Vrablecová et al. [7] developed the suitability of an online support vector regression (SVR) method to short-term power load forecasting and presented a comparison of 10 state-of-the-art forecasting methods in terms of accuracy for the public Irish Commission for Energy Regulation (CER) dataset. Tso and Yau [26] conducted weekly power consumption prediction for households in Hong Kong based on an artificial neural network (ANN), multiple regression (MR), and a decision tree (DT). They built the input variables of their prediction model by surveying the approximate power consumption for diverse electronic products, such as air conditioning, lighting, and dishwashing. Jain et al. [27] proposed a building electrical load forecasting model based on SVR. Electrical load data were collected from multi-family residential buildings located at the Columbia University campus in New York City. Grolinger et al. [28] proposed two electrical load forecasting models based on ANN and SVR to consider both events and external factors and performed electrical load forecasting by day, hour, and 15-min intervals for a large entertainment building. Amber et al. [29] proposed two forecasting models, genetic programming (GP) and MR, to forecast the daily power consumption of an administration building in London. Rodrigues et al. [30] performed forecasting methods of daily and hourly electrical load by using ANN. They used a database with consumption records, logged in 93 real households in Lisbon, Portugal. Efendi et al. [31] proposed a new approach for determining the linguistic out-sample forecasting by using the index numbers of the linguistics approach. They used the daily load data from the National Electricity Board of Malaysia as an empirical study.

Recently, a hybrid prediction scheme using multiple machine learning algorithms has shown a better performance than the conventional prediction scheme using a single machine learning algorithm [14]. The hybrid model aims to provide the best possible prediction performance by automatically managing the strengths and weaknesses of each base model. Xiao et al. [18] proposed two combination models, the no negative constraint theory (NNCT) and the artificial intelligence algorithm, and showed that they can always achieve a desirable forecasting performance compared to the existing traditional combination models. Jurado et al. [24] proposed a hybrid methodology that combines feature selection based on entropies with soft computing and machine learning approaches (i.e., fuzzy inductive reasoning, random forest, and neural networks) for three buildings in Barcelona. Abdoos et al. [20] proposed a new hybrid intelligent method for short-term load forecasting. They decomposed the electrical load signal into two levels using wavelet transform and then created the training input matrices using the decomposed signals and temperature data. After that, they selected the dominant features using the Gram–Schmidt method to reduce the dimensions of the input matrix. They used SVM as the classifier core for learning patterns of the training matrix. Dong et al. [21] proposed a novel hybrid data-driven "PEK" model for predicting the daily total load of the city of Shuyang, China. They constructed the model by using various function approximates, including partial mutual information (PMI)-based input variable selection, ensemble artificial neural network (ENN)-based output estimation, and K-nearest neighbor (KNN) regression-based output error estimation. Lee and Hong [22] proposed a hybrid model for forecasting the electrical load several months ahead based on a dynamic (i.e., air temperature dependency of power load) and a fuzzy time series approach. They tested their hybrid model using actual load data obtained from the Seoul metropolitan area, and compared its prediction performance with those of the other two dynamic models.

Previous studies on hybrid forecasting models comprise parameter selection and optimization technique-based combined approaches. This approach has the disadvantages that it is dependent on a designer's expertise and exhibits low versatility [18]. On the other hand, this paper proposes data post-processing technique combined approaches to construct a hybrid forecasting model by combining random forest and multilayer perceptron (MLP).

3. Hybrid Short-Term Load Forecasting

In this section, we describe our hybrid electrical load forecasting model. The overall steps for constructing the forecasting model are shown in Figure 1. First, we collect daily power consumption data, time series information, and weather information, which will be used as independent variables for our hybrid STLF model. After some preprocessing, we build a hybrid prediction model based on random forest and MLP. Lastly, we perform a seven-step-ahead (one week or 145 h ahead) time series cross-validation for the electrical load data.

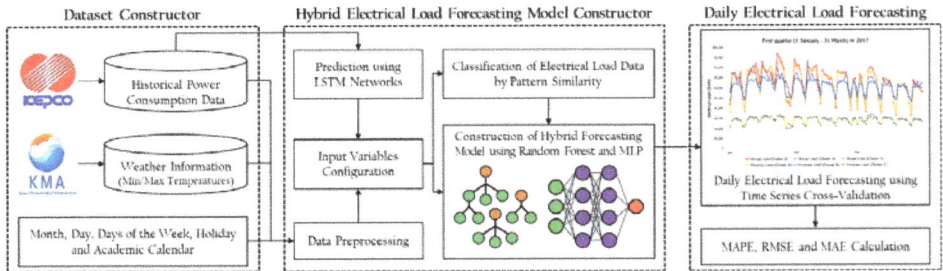

Figure 1. Our framework for hybrid daily electrical load forecasting.

3.1. Dataset

To build an effective STLF model for buildings or building clusters, it is crucial to collect their real power consumption data that show the power usage of the buildings in the real world. For this purpose, we considered three clusters of buildings with varied purposes and collected their daily power consumption data from a university in Korea. The first cluster is composed of 32 buildings with academic purposes, such as the main building, amenities, department buildings, central library, etc. The second cluster is composed of 20 buildings, with science and engineering purposes. Compared to other clusters, this cluster showed a much higher electrical load, mainly due to the diverse experimental equipment and devices used in the laboratories. The third cluster comprised 16 dormitory buildings, whose power consumption was based on the residence pattern. In addition, we gathered other data, including the academic schedule, weather, and event calendar. The university employs the i-Smart system to monitor the electrical load in real time. This is an energy portal service operated by the Korea Electric Power Corporation (KEPCO) to give consumers electricity-related data such as electricity usage and expected bill to make them use electricity efficiently. Through this i-Smart system, we collected the daily power consumption of six years, from 2012 to 2017. For weather information, we utilized the regional synoptic meteorological data provided by the Korea Meteorological Office (KMA). KMA's mid-term forecast provides information including the date, weather, temperature (maximum and minimum), and its reliability for more than seven days.

To build our hybrid STLF model, we considered nine variables; month, day of the month, day of the week, holiday, academic year, temperature, week-ahead load, year-ahead load, and LSTM Networks. In particular, the day of the week is a categorized variable and we present the seven days using integers 1 to 7 according to the ISO-8601 standard [32]. Accordingly, 1 indicates Monday and 7 indicates Sunday. Holiday, which includes Saturdays, Sundays, national holidays, and school anniversary [33], indicates whether the campus is closed or not. A detailed description of the input variables can be found in [12].

3.2. Data Preprocessing

3.2.1. Temperature Adjustment

Generally, the power consumption increases in summer and winter due to the heavy use of air conditioning and electric heating appliances, respectively. Since the correlations between the temperature and electrical load in terms of maximum and minimum temperatures are not that high, we need to adjust the daily temperature for more effective training based on the annual average temperature of 12.5 provided by KMA [34], using Equation (1) as follows:

$$Adjusted_{Temp} = \left| 12.5 - \frac{Minimum_{Temp} + Maximumm_{Temp}}{2} \right|. \tag{1}$$

To show that the adjusted temperature has a higher correlation than the minimum and maximum temperatures, we calculated the Pearson correlation coefficients between the electrical load and minimum, maximum, average, and adjusted temperatures, as shown in Table 1. In the table, the adjusted temperature shows higher coefficients for all building clusters compared to other types of temperatures.

Table 1. Comparison of Pearson correlation coefficients.

Temperature Type	Cluster #		
	Cluster A	Cluster B	Cluster C
Minimum temperature	−0.018	0.101	0.020
Maximum temperature	−0.068	0.041	−0.06
Average temperature	−0.043	0.072	−0.018
Adjusted temperature	0.551	0.425	0.504

3.2.2. Estimating the Week-Ahead Consumption

The electrical load data from the past form one of the perfect clues for forecasting the power consumption of the future and the power consumption pattern relies on the day of the week, workday, and holiday. Hence, it is necessary to consider many cases to show the electrical load of the past in the short-term load forecasting. For instance, if the prediction time is a holiday and the same day in the previous week was a workday, then their electrical loads can be very different. Therefore, it would be better to calculate the week-ahead load at the prediction time not by the electrical load data of the coming week, but by averaging the electrical loads of the days of the same type in the previous week. Thus, if the prediction time is a workday, we use the average electrical load of all workdays of the previous week as an independent variable. Likewise, if the prediction time is a holiday, we use the average electrical load of all holidays of the previous week. In this way, we reflect the different electrical load characteristics of the holiday and workday in the forecasting. Figure 2 shows an example of estimating the week-ahead consumption. If the current time is Tuesday, we already know the electrical load of yesterday (Monday). Hence, to estimate the week-ahead consumption of the coming Monday, we use the average of the electrical loads of workdays of the last week.

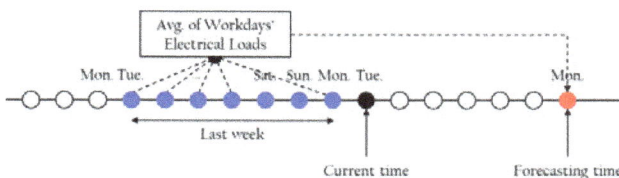

Figure 2. Example of estimating week-ahead consumption.

3.2.3. Estimating the Year-Ahead Consumption

The year-ahead load aims to utilize the trend of the annual electrical load by showing the power consumption of the same week of the previous year. However, the electrical load of the exact same week of the previous year is not always used because the days of the week are different and popular Korean holidays are celebrated according to the lunar calendar. Every week of the year has a unique week number based on ISO-8601 [32]. As mentioned before, the average of power consumptions of all holidays or workdays of the week are calculated to which the prediction time belongs and depending on the year, one year comprises 52 or 53 weeks. In the case of an issue such as the prediction time belongs to the 53rd week, there is no same week number in the previous year. To solve this problem, the power consumption of the 52nd week from the previous year is utilized since the two weeks have similar external factors. Especially, electrical loads show very low consumption on a special holiday like the Lunar New Year holidays and Korean Thanksgiving days [35]. To show this usage pattern, the average power consumption of the previous year's special holiday represents the year-ahead's special holiday's load. The week number can differ depending on the year, so representing the year-ahead's special holiday power consumption cannot be done directly using the week number of the holiday. This issue can be handled easily by exchanging the power consumption of the week and the week of the holiday in the previous year. Figure 3 shows an example of estimating the year-ahead consumption. If the current time is Monday of the 33rd week 2016, we use the 33rd week's electrical load of the last year. To estimate the year-ahead consumption of Sunday of the 33rd week, we use the average of the electrical loads of the holidays of the 33rd week of the last year.

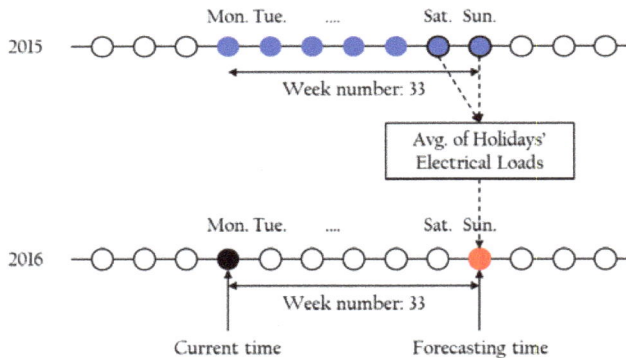

Figure 3. Example of estimating the year-ahead consumption.

3.2.4. Load Forecasting Based on LSTM Networks

A recurrent neural network (RNN) is a class of ANN where connections between units form a directed graph along a sequence. Unlike a feedforward neural network (FFNN), RNNs can use their internal state or memory to process input sequences [36]. RNNs can handle time series data in many applications, such as unsegmented, connected handwriting recognition or speech recognition [37]. However, RNNs have problems in that the gradient can be extremely small or large; these problems are called the vanishing gradient and exploding gradient problems. If the gradient is extremely small, RNNs cannot learn data with long-term dependencies. On the other hand, if the gradient is extremely large, it moves the RNN parameters far away and disrupts the learning process. To handle the vanishing gradient problem, previous studies [38,39] have proposed sophisticated models of RNN architectures. One successful model is long short-term memory (LSTM), which solves the RNN problem through a cell state and a unit called a cell with multiple gates. LSTM Networks use a method that influences the behind data by reflecting the learned information with the previous data as the learning progresses with time. Therefore, it is suitable for time series data, such as electrical load data. However, the LSTM networks can reflect yesterday's information for the next day's forecast. Since the daily load forecasting of a smart grid aims to be scheduled until after a week, LSTM networks are not suitable for application to the daily load forecasting because there is a gap of six days. Furthermore, if the prediction is not valid, the LSTM model method can give a bad result. For instance, national holidays, quick climate change, and unexpected institution-related events can produce unexpected power consumption. Therefore, the LSTM model alone is not enough for short-term load forecasting due to its simple structure and weakness in volatility. Eventually, a similar life pattern can be observed depending on the day of the week, which in return gives a similar electrical load pattern. This study uses the LSTM networks method to show the repeating pattern of power consumptions depending on the day of the week. The input variable period of the training dataset is composed of the electrical load from 2012 to 2015 and the dependent variable of the training set is composed of the electrical load from 2013 to 2015. We performed 10-fold cross-validation on a rolling basis for optimal hyper-parameter detection.

3.3. Discovering Similar Time series Patterns

So far, diverse machine learning algorithms have been proposed to predict electrical load [1,3,6,14]. However, they showed different prediction performances depending on the various factors. For instance, for time series data, one algorithm gives the best prediction performance on one segment, while for other segments, another algorithm can give the best performance. Hence, one way to improve the accuracy in this case is to use more than one predictive algorithm. We consider electrical load data as time series data and utilize a decision tree to classify the electrical load data by pattern similarity. Decision trees [26,40] can handle both categorical and numerical data, and are highly persuasive because they can be analyzed through each branch of the tree, which represents the process of classification or prediction. In addition, they exhibit a high explanatory power because they can confirm which independent variables have a higher impact when predicting the value of a dependent or target variable. On the other hand, continuous variables used in the prediction of values of the time series are regarded as discontinuous values, and hence, the prediction errors are likely to occur near the boundary of separation. Hence, using the decision tree, we divide continuous dependent variables into several classes with a similar electrical load pattern. To do this, we use the training dataset from the previous three years. We use the daily electrical load as the attribute of class label or dependent variable and the characteristics of the time series as independent variables, representing year, month, day, day of the week, holiday, and academic year. Details on the classification of time series data will be shown in the experimental section.

3.4. Building a Hybrid Forecasting Model

To construct our hybrid prediction model, we combine both a random forest model and multilayer perceptron model. Random forest is a representative ensemble model, while MLP is a representative deep learning model; both these models have shown an excellent performance in forecasting electrical load [5,12,15–17].

Random forest [41,42] is an ensemble method for classification, regression, and other tasks. It constructs many decision trees that can be used to classify a new instance by the majority vote. Each decision tree node uses a subset of attributes randomly selected from the original set of attributes. Random forest runs efficiently on large amounts of data and provides a high accuracy [43]. In addition, compared to other machine learning algorithms such as ANN and SVR, it requires less fine-tuning of its hyper-parameters [16]. The basic parameters of random forest include the total number of trees to be generated (nTree) and the decision tree-related parameters (mTry), such as minimum split and split criteria [17]. In this study, we find the optimal mTry and nTree for our forecasting model by using the training set and then verify their performance using the validation and test set. The authors in [42] suggested that a random forest should have 64 to 128 trees and we use 128 trees for our hybrid STLF model. In addition, the mTry values used for this study provided by scikit-learn are as follows.

- Auto: max features = n features.
- Sqrt: max features = sqrt (n features).
- Log2: max features = log2 (n features).

A typical ANN architecture, known as a multilayer perceptron, is a type of machine learning algorithm that is a network of individual nodes, called perceptrons, organized in a series of layers [5]. Each layer in MLP is categorized into three types: an input layer, which receives features used for prediction; a hidden layer, where hidden features are extracted; and an output layer, which yields the determined results. Among them, the hidden layer has many factors affecting performance, such as the number of layers, the number of nodes involved, and the activation function of the node [44]. Therefore, the network performance depends on how the hidden layer is configured. In particular, the number of hidden layers determines the depth or shallowness of the network. In addition, if there are more than two hidden layers, it is called a deep neural network (DNN) [45]. To establish our MLP, we use two hidden layers since we do not require many input variables in our prediction model. In addition, we use the same epochs and batch size as the LSTM model we described previously. Furthermore, as an activation function, we use an exponential linear unit (ELU) without the rectified linear unit (ReLU), which has gained increasing popularity recently. However, its main disadvantage is that the perceptron can die in the learning process. ELU [46] is an approximate function introduced to overcome this disadvantage, and can be defined by:

$$f(x) = \begin{cases} x & \text{if } x \geq 0 \\ \alpha(e^x - 1) & \text{if } x < 0 \end{cases} \quad . \tag{2}$$

The next important consideration is to choose the number of hidden nodes. Many studies have been conducted to determine the optimal number of hidden nodes for a given task [15,47,48], and we decided to use two different hidden node counts: the number of input variables and 2/3 of the number of input variables. Since we use nine input variables, the numbers of hidden nodes we will use are 9 and 6. Since our model has two hidden layers, we can consider three configurations, depending on the hidden nodes of the first and second layers: (9, 9), (9, 6), and (6, 6). As in the random forest, we evaluate these configurations using the training data for each building cluster and identify the configuration that gives the best prediction accuracy. After that, we compare the best MLP model with the random forest model for each cluster type.

3.5. Time series Cross-Validation

To construct a forecasting model, the dataset is usually divided into a training set and test set. Then, the training set is used in building a forecasting model and the test set is used in evaluating the resulting model. However, in traditional time series forecasting techniques, the prediction performance is poorer as the interval between the training and forecasting times increases. To alleviate this problem, we apply the time series cross-validation (TSCV) based on the rolling forecasting origin [49]. A variation of this approach focuses on a single prediction horizon for each test set. In this approach, we use various training sets, each containing one extra observation than the previous one. We calculate the prediction accuracy by first measuring the accuracy for each test set and then averaging the results of all test sets. This paper proposes a one-week (sum from 145 h to 168 h) look-ahead view of the operation for smart grids. For this, a seven-step-ahead forecasting model is built to forecast the power consumption at a single time (h + 7 + i − 1) using the test set with observations at several times (1, 2, ... , h + i − 1). If h observations are required to produce a reliable forecast, then, for the total T observations, the process works as follows.

For i = 1 to T − h − 6:

(1) Select the observation at time h + 7 + i − 1 for the test set;
(2) Consider the observations at several times 1, 2, $\cdots$, h + i − 1 to estimate the forecasting model;

(3) Calculate the 7-step error on the forecast for time $h + 7 + i - 1$;
(4) Compute the forecast accuracy based on the errors obtained.

4. Performance Metrics

To analyze the forecast model performance, several metrics, such as mean absolute percentage error (MAPE), root mean square error (RMSE), and mean absolute error (MAE), are used, which are well-known for representing the prediction accuracy.

4.1. Mean Absolute Percentage Error

MAPE is a measure of prediction accuracy for constructing fitted time series values in statistics, specifically in trend estimation. It usually presents accuracy as a percentage of the error and can be easier to comprehend than the other statistics since this number is a percentage. It is known that the MAPE is huge if the actual value is very close to zero. However, in this work, we do not have such values. The formula for MAPE is shown in Equation (3), where A_t and F_t are the actual and forecast values, respectively. In addition, n is the number of times observed.

$$\text{MAPE} = \frac{100}{n} \sum_{t=1}^{n} \left| \frac{A_t - F_t}{A_t} \right| \tag{3}$$

4.2. Root Mean Square Error

RMSE (also called the root mean square deviation, RMSD) is used to aggregate the residuals into a single measure of predictive ability. The square root of the mean square error, as shown in Equation (4), is the forecast value F_t and an actual value A_t. The mean square standard deviation of the forecast value F_t for the actual value A_t is the square root of RMSE. For an unbiased estimator, RMSE is the square root of the variance, which denotes the standard error.

$$\text{RMSE} = \sqrt{\frac{\sum_{i=1}^{n} (F_t - A_t)^2}{n}} \tag{4}$$

4.3. Mean Absolute Error

In statistics, MAE is used to evaluate how close forecasts or predictions are to the actual outcomes. It is calculated by averaging the absolute differences between the prediction values and the actual observed values. MAE is defined as shown in Equation (5), where F_t is the forecast value and A_t is the actual value.

$$\text{MAE} = \frac{1}{n} \sum_{i=1}^{n} |F_t - A_t| \tag{5}$$

5. Experimental Results

To evaluate the performance of our hybrid forecast model, we carried out several experiments. We performed preprocessing for the dataset in the Python environment and performed forecast modeling using scikit-learn [50], TensorFlow [51], and Keras [52]. We used six years of daily electrical load data from 2012 to 2017. Specifically, we used electrical load data of 2012 to configure input variables for a training set. Data from 2013 to 2015 was used as the training set, the data of 2016 was the validation set, and the data of 2017 was the test set.

5.1. Dataset Description

Table 2 shows the statistics of the electric consumption data for each cluster, including the number of valid cases, mean, and standard deviation. As shown in the table, Cluster B has a higher power consumption and wider deviation than clusters A and C.

Table 2. Statistics of power consumption data.

Statistics	Cluster #		
	Cluster A	Cluster B	Cluster C
Number of valid cases	1826	1826	1826
Mean	63,094.97	68,860.93	30,472.31
Variance	246,836,473	269,528,278	32,820,509
Standard deviation	15,711.03	16,417.31	5728.92
Maximum	100,222.56	109,595.52	46,641.6
Minimum	23,617.92	26,417.76	14,330.88
Lower quartile	52,202.4	56,678.88	26,288.82
Median	63,946.32	66,996.72	30,343.14
Upper quartile	76,386.24	79,209.96	34,719.45

5.2. Forecasting Model Configuration

In this study, we used the LSTM networks method to show the repeating pattern of power consumptions depending on the day of the week. We tested diverse cases and investigated the accuracy of load forecasting for the test cases to determine the best input data selection. As shown in Figure 4, the input variables consist of four electrical loads from one week ago to four weeks ago as apart at a weekly interval to reflect the cycle of one month. In the feature scaling process, we rescaled the range of the measured values from 0 to 1. We used tanh as the activation function and calculated the loss by using the mean absolute error. We used the adaptive moment estimation (Adam) method, which combines momentum and root mean square propagation (RMSProp), as the optimization method. The Adam optimization technique weighs the time series data and maintains the relative size difference between the variables. In the configuration of the remaining hyper-parameters of the model, we set the number of hidden units to 60, epochs to 300, and batch size to 12.

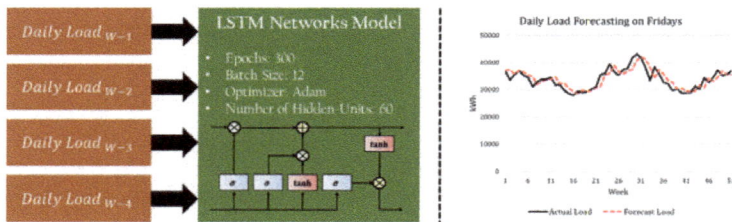

Figure 4. System architecture of LSTM networks.

We experimented with the LSTM model [52] by changing the time step from one cycle to a maximum of 30 cycles. Table 3 shows the mean absolute percentage error (MAPE) of each cluster for each time step. In the table, the predicted results with the best accuracy are marked in bold. Table 3 shows that the 27th time step indicates the most accurate prediction performance. In general, the electricity demand is relatively high in summer and winter, compared to that in spring and autumn. In other words, it has a rise and fall curve in a half-year cycle, and the 27th time step corresponds to a week number of about half a year.

We performed similar a time series pattern analysis based on the decision tree through 10-fold cross-validation for the training set. Among several options provided by scikit-learn to construct a decision tree, we considered the criterion, max depth, and max features. The criterion is a function for measuring the quality of a split. In this paper, we use the "mae" criterion for our forecasting model since it gives the smallest error rate between the actual and the classification value. Max depth is the maximum depth of the tree. We set max depth to 3, such that the number of leaves is 8. In other words, the decision tree classifies the training datasets into eight similar time series. Max features are the number of features to consider when looking for the best split. We have chosen the "auto" option to reflect all time variables. Figure 5 shows the result of the similar time series recognition for each cluster using the decision tree. Here, samples indicate the number of tuples in each leaf. The total number of samples is 1095, since we are considering the daily consumption data over three years. Value denotes the classification value of the similar time series. Table 4 shows the number of similar time series samples according to the decision tree for 2016 and 2017.

Table 3. MAPE results of LSTM networks.

Time Step	Cluster A	Cluster B	Cluster C	Average
1	9.587	6.989	6.834	7.803
2	9.169	6.839	6.626	7.545
3	8.820	6.812	6.463	7.365
4	8.773	6.750	6.328	7.284
5	8.686	6.626	6.191	7.168
6	8.403	6.695	5.995	7.031
7	8.405	6.700	6.104	7.070
8	8.263	6.406	5.846	6.839
9	8.260	6.583	5.648	6.830
10	8.286	6.318	5.524	6.709
11	8.095	6.438	5.666	6.733
12	8.133	6.469	5.917	6.840
13	7.715	6.346	5.699	6.587
14	7.770	6.263	5.399	6.477
15	7.751	6.139	5.306	6.399
16	7.561	5.974	5.315	6.283
17	7.411	5.891	5.450	6.251
18	7.364	6.063	5.398	6.275
19	7.466	6.089	5.639	6.398
20	7.510	5.892	5.627	6.343
21	7.763	5.977	5.451	6.397
22	7.385	5.856	5.460	6.234
23	7.431	5.795	5.756	6.327
24	7.870	6.089	5.600	6.520
25	7.352	5.923	5.370	6.215
26	**7.335**	5.997	5.285	6.206
27	7.405	**5.479**	5.371	**6.085**
28	7.422	5.853	**5.128**	6.134
29	7.553	5.979	5.567	6.366
30	7.569	5.601	5.574	6.248

Table 4. Similar time series patterns.

Pattern	Cluster A		Cluster B		Cluster C	
	2016	2017	2016	2017	2016	2017
1	62	62	62	62	62	62
2	14	14	14	14	140	138
3	107	111	107	111	20	20
4	64	58	64	58	25	25
5	14	15	14	15	1	2
6	53	52	53	52	10	9
7	16	16	5	5	99	98
8	36	37	47	48	9	11
Total	366	365	366	365	366	365

The predictive evaluation consists of two steps. Based on the forecast models of random forest and MLP, we used the training set from 2013 to 2015 and predicted the verification period of 2016. The objectives are to detect models with optimal hyper-parameters and then to select models with a better predictive performance in similar time series. Next, we set the training set to include data from 2013 to 2016 and predicted the test period of 2017. Here, we evaluate the predictive performance of the hybrid model we have constructed. Table 5 is the prediction result composed of MLP, and MAPE is used as a measure of prediction accuracy and the predicted results with the best accuracy are marked in bold. As shown in the table, overall, a model consisting of nine and nine nodes in each hidden layer showed the best performance. Although the nine and six nodes in each hidden layer showed a better performance in Cluster A, the model consisting of nine and nine nodes was selected to generalize the predictive model.

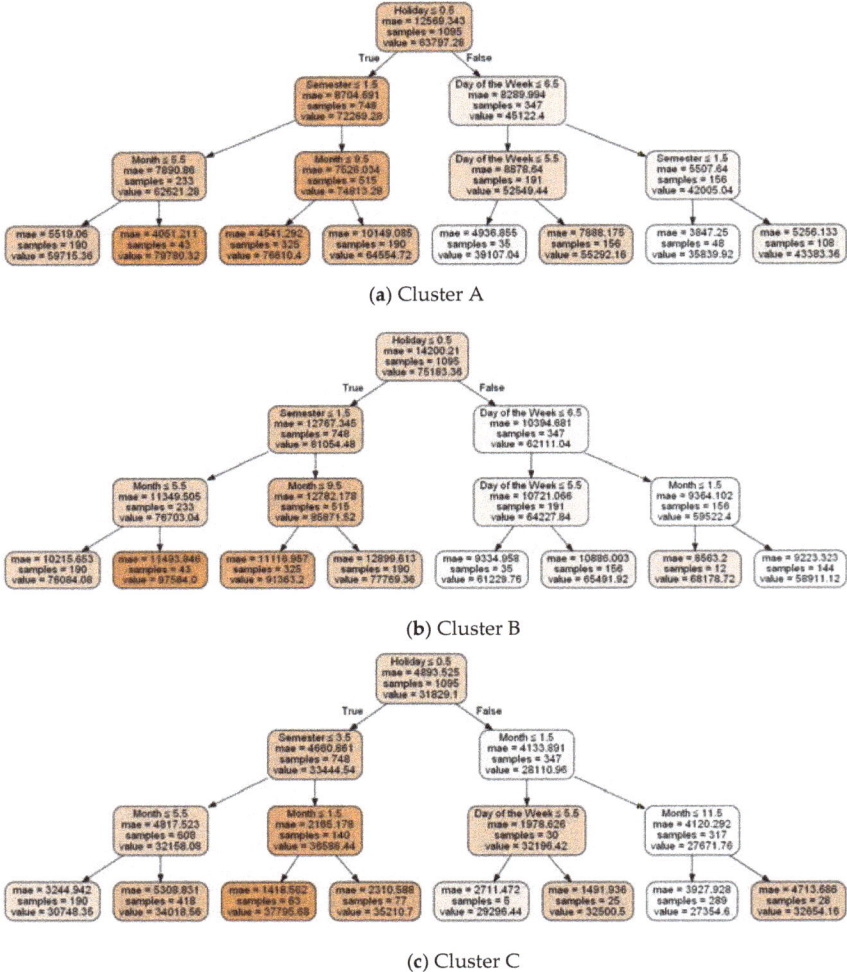

(a) Cluster A

(b) Cluster B

(c) Cluster C

Figure 5. Results of similar time series classifications using decision trees.

Table 5. MAPE results of the multilayer perceptron.

Cluster #	Number of Neurons in Each Layer		
	9-6-6-1	**9-9-6-1**	**9-9-9-1**
Cluster A	3.856	**3.767**	3.936
Cluster B	4.869	5.076	**4.424**
Cluster C	3.366	3.390	**3.205**

Table 6 shows the MAPE of random forest for each cluster under different mTry and the predicted results with the best accuracy are marked in bold. Since the input variable is 9, sqrt and log2 are recognized as 3 and the results are the same. We choose the sqrt that is commonly used [16,43].

Table 6. MAPE results of random forest.

Cluster #	Number of Features		
	Auto	sqrt	log2
Cluster A	3.983	**3.945**	**3.945**
Cluster B	4.900	**4.684**	**4.684**
Cluster C	3.579	**3.266**	**3.266**

Figure 6a–c show the use of forests of trees to evaluate the importance of features in an artificial classification task. The blue bars denote the feature importance of the forest, along with their inter-trees variability. In the figure, LSTM, which refers to the LSTM-RNN that reflects the trend of day of the week, has the highest impact on the model configuration for all clusters. Other features have different impacts, depending on the cluster type.

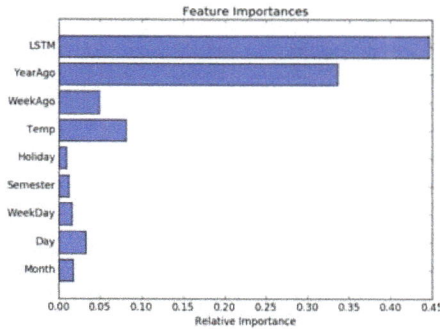

(a) Cluster A

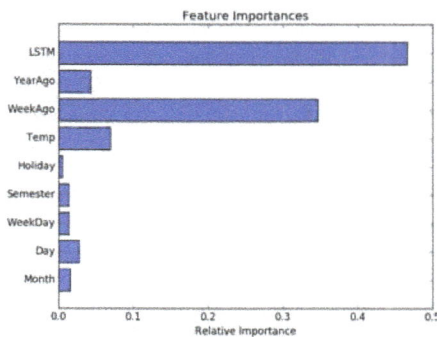

(b) Cluster B

Figure 6. *Cont.*

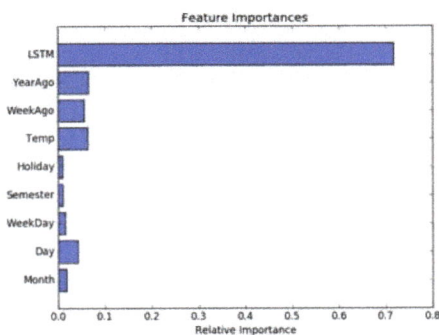

Feature Importances

(c) Cluster C

Figure 6. Feature importance in random forest.

Table 7 shows the electrical load forecast accuracy for the pattern classification of similar time series for 2016. In the table, the predicted results with a better accuracy are marked in bold. For instance, in the case of Cluster A, while random forest shows a better prediction accuracy for patterns 1 to 4, MLP shows a better accuracy for patterns 5 to 8. Using this table, we can choose a more accurate prediction model for the pattern and cluster type.

Table 7. MAPE results of load forecasting in 2016.

2016	Cluster A		Cluster B		Cluster C	
Pattern	MLP	RF	MLP	RF	MLP	RF
1	3.339	**3.092**	3.705	**2.901**	2.736	**2.475**
2	2.199	**1.965**	4.395	**3.602**	2.987	**2.731**
3	2.840	**2.712**	3.343	**2.990**	2.853	**2.277**
4	4.165	**3.472**	**3.794**	3.978	3.517	**2.568**
5	**7.624**	9.259	**8.606**	15.728	**4.229**	10.303
6	**4.617**	5.272	**5.404**	6.172	5.159	**4.894**
7	**3.816**	4.548	9.199	**8.860**	3.686	**4.718**
8	**6.108**	6.402	**5.844**	6.768	2.152	**2.595**

Table 8 shows prediction results of our model for 2017. Comparing Tables 7 and 8, we can see that MLP and random forest (RF) have a matched relative performance in most cases. There are two exceptions in Cluster A and one exception in Cluster B and they are underlined and marked in bold. In the case of Cluster C, MLP and RF gave the same relative performance. This is good evidence that our hybrid model can be generalized.

Table 8. MAPE results of load forecasting in 2017.

2017	Cluster A		Cluster B		Cluster C	
Pattern	MLP	RF	MLP	RF	MLP	RF
1	2.914	**2.709**	4.009	**3.428**	2.838	**2.524**
2	**_1.945_**	2.587	**_3.313_**	3.442	2.622	**2.474**
3	2.682	**2.629**	3.464	**3.258**	3.350	**2.583**
4	5.025	**4.211**	**4.005**	5.116	2.694	**2.391**
5	**7.103**	11.585	**9.640**	20.718	**3.300**	15.713
6	**4.503**	6.007	**5.956**	7.272	6.984	**6.296**
7	**3.451**	3.517	13.958	**12.386**	3.835	**4.443**
8	6.834	**_6.622_**	7.131	**8.106**	2.562	**3.722**

5.3. Comparison of Forecasting Techniques

To verify the validness and applicability of our hybrid daily load forecasting model, the predictive performance of our model should be compared with other machine learning techniques, including ANN and SVR, which are very popular predictive techniques [6]. In this comparison, we consider eight models, including our model, as shown in Table 9. In the table, GBM (Gradient Boosting Machine) is a type of ensemble learning technique that implements the sequential boosting algorithm. A grid search can be used to find optimal hyper-parameter values for the SVR/GBM [25]. SNN (Shallow Neural Network) has three layers of input, hidden, and output, and it was found that the optimal number of the hidden nodes is nine for all clusters.

Tables 9–11 compare the prediction performance in terms of MAPE, RMSE, and MAE, respectively. From the tables, the predicted results with the best accuracy are marked in bold and we observe that our hybrid model exhibits a superb performance in all categories. Figure 7 shows more detail of the MAPE distribution for each cluster using a box plot. We can deduce that our hybrid model has fewer outliers and a smaller maximum error. In addition, the error rate increases in the case of long holidays in Korea. For instance, during the 10-day holiday in October 2017, the error rate increased significantly. Another cause of high error rates is due to outliers or missing values because of diverse reasons, such as malfunction and surge. Figure 8 compares the daily load forecasts of our hybrid model and actual daily usage on a quarterly basis. Overall, our hybrid model showed a good performance in predictions, regardless of diverse external factors such as long holidays.

Table 9. MAPE distribution for each forecasting model.

Forecasting Model	Cluster #		
	Cluster A	Cluster B	Cluster C
MR	7.852	8.971	4.445
DT	6.536	8.683	6.004
GBM	4.831	6.896	3.920
SVR	4.071	5.761	3.135
SNN	4.054	5.948	3.181
MLP	3.961	4.872	3.139
RF	4.185	5.641	3.216
RF+MLP	**3.798**	**4.674**	**2.946**

Table 10. RMSE comparison for each forecasting model.

Forecasting Model	Cluster #		
	Cluster A	Cluster B	Cluster C
MR	5725.064	6847.179	1757.463
DT	6118.835	7475.188	2351.676
GBM	4162.359	5759.276	1495.751
SVR	3401.812	5702.405	1220.052
SNN	3456.156	4903.587	1236.606
MLP	3381.697	4064.559	1170.824
RF	4111.245	4675.762	1450.436
RF + MLP	**3353.639**	**3894.495**	**1143.297**

Nevertheless, we can see that there are several time periods when forecasting errors are high. For instance, from 2013 to 2016, Cluster B showed a steady increase in its power consumption due to building remodeling and construction. Even though the remodeling and construction are finished at the beginning of 2017, the input variable for estimating the year-ahead consumption is still reflecting such an increase. This was eventually adjusted properly for the third and fourth quarters by the time series cross-validation. On the other hand, during the remodeling, the old heating, ventilation, and air conditioning (HVAC) system was replaced by a much more efficient one and the new system started its operation in December 2017. Even though our hybrid model predicted much higher power consumption for the cold weather in the third week, the actual power consumption was quite low due to the new HVAC system. Lastly, Cluster A showed a high forecasting error on 29 November 2017. It turned out that at that time, there were several missing values in the actual power consumption. This kind of problem can be detected by using the outlier detection technique.

Table 11. MAE comparison for each forecasting model.

Forecasting Model	Cluster #		
	Cluster A	Cluster B	Cluster C
MR	4155.572	4888.821	1262.985
DT	3897.741	5054.069	1708.709
GBM	2764.128	3916.945	1122.530
SVR	2236.318	3956.907	898.963
SNN	2319.696	3469.775	919.014
MLP	2255.537	2795.246	910.351
RF	2708.848	3235.855	1063.731
RF+MLP	**2208.072**	**2742.543**	**860.989**

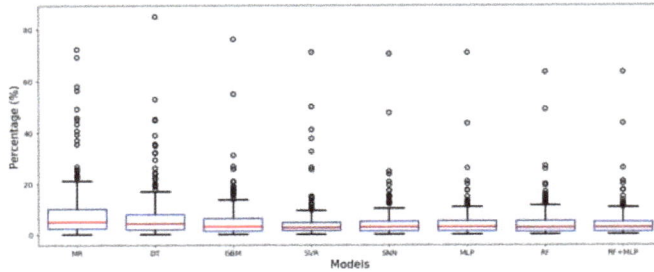

(a) Cluster A

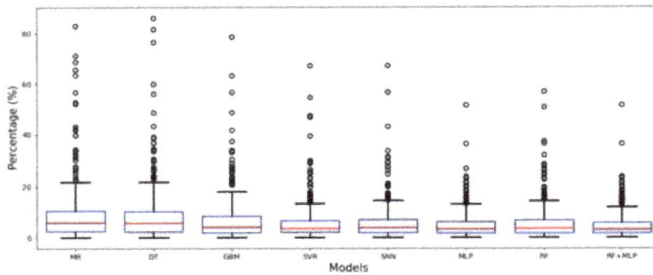

(b) Cluster B

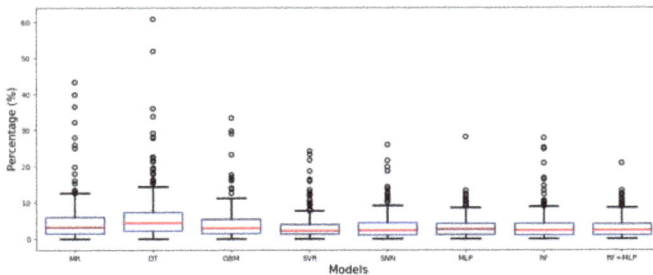

(c) Cluster C

Figure 7. Distribution of each model by MAPE.

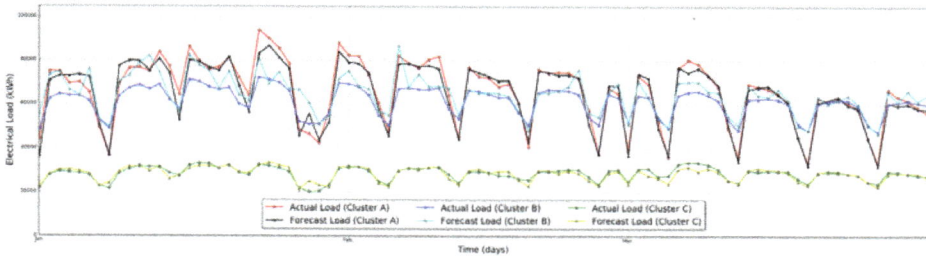

(**a**) First quarter (1 January–31 March) in 2017

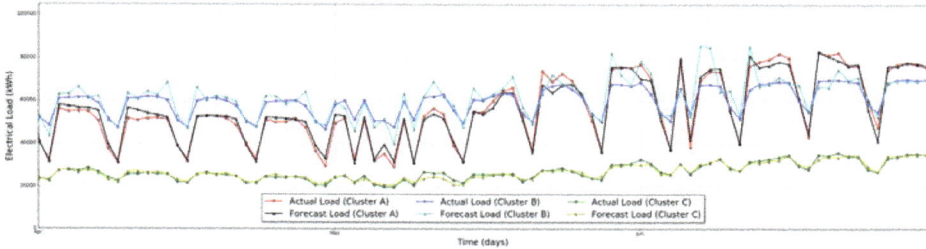

(**b**) Second quarter (1 April–30 June) in 2017

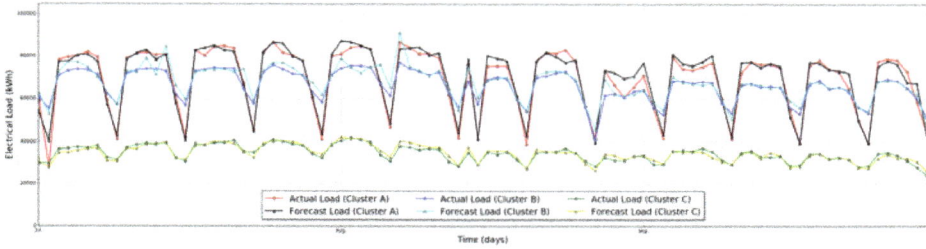

(**c**) Third quarter (1 July–30 September) in 2017

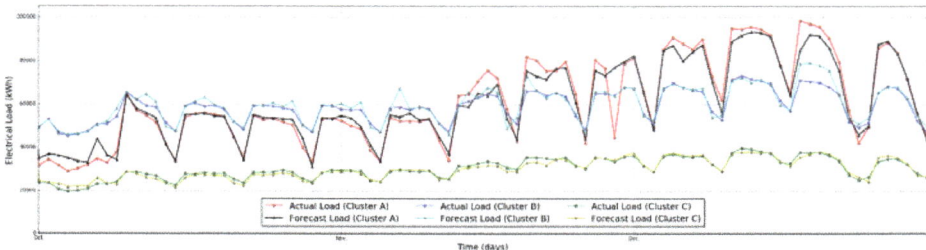

(**d**) Fourth quarter (1 October–31 December) in 2017

Figure 8. Daily electrical load forecasting for university campus.

6. Conclusions

In this paper, we proposed a hybrid model for short-term load forecasting for higher educational institutions, such as universities, using random forest and multilayer perceptron. To construct our forecast model, we first grouped university buildings into an academic cluster, science/engineering cluster, and dormitory cluster, and collected their daily electrical load data over six years. We divided the collected data into a training set, a validation set, and a test set. For the training set, we classified electrical load data by pattern similarity using the decision tree technique. We considered various configurations for random forest and multilayer perceptron and evaluated their prediction performance by using the validation set to select the optimal model. Based on this work, we constructed our hybrid daily electrical load forecast model by selecting models with a better predictive performance in similar time series. Finally, using the test set, we compared the daily electrical load prediction performance of our hybrid model and other popular models. The comparison results show that our hybrid model outperforms other popular models. In conclusion, we showed that LSTM networks are effective for reflecting an electrical load depending on the day of the week and the decision tree is effective in classifying time series data by similarity. Moreover, using these two forecasting models in a hybrid model can complement their weaknesses.

In order to improve the accuracy of electrical load prediction, we plan to use a supervised learning method reflecting various statistically significant data. Also, we will analyze the prediction performance in different look-ahead points (from the next day to a week) using probabilistic forecasting.

Author Contributions: J.M. designed the algorithm, performed the simulations, and prepared the manuscript as the first author. Y.K. analyzed the data and visualized the experimental results. M.S. collected the data, and developed and wrote the load forecasting based on the LSTM networks part. E.H. conceived and supervised the work. All authors discussed the simulation results and approved the publication.

Funding: This research was supported by Korea Electric Power Corporation (Grant number: R18XA05) and the Brain Korea 21 Plus Project in 2018.

Conflicts of Interest: The authors declare no conflict of interest.

References

1. Lindley, D. Smart grids: The energy storage problem. *Nat. News* **2010**, *463*, 18–20. [CrossRef] [PubMed]
2. Erol-Kantarci, M.; Mouftah, H.T. Energy-efficient information and communication infrastructures in the smart grid: A survey on interactions and open issues. *IEEE Commun. Surv. Tutor.* **2015**, *17*, 179–197. [CrossRef]
3. Raza, M.Q.; Khosravi, A. A review on artificial intelligence based load demand forecasting techniques for smart grid and buildings. *Renew. Sustain. Energy Rev.* **2015**, *50*, 1352–1372. [CrossRef]
4. Hernandez, L.; Baladron, C.; Aguiar, J.M.; Carro, B.; Sanchez-Esguevillas, A.J.; Lloret, J.; Massana, J. A survey on electric power demand forecasting: Future trends in smart grids, microgrids and smart buildings. *IEEE Commun. Surv. Tutor.* **2014**, *16*, 1460–1495. [CrossRef]
5. Kuo, P.-H.; Huang, C.-J. A High Precision Artificial Neural Networks Model for Short-Term Energy Load Forecasting. *Energies* **2018**, *11*, 213. [CrossRef]
6. Ahmad, A.; Hassan, M.; Abdullah, M.; Rahman, H.; Hussin, F.; Abdullah, H.; Saidur, R. A review on applications of ANN and SVM for building electrical energy consumption forecasting. *Renew. Sustain. Energy Rev.* **2014**, *33*, 102–109. [CrossRef]
7. Vrablecová, P.; Ezzeddine, A.B.; Rozinajová, V.; Šárik, S.; Sangaiah, A.K. Smart grid load forecasting using online support vector regression. *Comput. Electr. Eng.* **2017**, *65*, 102–117. [CrossRef]
8. Hong, T.; Fan, S. Probabilistic electric load forecasting: A tutorial review. *Int. J. Forecast.* **2016**, *32*, 914–938. [CrossRef]
9. Hahn, H.; Meyer-Nieberg, S.; Pickl, S. Electric load forecasting methods: Tools for decision making. *Eur. J. Oper. Res.* **2009**, *199*, 902–907. [CrossRef]
10. Moon, J.; Park, J.; Hwang, E.; Jun, S. Forecasting power consumption for higher educational institutions based on machine learning. *J. Supercomput.* **2018**, *74*, 3778–3800. [CrossRef]
11. Chung, M.H.; Rhee, E.K. Potential opportunities for energy conservation in existing buildings on university campus: A field survey in Korea. *Energy Build.* **2014**, *78*, 176–182. [CrossRef]
12. Moon, J.; Kim, K.-H.; Kim, Y.; Hwang, E. A Short-Term Electric Load Forecasting Scheme Using 2-Stage Predictive Analytics. In Proceedings of the IEEE International Conference on Big Data and Smart Computing (BigComp), Shanghai, China, 15–17 January 2018; pp. 219–226.

13. Palchak, D.; Suryanarayanan, S.; Zimmerle, D. An Artificial Neural Network in Short-Term Electrical Load Forecasting of a University Campus: A Case Study. *J. Energy Resour. Technol.* **2013**, *135*, 032001. [CrossRef]

14. Wang, Z.; Srinivasan, R.S. A review of artificial intelligence based building energy use prediction: Contrasting the capabilities of single and ensemble prediction models. *Renew. Sustain. Energy Rev.* **2016**, *75*, 796–808. [CrossRef]

15. Hippert, H.S.; Pedreira, C.E.; Souza, R.C. Neural networks for short-term load forecasting: A review and evaluation. *IEEE Trans. Power Syst.* **2001**, *16*, 44–55. [CrossRef]

16. Lahouar, A.; Slama, J.B.H. Day-ahead load forecast using random forest and expert input selection. *Energy Convers. Manag.* **2015**, *103*, 1040–1051. [CrossRef]

17. Ahmad, M.W.; Mourshed, M.; Rezgui, Y. Trees vs Neurons: Comparison between random forest and ANN for high-resolution prediction of building energy consumption. *Energy Build.* **2017**, *147*, 77–89. [CrossRef]

18. Xiao, L.; Wang, J.; Dong, Y.; Wu, J. Combined forecasting models for wind energy forecasting: A case study in China. *Renew. Sustain. Energy Rev.* **2015**, *44*, 271–288. [CrossRef]

19. Pinson, P.; Kariniotakis, G. Conditional prediction intervals of wind power generation. *IEEE Trans. Power Syst.* **2010**, *25*, 1845–1856. [CrossRef]

20. Abdoos, A.; Hemmati, M.; Abdoos, A.A. Short term load forecasting using a hybrid intelligent method. *Knowl. Based Syst.* **2015**, *76*, 139–147. [CrossRef]

21. Dong, J.-r.; Zheng, C.-y.; Kan, G.-y.; Zhao, M.; Wen, J.; Yu, J. Applying the ensemble artificial neural network-based hybrid data-driven model to daily total load forecasting. *Neural Comput. Appl.* **2015**, *26*, 603–611. [CrossRef]

22. Lee, W.-J.; Hong, J. A hybrid dynamic and fuzzy time series model for mid-term power load forecasting. *Int. J. Electr. Power. Energy Syst.* **2015**, *64*, 1057–1062. [CrossRef]

23. Zhang, T.; Wang, J. K-nearest neighbors and a kernel density estimator for GEFCom2014 probabilistic wind power forecasting. *Int. J. Forecast.* **2016**, *32*, 1074–1080. [CrossRef]

24. Jurado, S.; Nebot, À.; Mugica, F.; Avellana, N. Hybrid methodologies for electricity load forecasting: Entropy-based feature selection with machine learning and soft computing techniques. *Energy* **2015**, *86*, 276–291. [CrossRef]

25. Feng, C.; Cui, M.; Hodge, B.-M.; Zhang, J. A data-driven multi-model methodology with deep feature selection for short-term wind forecasting. *Appl. Energy* **2017**, *190*, 1245–1257. [CrossRef]

26. Tso, G.K.; Yau, K.K. Predicting electricity energy consumption: A comparison of regression analysis, decision tree and neural networks. *Energy* **2007**, *32*, 1761–1768. [CrossRef]

27. Jain, R.K.; Smith, K.M.; Culligan, P.J.; Taylor, J.E. Forecasting energy consumption of multi-family residential buildings using support vector regression: Investigating the impact of temporal and spatial monitoring granularity on performance accuracy. *Appl. Energy* **2014**, *123*, 168–178. [CrossRef]

28. Grolinger, K.; L'Heureux, A.; Capretz, M.A.; Seewald, L. Energy forecasting for event venues: Big data and prediction accuracy. *Energy Build.* **2016**, *112*, 222–233. [CrossRef]

29. Amber, K.; Aslam, M.; Hussain, S. Electricity consumption forecasting models for administration buildings of the UK higher education sector. *Energy Build.* **2015**, *90*, 127–136. [CrossRef]

30. Rodrigues, F.; Cardeira, C.; Calado, J.M.F. The daily and hourly energy consumption and load forecasting using artificial neural network method: A case study using a set of 93 households in Portugal. *Energy* **2014**, *62*, 220–229. [CrossRef]

31. Efendi, R.; Ismail, Z.; Deris, M.M. A new linguistic out-sample approach of fuzzy time series for daily forecasting of Malaysian electricity load demand. *Appl. Soft Comput.* **2015**, *28*, 422–430. [CrossRef]

32. ISO Week Date. Available online: https://en.wikipedia.org/wiki/ISO_week_date (accessed on 19 October 2018).

33. Holidays and Observances in South Korea in 2017. Available online: https://www.timeanddate.com/holidays/south-korea/ (accessed on 28 April 2018).

34. Climate of Seoul. Available online: https://en.wikipedia.org/wiki/Climate_of_Seoul (accessed on 28 April 2018).

35. Son, S.-Y.; Lee, S.-H.; Chung, K.; Lim, J.S. Feature selection for daily peak load forecasting using a neuro-fuzzy system. *Multimed. Tools Appl.* **2015**, *74*, 2321–2336. [CrossRef]

36. Kong, W.; Dong, Z.Y.; Jia, Y.; Hill, D.J.; Xu, Y.; Zhang, Y. Short-term residential load forecasting based on LSTM recurrent neural network. *IEEE Trans. Smart Grid* **2017**. [CrossRef]

37. Kanai, S.; Fujiwara, Y.; Iwamura, S. Preventing Gradient Explosions in Gated Recurrent Units. In Proceedings of the Neural Information Processing Systems, Long Beach, CA, USA, 4–9 December 2017; pp. 435–444.

38. Cho, K.; van Merriënboer, B.; Gulcehre, C.; Bahdanau, D.; Bougares, F.; Schwenk, H.; Bengio, Y. Learning phrase representations using RNN encoder-decoder for statistical machine translation. *arXiv* **2014**, arXiv:1406.1078.

39. Hochreiter, S.; Schmidhuber, J. Long short-term memory. *Neural Comput.* **1997**, *9*, 1735–1780. [CrossRef] [PubMed]

40. Rutkowski, L.; Jaworski, M.; Pietruczuk, L.; Duda, P. The CART decision tree for mining data streams. *Inf. Sci.* **2014**, *266*, 1–15. [CrossRef]

41. Breiman, L. Random forests. *Mach. Learn.* **2001**, *45*, 5–32. [CrossRef]

42. Oshiro, T.M.; Perez, P.S.; Baranauskas, J.A. How many trees in a random forest? In Proceedings of the International Conference on Machine Learning and Data Mining in Pattern Recognition, Berlin, Germany, 13–20 July 2012; pp. 154–168.

43. Díaz-Uriarte, R.; De Andres, S.A. Gene selection and classification of microarray data using random forest. *BMC Bioinform.* **2006**, *7*, 3. [CrossRef] [PubMed]

44. Suliman, A.; Zhang, Y. A Review on Back-Propagation Neural Networks in the Application of Remote Sensing Image Classification. *J. Earth Sci. Eng. (JEASE)* **2015**, *5*, 52–65. [CrossRef]

45. Bengio, Y. Learning deep architectures for AI. *Found. Trends® Mach. Learn.* **2009**, *2*, 1–127. [CrossRef]

46. Clevert, D.-A.; Unterthiner, T.; Hochreiter, S. Fast and accurate deep network learning by exponential linear units (elus). *arXiv* **2015**, arXiv:1511.07289.

47. Sheela, K.G.; Deepa, S.N. Review on methods to fix number of hidden neurons in neural networks. *Math. Probl. Eng.* **2013**, 425740. [CrossRef]

48. Xu, S.; Chen, L. A novel approach for determining the optimal number of hidden layer neurons for FNN's and its application in data mining. In Proceedings of the 5th International Conference on Information Technology and Application (ICITA), Cairns, Australia, 23–26 June 2008; pp. 683–686.

49. Hyndman, R.J.; Athanasopoulos, G. *Forecasting: Principles and Practice*; Otexts: Melbourne, Australia, 2014; ISBN 0987507117.

50. Pedregosa, F.; Varoquaux, G.; Gramfort, A.; Michel, V.; Thirion, B.; Grisel, O.; Blondel, M.; Prettenhofer, P.; Weiss, R.; Dubourg, V. Scikit-learn: Machine learning in Python. *J. Mach. Learn. Res.* **2011**, *12*, 2825–2830.

51. Abadi, M.; Barham, P.; Chen, J.; Chen, Z.; Davis, A.; Dean, J.; Devin, M.; Ghemawat, S.; Irving, G.; Isard, M. TensorFlow: A System for Large-Scale Machine Learning. In Proceedings of the 12th USENIX Symposium on Operating Systems Design and Implementation (OSDI '16), Savannah, GA, USA, 2–4 November 2016; pp. 265–283.

52. Ketkar, N. Introduction to Keras. In *Deep Learning with Python*; Apress: Berkeley, CA, USA, 2017; pp. 97–111.

energies

MDPI

Article

Empirical Comparison of Neural Network and Auto-Regressive Models in Short-Term Load Forecasting

Miguel López *, Carlos Sans, Sergio Valero and Carolina Senabre

Department of Mechanic Engineering and Energy, Universidad Miguel Hernández, 03202 Elx, Alacant, Spain; carsantr@gmail.com (C.S.); svalero@umh.es (S.V.); csenabre@umh.es (C.S.)
* Correspondence: m.lopezg@umh.es; Tel.: +34-965-22-2407

Received: 26 June 2018; Accepted: 1 August 2018; Published: 10 August 2018

Abstract: Artificial Intelligence (AI) has been widely used in Short-Term Load Forecasting (STLF) in the last 20 years and it has partly displaced older time-series and statistical methods to a second row. However, the STLF problem is very particular and specific to each case and, while there are many papers about AI applications, there is little research determining which features of an STLF system is better suited for a specific data set. In many occasions both classical and modern methods coexist, providing combined forecasts that outperform the individual ones. This paper presents a thorough empirical comparison between Neural Networks (NN) and Autoregressive (AR) models as forecasting engines. The objective of this paper is to determine the circumstances under which each model shows a better performance. It analyzes one of the models currently in use at the National Transport System Operator in Spain, Red Eléctrica de España (REE), which combines both techniques. The parameters that are tested are the availability of historical data, the treatment of exogenous variables, the training frequency and the configuration of the model. The performance of each model is measured as RMSE over a one-year period and analyzed under several factors like special days or extreme temperatures. The AR model has 0.13% lower error than the NN under ideal conditions. However, the NN model performs more accurately under certain stress situations.

Keywords: short-term load forecasting (STLF); neural networks; artificial intelligence (AI)

1. Introduction

The development of Short-Term Load Forecasting (STLF) tools has been a common topic in the late years [1–3]. STLF is defined as forecasting from 1 h to several days ahead, and it is usually done hourly or half-hourly. The application of STLF include transport and system operators that need to ensure reliability and efficiency of the system and networks and producers that require to establish schedules and utilization of their power facilities. In addition, STLF is required for the optimization of market bidding for both buyers and sellers in the market. The ability to foresee the electric demand will reduce the costs of deviations from the committed offers. These aspects have been especially relevant in the last decade in which the deregulation of the Spanish market following European directives has been enforced. In addition, the increasing availability of renewable energy sources, makes the balancing of the system more unstable as it adds more uncertainty on the producing end. All of these reasons make STLF a critical aspect to ensure reliability and efficiency of the power system.

Forecasting models use several techniques that can be grouped in Statistical, Artificial Intelligence and Hybrid techniques. Statistical methods require a mathematical model that provide the relationship between load and other input factors. These methods were the first ones used and are still currently relevant. They include multiple linear regression models [4–6], time-series [7–10] and exponential smoothing techniques [11]. Pattern recognition is a key aspect of load forecasting.

Determining the daily, weekly and seasonal patterns of consumers is at the root of the load-forecasting problem. Pattern recognition techniques stem from the context of computer vision and from there, they have evolved to applications in all fields of engineering in forms of different types of Artificial Intelligence. These techniques (AI) have gained attention over the last 20 years. AI offers a variety of techniques that generally require the selection of certain aspects of their topology but they are able to model non-linear behavior from observing past instances. The term refers to methods that employ Artificial Neural Networks [12–16], Fuzzy Logic [13,15,17–20], Support Vector Machines [21] or Evolutionary Algorithms [15,17,22–24]. Hybrid models are those that combine the use of two or more techniques in the forecasting process. These are some examples that include some of the already mentioned [15,23,25–27]. Other application of pattern recognition and AI techniques to STLF include smaller scale systems, which present their own specificities [28,29].

The previous paragraph focused solely on the forecasting engine used to calculate the actual forecast, as this part usually receives the most attention. However, it is not the only key aspect of the forecasting problem. In [30], it is proposed a standard that includes 5 stages that need to be properly addressed in order to obtain accurate forecasts:

- Data Pre-processing: Data normalizing, filtering of outliers and decomposition of signals by transforms. This last aspect has received significant attention recently [23,24,31,32].
- Input Selection: Analysis of the available information and of how the forecasting engines process this information best. In [33], an example of how to determine which variable should be included is shown. The information about special days is also included in this stage, relevant attempts to determine the best way to convert type of day information to valid input to the forecasting engine are found in [18–20,34–36].
- Time Frame Selection: Refers to determining which period should be used for training. In [16], a time scheme including similar days is proposed. In this paper, this issue will be addressed by determining how the availability of historical data affects the accuracy of forecasts carried out by different forecasting engines.
- Load Forecasting: Refers to the forecasting engine.
- Data Post-Processing: De-normalizing, re-composition, etc.

To sum up, it is also relevant to mention examples of real world applications [37–39]. The publishing of models that are validated through actual use by the industry instead of through lab conditions is especially important for the advancement of the field [2].

The referred examples contain descriptions of particular forecasting models that are usually described by defining their input and the inner workings, topology, configuration and other characteristics of the forecasting engine. They also include the results of the model when it has been tested on a specific database and for a certain period of time. This methodology has provided a wide variety of models for the industry and scientific community to choose from for any particular application. However, it has provided very little information on how to compare each method and how to determine the strong and weak suits of each technique. The lack of analysis of the characteristics of the database, and in some cases the use of testing periods shorter than a full year, makes it very difficult for the reader to *a priori* determine which of the proposed models would suit best their own personal case.

This issue has been treated in [40,41], in which the authors propose a certain methodology to adopt different techniques depending on the forecasting problem. These papers include an analysis of the load prior to the actual forecasting process. However, they only test one technique for the forecasting engine. In [42] the issue of predictability of databases is addressed to provide a benchmark indicator that could provide a fair comparison among results of different models on different databases that may or may not be similarly affected by the same factors (temperature, social activities . . .). This type of information along with the standardization proposed in [30] would be useful to determine the characteristics of a specific problem and the features of each model available that best addresses the subject at hand.

Consequently, there is consensus that a general solution does not exist and that the STLF problem does not have a "one-size-fits-all" fix. Nevertheless, the objective of this paper is to provide a comparison between two of the most common forecasting engines: the autoregressive model (AR) and the Neural Network (NN). The goal is to determine how a given set of conditions and configuration parameters affect the accuracy of each technique (AR and NN) and use this information to define their strong and weak points.

The methodology aims to determine the circumstances under which each of the forecasting engines performs more accurately. The conditions of the forecasts: historical data available, sources of temperature information, computational burden, maintenance needed . . . are modified to determine how each of them affects each technique. In addition, the performance results are analyzed in terms of type of days (cold, hot, special days) in order to better assess whether one of the forecasting engines performs better on a certain type of day.

This paper provides results from a real application using two different techniques under the same set of conditions. These results are classified by the type of day to facilitate the analysis. The obtained results provide proof that NN models are more reliable when meteorological information is scarce (only few locations are available) or when it is not properly pre-processed. Nevertheless, the NN requires a larger historical database to match the accuracy of the AR model. The overall results show that each technique is better suited for specific types of days, but more importantly, that there are conditions under which one technique clearly outperforms the other.

Section 2 contains the description of the forecasting engines that are compared, the parameters and conditions under which the forecasting engines are tested and the categorization of type of days used to compare the results. On Section 3, the characteristics of the data used are explained: characteristics of the load, meteorological variables and their treatment and information to determine the type of day. Section 4 includes the results obtained on the tests: a revision of each parameter and how its variation affects the performance to both forecasting engines. Finally, Section 5 includes a brief conclusion that summarizes the most relevant aspects of the results.

2. Methodology

This section provides a detailed description of the analyzed forecasting techniques, the conditions under which they are tested and the classification of the results used to draw conclusions.

2.1. Forecasting Models

Both forecasting models under analysis are extracted from the STLF system currently working at Red Eléctrica de España (REE), the Transport System Operator in Spain. They have been thoroughly described in [39], and have been running on the REE headquarters for over two years now. Both forecasting engines use the same data filtering system to discard outliers, usually caused by malfunctioning of the data acquisition systems. The forecasting scheme provides a forecast every hour that contains the forecasted hourly profile for the current day and the next nine days. Internally, each hour is forecasted separately by different sub-models. Therefore, each full model includes 24 sub-models to forecast the load profile of a full day, and different submodels are used depending on how distant in the future the forecasted day is.

To simplify the comparison, the metric that will be used as reference is the error of the forecast made at 9 a.m. for the full 24 h of the next day. This forecast is the most relevant for REE as it is the one that serves as a base for operation and planning.

The input for any of the submodels is a vector that contains the latest load information available, temperature forecasts, annual trends and calendar information. This data will be further discussed on the next section, but it is the same for both techniques AR and NN that are now explained.

2.1.1. Auto-Regressive Model

The auto-regressive model is actually an auto-regressive model with errors that includes exogenous variables. Regression models with time series errors describe the behavior of a series by accounting for linear effects of exogenous variables. However, the errors are not considered white noise but a time series. This type of model is described in Equation (1).

$$y_t = \sum_{i=1}^{p} \varphi_i \cdot e_{t-i} + X_t \cdot \theta + \varepsilon_t \tag{1}$$

where, the output y_t is expressed as a linear combination of previous known errors, e_{t-i}, exogenous variables X_t and a random shock, ε_t. The coefficients φ_i and vector θ are calculated from the training data by a maximum likelihood method. The parameter p expresses the number of lags of the error that are included in the model.

2.1.2. Neural Network

The Neural Network model uses a non-linear auto-regressive system with exogenous input. mathematically expressed in Equation (2):

$$y_t = f\left(y_{t-1}, \ldots, y_{t-n_y}, u_{t-1}, \ldots, u_{t-n_u}\right) \tag{2}$$

where, the output value y_t is a non-linear function of n_y previous outputs and n_u inputs. This non-linear function is, in our case, a feedforward neural network. Further description of this model can be found in [39]. Figure 1 shows a visualization of this type of networks working online. The figure shows a feedforward neural network with 119 exogenous inputs and a feedback of 14 previous values, 10 neurons in the hidden layer and 1 output.

The random nature of the training process of the NARX systems requires certain redundancy to estabilize the output. This is achieved by using a number NN in parallel. Also, the ability of the NN to capture non-linear behavior depends on the size of the hidden layer. Both of these parameters affect the computational burden imposed on the system, which is one of the conditions under which the models are tested.

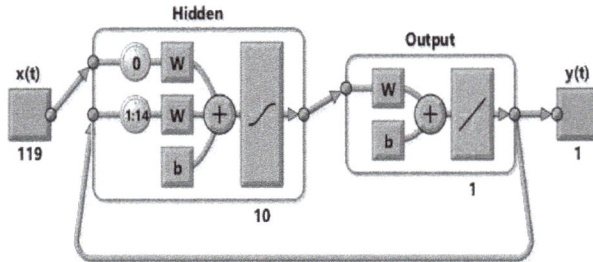

Figure 1. Schematic view of the NARX system as shown on a Matlab Mathworks visualization.

2.2. Parameters and Forecasting Conditions

The forecasting engines described above have been tested with different configuration parameters and external conditions to determine how they adapt to different situations. External conditions are historical load availability, temperature locations availability and response timeliness, which is related to computational burden. Configuration parameters are temperature treatment, frequency of training and number of auto-regressive lags.

2.2.1. Historical Load Availability

The most important input of a load-forecasting model is its past behavior. A persistent model that only takes into account previous values may provide, in some cases, a valid baseline to start developing a more complex one. However, in many situations, and especially in industry applications, the availability of such historical data is not as deep as desired and it is restricted due to the quantity or the quality of the stored data. In some cases, the data acquisition system has not been running long enough, or a change in its configuration may cause old data to be obsolete.

The question of how old the data that we use in our forecasting system should be is a valid one. The inclusion of data from too far back may cause the model to learn obsolete behavior that has changed over the years and that is not currently accurate: the increment of air conditioning systems may increase the sensitivity of load to temperature increase while the use of more efficient lighting may decrease the load in after-sunset hours. On the other hand, there are certain phenomena like extreme temperatures or special days that do not happen for long periods of time and, therefore, if the database is not deep enough, it may not have enough examples to shape this type of behavior.

Our research proposes using data from the last 3, 5 and 7 years to train both models. The goal of these experiments is to determine which one of them requires a deeper database, or which one can benefit the most from such data availability. The data will be broken down into separate types of days in order to determine which category is affected by this condition.

2.2.2. Temperature Locations Availability

Temperature is the most important exogenous factor for load forecasting of regular days as both extremes of the temperature range increase electricity consumption. Load forecasting of small areas in which temperature is homogenous may require only one series of temperature data to learn the area's behaviors that are related to temperature. However, if the region is larger and the weather presents higher variability, it is necessary to determine which locations provide a relevant temperature series that could model the local area's behavior related to weather. Needless to say, not all local areas will be equally affected by temperature and the relevance of each area within the overall load for the region will vary depending on the lower or higher electricity capacity of each area. The electricity capacity normally relates to the area's gross product.

In our case, Spain is a large country with a wide weather diversity. In addition, the population distribution also causes a high variability of power consumption among areas. According to this, the model used at REE includes data from five locations that represent the five weather regions: North-Atlantic (Bilbao), Mediterranean (Barcelona), Upper-Center (Zaragoza), Lower-Center (Madrid) and South (Sevilla). These cities, shown in Figure 2 are the most power demanding areas in each weather region.

Figure 2. Location of the five temperature series and distribution of the weather regions in Spain.

The lack of availability of all temperature series affects the accuracy of the system. Both models have been tested by including only one series of data and then adding the rest one at a time. This experiment allows to determine which model can perform best under scarce information and which can benefit the most from a richer dataset.

2.2.3. Temperature Treatment

As it was aforementioned, temperature has a non-linear relation with electricity consumption, as both high and low temperature causes an increase in demand. To illustrate this, Figure 3 shows a plot of the average load on regular days at 18 h against the average temperature of the day. Therefore, in order for the forecasting engine to capture such behavior, it may require a pre-processing of the data.

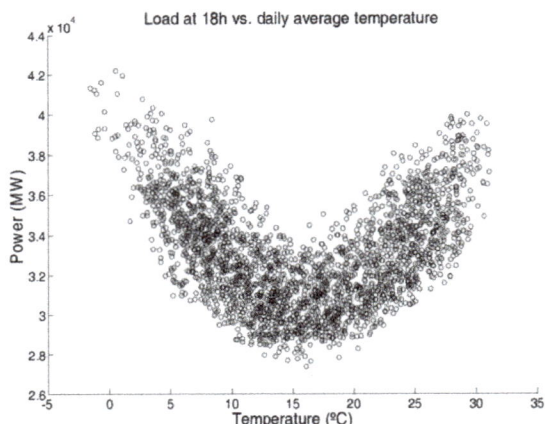

Figure 3. Scatter plot of national load at 18 h against the daily average temperature in Madrid.

One common approach to this is using a technique called Heating Degree Days (HDD) and Cooling Degree Days (CDD). This technique linearizes the temperature load relation by defining threshold for high and low temperatures and splitting the series into one that accounts for cold days and another that does for hot days. The CDD and HDD series are described in Equations (3) and (4) and they are further discussed in [34].

$$CDD_d = \begin{cases} T_{med,d} - TH_{hot}, & \text{if } T_{med} > TH_{hot} \\ 0, & \text{otherwise} \end{cases} \tag{3}$$

$$HDD_d = \begin{cases} TH_{cold} - T_{med,d}, & \text{if } T_{med} < TH_{cold} \\ 0, & \text{otherwise} \end{cases} \tag{4}$$

where $T_{med,d}$ is the average temperature of day d, TH_{hot} and TH_{cold} are the thresholds for hot and cold days and CDD_d and HDD_d are the values of each series for day d.

This technique requires the thresholds to be properly tuned to each location's effect on the load. This optimization process is described in [39] and the optimal threshold for each zone has been calculated. However, the robustness of each model against the variation of these values has been tested by introducing variations of up to 12 degrees on each threshold.

2.2.4. Neural Network Size, Redundancy and Computational Burden

According to the selected topology shown in Figure 1, part of the configuration of the network is the selection of the number of neurons in the hidden layer. The complexity of the network is related to this parameter, as it is associated to its ability to model non-linear behaviors. A network with a low number of neurons in its hidden layer would fail to learn complex, non-linear relations between input and output. On the other hand, the number of neurons increases the computational burden of the training and forecasting process and, therefore it should be minimized if the system is working online and has a response time limit.

In addition, the neural network training algorithm relies on a random initialization of the neurons' weights. The randomness causes the network's output to contain a random component. In order to minimize the effect of this randomness, the working model includes a redundant design. Each network is replicated n times to obtain n different outputs for each forecast. The final output is obtained then by discarding the lowest and highest values and averaging the rest. Increasing the number of replicas costs a linear increase of computational burden while it reduces the randomness of the output and reduces the variability of the output, minimizing the maximum error of a forecasted period.

The response time of the system is a critical feature. If the forecast is not produced on time, then the whole effort could be useless. In order to test how the limit of time response affects the models the number of neurons is set from 3 to 20 and the number of redundant networks from 3 to 25. As the neural network model is the one with higher computational burden it is the only one affected by this limitation.

2.2.5. Frequency of Training

As it will be further discussed in Section 3, the load series evolve over time due to changes in factors like economic growth or shifts in consumer behaviors. This causes forecasting models to become obsolete if the data used during training no longer follows the current trends. Therefore, in order to keep up with load shifting behavior, forecasting models need to be frequently retrained with new data.

The training process may have heavy computational requirements that make it unpractical to increase frequency needlessly. Therefore, the period in between trainings is a factor that may alter the accuracy of the model.

In this research, both AR and NN models have been tested with training frequencies of 3, 6, 12 and 24 months. In each of these tests, all sub-models were retrained using the most recent data. In accordance with this, for frequencies higher than 12 months, the simulation period of one year was split into separate blocks as the Table 1 shows.

To evaluate the results, all blocks from each frequency are added together into a single one-year period and the corresponding Root Mean Square Error (RMSE) is calculated for both AR and NN models.

Table 1. Training and simulation periods used for testing the effect of training frequency.

Frequency (Months)	Block	Training Period		Simulation Period	
		Start	End	Start	End
3	1	1 January 2010	31 December 2016	1 January 2017	31 March 2017
	2	1 April 2010	31 March 2017	1 April 2017	30 June 2017
	3	1 July 2010	30 June 2017	1 July 2017	30 September 2017
	4	1 October 2010	30 September 2017	1 October 2017	31 December 2017
6	1	1 January 2010	31 December 2016	1 January 2017	30 June 2017
	2	1 July 2010	30 June 2017	1 July 2017	31 December 2017
12	1	1 January 2010	31 December 2016	1 January 2017	31 December 2017
24	1	1 January 2009	31 December 2015	1 January 2017	31 December 2017

2.2.6. Number of Auto-Regressive Lags

As it was aforementioned, both models present an auto-regressive component. This part of the model introduces the previous values as a feedback in order to enable to forecasting engine to reduce errors due to unaccounted factors that are persistent in time.

The key parameter to configure this aspect of the models is the number of lags, which represents how many previous values are fed back into the model. Intuitively, the most recent values carry the most information while the further back in time that we reach, the less relevant the data become. In addition, the AR model uses a linear relation to capture the lagged results while the NN model allows non-linearity. Therefore, it is possible that one model is able to use a different amount of lags than the other.

The auto-regressive order of each model has been tested from 0 to 25. The load series is highly self-correlated on lags multiple of seven due to the weekly patterns, as it is shown in Figure 4. Therefore, lags around 7, 14 and 21 were explored. Auto-correlation measures the correlation between y_t and y_{t+k}, and its calculation is described in [43].

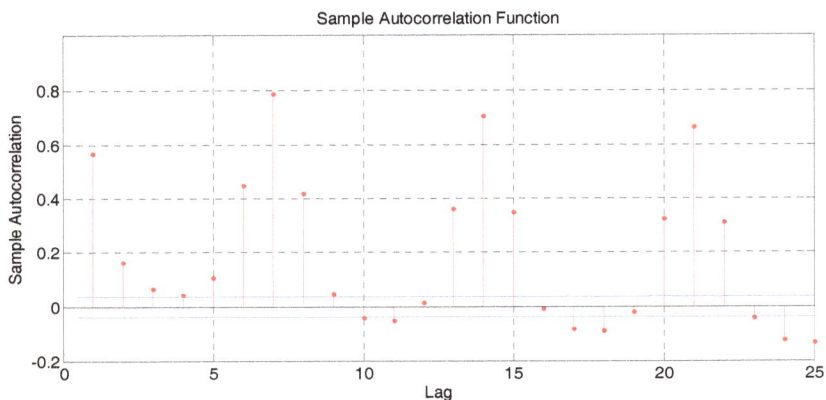

Figure 4. Sample autocorrelation function for National load at 18 h.

It is worth mentioning that the objective of this paper is not to provide or suggest analytical or statistical methods to determine the order of auto-regressive models like [44,45] but to offer a comparison between AR and NN based models to understand the effect that the auto-regressive order has on the forecasting accuracy.

2.3. *Types of Days*

Each of the proposed parameters and conditions under which the forecasting models are tested will cause the forecasting accuracy to change over the whole one-year simulating period. This variation, however, may affect some type of days more than other and, therefore, it may seem irrelevant when it is averaged over the whole testing period. In order to avoid this error, it is important to dissect the results and analyze the accuracy of the models on different categories of days to determine which conditions affect which type of days and how they do it.

There are two aspects to classify the days: social character and temperature. The first one considers days as special if they are a holiday, are in between two holidays or weekend, or are affected by Daylight Saving Time or the vacational periods at Christmas or Easter. A more detailed description of the days considered special is found in Section 3.

Temperature is used to classify days as hot and cold. For each category, the top 20 and bottom 20 days from the temperature series are considered. If one of the 20 days is also a special day, then it is discarded as either hot or cold. All days that do not belong to one of the categories (special, hot or cold) are considered as regular days.

3. Data Analysis

It is important to describe the characteristics of the data series relevant to the forecasting process in order to understand the forecasting problem and whether or not its conclusions may apply to a different case:

3.1. Load

The load data series covers from 2010 to 2017 and it includes hourly values of electricity consumption in the Spanish inland system. The long-term trend of the series shown in Figure 5 is related to economic growth, efficiency improvements and behavioral shifts like the use of AC systems.

On a shorter term scale, the factors driving the load in Spain are temperature and social events and holidays, which are explained in the following subsections.

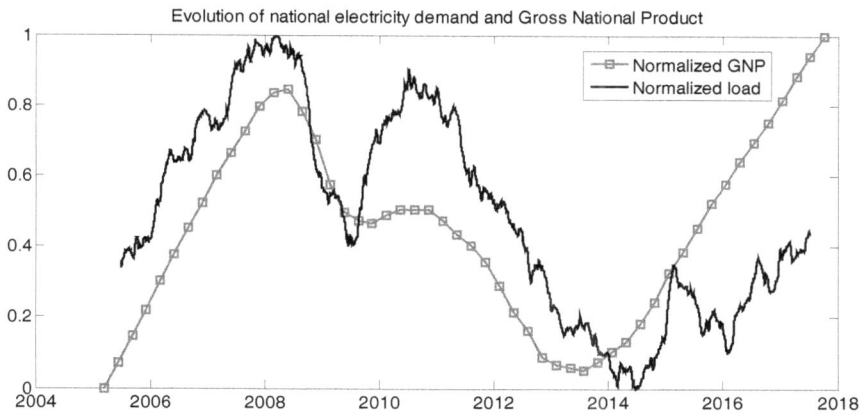

Figure 5. Evolution of 52 weeks moving average load and Gross National Product. Both series are normalized [0, 1].

3.2. Temperature

The temperature data available includes series from 59 stations scattered across the country. Real data of daily maximum and minimum data is collected along with daily forecasts of up to ten days ahead. Therefore, it is possible to simulate real time conditions if forecasts are used instead of real data.

As it was explained before, the national forecast only uses information from five stations selected from the 59 available. This selection is made through an empirical evaluation. In addition, the temperature from up to four previous days is also used in order to capture the dynamics of the temperature-load relation. The non-linearity of the relation is modeled using the CDD and HDD approach already discussed. Figure 6 shows the scatter plot of national load at 18 h on weekday against temperature at the three most relevant locations. The HDD and CDD linearization is also plotted for each location along with the Mean Average Percentage Error (MAPE) between the actual load and the linearized one.

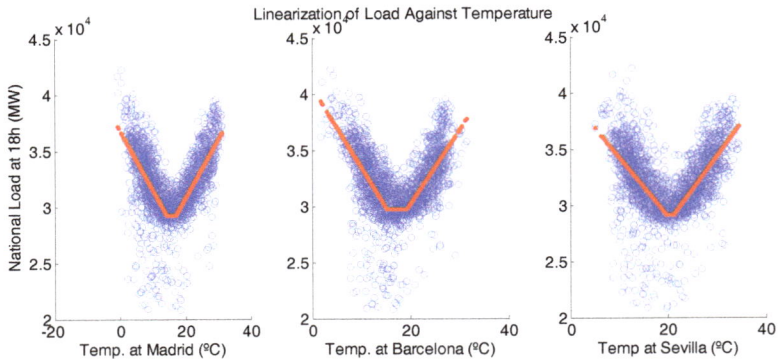

Figure 6. Scatter plot of national load and its linearization against temperature at Madrid, Barcelona and Sevilla.

3.3. Calendar

The type of day is determined by the official national calendar published in the Official Gazette [46]. The days are classified into 34 exclusive categories some of which include several days under a general rule: Mondays, Wednesdays, national holidays, Mondays before a holiday … and others for specific and unique days: 1 May, 25 December, 1 January … In addition to the exclusive categories (each day can only be assigned to one of these), there are also 18 modifying categories that may be simultaneously active with an exclusive one. These include regional holidays, days affected by DST … The complete classification can be found in [39].

The relevance of a proper day categorization is shown in Figure 7. The graph represents the average load profiles for 8 December, which is a national holiday, 7 December, before a holiday, and 30 November, regular day 7 days prior to 7 December. The years considered are the ones in which 7 December was not Saturday, Sunday or Monday. Figure 7 also includes the profile for 7 December on a Saturday. The graph shows how depending on the calendar (effects of temperature are averaged out), the profile not only shows variation of up to 20% from a regular weekday to a national holiday but it also shows different profiles in between.

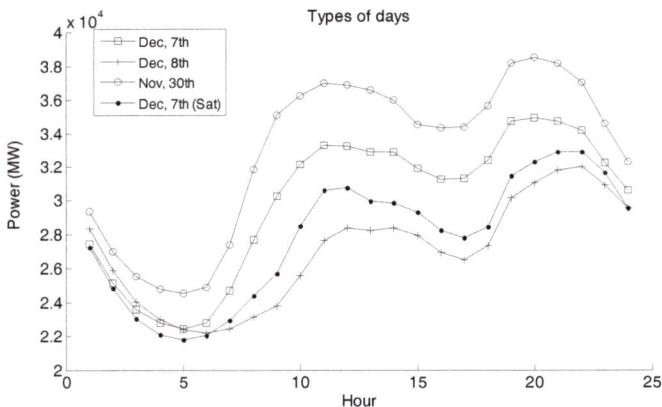

Figure 7. Different average profiles for national holidays, normal days, Saturdays and adjacent to holiday.

4. Results

The results expressed in this section correspond to the forecasting period of 2017. Each subsection presents the accuracy of both techniques (AR and NN) when the correspondent parameter or external condition changes. In addition, these results have been analyzed under the categories described in Section 2.3.

4.1. Historical Load Availability

The results shown in Table 2 represent the effect of increasing the number of previous years considered in the training of the model from 3 to 7 for both models. The results show a generally more accurate performance by the AR model especially with fewer years of data (1.50% vs. 2.17%). The NN model, however, benefits more from the availability of more data and this difference is reduced to 0.1% when seven years are used. The AR model shows very little improvement from 3 to 7 years while the NN model appears to be able to benefit from even longer training data as its performance on all categories continues to improve from 5 to 7 years (see Figure 8). Unfortunately, the available data base is not yet deep enough to test this.

Table 2. Forecasting error (RMSE) with training periods from 3 to 7 years.

Type of Day	3-Years		5-Years		7-Years	
	AR	NN	AR	NN	AR	NN
Overall	1.50%	2.17%	1.52%	1.72%	1.45%	1.55%
Regular	1.44%	1.96%	1.47%	1.57%	1.40%	1.44%
Special	1.91%	3.62%	1.81%	2.71%	1.81%	2.31%
Hot	1.63%	2.65%	1.53%	2.08%	1.55%	1.93%
Cold	1.61%	2.79%	1.73%	1.81%	1.72%	1.48%

Test conditions: 10 neurons (10N), 10 redundant networks (10RN), 5 temperature locations (5TL), 12 month training freq (12MF), 7 lags for AR and 14 for NN (7/14LAG).

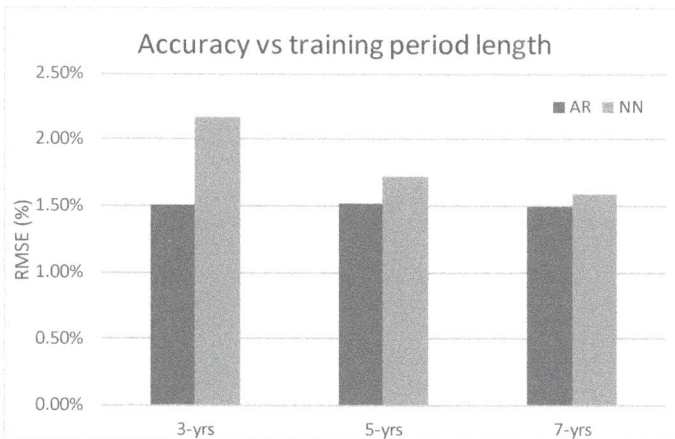

Figure 8. Overall forecasting error (RMSE) with training periods from 3 to 7 years.

Regarding the categorized results, regular days obtain almost the same result while in hot and special days the AR outperforms the NN model. However, cold days are clearly forecasted more accurately by the NN model. This could imply that the linear restriction present in the AR model limits its capacity to model the behavior of the load with the data treatment used.

4.2. Temperature Locations

The results for testing the availability of temperature data series from different locations are included in Table 3. In addition, Figure 9 shows the evolution of the overall RMSE of both models from having only location to including all five. Locations are included sequentially from most to least relevant.

Table 3. Forecasting error (RMSE) with available temperature location from 1 to 5.

Type of Day	MAD		MAD, BAR		MAD, BAR, VIZ		MAD, BAR, VIZ, SEV		MAD, BAR, VIZ, SEV, ZAR	
	AR	NN	AR	NN	AR	NN	AR	NN	AR	NN
Overall	1.63%	1.61%	1.53%	1.59%	1.48%	1.54%	1.46%	1.54%	1.45%	1.55%
Regular	1.59%	1.53%	1.48%	1.50%	1.43%	1.45%	1.41%	1.44%	1.40%	1.44%
Special	1.84%	2.22%	1.86%	2.21%	1.81%	2.15%	1.80%	2.17%	1.81%	2.31%
Hot	1.83%	2.02%	1.63%	1.91%	1.52%	1.84%	1.55%	1.94%	1.55%	1.93%
Cold	2.00%	1.61%	1.81%	1.49%	1.83%	1.47%	1.76%	1.48%	1.72%	1.48%

Test conditions: 7 Years Training (7YT), 10N, 10RN, 12MF, 7/14LAG.

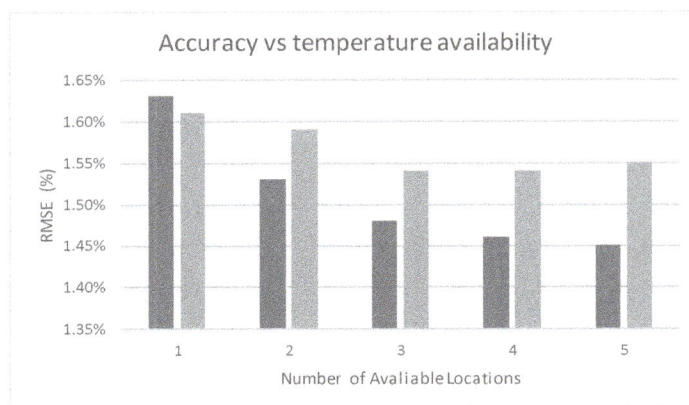

Figure 9. Overall forecasting error (RMSE) with available temperature location from 1 to 5.

The NN outperforms the AR model when only one location is available. Both models benefit from having more data series included, but the AR model obtains a more accurate forecast with five locations. In fact, the NN model obtains a larger error with five locations than it does with four. This could imply that the linear restriction on the AR model allows it to correctly include this information in the model. The excessive availability of information, however, seems to increase the risk of NN model overfitting the training data and, therefore, losing forecasting capabilities.

4.3. Temperature Treatment

The preprocessing of the temperature data is a key aspect of the forecasting system. The thresholds need to be properly tuned so that the linearization of the relation is correct. However, these thresholds may shift over time as consumers' behavior regarding temperature changes. Therefore, robustness to this configuration is also important.

The results were obtained using one location each time and varying HDD and CDD thresholds from 13 to 25 °C. Table 4 shows the overall results for shifting the HDD threshold for Barcelona along with the hot and cold categories as the special days are not relevant to this test.

The effect of adjusting the threshold is more clearly shown in Figure 10, in which forecasting accuracy of both models using temperature from Zaragoza and Barcelona is plotted. The graph shows how the NN is much less dependent on the chosen threshold while the AR performance is clearly thrown off by a misadjusted threshold.

Table 4. Forecasting error (RMSE) with different HDD threshold adjustment in Barcelona.

Type of Day and Model		HDD Threshold											
		13	14	15	16	17	18	19	20	21	22	23	24	25
Overall	AR	2.00%	1.98%	1.96%	1.95%	1.94%	1.92%	1.91%	1.91%	1.91%	1.91%	1.92%	1.93%	1.94%
	NN	1.92%	1.93%	1.93%	1.93%	1.94%	1.92%	1.92%	1.93%	1.92%	1.92%	1.93%	1.93%	1.94%
Hot	AR	2.62%	2.63%	2.63%	2.62%	2.60%	2.58%	2.55%	2.53%	2.50%	2.48%	2.46%	2.43%	2.45%
	NN	2.54%	2.55%	2.55%	2.53%	2.56%	2.54%	2.55%	2.54%	2.52%	2.51%	2.51%	2.46%	2.47%
Cold	AR	2.16%	2.11%	2.07%	2.03%	2.00%	1.99%	1.99%	1.99%	1.99%	2.00%	2.01%	2.02%	2.02%
	NN	1.65%	1.67%	1.69%	1.71%	1.71%	1.68%	1.71%	1.68%	1.72%	1.67%	1.71%	1.69%	1.69%

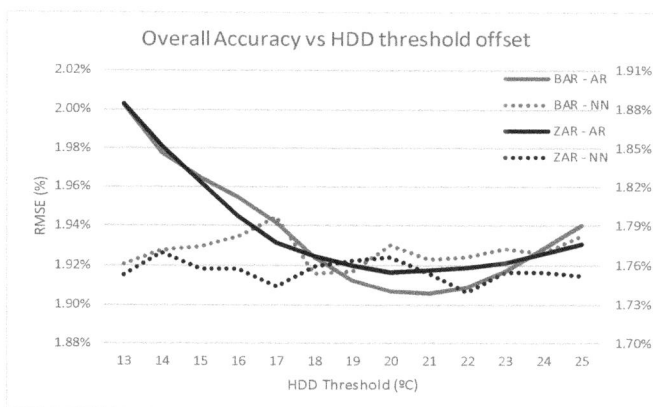

Figure 10. Overall forecasting error (RMSE) with different HDD threshold adjustment in Barcelona and Zaragoza.

4.4. Number of Neurons

The number of neurons in the hidden layer affects both computational burden and the NN's performance. Therefore, both aspects are reported as results on this test. Table 5 shows the accuracy of the neural network as the number of neurons is increased. In addition, the forecasting time for a single 24-h profile is included. It is worth noticing that the rest of forecasting processes like data access or treatment also consume time and, therefore, the reported time is not the only concern in order to obtain a timely forecast.

Table 5. Forecasting error (RMSE) and execution time with different number of neurons.

Type of Day	Number of Neurons					
	3	4	5	10	15	20
Overall	1.56%	1.55%	1.56%	1.55%	1.58%	1.62%
Regular	1.49%	1.46%	1.45%	1.44%	1.46%	1.50%
Special	2.00%	2.10%	2.28%	2.31%	2.36%	2.46%
Hot	2.00%	1.93%	1.95%	1.93%	2.00%	2.04%
Cold	1.55%	1.45%	1.51%	1.48%	1.51%	1.58%
Time (s)	1.610	1.615	1.620	1.630	1.639	1.643

Test conditions: 7YT, 10RN, 5TL, 12MF, 7/14LAG.

Figure 11 shows the evolution of accuracy and simulation time against the number of neurons in the hidden layer. It can be seen that the execution time is almost constant and therefore the number of neurons is not an issue regarding computational burden. In addition, accuracy on regular days does not improve with more complex networks. Special days, however, show a deterioration as the number of neurons increases. A possible explanation to this is that a more complex network is able to overfit the training data and lose generality. This is especially obvious on the special-day category due to the scarcity of data.

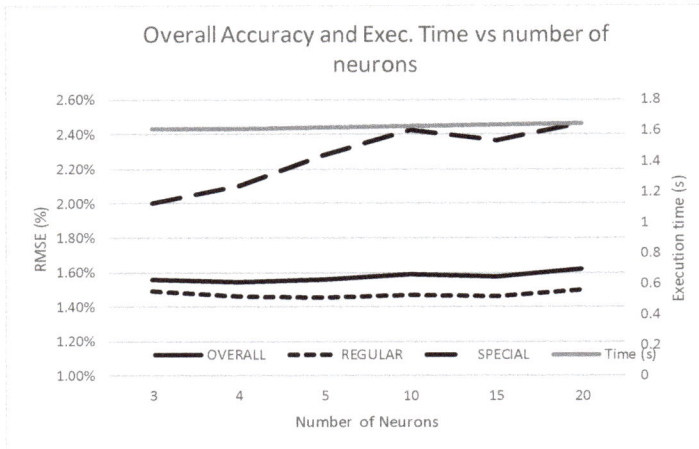

Figure 11. Overall forecasting error (RMSE) and execution time with different number of neurons.

4.5. Redundancy of Neural Networks

The use of a redundant number of NN reduces the model's dependency of random initial conditions. Furthermore, eliminating extreme values also reduces the overall error. Table 6 shows the results of using from 3 to 25 redundant networks for the NN model.

Table 6. Forecasting error (RMSE) and execution time with different redundant networks.

Type of Day	Number of Networks									
	3	5	10	11	12	14	15	16	20	25
Overall	1.67%	1.62%	1.55%	1.56%	1.54%	1.54%	1.54%	1.55%	1.55%	1.56%
Regular	1.55%	1.51%	1.46%	1.48%	1.46%	1.45%	1.46%	1.46%	1.45%	1.46%
Special	2.51%	2.30%	2.10%	2.08%	2.08%	2.11%	2.12%	2.13%	2.18%	2.22%
Hot	2.02%	1.97%	1.93%	1.98%	1.92%	1.94%	1.92%	1.93%	1.92%	1.94%
Cold	1.54%	1.57%	1.45%	1.43%	1.45%	1.43%	1.44%	1.46%	1.47%	1.44%
Time (s)	0.839	1.09	1.686	1.746	1.868	2.039	2.228	2.333	2.713	3.363

Test conditions: 7YT, 4N, 5TL, 12MF, 7/14LAG.

There is an improvement using up to 10 redundant networks. However, there is not significant error reduction from 10 to 25 networks. The execution time shows an increase, although for the optimum amount of 10 networks the computational burden is still manageable. As a reference, we have used the execution time for the AR model, which is 0.835 s. In addition, in Figure 12 it can be seen that the type of days that benefit the most from increasing number of networks from 3 to 10 are special days. Again, this is probably due to the higher variability in the output from different networks for this scarcer type of days.

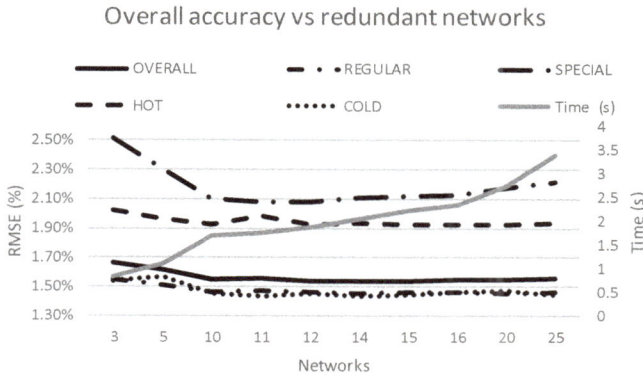

Figure 12. Overall forecasting error (RMSE) and execution time with different redundant networks.

4.6. Frequency of Training

The results from Table 7 show the performance of both models when the training period is changed from 3 months to 24 months. The testing period remains the same as described in Table 1, but the data used to train the model that forecasted each block changes. There appear to be no significant improvement from retraining the models more frequently than annually, as seen on Figure 13. However, a training period longer than a year seems to cause an increase in the forecasting error. Both models are affected very similarly by this parameter, with an increase in the error of about 23% for both models when increasing the time in between trainings from 12 to 24 months.

Table 7. Forecasting error (RMSE) with different training frequency.

Type of Day	3 Months		6 Months		12 Months		24 Months	
	AR	NN	AR	NN	AR	NN	AR	NN
Overall	1.44%	1.54%	1.44%	1.56%	1.45%	1.55%	1.78%	2.07%
Regular	1.39%	1.45%	1.40%	1.47%	1.40%	1.46%	1.74%	2.01%
Special	1.78%	2.09%	1.79%	2.10%	1.81%	2.10%	2.13%	2.35%
Hot	1.57%	1.96%	1.57%	1.97%	1.55%	1.93%	2.26%	3.11%
Cold	1.70%	1.45%	1.69%	1.46%	1.72%	1.45%	1.96%	1.96%

Test conditions: 7YT, 4N, 10RN, 5TL, 7/14LAG.

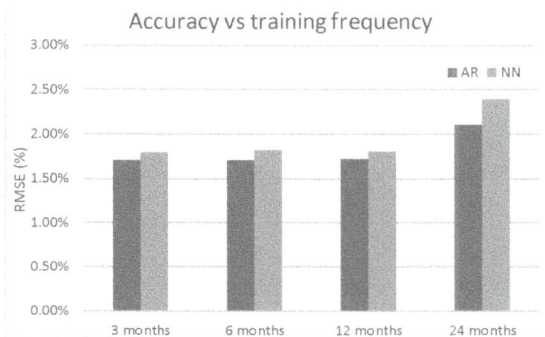

Figure 13. Forecasting error (RMSE) for AR and NN models with training frequency from 3 to 24 months.

4.7. Number of Lags

The number of lags in each model is changed from 0 to 25 in order to expose how this parameter affects the accuracy of each model. The results are categorized by type of day on Table 8. The AR model obtains a less accurate forecast than the NN when the lags are below 7 days. However, the results beyond this threshold benefit the AR model clearly. The AR model seems to continue its improvement up to lag number 21 (three weeks) but the NN reaches a plateau at lag 7. Once again, the NN model performs more accurately when little information (in this case lags) is available but it is outperformed by the AR model when the limitation is lifted. Figure 14 represents the overall accuracy of both models as the number of lags is increased. It is worth noticing how the AR model improves specially at lags 7, 14 and 21.

Table 8. Forecasting error (RMSE) with different lagged feedback.

LAG	Overall		Regular		Special		Hot		Cold	
	AR	NN	AR	NN	AR	NN	AR	NN	AR	NN
0	1.80%	1.73%	1.76%	1.64%	2.08%	2.31%	1.94%	1.89%	2.00%	1.57%
1	1.73%	1.65%	1.68%	1.55%	2.04%	2.34%	1.86%	1.81%	1.85%	1.47%
3	1.64%	1.61%	1.59%	1.49%	1.96%	2.36%	1.81%	1.82%	1.87%	1.53%
5	1.62%	1.57%	1.57%	1.48%	1.95%	2.16%	1.81%	1.91%	1.86%	1.44%
6	1.56%	1.56%	1.51%	1.47%	1.89%	2.11%	1.75%	1.91%	1.81%	1.41%
7	1.45%	1.56%	1.40%	1.47%	1.81%	2.14%	1.55%	1.89%	1.72%	1.47%
8	1.45%	1.55%	1.39%	1.46%	1.81%	2.15%	1.55%	1.90%	1.72%	1.45%
13	1.45%	1.57%	1.39%	1.47%	1.84%	2.16%	1.52%	1.97%	1.72%	1.45%
14	1.43%	1.55%	1.37%	1.46%	1.83%	2.10%	1.51%	1.93%	1.71%	1.45%
15	1.43%	1.57%	1.37%	1.48%	1.83%	2.15%	1.51%	1.92%	1.71%	1.50%
20	1.43%	1.58%	1.37%	1.48%	1.83%	2.21%	1.52%	1.95%	1.70%	1.48%
21	1.42%	1.56%	1.35%	1.48%	1.83%	2.08%	1.49%	1.91%	1.68%	1.48%
22	1.42%	1.54%	1.35%	1.46%	1.83%	2.03%	1.50%	1.92%	1.68%	1.45%
24	1.42%	1.58%	1.36%	1.48%	1.84%	2.17%	1.50%	1.93%	1.68%	1.47%
25	1.42%	1.59%	1.35%	1.49%	1.84%	2.25%	1.50%	1.92%	1.68%	1.49%

Test conditions: 7YT, 4N, 10RN, 5TL, 12MF.

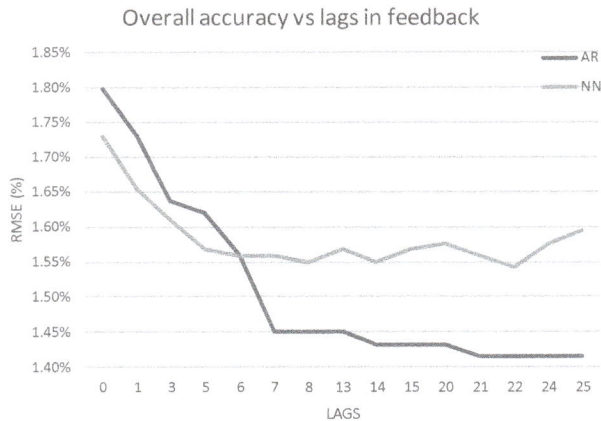

Figure 14. Overall forecasting error (RMSE) with different lagged feedback.

4.8. Overall Results

The previous subsections show how there is not a single solution for the load-forecasting problem. The conditions under which the forecast is done due to availability or data or time constraints affect the accuracy of each technique differently and, therefore, these conditions need to be taken into consideration when designing a forecasting system. As a general result, the AR model appears to be slightly more accurate but requires a finer tuning when treating the temperature data and requires a larger amount of temperature data sources.

5. Conclusions

Many different short-term load forecasting models have been proposed in the recent years. However, it is difficult to compare the accuracy or the general performance of each model when each one is tested under different conditions, testing periods and databases. The goal of this paper is to provide a series of comparisons between two of the most used forecasting engines: auto-regressive models and neural networks. The starting point is a forecasting system currently in use by REE that includes both techniques. Several tests have been run in order to determine the conditions under which each model performs best.

The results show that both models obtain very similar accuracy and, therefore both of them should remain in use. The AR model obtained a better overall result under the best possible condition but the NN model was superior when fewer temperature locations are available, the treatment of the temperature data is not properly adjusted or the feedback is limited to less than 7 lagged days. The AR showed higher accuracy when historical data is limited to less than 7 years. Both models have the same needs in terms of training frequency: a one-year period in between trainings is sufficient.

Regarding computational burden, the AR model is less computationally intense than the NN. However, the optimum configuration found at 4 neurons in the hidden layer and 10 redundant networks only costs twice as much as the AR model. Therefore, neither model has a definite advantage on this front.

To sum up, this paper enables the researcher to establish a set of rules to guide them in the process of selecting or designing a forecasting system. The results of this research offer very practical information that responds to actual empirical implementations of the system rather than to theoretical experiments. Further research in this area should include the analysis of different databases from other systems. The use of information from other systems would help determine if the conclusions drawn are general or database specific, in which case, studying the specificities of each database and determining why they behave differently would also be of value to the field.

Author Contributions: M.L. conceived and designed the experiments; C.S. (Carlos Sans) performed the experiments; M.L. and C.S. (Carlos Sans) analyzed the data; M.L.; S.V. and C.S. (Carolina Senabre) wrote the paper.

Acknowledgments: This research is a byproduct of a collaboration project between REE and Universidad Miguel Hernández. Open access costs will be funded by this project.

Conflicts of Interest: The authors declare no conflict of interest. The funding sponsors had no role in the design of the study; in the collection, analyses, or interpretation of data; in the writing of the manuscript, and in the decision to publish the results.

References

1. Hippert, H.S.; Pedreira, C.E.; Souza, R.C. Neural networks for short-term load forecasting: A review and evaluation. *IEEE Trans. Power Syst.* **2001**, *16*, 44–55. [CrossRef]
2. Hong, T.; Fan, S. Probabilistic electric load forecasting: A tutorial review. *Int. J. Forecast.* **2016**, *32*, 914–938. [CrossRef]
3. Kuster, C.; Rezgui, Y.; Mourshed, M. Electrical load forecasting models: A critical systematic review. *Sustain. Cities Soc.* **2017**, *35*, 257–270. [CrossRef]
4. Papalexopoulos, A.D.; Hesterberg, T.C. A regression-based approach to short-term system load forecasting. *IEEE Trans. Power Syst.* **1990**, *5*, 1535–1547. [CrossRef]
5. Charlton, N.; Singleton, C. A refined parametric model for short term load forecasting. *Int. J. Forecast.* **2014**, *30*, 364–368. [CrossRef]

6. Wang, P.; Liu, B.; Hong, T. Electric load forecasting with recency effect: A big data approach. *Int. J. Forecast.* **2016**, *32*, 585–597. [CrossRef]

7. Hagan, M.T.; Behr, S.M. The Time Series Approach to Short Term Load Forecasting. *IEEE Trans. Power Syst.* **1987**, *2*, 785–791. [CrossRef]

8. Amjady, N. Short-term hourly load forecasting using time-series modeling with peak load estimation capability. *IEEE Trans. Power Syst.* **2001**, *16*, 498–505. [CrossRef]

9. Amini, M.H.; Kargarian, A.; Karabasoglu, O. ARIMA-based decoupled time series forecasting of electric vehicle charging demand for stochastic power system operation. *Electr. Power Syst. Res.* **2016**, *140*, 378–390. [CrossRef]

10. Boroojeni, K.G.; Amini, M.H.; Bahrami, S.; Iyengar, S.S.; Sarwat, A.I.; Karabasoglu, O. A novel multi-time-scale modeling for electric power demand forecasting: From short-term to medium-term horizon. *Electr. Power Syst. Res.* **2017**, *142*, 58–73. [CrossRef]

11. Taylor, J.W. Short-Term Load Forecasting with Exponentially Weighted Methods. *IEEE Trans. Power Syst.* **2012**, *27*, 458–464. [CrossRef]

12. Mandal, P.; Senjyu, T.; Urasaki, N.; Funabashi, T. A neural network based several-hour-ahead electric load forecasting using similar days approach. *Int. J. Electr. Power Energy Syst.* **2006**, *28*, 367–373. [CrossRef]

13. Zhang, Y.; Zhou, Q.; Sun, C.; Lei, S.; Liu, Y.; Song, Y. RBF Neural Network and ANFIS-Based Short-Term Load Forecasting Approach in Real-Time Price Environment. *IEEE Trans. Power Syst.* **2008**, *23*, 853–858. [CrossRef]

14. Kalaitzakis, K.; Stavrakakis, G.S.; Anagnostakis, E.M. Short-term load forecasting based on artificial neural networks parallel implementation. *Electr. Power Syst. Res.* **2002**, *63*, 185–196. [CrossRef]

15. Liao, G.-C.; Tsao, T.-P. Application of a fuzzy neural network combined with a chaos genetic algorithm and simulated annealing to short-term load forecasting. *IEEE Trans. Evol. Comput.* **2006**, *10*, 330–340. [CrossRef]

16. López, M.; Valero, S.; Senabre, C.; Aparicio, J.; Gabaldon, A. Application of SOM neural networks to short-term load forecasting: The Spanish electricity market case study. *Electr. Power Syst. Res.* **2012**, *91*, 18–27. [CrossRef]

17. Hinojosa, V.H.; Hoese, A. Short-Term Load Forecasting Using Fuzzy Inductive Reasoning and Evolutionary Algorithms. *IEEE Trans. Power Syst.* **2010**, *25*, 565–574. [CrossRef]

18. Srinivasan, D.; Chang, C.S.; Liew, A.C. Demand forecasting using fuzzy neural computation, with special emphasis on weekend and public holiday forecasting. *IEEE Trans. Power Syst.* **1995**, *10*, 1897–1903. [CrossRef]

19. Kim, K.-H.; Youn, H.-S.; Kang, Y.-C. Short-term load forecasting for special days in anomalous load conditions using neural networks and fuzzy inference method. *IEEE Trans. Power Syst.* **2000**, *15*, 559–565.

20. Song, K.-B.; Baek, Y.-S.; Hong, D.H.; Jang, G. Short-term load forecasting for the holidays using fuzzy linear regression method. *IEEE Trans. Power Syst.* **2005**, *20*, 96–101. [CrossRef]

21. Chen, Y.; Yang, Y.; Liu, C.; Li, C.; Li, L. A hybrid application algorithm based on the support vector machine and artificial intelligence: An example of electric load forecasting. *Appl. Math. Model.* **2015**, *39*, 2617–2632. [CrossRef]

22. Wang, J.; Jin, S.; Qin, S.; Jiang, H. Swarm Intelligence-based hybrid models for short-term power load prediction. *Math. Probl. Eng.* **2014**, *2014*, 17. [CrossRef]

23. Bashir, Z.A.; El-Hawary, M.E. Applying wavelets to short-term load forecasting using pso-based neural networks. *IEEE Trans. Power Syst.* **2009**, *24*, 20–27. [CrossRef]

24. Amjady, N.; Keynia, F. Short-term load forecasting of power systems by combination of wavelet transform and neuro-evolutionary algorithm. *Energy* **2009**, *34*, 46–57. [CrossRef]

25. Ho, K.-L.; Hsu, Y.-Y.; Chen, C.-F.; Lee, T.-E.; Liang, C.-C.; Lai, T.-S.; Chen, K.-K. Short term load forecasting of Taiwan power system using a knowledge-based expert system. *IEEE Trans. Power Syst.* **1990**, *5*, 1214–1221.

26. Huang, C.-M.; Huang, C.-J.; Wang, M.-L. A particle swarm optimization to identifying the ARMAX model for short-term load forecasting. *IEEE Trans. Power Syst.* **2005**, *20*, 1126–1133. [CrossRef]

27. Niu, D.; Xing, M. Research on neural networks based on culture particle swarm optimization and its application in power load forecasting. In Proceedings of the Third International Conference on Natural Computation (ICNC 2007), Haikou, China, 24–27 August 2007; Volume 1, pp. 270–274.

28. Gajowniczek, K.; Ząbkowski, T. Electricity forecasting on the individual household level enhanced based on activity patterns. *PLoS ONE* **2017**, *12*, e0174098. [CrossRef] [PubMed]

29. Singh, S.; Yassine, A. Big data mining of energy time series for behavioral analytics and energy consumption forecasting. *Energies* **2018**, *11*, 452. [CrossRef]

30. López, M.; Valero, S.; Senabre, C.; Aparicio, J.; Gabaldon, A. Standardization of short-term load forecasting models. In Proceedings of the 2012 9th International Conference on the European Energy Market, Florence, Italy, 10–12 May 2012; pp. 1–7.

31. Kim, C.; Yu, I.; Song, Y.H. Kohonen neural network and wavelet transform based approach to short-term load forecasting. *Electr. Power Syst. Res.* **2002**, *63*, 169–176. [CrossRef]

32. Chen, Y.; Luh, P.B.; Rourke, S.J. Short-term load forecasting: Similar day-based wavelet neural networks. In Proceedings of the 2008 7th World Congress on Intelligent Control and Automation, Chongqing, China, 25–27 June 2008; pp. 3353–3358.

33. López, M.; Valero, S.; Senabre, C.; Gabaldón, A. Analysis of the influence of meteorological variables on real-time Short-Term Load Forecasting in Balearic Islands. In Proceedings of the 2017 11th IEEE International Conference on Compatibility, Power Electronics and Power Engineering (CPE-POWERENG), Cadiz, Spain, 4–6 April 2017; pp. 10–15.

34. Cancelo, J.R.; Espasa, A.; Grafe, R. Forecasting the electricity load from one day to one week ahead for the Spanish system operator. *Int. J. Forecast.* **2008**, *24*, 588–602. [CrossRef]

35. Lamedica, R.; Prudenzi, A.; Sforna, M.; Caciotta, M.; Cencellli, V.O. A neural network based technique for short-term forecasting of anomalous load periods. *IEEE Trans. Power Syst.* **1996**, *11*, 1749–1756. [CrossRef]

36. Arora, S.; Taylor, J.W. Short-term forecasting of anomalous load using rule-based triple seasonal methods. *IEEE Trans. Power Syst.* **2013**, *28*, 3235–3242. [CrossRef]

37. Khotanzad, A.; Afkhami-Rohani, R.; Maratukulam, D. ANNSTLF-artificial neural network short-term load forecaster generation three. *IEEE Trans. Power Syst.* **1998**, *13*, 1413–1422. [CrossRef]

38. Fan, S.; Methaprayoon, K.; Lee, W.-J. Multiregion load forecasting for system with large geographical area. *IEEE Trans. Ind. Appl.* **2009**, *45*, 1452–1459. [CrossRef]

39. López, M.; Valero, S.; Rodriguez, A.; Veiras, I.; Senabre, C. New online load forecasting system for the Spanish transport system operator. *Electr. Power Syst. Res.* **2018**, *154*, 401–412. [CrossRef]

40. Mares, J.J.; Mercado, K.D.; Quintero M., C.G. A Methodology for short-term load forecasting. *IEEE Lat. Am. Trans.* **2017**, *15*, 400–407. [CrossRef]

41. Almeshaiei, E.; Soltan, H. A methodology for electric power load forecasting. *Alex. Eng. J.* **2011**, *50*, 137–144. [CrossRef]

42. García, M.L.; Valero, S.; Senabre, C.; Marín, A.G. Short-term predictability of load series: characterization of load data bases. *IEEE Trans. Power Syst.* **2013**, *28*, 2466–2474. [CrossRef]

43. Box, G.E.P.; Jenkins, G.M.; Reinsel, G.C.; Ljung, G.M. *Time Series Analysis: Forecasting and Control*, 5th ed.; John Wiley & Sons, Inc.: Hoboken, NJ, USA, 2016.

44. Akaike, H. Statistical predictor identification. *Ann. Inst. Stat. Math.* **1970**, *22*, 203–217. [CrossRef]

45. Akaike, H. Information theory and an extension of the maximum likelihood principle. In Proceedings of the Second International Symposium on Information Theory, Tsahkadsor, Armenia, 2–8 September 1971; Petrov, B.N., Caski, F., Eds.; Akadémiai Kiado: Budapest, Hungary; pp. 267–281.

46. Gobierno de España. Boletín Oficial del Estado. Available online: www.boe.es (accessed on 10 April 2018).

energies **MDPI**

Article

Load Forecasting for a Campus University Using Ensemble Methods Based on Regression Trees

María del Carmen Ruiz-Abellón [1], Antonio Gabaldón [2,*] and Antonio Guillamón [1]

[1] Department of Applied Mathematics and Statistics, Universidad Politécnica de Cartagena, 30202 Cartagena, Spain; maricarmen.ruiz@upct.es (M.d.C.R.-A.); antonio.guillamon@upct.es (A.G.)
[2] Department of Electrical Engineering, Universidad Politécnica de Cartagena, 30202 Cartagena, Spain
* Correspondence: antonio.gabaldon@upct.es; Tel.: +34-968-338944

Received: 6 July 2018; Accepted: 1 August 2018; Published: 6 August 2018

Abstract: Load forecasting models are of great importance in Electricity Markets and a wide range of techniques have been developed according to the objective being pursued. The increase of smart meters in different sectors (residential, commercial, universities, etc.) allows accessing the electricity consumption nearly in real time and provides those customers with large datasets that contain valuable information. In this context, supervised machine learning methods play an essential role. The purpose of the present study is to evaluate the effectiveness of using ensemble methods based on regression trees in short-term load forecasting. To illustrate this task, four methods (bagging, random forest, conditional forest, and boosting) are applied to historical load data of a campus university in Cartagena (Spain). In addition to temperature, calendar variables as well as different types of special days are considered as predictors to improve the predictions. Finally, a real application to the Spanish Electricity Market is developed: 48-h-ahead predictions are used to evaluate the economical savings that the consumer (the campus university) can obtain through the participation as a direct market consumer instead of purchasing the electricity from a retailer.

Keywords: Electricity Markets; load forecasting models; regression trees; ensemble methods; direct market consumers

---

1. Introduction

Load forecasting has been a topic of interest for many decades and the literature is plenty with a wide variety of techniques. Forecasting methods can be divided into three different categories: time-series approaches, regression based, and artificial intelligence methods (see [1]).

Among the classical time-series approaches, the ARIMA model is one of the most utilized (see [2–5]). Regression approaches, see [2,6], are also widely used in the field of short-term and medium-term load forecasting, including non-linear regression [7] and nonparametric regression [8] methods. Recently, in [9] the authors use linear multiple regression to predict the daily electricity consumption of administrative and academic buildings located at a campus of London South Bank University.

Several machine learning or computational intelligence techniques have been applied in the field of Short Term Load Forecasting. For example, decision trees [10], Fuzzy Logic systems [11,12], and Neural Networks [13–20]. In this paper, we propose the using of a particular set of supervised machine learning techniques (called ensemble methods based on decision trees) to predict the hourly electricity consumption of university buildings. In general, an ensemble method combines a set of weak learners to obtain a strong learner that provides better performance than a single one. Four particular cases of ensemble methods are bagging, random forest, conditional forest, and boosting, which are described in Section 2. There some recent literature regarding random forest and short-term load forecasting: for example, in [21] the authors use random forest to predict the hourly electrical load data of the Polish power system, whereas in [22] the same method is used to predict residential energy consumption. In [23], the authors propose three different methods for ensemble probabilistic forecasting. The ensemble methods are derived from seven individual machine learning models, which include random forest, among others, and it is tested in the field of solar power forecasts. On the other hand, in [24] the authors establish a novel ensemble model that is based on variational mode decomposition and the extreme learning machine. The proposed ensemble model is illustrated while using data from the Australian electricity market.

The main objective of this paper is to illustrate the performance of different ensemble methods for predicting the electricity consumption of some university buildings, analyzing their accuracies, relevant predictors, computational times, and parameter selection. Besides, we apply the prediction results to the context of Direct Market Consumers (DMC) in the Spanish Electricity Market.

In Spain, electricity price seems to be above our European neighbors, mainly due to the energy production mix and the weak electrical interconnections with the Central European Electricity System and Markets, but consumers can do little about that. Therefore, it is quite challenging for Spanish consumers to reduce this cost. Currently, a high voltage consumer (voltage supply greater than 1 kV), which is the case of a small campus university, can opt for two types of supply: captive customer (price freely agreed with a retailer or a provider) and Direct Market Consumer (also called qualified customer), taking advantage of the operation of the wholesale markets that are involved in the Spanish Electricity System. The literature concerning the topic of DMC is nearly non-existent and it reduces to some official web pages, such as [25,26].

In order to participate as a DMC in the Electricity Market, the customer needs to evaluate his load requirements, with roughly two days in advance. Another objective of this paper is to evaluate the savings that the university would have participating as a DMC, taking the 48-h-ahead predictions of one of the ensemble methods analyzed.

The main differences among the present paper and the previous ones dealing with the using of ensemble methods for forecasting porpoises (for example, ref. [27] employs the gradient boosting method for modeling the energy consumption of commercial buildings) are the following: in the present paper, we propose the XGBoost method as a useful tool for a medium-size consumer to purchase the electricity directly in the wholesale market. For that, a different prediction horizon (48 h ahead) is considered and the new predictors are needed. Indeed, we highlight the importance of calendar variables (distinguishing different types of festivities) for the case of electricity consumption in university buildings. This approach allows us to evaluate the savings of this kind of customers participating as Direct Market Consumers. Another novelty respect to previous papers is the using of conditional forest as an ensemble method to get load predictions, as well as the conditional importance measure to evaluate the relevance of each feature.

Firstly, in Section 2, four ensemble methods based on regression trees are described. Section 3 depicts the customer in study (a small campus university) and the data, discusses the parameter selection for each ensemble method as well as other relevant aspects and it shows the prediction results. Finally, in Section 4, the economic saving of a small campus university is computed when it participates as a Direct Market Consumer instead of acquiring the electric power from a traditional retailer. Note that it is not an energy efficiency study, the economic saving is given just by the type of supply: retail or wholesale market.

2. Ensemble Methods Based on Regression Trees

Taking into account the type of data in the analysis (continuous data corresponding to electricity consumption), in this section, we will focus on describing tree-based methods for regression and some related ensemble techniques. However, decision trees and ensemble methods can be applied to both regression and classification problems.

The process of building a regression tree can be summarized in two steps: firstly, we divide the predictor space into a number of non-overlapping regions (for example J regions), and secondly, the prediction for a new observation is given by the mean of the response values of the training data belonging to the same region as the new observation.

The criterion to construct the regions or "boxes" is to minimize the residual sum of squares (RSS), but not considering every possible partition of the feature space into J boxes because it would be computationally infeasible. Instead, a recursive binary splitting is used: at each step, the algorithm chooses the predictor and cutpoint, such that the resulting tree has the lowest RSS. The process is repeated until a stopping criterion is reached, see [28].

Let $\{(x_1, y_1), (x_2, y_2), \ldots, (x_n, y_n)\}$ be the training dataset, where each y_i denotes the i-th output (response variable) and $x_i = (x_{i_1}, x_{i_2}, \ldots, x_{i_s})$ the corresponding input of the "s" predictors (features) in study. The objective in a regression tree is to find boxes $B_1, B_2, \ldots, B_j$ that minimize the RSS, given by (1):

$$\sum_{j=1}^{J} \sum_{i \in B_j} (y_i - \hat{y}_{B_j})^2 \tag{1}$$

where $\hat{y}_{B_j}$ is the mean response for the training observations within the jth box.

A regression tree can be considered as a base learner in the field of machine learning. The main advantage of regression trees against lineal regression models is that in the case of highly non-linear and complex relationship between the features and the response, decision trees may outperform classical approaches. Although regression trees can be very non-robust and can generally provide less predictive accuracy than some of the other regression methods, these drawbacks can be easily improved by aggregating many decision trees, using methods, such as bagging, random forests, conditional forest, and boosting. These four methods have in common that can be considered as ensemble learning methods.

An ensemble method is a Machine Learning concept in which the idea is to build a prediction model by combining a collection of "N" simpler base learners. These methods are designed to reduce bias and variance with respect to a single base learner. Some examples of ensemble methods are bagging, random forest, conditional forest, and boosting.

2.1. Bagging

In the case of bagging (bootstrap aggregating), the collection of "N" base learners to ensemble is produced by bootstrap sampling on the training data. From the original training data set, N new training datasets are obtained by random sampling with replacement, where each observation has the same probability to appear in the new dataset. The prediction of a new observation with bagging is computed by averaging the response of the N learners for the new input (or majority vote in case of classification problems). In particular, when we apply bagging to regression trees, each individual tree has high variance, but low bias. Averaging the resulting prediction of these N trees reduces the variance and substantially improves in accuracy (see [28]).

The efficiency of the bagging method depends on a suitable selection of the number of trees N, which can be obtained by plotting the out-of-bag (OOB) error estimation with respect to N. Note that the bootstrap sampling step with replacement involves that each observation of the original training dataset is included in roughly two-thirds of the N bagged trees and it is out of the remaining ones. Then, the prediction of each observation of the original training dataset can be obtained by averaging the predictions of the trees that were not fit using that observation. This is a simple way, called OOB, to get a valid estimate of the test error for the bagged model avoiding a validation dataset or cross-validation.

Some other parameters that can also vary are the node size (minimum number of observations of the terminal nodes, generally five by default) and the maximum number of terminal nodes in the forest (generally trees are grown to the maximum possible, subject to limits by node size).

In this paper, the bagging method has been applied by means of the R package "randomForest", see [28]. The package also includes two measures of predictor importance that help to quantify the importance of each predictor in the final forecasting model and might suggest a reduced set of predictors.

2.2. Random Forest

Random forests are indeed a generalization of bagging. Instead of considering all of the predictors at each split of the tree, only a random sample of "*mtry*" predictors can be chosen each time. The main advantage of random forests respect to bagging can be noticed in the case of correlated predictors, as it is stated in [28]: predictions from the bagged trees will be highly correlated so that bagging will not reduce the variance so much, whereas random forests overcome this problem by forcing each split to consider only a subset of the predictors.

In the case of random forest, the efficiency of the method depends on a suitable selection of the number of trees N and the number of predictors *mtry* tested at each split. Again, the OOB error can be used for searching a suitable N as well as a suitable *mtry*. As with bagging, random forests will not overfit if we increase N, so the goal is to choose a value that is sufficiently large. The random forest method that is used in this paper has been implemented throughout the R package "randomForest", see [29].

2.3. Conditional Forest

Conditional forests consist in an implementation of the bagging and random forest ensemble algorithms, but utilizing conditional inference trees as base learners. Conditional inference trees are not only suitable for prediction (its partitioning algorithm avoid overfitting), but also for explanation purposes because they select variables in an unbiased way. They are especially useful in the presence of high-order interactions and when the number of predictors is large when compared to the sample size.

In conditional forests, each tree is obtained by binary recursive partitioning, as follows (see [30]): firstly, the algorithm tests whether any predictor is associated with the response, and it chooses the one that has the strongest association; secondly, the algorithm makes a binary split in this variable; finally, the previous two steps are repeated for each subset until there are no predictors that are associated with the response. The first step uses the permutation tests for conditional inference developed in [31].

As with random forest, in the case of conditional forest, we need a suitable selection of the number *mtry* of features tested at each split (the total number of predictors might be preferred) and the number of trees N (generally a lower value than for random forest is required). In this paper, the conditional forest method has been implemented throughout the R package "party", see [32].

2.4. Boosting

In contrast to the above ensemble methods, in boosting the "N" base, learners are obtained sequentially, that is, each base learner is determined while taking into account the success and errors of the previous base learners.

The first boosting algorithm was Adaptive Boosting (AdaBoost), as introduced in [33]. Instead of using bootstrap sampling, the original training sample is weighted at each step, giving more importance to those observations that provided large errors at previous steps. Besides, the prediction for a new observation is given by a weighted average (instead of a simple average) of the responses of the N base learners.

AdaBoost was later recast in a statistical framework as a numerical optimization problem where the objective is to minimize a loss function using a gradient descent procedure, see [34]. This new approach was called "gradient boosting", and it is considered one of the most powerful techniques for building predictive models.

Gradient boosting involves three elements: a loss function to be optimized, a weak learner to make predictions (in this case, decision trees obtained in a greedy manner), and an additive model to add weak learners (the output for each new tree is added to the output of the existing sequence of trees). The loss function used depends on the type of problem. For example, a regression problem may use a squared error loss function, whereas a classification problem may use logarithmic loss. Indeed, any differentiable loss function can be used.

Although boosting methods reduces bias more than bagging, they are more likely to overfit a training dataset. To overcome this task, several regularization techniques can be applied.

- Tree constraints: there are several ways to introduce constraints when constructing regression trees. For example, the following tree constraints can be considered as regularization parameters:

  ○ The number of gradient boosting iterations N: increasing N reduces the error on the training dataset, but may lead to overfitting. An optimal value of N is often selected by monitoring prediction error on a separate validation data set.

  ○ Tree depth: the size of the trees or number of terminal nodes in trees, which controls the maximum allowed level of interaction between variables in the model. The weak learners need to have skills but they should remain weak, thus shorter trees are preferred. In general, values of tree depth between 4 and 8 work well and values greater than 10 are unlikely to be required, see [35].

  ○ The minimum number of observation per split: the minimum number of observations needed before a split can be considered. It helps to reduce prediction variance at leaves.

- Shrinkage or learning rate: in regularization by shrinkage, each update is scaled by the value of the learning rate parameter "*eta*" in (0,1]. Shrinkage reduces the influence of each individual tree and leaves space for future trees to improve the model. As it is stated in [28], small learning rates provide improvements in model's generalization ability over gradient boosting without shrinking (*eta* = 1), but the computational time increases. Besides, the number of iterations and learning rate are tightly related: for a smaller learning rate "*eta*", a greater N is required.

- Random sampling: to reduce the correlation between the trees in the sequence, at each step, a subsample of the training data is selected without replacement to fit the base learner. This modification prevent overfitting and it was first introduced in [36], which is also called stochastic gradient boosting. Friedman observed an improvement in gradient boosting's accuracy with samplings of around one half of the training datasets. An alternative to row sampling is column sampling, which indeed prevents over-fitting more efficiently, see [37].

- Penalize tree complexity: complexity of a tree can be defined as a combination of the number of leaves and the L2 norm of the leaf scores. This regularization not only avoids overfitting, it also tends to select simple and predictive models. Following this approach, ref. [37] describes a scalable tree boosting system called XGBoost. In that paper, the objective to be minimized is a combination of the loss function and the complexity of the tree. In contrast to the previous ensemble methods, XGBoost requires a minimal amount of computational resources to solve real-world problems.

In XGBoost, the model is trained in an additive manner and it considers a regularized objective that includes a loss function and penalizes the complexity of the model. Following [37], if we denote by $\hat{y}_i^{(t)}$, the prediction of the i-th instance of the response at the t-th iteration, we need to find the tree structure f_t that minimizes the following objective:

$$\mathcal{L}^{(t)} = \sum_{i=1}^{n} l\left(y_i, \hat{y}_i^{(t-1)} + f_t(x_i)\right) + \Omega(f_t) \tag{2}$$

In the first term of (2), l is a differentiable convex loss function that measures the difference between the observed response y_i and the resulting prediction $\hat{y}_i$. The second term of (2) penalizes the complexity of the model, as follows:

$$\Omega(f) = \gamma T + \frac{1}{2}\lambda \| w \|^2 \tag{3}$$

where T is the number of leaves in the tree with leaf weights $w = (w_1, w_2, \ldots, w_T)$. Using the second order Taylor expansion, (3) can be simplified to:

$$\tilde{\mathcal{L}}^{(t)} = \sum_{i=1}^{n} \left[g_i f_t(x_i) + \frac{1}{2} h_i f_t^2(x_i) \right] + \gamma T + \frac{1}{2} \lambda \sum_{j=1}^{T} w_j^2 \tag{4}$$

where $g_i = \partial_{\hat{y}^{(t-1)}} l\left(y_i, \hat{y}^{(t-1)}\right)$ and $h_i = \partial_{\hat{y}^{(t-1)}}^2 l\left(y_i, \hat{y}^{(t-1)}\right)$.

Denoting by $I_j = \{i | q(x_i) = j\}$ the instance set of leaf j, we can rewrite (4), as follows:

$$\tilde{\mathcal{L}}^{(t)} = \sum_{j=1}^{T} \left[\left(\sum_{i \in I_j} g_i \right) w_j + \frac{1}{2} \left(\sum_{i \in I_j} h_i + \lambda \right) w_j^2 \right] + \gamma T \tag{5}$$

Therefore, the optimal weight is given by:

$$w_j^* = -\frac{\sum_{i \in I_j} g_i}{\sum_{i \in I_j} h_i + \lambda} \tag{6}$$

and the corresponding optimal objective by:

$$\tilde{\mathcal{L}}^{(t)}(q) = -\frac{1}{2} \sum_{j=1}^{T} \frac{\left(\sum_{i \in I_j} g_i \right)^2}{\sum_{i \in I_j} h_i + \lambda} + \gamma T \tag{7}$$

where q represents the optimal tree structure with T leaves and leaf weights $w^* = (w_1^*, w_2^*, \ldots, w_T^*)$.

Due to the impossibility of enumerating all the possible tree structures q, a greedy algorithm is used (it starts with a single leaf and adds branches iteratively). Denoting by I_L and I_R the instance sets of left and right nodes after the split, $I = I_L \cup I_R$, the reduction in the objective after the split is given by:

$$\mathcal{L}_{split} = \frac{1}{2} \left[\frac{\left(\sum_{i \in I_L} g_i \right)^2}{\sum_{i \in I_L} h_i + \lambda} + \frac{\left(\sum_{i \in I_R} g_i \right)^2}{\sum_{i \in I_R} h_i + \lambda} - \frac{\left(\sum_{i \in I} g_i \right)^2}{\sum_{i \in I} h_i + \lambda} \right] - \gamma \tag{8}$$

The task of searching the best split has been developed in two scenarios: an exact greedy algorithm (it enumerates all the possible splits on all the features, which is computational demanding) and an approximate greedy algorithm for big data sets, see [37] for more details.

The main difference between random forest and boosting is that the former builds the base learners independently through bootstrap sampling on the training dataset, while the latter obtains them sequentially focusing on the errors of the previous iteration and using gradient descent methods. Some strengths of the XGBoost implementation comparing to other methods are:

- An exact greedy algorithm is available.
- Approximate global and approximate local algorithms are available for big datasets.
- It performs parallel learning. Besides, an effective cache-aware block structure is available for out-of-core tree learning.
- It is efficient in case of sparse input data (including the presence of missing values).

The extreme gradient boosting method (XGBoost) has been implemented by means of the R package "xgboost", see [38].

Apart from its highly computational efficiency, the XGBoost offers a great flexibility, but it requires setting up more than the ten parameters that could not be learned from the data. Taking into account that R package "xgboost" does not have any hyperparameter tuning, the parameter tuning can be done by means of cross validation. However, creating a grid for all of the parameters to be tuned implies an extremely high computational cost.

3. Prediction Results for the University Buildings

In this section, the four ensemble methods that are described above are applied to the electricity consumption of a small campus university to evaluate the adequacy of each technique in this type of customers. Specifically, we will focus on 48-h-ahead predictions in order to apply them to the context of Direct Market Consumers, although different prediction horizons will be also considered for the case of XGBoost method. Some other aspects, such us predictors importance or parameter selection, for each method are also developed.

Firstly, in this section, the customer in study is introduced. Secondly, the load data, predictors, and some goodness of fit measurements are depicted. Finally, the forecasting results for the case study are shown.

3.1. Customer Description: A Campus University

The campus "Alfonso XIII" of the Technical University of Cartagena (UPCT, Spain) comprises seven buildings ranging from 2000 m^2 to 6500 m^2 and a meeting zone (10,000 m^2). Buildings are of two kinds: naturally ventilated cellular (individual windows, local light switches, and local heating control) and naturally ventilated open-plan (office equipment, light switched in longer groups, and zonal heating control). This campus has an overall surface larger than 35,500 m^2 to fulfill the needs of different Faculties for classrooms, departmental offices, administrative offices, and laboratories for 1800 students and 200 professors. Unfortunately, the age of buildings (50 years old in four cases) and architectural conditioning works are far from actual energy efficiency standards, specifically in the two main electrical end-uses of the building: air conditioning/space heating (low performance, insufficient heat insulation, and an important cluster of individual appliances for offices and small laboratories) and lighting (where conventional magnetic ballasts and fluorescent are still used at a great extend).

With respect to the share of end-uses in the "Campus Alfonso XIII" of UPCT, heating, ventilation, and air conditioning (HVAC) is the largest energy end-use (this trend is the same both in the residential and non-residential buildings in Spain and other countries, see Table 1) with 40–50% of overall demand; lighting follows with 25–30%, electronics and office equipment 7–12% and other appliances with 8–10% (i.e., vending machines, refrigeration, water heaters WH, laboratory equipment, etc.). Notice that building type is critical in how energy end uses are distributed in each specific building. Table 1 shows a comparative of end-uses in office buildings in three countries [39] and in the analysed case, campus "Alfonso XIII".

Table 1. Energy demand in office buildings by end-use.

End-Use	USA (%)	UK (%)	Spain (%)	University Buildings (%) (UPCT)
HVAC	48	55	52	40–50
Lighting	22	17	33	25–30
Equipment (appliances)	13	5	10	7–12
Other (WH, refrigeration)	17	23	5	8–10

3.2. Data Description

Data used in this paper correspond to the campus Alfonso XIII of the Technical University of Cartagena, as described in the previous subsection. Hourly load data from 2011 to 2016 (both included) were analyzed, obtained from the retailer electric companies (Nexus Energía S.A. and Iberdrola S.A.). It is well known that electricity consumption is related to several exogenous factors, such as the hour of the day, the day of the week, or the month of the year, and therefore these factors must be taken into account in the design of the prediction model. Temperature is a factor that might affect the electricity consumption (cooling and heating of the university buildings). Thus, the hourly temperature was considered as an input in the forecasting model, as provided by AEMET (Agencia Española de Meteorología) for the city of Cartagena (where the campus university is located), from 2011 to 2016. Besides, depending on the end-uses of the customer in study, some other features can be relevant for the load. For example, in this case study, different types of holidays or special days have been distinguished throughout binary variables (see Table 2 for a detailed description).

Table 2. Description of the predictors.

Predictors	Description
H2, H3, ... H24	Hourly dummy variables corresponding to the hour of the day
WH2, WH3, ... WH7	Hourly dummy variables corresponding to the day of the week
MH2, MH3, ... , MH12	Hourly dummy variables corresponding to the month of the year
FH1	Hourly dummy variables corresponding to the month of the year
FH2	Hourly dummy variable corresponding to Christmas and Eastern days
FH3	Hourly dummy variable corresponding to academic holidays (patron saint festivities)
FH4	Hourly dummy variable corresponding to national, regional or local holidays
FH5	Hourly dummy variable corresponding to academic periods with no-classes and no-exams (tutorial periods)
Temperature_lag_i	Hourly external temperature lagged "*i*" hours. Depending on the prediction horizon, different lags will be considered.
LOAD_lag_i	Hourly load lagged "*i*" hours. Depending on the prediction horizon, different lags will be considered.

Three different measurements given in (9), (10), and (11) were used to obtain the accuracy of the forecasting models: the root mean square error (*RMSE*), the *R-squared* (percentage of the variability explained by the forecasting model), and the mean absolute percentage error (*MAPE*). Although the *MAPE* is the most used error measure, see [1], the squared error measures might be more fitting because the loss function in Short Term Load Forecasting is not linear, see [13]. Some descriptive measures of the errors (such as the mean, skewness, and kurtosis) were also considered to evaluate the performance of the forecasting methods.

The root mean square error is defined by:

$$RMSE = \sqrt{\sum_{t=1}^{n} \frac{(y_t - \hat{y}_t)^2}{n}} \tag{9}$$

the *R-squared* is given by:

$$R - squared = 1 - \frac{\sum_{t=1}^{n} (y_t - \hat{y}_t)^2}{\sum_{t=1}^{n} (y_t - \bar{y})^2} \tag{10}$$

and the mean absolute percentage error is defined by:

$$MAPE = \frac{100}{n} \sum_{t=1}^{n} \left| \frac{y_t - \hat{y}_t}{y_t} \right| \tag{11}$$

where n is the number of data, y_t is the actual load at time t, and $\hat{y}_t$ is the forecasting load at time t.

3.3. Forecasting Results

Data from 1 January 2011 to 31 December 2015 were selected as the training period in all methods, whereas data from 1 January 2016 to 31 December 2016 constituted the test period. In this subsection, firstly a prediction horizon of 48 h is established, whose forecasting results will be used in the next section dealing with Direct Market Consumers. In this case, we consider 53 predictors (see Table 2):

23 dummies for the hour of the day, six dummies for the day of the week, 11 dummies for the month of the year, five dummies for special days (FH1, ... , FH5), two predictors of historic temperatures (lags 48 h and 72 h), and six predictors of historic loads (lags 48 h, 72 h, 96 h, 120 h, 144 h, and 168 h).

For each ensemble method, the parameter selection has been developed and measures of variable importance have been obtained (see Table 3 for the meaning of each term). In order to have reproducible models and comparable results, the same *seed* was selected in all procedures that require random sampling. In the case of bagging and random forest, we have selected an optimal number of trees (*ntree*) through the OOB error estimate and we have ordered the predictors according to the node impurity importance measure, see [28]. For bagging, the number of predictors that are considered at each split must be the total number of predictors, whereas in the case of random forest, the optimal parameter has been selected using the OOB error estimate for different values of *mtry*. In the case of conditional forest, the conditional variable importance measure introduced in [40] has been considered, which better reflects the true impact of each predictor in presence of correlated predictors.

While in bagging and random forest the OOB error was used to tune the parameters, in the case of conditional forest and XGBoost the parameters were tuned by means of cross validation with five folds (approximately one year in each fold). As for conditional forest, only two parameters need to be tuned (*ntree* and *mtry*), but in XGBoost, there are more parameters to tune. Although one can apply cross validation taking into account a multi-dimension grid with all of the parameters to tune (this approach would imply a high computational cost), we considered a simplification of the search selecting subsample = 0.5, max depth = 6 (appropriate in most problems) and looking for a good combination of "*eta*" and "*nrounds*", see Table 4. The rest of parameters of the method were set up by default, according to the R package [38]. In the case of XGBoost, features have been ordered by decreasing importance while using the gain measure defined in [36].

Table 3. Notation.

Term	Description
ntree (N)	Number of trees or iterations in bagging, random forest and conditional forest
mtry	Number of predictors considered at each split in bagging, random forest and conditional forest
node impurity	Importance measure in random forest
max_depth	Maximum depth of a tree
subsample	Subsample ratio of the training instance
eta	Shrinkage or learning rate
nrounds	Number of boosting iterations
gain	Fractional contribution of each feature to the model

Table 4 shows the results of the parameter selection for the XGBoost method. Recall that a lower learning rate *eta* implies a greater number of iterations *nround*, but a too large *nround* can lead to overfitting. Combination (*eta* = 0.02, *nrounds* = 3400) provided the lowest *RMSE* and the highest *R-squared* scores for the test data, whereas (*eta* = 0.01, *nrounds* = 5700) got the lowest *MAPE*. However, any pair of parameters in Table 4 could be appropriate because they lead similar accuracy.

Table 4. Results of the parameter selection for the XGBoost method.

XGBoost Pred. Horizon = 48 h	eta = 0.01, nrounds = 5700	eta = 0.02, nrounds = 3400	eta = 0.05, nrounds = 1700	eta = 0.10, nrounds = 566
RMSE_train (kWh)	11.91	11.02	10.02	12.50
RMSE_test (kWh)	23.74	23.65	23.92	24.26
R-squared_train	0.988	0.989	0.991	0.986
R-squared_test	0.946	0.946	0.945	0.943
MAPE_train (%)	5.03	4.76	4.45	5.28
MAPE_test (%)	8.98	9.00	9.12	9.23
E_mean_train (kWh)	0.00	0.00	0.00	0.00
E_mean_test (kWh)	−0.16	−0.35	−0.47	−0.09
E_skewness_train	0.31	0.29	0.24	0.28
E_skewness_test	0.14	0.02	0.09	0.04
E_kurtosis_train	6.63	6.28	5.64	6.51
E_kurtosis_test	7.58	7.68	7.41	7.53
Computational time	13 min	8 min	4 min	1.5 min

Tables 5 and 6 show the results that were obtained for the best parameter selection of each ensemble method. They also include the comparison with traditional and simple forecasting models, such as naïve (prediction at hour h is given by the real consumption at hour h-168) and multiple linear regression (MLR) with the same predictors, as used in the ensemble methods. According to Table 5, XGBoost method provides nearly null bias, more symmetry of the errors than the other ensemble methods and the traditional ones, as well as values of the kurtosis that are closer to zero (considered desired properties for residual in forecasting techniques).

Table 5. Descriptive measures of the errors for each ensemble method.

Pred. Horizon = 48 h	Bagging	RForest	CForest	XGBoost	MLR	Naïve
Optimal parameters	ntree = 200, mtry = 53	ntree = 200, mtry = 20	ntree = 3, mtry = 53	max_depth = 6, subsample = 0.5, eta = 0.02, nrounds = 3400	number of predictors = 53	lag = 168 h
Error_mean_train (kWh)	0.056	0.04	0.50	0.00	0.00	0.36
Error_mean_test (kWh)	0.25	−0.13	1.41	−0.35	−3.41	1.31
Error_skewness_train	1.48	1.46	1.54	0.29	0.64	0.49
Error_skewness_test	1.19	0.61	1.81	0.02	0.61	0.35
Error_kurtosis_train	31.12	27.12	23.44	6.28	8.14	13.39
Error_kurtosis_test	13.18	10.19	15.68	7.68	7.13	12.28

Although bagging and random forest provide the best accuracy in the training dataset (see Table 6), XGBoost fits better in the test dataset (in this case, gradient boosting avoid more overfitting than the others ensemble methods due to a suitable selection of the parameters). Furthermore, when comparing the results of random forest and XGBoost, we can state that the latter fits lightly better and it is twelve times faster to compute. Table 6 also shows that all ensemble methods significantly improve the accuracy of the predictions with respect to MLR and naïve models.

It is also important to remark that, for all methods, roughly half of the predictors accumulate more that 99% of the relative importance. In the case of ensemble methods, the corresponding importance measure has been computed (for example, the node impurity for random forest and the gain for XGBoost), whereas in the case of MLR, the forward stepwise selection method and R-squared were used to evaluate the relative importance of each predictor. We can also highlight the following aspects: the electricity consumption at the same hour of the previous week (predictor LOAD_lag_168) results the most important feature in all methods, the electricity load with lags 48 h and 144 h appear among the five most important predictors in all of the ensemble methods, and finally, the presence of the features WH6, WH7, FH1, and FH3 among the five most important predictors for different methods evidences that calendar variables and types of holidays are essential for this kind of customer. However, the temperature has a reduced effect on the response because it appears between the 10th and 12th position of importance (depending on the method), with a relative importance of around 1%.

Table 6. Accuracy results for the ensemble methods.

Pred. Horizon = 48 h	Bagging	RForest	CForest	XGBoost	MLR	Naïve
Optimal parameters	*ntree* = 200, *mtry* = 53	*ntree* = 200, *mtry* = 20	*ntree* = 3, *mtry* = 53	*max_depth* = 6, *subsample* = 0.5, *eta* = 0.02, *nrounds* = 3400	*number of predictors* = 53	*lag* = 168 h
RMSE_train (kWh)	8.83	8.79	25.86	11.02	42.34	52.6
RMSE_test (kWh)	27.65	25.45	33.09	23.65	40.67	50.87
R-squared_train	0.99	0.99	0.94	0.99	0.844	0.76
R-squared_test	0.93	0.94	0.89	0.95	0.841	0.75
MAPE_train (%)	2.6	2.69	8	4.76	18.05	14.6
MAPE_test (%)	9.6	9.33	10.82	8.99	19.33	15.63
Total number of predictors	53	53	53	53	53	0
Number important predictors (cumulative importance >99%)	24	28	25	25	30	Not applicable
Top 5 important predictors	LOAD_lag_168 (77.99%) FH1 (2.88%) LOAD_lag_144 (2.63%) FH3 (1.97%)	LOAD_lag_168 (42.55%) LOAD_lag_144 (16.37%) LOAD_lag_48 (9.26%) LOAD_lag_120 (6.67%) WH7 (5.13%)	LOAD_lag_168 (58.15%) WH7 (10.57%) WH6 (6.37%) LOAD_lag_48 (5.23%) LOAD_lag_144 (3.89%)	LOAD_lag_168 (75.85%) LOAD_lag_48 (3.76%) LOAD_lag_144 (3.52%) FH1 (2.86%) FH3 (1.93%)	LOAD_lag_168 (91.81%) FH3 (1.52%) LOAD_lag_48 (0.93%) FH1 (0.69%) WH7 (0.46%)	Not applicable
Computational time	46 min	18 min	1.5 min	8 min	0 min	0 min

In order to compare the accuracy for the different types of day, the days of the test data (2016) were divided in two groups: special days, which include weekends, August (official academic holidays), and all days that are determined by the dummy variables FH1, . . . , FH5 in Table 2; and, regular days, which include the rest of the days. Results are exposed in Table 7. Notice that the lowest *MAPE* scores are always reached for regular days.

Table 7. MAPE (%) for regular and special days in 2016.

Pred. Horizon = 48h	Bagging	RForest	CForest	XGBoost
MAPE regular days (149)	9.07	8.60	10.44	*8.15*
MAPE special days (217)	9.97	9.83	11.08	9.57
MAPE total days (366)	9.60	9.33	10.82	8.99

Figure 1a,b show the monthly evolution of two goodness-of-fit measures (*RMSE* and *MAPE*). Remark that accuracies of random forest and XGBoost are quite similar, with greatest differences in January and March (due to lack of accuracy in Christmas and Eastern days). Also, the models fit better for night hours (from 10 p.m. to 5 a.m.) due to the absence of activity during that period (see Figure 2a,b).

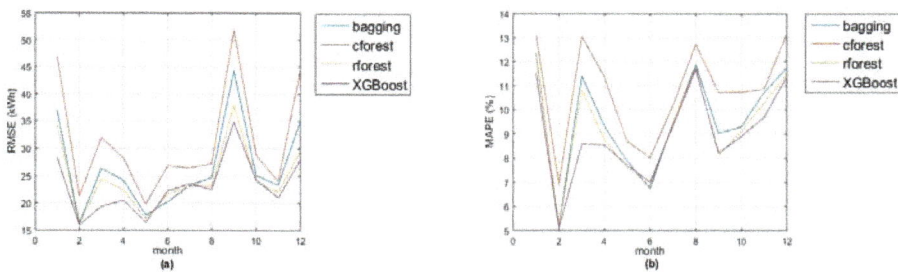

Figure 1. Goodness-of-fit measures for each month in 2016 and each ensemble method: (**a**) using root mean square error (RMSE) (kWh); and, (**b**) using mean absolute percentage error (MAPE) (%).

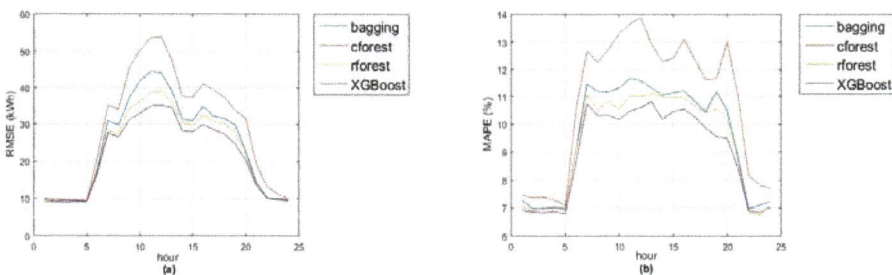

Figure 2. Goodness-of-fit measures for each ensemble method by hour of the day in 2016: (**a**) using RMSE (kWh); (**b**) using MAPE (%).

As an example, Figure 3 shows the actual and prediction load for a complete week in May 2016.

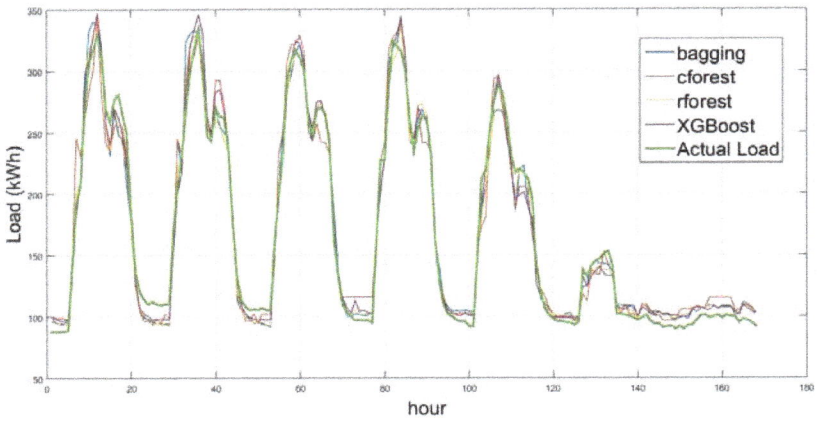

Figure 3. Actual and forecasting load (kWh) for a week (9–15 May 2016).

Finally, in this section, we analyze the adecuacy of the forecasting method XGBoost for the case study when considering different prediction horizons (1 h, 2 h, 12 h, 24 h, and 48 h). In all cases, we selected the same parameters: *subsample* = 0.5, *max_depth* = 6, *eta* = 0.05 and *nrounds* = 1700. Accuracy results for the training and test datasets are given in Table 8 as well as the most important predictors in each case.

Table 8. Results for XGBoost method and different prediction horizons.

XGBoost	Pred. Horizon = 1 h	Pred. Horizon = 2 h	Pred. Horizon = 12 h	Pred. Horizon = 24 h	Pred. Horizon = 48 h
RMSE_train (kWh)	4.37	5.45	7.51	8.59	10.02
RMSE_test (kWh)	10.77	13.91	22.15	22.44	23.92
R-squared_train	0.998	0.997	0.995	0.993	0.991
R-squared_test	0.989	0.981	0.953	0.951	0.945
MAPE_train (%)	1.85	2.37	3.38	7.09	4.45
MAPE_test (%)	3.50	4.67	7.51	8.14	9.12
Total number of predictors	99	97	77	55	53
Number important predictors (cumulative importance >99%)	34	47	38	20	26
Top 5 important predictors	LOAD_lag_1 (85.24%) LOAD_lag_168 (8.22%) LOAD_lag_24 (7.94%) H7 (3.92%) LOAD_lag_5 (3.60%)	LOAD_lag_168 (60.92%) LOAD_lag_2 (23.81%) LOAD_lag_24 (2.59%) LOAD_lag_12 (1.47%) FH1 (0.85%)	LOAD_lag_168 (72.87%) LOAD_lag_24 (8.72%) WH6 (2.58%) FH1 (1.92%) FH3 (1.83%)	LOAD_lag_168 (73.40%) LOAD_lag_24 (9.54%) WH6 (2.57%) FH1 (2.07%) FH3 (1.85%)	LOAD_lag_168 (75.63%) LOAD_lag_48 (3.79%) LOAD_lag_144 (3.33%) FH1 (2.86%) FH3 (1.95%)
Computational time	7 min	7 min	5 min	4.5 min	4 min

Obviously, the best accuracies are obtained for the shortest prediction horizon (1 h), where the most important feature is the consumption at the previous hour (lag = 1) with more than 85% of relative importance. However, for the rest of prediction horizons, the most important predictor is, once again, the load with lag 168 h.

4. Direct Market Consumers

Direct Consumers in the market by point of supply or installation are those consumers of electric energy who purchase electricity directly on the production market for their own consumption and who meet some specific conditions. Firstly, this section is dealing with the performance of the Spanish Market, specifically the aspects that are related to DMC type of supply. Secondly, the components that define the price of the energy as a DMC are introduced. Finally, the results for the case of a campus university are shown.

4.1. Law Framework for DMC and Market Performance

Law 24/2013 of 26 December [41] defines the Direct Consumers in the Spanish Market in article 6.g) and establishes its rights and obligations. The activity of these subjects is regulated in Royal Decree 1955/2000 of 1 December [42], which regulates the activities of transportation, distribution, retailing, supply, and authorization procedures of Power Systems. To start the activity of qualified consumer in the market, the interested party must send several documents to different official bodies and fulfill a series of requirements, such as: have provided the System Operator with sufficient guarantee to cover economic obligations and to have the status of market agent, among others. Currently, the list of DMC includes around 200 consumers, most of them small and medium companies (see [43]).

The Day-ahead Market, as part of the electric power production market, aims to carry out electric power transactions for the next day by resolving offers and bids that are offered by market agents. The Market Operator ranks and matches selling offers with buying bids for electricity (received before 10:00 a.m. on the day before the dispatch), using the simple or complex matching method, according to simple or there are offers that incorporate complex conditions.

According to that, DMC must make their bids for the day D (day of dispatch) before 10:00 a.m. of day D-1 (day before the dispatch), so nearly two-day-ahead forecasting models for the demand are needed. After this process, the System Operator established the Daily Base Program, which is published at 12:00, based on the program resulting from the Market Operator program for the Day-Ahead Market and the communication of the execution of bilateral contracts. The Intraday Market aims to meet the Definitive Viable Daily Program through the presentation of energy offers and bids by the markets agents. The final scheduling is the result of the aggregation of all the firm transactions that are formalized for each programming period as a consequence of the viable daily program and market matching intraday once the technical restrictions identified have been resolved and the subsequent rebalancing has taken place. Finally, generation and demand deviations arising from the closing of the final scheduling are managed by the System Operator through balance management procedures and the provision of secondary and tertiary regulation services.

4.2. Price of the Energy Participating as a DMC

The final price of the energy consumed as a DMC consists of three clearly differentiated components, as described below.

- Regulated prices: these are prices set by the State and also depend on the supply rate. This component includes access fees, capacity payments and loss coefficients. This component does not depend on the type of supply, thus the corresponding cost would be the same for consumers through retailers and DMC.

- Taxes: they are also regulated prices, although of a different nature from the previous ones. This component is given by the special tax on electricity (currently 5113%) and VAT (currently 21%). This component is also common for all consumers.
- Unregulated prices: this component of the billing contemplates the price for the energy consumed in wholesale market and therefore it is not regulated by the State. It includes the price of energy in the Day-ahead and Intraday Market, costs for bilateral contracts, costs for measured deviations (difference between energy consumed and programmed energy), and costs for ancillary services.

Therefore, the Final Cost of the energy for a Direct Market Consumer is given by:

$$
\begin{aligned}
FinalCostDMC = \ &RegulatedPricesComponent \\
&+ UnregulatedPricesComponent + Taxes
\end{aligned}
\tag{12}
$$

The price of energy in the Day-ahead or Daily Market, which is also called the marginal price, is the result of matching sales and purchase offers managed the day before the energy dispatch. It is therefore a non-regulated component of the billing. The price of energy in the Day-ahead Market is determined for each of the 24 h of the day as a result of the matching, values that are available on the website of the System Operator [44] (Red Eléctrica Española, REE). It is the largest component (more than 80%) of the average final price, as it is shown in Figure 4.

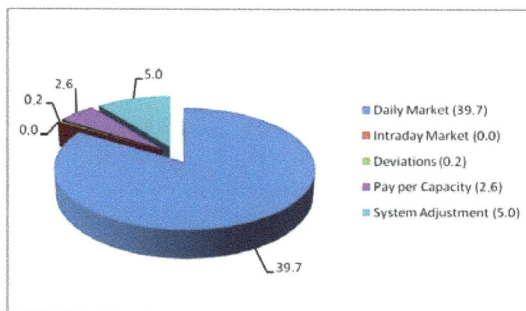

Figure 4. Components of the Average Final Price in 2016, price for 1 MWh in euros.

As in the Daily Market, the price of energy in the Intraday Market is the result of the negotiation of sales and purchase offers managed in the sessions held a few hours before the dispatch of energy (intraday sessions), and both are variable and unregulated prices. The price for each hour of the day and each intraday session are posted on the website of the System Operator, in this case REE [44].

Once the daily scope of the agents, consumers, and generators programs has been reached, the processes of liquidation of their energies (charges and payments) actually produced and consumed are entered, with each passing the costs of the deviation that they have incurred by have "failed" their respective programs of production and consumption. Thus, those who have deviated to rise at a certain time (generators that have produced more than their program and consumers who have consumed less than their programs) are passed on the corresponding cost in case that deviation has gone in the opposite direction (the generators charge a price lower than the marginal price of the hour for their additional production, and consumers receive a price lower than the marginal price they paid in that hour for their lower consumption), while if their deviation was in the same sense of the needs of the system, no cost is passed on to them (generators charge the marginal and consumers receive the marginal). Identical reasoning governs the case of deviations to go down, in which producers have generated less energy than their program and consumers have consumed more than what is established in their schedule.

In order to compare the real electricity bill in 2016 of the customer (the campus university) with the one that it would have had acting as a DMC, the parts that are different in both of the bills have been emphasized. The real electricity bill emitted by the supplier in 2016 (a retailer) consists of three components: the access fees (regulated price), the taxes, and the "referenced energy" (which includes some regulated prices such as the capacity payments or loss coefficients and all of the unregulated prices).

$$FinalCostRetailer = AccessFees + ReferencedEnergy + Taxes \qquad (13)$$

Taking into account that the access fees and taxes are the same for the two types of supply (retailer and DMC), the cost of the "referenced energy" for 2016 is analyzed.

For a DMC, the hourly cost of the (referenced) energy is given by the following sum of costs:

$$E(h) = ECBC(h) + DMP(h) \cdot EDM(h) + IMP(h) \cdot EIM(h) + SAC(h) \cdot EMCB(h) \\ + MDP(h) \cdot EMD(h) + CPP(h) \cdot EMCB(h) \qquad (14)$$

where:

- $E(h)$ = Energy cost in the hour "h", in €.
- $ECBC(h)$ = Energy cost in the hour "h" from bilateral contracts, in €.
- $DMP(h)$ = Daily Market price in the hour "h", in €/kWh.
- $EDM(h)$ = Energy bought in the Daily Market in the hour "h", in kWh.
- $IMP(h)$ = Intraday Market price in the hour "h", in €/kWh.
- $EIM(h)$ = Energy bought in the Intraday Market in the hour "h", in kWh.
- $SAC(h)$ = System adjustment cost passed on to the DMC in the hour "h", in €/kWh.
- $EMCB(h)$ = Energy measured in Central Bars in the hour "h", in kWh.
- $MDP(h)$ = Measured Deviations price in the hour "h", in €/kWh.
- $EMD(h)$ = Measured Deviation of Energy in the hour "h" = Difference between consumed energy and programmed energy in the hour "h", in kWh.
- $CPP(h)$ = Capacity payment price in the hour "h", in €/kWh.

In this paper, it is assumed that the DMC in the study (the campus university) does not participate in bilateral contracts nor in the Intraday Market, thus the hourly cost of the energy reduces to:

$$E(h) = DMP(h) \cdot EDM(h) + SAC(h) \cdot EMCB(h) + MDP(h) \cdot EMD(h) + CPP(h) \cdot EMCB(h) \quad (15)$$

It is mandatory for the Spanish Regulator (Comisión Nacional del Mercado y la Competencia, CNMC) to publish on its website a document with the criterion used to calculate the average final price (AFP) of energy in the market. The AFP (see Figure 4) represents an approximate value of the cost of electric energy per kWh, being only a reference that can vary to a greater or lesser extent from the actual final price, depending on the consumer. Specifically, the capacity payments and the deviations between energy consumed and programmed, are those that can mark greater differences between the real cost of the invoicing and the cost resulting from using the average final price. As an additional objective, we compare the real cost acting as a DMC with the resulting cost using the AFP.

4.3. Case Study: A Campus University as a DMC

To date, all the dependencies of the Technical University of Cartagena have contracted supply with a retailer, which is the modality of supplying of almost all consumers in high voltage of the Spanish electrical system. Only around 200 consumers have dared to participate in the Market as DMC, see the list in [43]. In 2016, the contracted tariff for the Alfonso XIII campus was the ATR 6.1, 6-period high voltage tariff, with a supply voltage of 20 kV.

As it has been stated before, the final price of the campus university's invoice is composed of the access fees (which refers to the use of the network), the taxes, and the price of the energy freely agreed with the retailer (which refers to the value of the energy consumed). Note that the concepts corresponding to access fees (power and energy terms) and taxes are independent of the mode of supply, so they do not change for a Direct Consumer. Therefore, the calculation of the cost of the referenced energy for the DMC and its comparison with the retailer cost, is the main concern for this study.

Recall that, under the assumptions of this study, the hourly energy cost as a DMC is given by the sum of four components: the cost in the Daily Market (DM cost), the adjustment services (AS cost), the measured deviations (MD cost), and the capacity payments (CP cost). Table 9 shows the value of each component when the cost of the energy as a DMC is evaluated. In this section, 48-h-ahead predictions obtained with the XGBoost method (*eta* = 0.02, *nrounds* = 3700) were used, although any of the other ensemble methods would lead to similar results. It is worth to mention that the cost of deviations is quite limited due to the accuracy of the load forecasting method.

Table 9. Monthly cost acting as a Direct Market Consumers (DMC) and its components.

Month	DM Cost (in €)	AS Cost (in €)	MD Cost (in €)	CP Cost (in €)	DMC Cost (in €)
January	5478	685	91	313	6567
February	4492	815	56	409	5772
March	3644	763	45	99	4551
April	2980	649	70	105	3804
May	3976	801	43	127	4948
June	6692	682	42	336	7752
July	7151	610	28	524	8313
August	4450	430	56	0	4936
September	8013	724	57	195	8989
October	8289	708	48	130	9176
November	7960	474	91	140	8665
December	8727	492	46	285	9575
Total 2016	71,853 (86.52%)	7834 (9.44%)	697 (0.84%)	2664 (3.2%)	83,048

Table 10 shows the electricity consumption (in kWh) of the campus university in 2016 and the cost of the referenced energy (consumption) in four cases: the real cost paid to the retailer, the cost using the Average Final Price (AFP), acting as a DMC, and what we call the pessimist price (a Direct Consumer with all the deviations against the system). According to the results, it can be established that DMC modality would have produced savings of around 11% in the energy term of the invoice when compared to the retail price. Note also that the cost using the AFP does not coincide with the cost of the DMC because the cost due to deviations and the capacity payments components depend on the consumer. On the other hand, the results show that, even in the pessimistic case (all deviations of the predictions against the system), the DMC type of supply is worthy against the retailer.

It is important to highlight that the economic benefits of the DMC type of supply depend on two main aspects: the magnitude of the deviations and the direction of the deviations (towards or against the system). The first aspect (magnitude of the deviations) is determined by the accuracy of the forecasting method. However, the second aspect (direction of the deviations) is out of our control and it depends on the whole Electric System. In particular, some worse forecasting methods could lead to greater benefits than more accuracy methods, but only by chance and assuming that the forecasting values are good enough (moderate deviations). Therefore, the load forecasting method is important to some extent, but obviously lower deviations are preferable to greater deviations.

Table 10. Comparison of costs in four cases: average final price (AFP), pessimist, DMC, and retailer.

Month	Consumption kWh	AFP (in €)	Pessimist (in €)	DMC (in €)	Retailer (in €)	Saving %
January	125,702	6643	6677	6567	7434	12
February	136,620	5834	5821	5772	6760	15
March	119,103	4778	4628	4551	5338	15
April	108,475	3965	3874	3804	4346	12
May	130,149	5164	5001	4948	5571	11
June	157,785	7953	7802	7752	8815	12
July	160,212	8423	8361	8313	9315	11
August	100,343	5133	4957	4936	5477	10
September	167,116	9272	9040	8989	10,036	10
October	141,077	9410	9213	9176	9953	8
November	127,613	8818	8691	8665	9534	9
December	130,583	9717	9634	9575	10,524	9
Total 2016	1,604,778	85,111	83,698	83,048	93,103	11

5. Conclusions

Load forecasting has been an important concern to provide accurate estimates for the operation and planning of Power Systems, but it can also arise as an important tool to engage and empower customers in markets, for example for decision making in electricity markets.

In this paper, we propose the using of different ensemble methods that are based on regression trees as alternative tools to obtain short-term load predictions. The main advantages of this approach are the flexibility of the model (suitable for linear and non-linear relationships), they take into account interactions among the predictors at different levels, no assumption or transformations on the data are needed, and they provide very accurate predictions.

Four ensemble methods (bagging, random forest, conditional forest, and boosting) were applied to the electricity consumption of the campus Alfonso XIII of the Technical University of Cartagena (Spain). In addition to historical load data, some calendar variables and historical temperatures were considered, as well as dummy variables representing different types of special days in the academic context (such as exams periods, tutorial periods, or academic festivities). The results show the effectiveness of the ensemble methods, mainly random forest, and a recent variant of gradient boosting called the XGBoost method. It is also worth to mention the fast computational time of the latter.

To illustrate the utility of this load-forecasting tool for a medium-size customer (a campus university), predictions with a horizon of 48h were obtained to evaluate the benefits that are involved in the change from tariffs to price of wholesale markets in Spain. This possibility provides an interesting option for the customer (a reduction of around 11% in electricity costs).

Author Contributions: M.d.C.R.-A. and A.G. (Antonio Gabaldón) conceived, designed the experiments and wrote the part concerning load forecasting. A.G. (Antonio Guillamón) and M.d.C.R.-A. collected the data, developed and wrote the part concerning the direct market consumer. All authors have approved the final manuscript.

Funding: This research was funded by the Ministerio de Economía, Industria y Competitividad (Agencia Estatal de Investigación, Spanish Government) under research project ENE-2016-78509-C3-2-P, and EU FEDER funds. The third author is also partially funded by the Spanish Government through Research Project MTM2017-84079-P (Agencia Estatal de Investigación and Fondo Europeo de Desarrollo Regional). Authors have also received funds from these grants for covering the costs to publish in open access.

Acknowledgments: This work was supported by the Ministerio de Economía, Industria y Competitividad (Agencia Estatal de Investigación, Spanish Government) under research project ENE-2016-78509-C3-2-P, and EU FEDER funds. The third author is also partially funded by the Spanish Government through Research Project MTM2017-84079-P (Agencia Estatal de Investigación and Fondo Europeo de Desarrollo Regional). Authors have also received funds from these grants for covering the costs to publish in open access.

Conflicts of Interest: The authors declare no conflict of interest.

References

1. Hahn, H.; Meyer-Nieberg, S.; Pickl, S. Electric load forecasting methods: Tools for decision making. *Eur. J. Oper. Res.* **2009**, *199*, 902–907. [CrossRef]
2. Alfares, H.K.; Nazeeruddin, M. Electric load forecasting: Literature survey and classification of methods. *Int. J. Syst. Sci.* **2002**, *33*, 23–34. [CrossRef]
3. Yang, H.T.; Huang, C.M.; Huang, C.L. Identification of ARMAX model for short term load forecasting: An evolutionary programming approach. *IEEE Trans. Power Syst.* **1996**, *11*, 403–408. [CrossRef]
4. Taylor, W.; Menezes, L.M.; McSharry, P.E. A comparison of univariate methods for forecasting electricity demand up to a day ahead. *Int. J. Forecast.* **2006**, *22*, 1–16. [CrossRef]
5. Newsham, G.R.; Birt, B.J. Building-level occupancy data to improve arima-based electricity use forecasts. In Proceedings of the 2nd ACM Workshop on Embedded Sensing Systems for Energy-Efficiency in Building, Zurich, Switzerland, 3–5 November 2010; pp. 13–18.
6. Massana, J.; Pous, C.; Burgas, L.; Melendez, J.; Colomer, J. Shortterm load forecasting in a non-residential building contrasting models and attributes. *Energy Build.* **2015**, *92*, 322–330. [CrossRef]
7. Bruhns, A.; Deurveilher, G.; Roy, J.S. A nonlinear regression model for midterm load forecasting and improvements in seasonality. In Proceedings of the 15th Power Systems Computation Conference, Liege, Belgium, 22–26 August 2005.
8. Charytoniuk, W.; Chen, M.S.; Van Olinda, P. Nonparametric regression based short-term load forecasting. *IEEE Trans. Power Syst.* **1998**, *13*, 725–730. [CrossRef]
9. Amber, K.P.; Aslam, M.W.; Mahmood, A.; Kousar, A.; Younis, M.Y.; Akbar, B.; Chaudhary, G.Q.; Hussain, S.K. Energy Consumption Forecasting for University Sector Buildings. *Energies* **2017**, *10*, 1579. [CrossRef]
10. Tso, G.K.F.; Yau, K.K.W. Predicting electricity energy consumption: A comparison of regression analysis, decision tree and neural networks. *Energy* **2007**, *32*, 1761–1768. [CrossRef]
11. Li, K.; Su, H.; Chu, J. Forecasting building energy consumption using neural networks and hybrid neuro-fuzzy system: A comparative study. *Energy Build.* **2011**, *43*, 2893–2899. [CrossRef]
12. Liao, G.C.; Tsao, T.P. Application of a fuzzy neural network combined with a chaos genetic algorithm and simulated annealing to short-term load forecasting. *IEEE Trans. Evol. Comput.* **2016**, *10*, 330–340. [CrossRef]
13. Hippert, H.S.; Pedreira, C.E.; Souza, R.C. Neural networks for short-term load forecasting: A review and evaluation. *IEEE Trans. Power Syst.* **2001**, *16*, 44–55. [CrossRef]
14. Karatasou, S.; Santamouris, M.; Geros, V. Modeling and predicting building's energy use with artificial neural networks: Methods and results. *Energy Build.* **2006**, *38*, 949–958. [CrossRef]
15. Metaxiotis, K.; Kagiannas, A.; Askounis, D.; Psarras, J. Artificial intelligence in short term electric load forecasting: A state-of-the-art survey for the researcher. *Energy Convers. Manag.* **2003**, *44*, 1525–1534. [CrossRef]
16. Buitrago, J.; Asfour, S. Short-term forecasting of electric loads using nonlinear autoregressive artificial neural networks with exogenous vector inputs. *Energies* **2017**, *10*, 40. [CrossRef]
17. Hashmi, M.U.; Arora, V.; Priolkar, J.G. Hourly electric load forecasting using Nonlinear Auto Regressive with eXogenous (NARX) based neural network for the state of Goa, India. In Proceedings of the International Conference on Industrial Instrumentation and Control. (ICIC), Pune, India, 28–30 May 2015; pp. 1418–1423. [CrossRef]
18. Hanshen, L.; Yuan, Z.; Jinglu, H.; Zhe, L. A localized NARX Neural Network model for Short-term load forecasting based upon Self-Organizing Mapping. In Proceedings of the IEEE 3rd International Future Energy Electronics Conference and ECCE Asia (IFEEC 2017—ECCE Asia), Kaohsiung, Taiwan, 3–7 June 2017; pp. 749–754. [CrossRef]
19. Fan, G.-F.; Qing, S.; Wang, H.; Hong, W.-C.; Li, H.-J. Support Vector Regression Model Based on Empirical Mode Decomposition and Auto Regression for Electric Load Forecasting. *Energies* **2013**, *6*, 1887–1901. [CrossRef]
20. Dong, Y.; Ma, X.; Ma, C.; Wang, J. Research and Application of a Hybrid Forecasting Model Based on Data Decomposition for Electrical Load Forecasting. *Energies* **2016**, *9*, 1050. [CrossRef]
21. Dudek, G. Short-Term Load Forecasting Using Random Forests. In *Intelligent Systems'2014. Advances in Intelligent Systems and Computing*; Springer: Cham, Switzerland, 2015; pp. 821–828. ISBN 978-3-319-11310-4.

22. Hedén, W. Predicting Hourly Residential Energy Consumption Using Random Forest and Support Vector Regression: An Analysis of the Impact of Household Clustering on the Performance Accuracy, Degree-Project in Mathematics (Second Cicle). Royal Institute of Technology SCI School of Engineering Sciences. 2016. Available online: https://kth.diva-portal.org/smash/get/diva2:932582/FULLTEXT01.pdf (accessed on 4 July 2018).

23. Ahmed Mohammed, A.; Aung, Z. Ensemble Learning Approach for Probabilistic Forecasting of Solar Power Generation. *Energies* **2016**, *9*, 1017. [CrossRef]

24. Lin, Y.; Luo, H.; Wang, D.; Guo, H.; Zhu, K. An Ensemble Model Based on Machine Learning Methods and Data Preprocessing for Short-Term Electric Load Forecasting. *Energies* **2017**, *10*, 1186. [CrossRef]

25. Sistema de Información del Operador del Sistema (Esios); Red Eléctrica de España. Alta Como Consumidor Directo en Mercado Peninsular. Available online: https://www.esios.ree.es/es/documentacion/guia-alta-os-consumidor-directo-mercado-peninsula (accessed on 4 July 2018).

26. MINETAD. Ministerio de Energía, Turismo y Agenda Digital. Gobierno de España. Available online: http://www.minetad.gob.es/ENERGIA/ELECTRICIDAD/DISTRIBUIDORES/Paginas/ConsumidoresDirectosMercado.aspx (accessed on 4 July 2018).

27. Touzani, S.; Granderson, J.; Fernandes, S. Gradient boosting machine for modeling the energy consumption of commercial buildings. *Energy Build.* **2018**, *158*, 1533–1543. [CrossRef]

28. James, G.; Witten, D.; Hastie, T.; Tibshirani, R. *An Introduction to Statistical Learning*; Springer: New York, NY, USA, 2013; ISBN 978-1-4614-7138-7.

29. R Package: RandomForest. Repository CRAN. Available online: https://cran.r-project.org/web/packages/randomForest/randomForest.pdf (accessed on 4 July 2018).

30. Hothorn, T.; Hornik, K.; Zeileis, A. Unbiased Recursive Partitioning: A Conditional Inference Framework. *J. Comput. Graph. Stat.* **2006**, *15*, 651–674. [CrossRef]

31. Strasser, H.; Weber, C. On the asymptotic theory of permutation statistics. *Math. Methods Stat.* **1999**, *8*, 220–250.

32. R Package: Party. Repository CRAN. Available online: https://cran.r-project.org/web/packages/party/party.pdf (accessed on 4 July 2018).

33. Freund, Y.; Schapire, R.E. A decision-theoretic generalization of on-line learning and an application to boosting. *J. Comput. Syst. Sci.* **1997**, *55*, 119–139. [CrossRef]

34. Friedman, J.H. Greedy function approximation: A gradient boosting machine. *Ann. Stat.* **2001**, *19*, 1189–1232. [CrossRef]

35. Hastie, T.; Tibshirani, R.; Friedman, J.H. 10. Boosting and Additive Trees. In *The Elements of Statistical Learning*, 2nd ed.; Springer: New York, NY, USA, 2009; pp. 337–384. ISBN 0-387-84857-6.

36. Friedman, J.H. Stochastic Gradient Boosting. *Comput. Stat. Data Anal.* **2002**, *38*, 367–378. [CrossRef]

37. Chen, T.; Guestrin, C. XGBoost: A scalable tree boosting system. In Proceedings of the 22nd ACM SIGKDD International Conference on Knowledge Discovery and Data Mining, San Francisco, CA, USA, 13–17 August 2016; pp. 785–794.

38. R Package: Xgboost. Repository CRAN. Available online: https://cran.r-project.org/web/packages/xgboost/xgboost.pdf (accessed on 4 July 2018).

39. Pérez-Lombard, L.; Ortiz, J.; Pout, C. A review on buildings consumption information. *Energy Build.* **2008**, *40*, 394–398. [CrossRef]

40. Strobl, C.; Boulesteix, A.L.; Kneib, T.; Augustin, T.; Zeileis, A. Conditional Variable Importance for Random Forests. *BMC Bioinf.* **2008**, *9*, 307. [CrossRef] [PubMed]

41. Boletín Oficial del Estado. Ley 24/2013, de 26 de Diciembre, del Sector Eléctrico. Available online: https://www.boe.es/buscar/doc.php?id=BOE-A-2013-13645 (accessed on 4 July 2018).

42. Boletín Oficial del Estado. Real Decreto 1955/2000, de 1 de Diciembre, Por el Que se Regulan las Actividades de Transporte, Distribución, Comercialización, Suministro y Procedimientos de Autorización de Instalaciones de Energía Eléctrica. Available online: http://www.boe.es/buscar/act.php?id=BOE-A-2000-24019&tn=1&p=20131230&vd=#a70 (accessed on 4 July 2018).

Energies **2018**, *11*, 2038

43. Comisión Nacional de Los Mercados y la Competencia (CNMC). List of Direct Market Consumers. Available online: https://www.cnmc.es/ambitos-de-actuacion/energia/mercado-electrico#listados (accessed on 4 July 2018).

44. Red Eléctrica Española (REE). Electricity Market Data. Available online: https://www.esios.ree.es/es/analisis (accessed on 4 July 2018).

energies

MDPI

Article

Short-Term Load Forecasting of Natural Gas with Deep Neural Network Regression [†]

Gregory D. Merkel, Richard J. Povinelli * [iD] and Ronald H. Brown

Opus College of Engineering, Marquette University, Milwaukee, WI 53233, USA;
gregory.merkel@marquette.edu (G.D.M.); ronald.brown@marquette.edu (R.H.B.)
* Correspondence: richard.povinelli@marquette.edu; Tel.: +1-414-288-7088
† This work is an extension of the paper "Deep neural network regression for short-term load forecasting of natural gas" presented at the International Symposium on Forecasting, 17–20 June 2015, Cairns, Australia, and is published in their proceedings.

Received: 29 June 2018; Accepted: 1 August 2018; Published: 2 August 2018

Abstract: Deep neural networks are proposed for short-term natural gas load forecasting. Deep learning has proven to be a powerful tool for many classification problems seeing significant use in machine learning fields such as image recognition and speech processing. We provide an overview of natural gas forecasting. Next, the deep learning method, contrastive divergence is explained. We compare our proposed deep neural network method to a linear regression model and a traditional artificial neural network on 62 operating areas, each of which has at least 10 years of data. The proposed deep network outperforms traditional artificial neural networks by 9.83% weighted mean absolute percent error (*WMAPE*).

Keywords: short term load forecasting; artificial neural networks; deep learning; natural gas

---

1. Introduction

This manuscript presents a novel deep neural network (DNN) approach to forecasting natural gas load. We compare our new method to three approaches—a state-of-the-art linear regression algorithm and two shallow artificial neural networks (ANN). We compare our algorithm on 62 datasets representing many areas of the U.S. Each dataset consists of 10 years of training data and 1 year of testing data. Our new approach outperforms each of the existing approaches. The remainder of the introduction overviews the natural gas industry and the need for accurate natural gas demand forecasts.

The natural gas industry consists of three main parts; production and processing, transmission and storage, and distribution [1]. Like many fossil fuels, natural gas (methane) is found underground, usually near or with pockets of petroleum. Natural gas is a common byproduct of drilling for petroleum. When natural gas is captured, it is processed to remove higher alkanes such as propane and butane, which produce more energy when burned. After the natural gas has been processed, it is transported via pipelines directly to local distribution companies (LDCs) or stored either as liquid natural gas in tanks or back underground in aquifers or salt caverns. The natural gas is purchased by LDCs who provide natural gas to residential, commercial, and industrial consumers. Subsets of the customers of LDCs organized by geography or municipality are referred to as operating areas. Operating areas are defined by the individual LDCs and can be as large as a state or as small as a few towns. The amount of natural gas used often is referred to as the load and is measured in dekatherms (Dth), which is approximately the amount of energy in 1000 cubic feet of natural gas.

For LDCs, there are several uses of natural gas, but the primary use is for heating homes and business buildings, which is called heatload. Heatload changes based on the outside temperature.

During the winter, when outside temperatures are low, the heatload is high. When the outside temperature is high during the summer, the heatload is approximately zero. Other uses of natural gas, such as cooking, drying clothes, and heating water and other household appliances, are called baseload. Baseload is generally not affected by weather and typically remains constant throughout the year. However, baseload may increase with a growth in the customer population.

Natural gas utility operations groups depend on reliable short-term natural gas load forecasts to make purchasing and operating decisions. Inaccurate short-term forecasts are costly to natural gas utilities and customers. Under-forecasts may require a natural gas utility to purchase gas on the spot market at a much higher price. Over-forecasts may require a natural gas utility to store the excess gas or pay a penalty.

In this paper, we apply deep neural network techniques to the problem of short term load forecasting of natural gas. We show that a moderately sized neural network, trained using a deep neural network technique, outperforms neural networks trained with older techniques by an average of 0.63 (9.83%) points of weighted mean absolute percent error (*WMAPE*). Additionally, a larger network architecture trained using the discussed deep neural network technique results in an additional improvement of 0.20 (3.12%) points of *WMAPE*. This paper is an extension of Reference [2].

The rest of the manuscript is organized as follows. Section 2 provides an overview of natural gas forecasting, including the variables used in typical forecasting models. Section 3 discusses prior work. Section 4 provides an overview of ANN and DNN architecture and training algorithms. Section 5 discusses the data used in validating our method. Section 6 describes the proposed method. Section 7 explains the experiments and their results. Section 8 provides conclusions.

2. Overview of Natural Gas Forecasting

The baseload of natural gas consumption, which does not vary with temperature for an operating area, typically changes seasonally and slowly as the number of customers, or their behavior, changes. Given the near steady nature of baseload, most of the effort in forecasting natural gas load focuses on predicting the heatload (load which varies with temperature). Hence, the most important factor affecting the natural gas load is the weather.

Figure 1 shows that natural gas load has a roughly linear relationship with temperatures above 65 °F. For this reason, it is important to consider a variety of temperature-related exogenous variables as potential inputs to short-term load forecasting models. This section discusses a few of these exogenous variables, which include heating degree day (HDD), dew point (DPT), cooling degree day (CDD), day of the week (DOW), and day of the year (DOY).

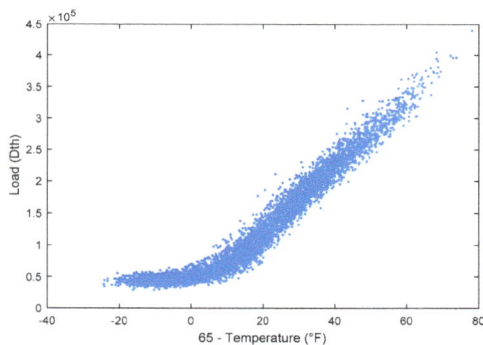

Figure 1. Weighted combination of several midwestern U.S. operating areas, including Illinois, Michigan, and Wisconsin. Authors obtained data directly from local distribution companies. The data is from 1 January 2003 to 19 March 2018.

Note the kink in the trend of Figure 1 at about 65 °F. At temperatures greater than 65 °F, residential and commercial users typically stop using natural gas for heating. At temperatures greater than 65 °F, only the baseload remains. Thus, heating degree days (HDD) are used as inputs to forecasting models,

$$HDD = \max(0, T_{ref} - T),\tag{1}$$

where T is the temperature, and T_{ref} is the reference temperature [3]. Reference temperature is indicated by concatenating it to HDD, i.e., HDD65 indicates a reference temperature of 65 °F.

Several other weather-based inputs can be used in forecasting natural gas, such as wind-adjusted heating degree day (HDDW); dew point temperature (DPT), which captures humidity; and cooling degree days (CDD),

$$CDD = \max(0, T - T_{ref})\tag{2}$$

and is used to model temperature-related effects above T_{ref} as seen in Figure 1.

In addition to weather inputs, time variables are important for modeling natural energy demand [4]. Figure 2 illustrates the day of the week (DOW) effect. Weekends (Friday–Sunday) have less demand than weekdays (Monday–Thursday). The highest demand typically occurs on Wednesdays, while the lowest demand generally occurs on Saturdays. A day of the year (DOY) variable is also important. This allows homeowner behaviors between seasons to be modeled. In September, a 50 °F temperature will cause few natural gas customers to turn on their furnaces, while in February at 50 °F all furnaces will be on.

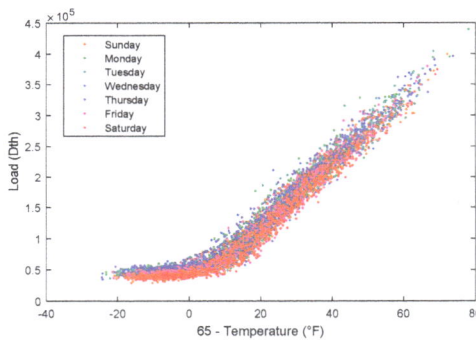

Figure 2. The same data as in Figure 1 colored by day of the week.

3. Prior Work

Multiple linear regression (LR) and autoregressive integrated moving average (ARIMA) are common models for forecasting short-term natural gas demand [5]. Vitullo et al. propose a five-parameter linear regression model [5]. Let $\hat{s}$ be the day ahead forecasted natural gas demand, HDD65 be the forecasted HDD with a reference temperature of 65 °F, HDD55 be the forecasted HDD with a reference temperature 55 °F, and CDD65 be the forecasted CDD with a reference temperature 65 °F. Let ΔHDD65 be the difference between the forecasted HDD65 and the prior day's actual HDD65. Then, Vitullo's model is described as

$$\hat{s} = \beta_0 + \beta_1 HDD65 + \beta_2 HDD55 + \beta_3 \Delta HDD65 + \beta_4 CDD65.\tag{3}$$

β_0 is the natural gas load not dependent on temperature. The natural gas load dependent on temperature is captured by the sum of β_1 and β_2. The two reference temperatures better model the smooth transition from heating to non-heating days. β_3 accounts for recency effects [5,6]. Finally, β_4 models small, but not insignificant, temperature effects during non-heating days.

While the Vitullo model and other linear models perform well on linear stationary time-series, they assume that load has roughly a linearly relationship with temperature [7]. However, natural gas demand time series is not purely linear with temperature. Some of the nonlinearities can be modeled using heating and cooling degrees, but natural gas demand also contains many smaller nonlinearities that cannot be captured easily with linear or autoregressive models even with nonlinear transformations of the data.

To address these nonlinearities, forecasters have used artificial neural networks (ANNs) in place of, or in conjunction with, linear models [5,8,9]. ANNs are universal approximators, meaning that with the right architecture, they can be used to model almost any regression problem [8]. Artificial neural networks are composed of processing nodes that take a weighted sum of their inputs and then output a nonlinear transform of that sum.

Recently, new techniques for increasing the depth (number of layers) of ANNs have yielded deep neural networks (DNN) [10]. DNNs have been applied successfully to a range of machine learning problems, including video analysis, motion capture, speech recognition, and image pattern detection [10,11].

As will be described in depth in the next section, DNNs are just large ANNs with the main difference being the training algorithms. ANNs are typically trained using gradient descent. Large neural networks trained using gradient descent suffer from diminishing error gradients. DNNs are trained using the contrastive divergence algorithm, which pre-trains the model. The pre-trained model is fine-tuned using gradient descent [12].

This manuscript adapts the DNNs to short-term natural gas demand forecasting and evaluates DNNs' performance as a forecaster. Little work has been done in the field of time series regression using DNNs, and almost no work has been done in the field of energy forecasting with DNNs. One notable example of literature on these subjects is Qui et al., who claim to be the first to use DNNs for regression and time series forecasting [13]. They show promising results on three electric load demand time series and several other time series using 20 DNNs ensembled with support vector regression. However, the DNNs they used were quite small; the largest architecture consists of two hidden layers of 20 neurons each. Because of their small networks, Qui et al. did not take full advantage of the DNN technology.

Another example of work in this field is Busseti et al. [14], who found that deep recurrent neural networks significantly outperformed the other deep architectures they used for forecasting energy demand. These results are interesting but demonstrated poor performance when compared to the industry standard in energy forecasting, and they are nearly impossible to replicate given the information in the paper.

Some good examples of time series forecasting using DNNs include Dalto, who used them for ultra-short-term wind forecasting [15], and Kuremoto et al. [16], who used DNNs on the Competition on Artificial Time Series benchmark. In both applications, DNNs outperformed neural networks trained by backpropagation. Dalto capitalized on the work of Glorot and Bengio when designing his network and showed promising results [17]. Meanwhile, Kuremoto successfully used Kennedy's particle swarm optimization in selecting their model parameters [18]. The work most similar to ours is Ryu et al., who found that two different types of examined DNNs performed better on short-term load forecasting of electricity than shallow neural networks and a double seasonal Holt-Winters model [19].

Other, more recent examples of work in this field include Kuo and Huang [20], who use a seven-layer convolutional neural network for forecasting energy demand with some success. Unfortunately, they do not use any weather information in their model which results in poor forecasting accuracy compared to those who do account for weather. Li et al. used a DNN combined with hourly consumption profile information to do hourly electricity demand forecasting [21]. Chen et al. used a deep residual network to do both point and probabilistic short-term load forecasting of natural gas [22]. Perhaps the most similar recent work to that which is presented in this paper is Hosein and Hosein, who compared a DNN without RBM pretraining to one with RBM pretraining on short-term load forecasting of electricity. They found that the pretrained DNN performed better, especially as network size increased [23].

Given the successful results of these deep neural network architectures on similar problems, it is expected that DNNs will surpass ANNs in many regression problems, including the short-term load forecasting of natural gas. This paper explores the use of DNNs to model a natural gas system by comparing the performance of the DNN to various benchmark models and the current state-of-the-art models.

4. Artificial and Deep Neural Networks

This section provides an overview of ANNs and DNNs and how to train them to solve regression problems. An ANN is a network of nodes. Each node sums its inputs and then nonlinearly transforms them. Let x_i represent the ith input to the node of a neural network, w_i the weight of the ith input, b the bias term, n the number of inputs, and o the output of the node. Then

$$o = \sigma\left(\sum_{i=1}^{n} w_i x_i + b\right), \tag{4}$$

where

$$\sigma(x) = \frac{1}{1 + e^{-x}}. \tag{5}$$

This type of neural network node is a sigmoid node. However other nonlinear transforms may be used. For regression problems, the final node of the network is typically a linear node where

$$o = \sum_{i=1}^{n} w_i x_i + b. \tag{6}$$

A network of nodes is illustrated in Figure 3 below for a feedforward ANN, whose outputs always connect to nodes further in the network. The arrows in Figure 3 indicate how the outputs of nodes in one layer connect to the inputs in the next layer. The visible nodes are labelled with a V. The hidden nodes are labelled with an Hx.y, where x indicates the layer number and y indicates the node number. The output node is labeled O.

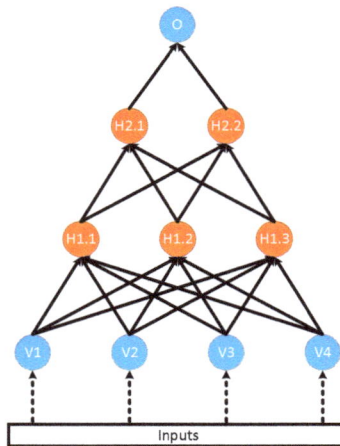

Figure 3. A feedforward ANN with four visible nodes, three nodes in the first hidden layer, two nodes in the second hidden layer, and a single node in the output layer.

The ANN is trained using the backpropagation algorithm [24]. The backpropagation algorithm is run over all the training data. This is called an epoch. When training an ANN, many epochs are performed with a termination criterion such as a maximum number of epochs or the error falling below a threshold.

Next, we describe a DNN. A DNN is essential an ANN with many hidden layers. The difference is in the training process. Rather than training the network using only the backpropagation algorithm, an initialization phase is done using the contrastive divergence algorithm [25,26]. The contrastive divergence algorithm is performed on a restricted Boltzmann machine (RBM). Figure 4 illustrates a RBM with four visible nodes and three hidden nodes. Important to note is that unlike the ANN, the arrows point in both directions. This is to indicate that the contrastive divergence algorithm updates the weights by propagating the error in both directions.

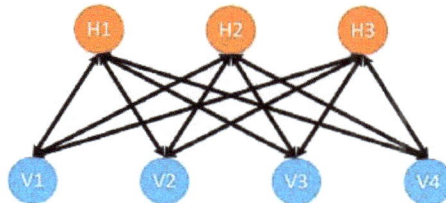

Figure 4. A restricted Boltzmann machine with four visible units and three hidden units. Note the similarity with a single layer of a neural network.

Similar to an ANN, a RBM has bias terms. However, since the error is propagated in both directions there are two bias terms, b and c. The visible and hidden nodes are calculated from one another [26]. Let v_i represent the ith visible node, w_i the weight of the ith visible node, c the bias term, n the number of visible nodes, and h the hidden node.

$$h = \sigma\left(\sum_{i=1}^{n} w_i v_i + c\right), \tag{7}$$

which can be rewritten in vector notation for all hidden units as

$$\mathbf{h} = \sigma(\mathbf{W}\mathbf{v} + \mathbf{c}). \tag{8}$$

Similarly, the visible node can be calculated in terms of the hidden nodes. Let h_j represent the jth hidden node, w_j the weight of the jth hidden node, b the bias term, m the number of hidden nodes, and v the visible node. Then

$$v = \sigma\left(\sum_{j=1}^{m} w_j h_j + b\right), \tag{9}$$

which can be rewritten in vector notation for all visible units as

$$\mathbf{v} = \sigma\left(\mathbf{W}^T \mathbf{h} + \mathbf{b}\right), \tag{10}$$

where $\mathbf{W}^T$ is the transpose of $\mathbf{W}$.

Training a RBM is done in three phases as described in Algorithm 1 for training vector $\mathbf{v}_0$ and a training rate ε. Algorithm 1 is performed on iterations (epochs) of all input vectors.

Algorithm 1: Training restricted Boltzmann machines using contrastive divergence

---

1 //Positive Phase
2 $\mathbf{h}_0 = \sigma\,(\mathbf{Wv}_0 + \mathbf{c})$
3 for each hidden unit h_{0i}:
4 if $h_{0i} >$ rand(0,1)//rand(0,1) represents a sample drawn from the uniform distribution
5 $h_{0i} = 1$
6 else
7 $h_{0i} = 0$
8 //Negative Phase
9 $\mathbf{v}_1 = \sigma\,(\mathbf{W}^T\mathbf{h}_0 + \mathbf{b})$
10 for each visible units v_{1j}:
11 if $v_{1j} >$ rand(0,1)
12 $v_{1j} = 1$
13 else
14 $v_{1j} = 0$
15 //Update Phase
16 $\mathbf{h}_1 = \sigma\,(\mathbf{Wv}_1 + \mathbf{c})$
17 $\mathbf{W} = \varepsilon\,(\mathbf{h}_0\mathbf{v}_0{}^T - \mathbf{h}_1\mathbf{v}_1{}^T)$
18 $\mathbf{b} = \varepsilon\,(\mathbf{h}_0 - \mathbf{h}_1)$
19 $\mathbf{c} = \varepsilon\,(\mathbf{v}_0 - \mathbf{v}_1)$

As can be seen in Figure 4, a trained RBM closely resembles a single layer of an ANN. We stack RBMs to form an ANN. First, RBM1 is trained based on our input data using Algorithm 1. Then, the entire input set is fed into the visible layer of a now fixed RBM1, and the outputs at the hidden layer are collected. These outputs are used as the inputs to train RBM2. This process is repeated after RBM2 is fully trained to generate the inputs for RBM3, and so on, as shown in Figure 5. This training is unsupervised, meaning that no target outputs are given to the model. It has information about the inputs and how they are related to one another, but the network is not able to solve any real problem yet.

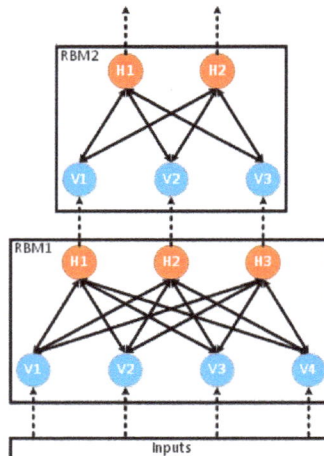

Figure 5. Graphical representation of how RBMs are trained and stacked to function as an ANN.

The next step is training a DNN. Backpropagation is used to train the neural network to solve a particular problem. Since our problem is short-term load forecasting, natural gas load values are used as target outputs, and a set of features such as temperature, wind speed, day of the week, and previous loads are used as the inputs. After the backpropagation training, the DNN functions identically to a large ANN.

5. Data

One common problem with training any type of neural network is that there is always some amount of randomness in the results [27]. This means that it is difficult to know whether a single trained model is performing well because the model parameters are good or because of randomness. Hanson and Salamon mitigated this problem using cross validation and an ensemble of similar neural networks [27]. They trained many models on the different parts of the same set of data so that they could test their models on multiple parts of the data.

This paper mitigates this problem by using data sets from 62 operating areas from local distribution companies around the United States. These operating areas come from many different geographical regions including the Southwest, the Midwest, West Coast, Northeast, and Southeast and thus represent a variety of climates. The data sets also include a variety of urban, suburban and rural areas. This diverse data set allows for broader conclusions to be made about the performance of the forecasting techniques.

For each of the 62 operating areas, several models are trained using at least 10 years of data for training and 1 year for testing. The inputs to these models are those discussed in Section 2. The natural gas flow is normalized using the method proposed by Brown et al. [28]. All the weather inputs in this experiment are observed weather as opposed to forecasted weather for the sake of simplicity.

6. Methods

This section discusses the models at the core of this paper. Four models are compared: a linear regression (LR) model [5], an ANN trained as described in Reference [26], and two DNNs trained as described in Section 3. The first DNN is a shallow neural network with the same size and shape as the ANN. The other DNN is much larger.

The ANN has two hidden layers of 12 and four nodes each and is trained using a Kalman filter-based algorithm [29]. The first DNN has the same architecture as the ANN but is pretrained using contrastive divergence. The purpose of using this model is to determine if the contrastive divergence algorithm can outperform the Kalman filter-based algorithm on these 62 data sets when all other variables are equal. Each RBM is trained for 1000 epochs, and 20 epochs of backpropagation are performed. Despite its small size, the contrastive divergence trained neural network is referred to as a DNN to simplify notation.

In addition to these models, which represent the state-of-the-art in short-term load forecasting of natural gas, a large DNN with hidden layers of 60, 60, 60, and 12 neurons, respectively, is studied. The purpose of this model is to show how much improvement can be made by using increasingly complex neural network architectures. All forecasting methods are provided with the same inputs to ensure a fair comparison.

7. Results

To evaluate the performance of the respective models, we considered several metrics to evaluate the performance of each model. The first of these is the root mean squared error:

$$RMSE = \sqrt{\frac{1}{N}\sum_{n=1}^{N}[\hat{s}(n) - s(n)]^2}, \tag{11}$$

for a testing vector of length N, actual demand s, and forecasted demand $\hat{s}$. *RMSE* is a powerful metric for short-term load forecasting of natural gas because it naturally places more value on the days with higher loads. These days are important, as they are when natural gas is the most expensive, which means that purchasing gas on the spot market or having bought too much gas can be costly. Unfortunately, *RMSE* is magnitude dependent, meaning that larger systems have larger *RMSE* if the percent error is constant, which makes it a poor metric for comparing the performance of a model across different systems.

Another common metric for evaluating forecasts is mean absolute percent error,

$$MAPE = 100\frac{1}{N}\sum_{n=1}^{N}\frac{|\hat{s}(n) - s(n)|}{s(n)}. \tag{12}$$

Unlike *RMSE*, *MAPE* is unitless and not dependent on the magnitude of the system. This means that it is more useful for comparing the performance of a method between operating areas. It does, however, put some emphasis on the lowest flow days, which, on top of being the least important days to forecast correctly, are often the easiest days to forecast. As such, *MAPE* is not the best metric for looking at the performance of the model across all the days in a year, but can be used to describe the performance on a subset of similar days.

The error metric used in this paper is weighted *MAPE*:

$$WMAPE = 100\frac{\sum_{n=1}^{N}|\hat{s}(n) - s(n)|}{\sum_{n=1}^{N}s(n)} \tag{13}$$

This error metric does not emphasize the low flow and less important days while being unitless and independent of the magnitude of the system. This means that it is the most effective error metric for comparing the performance of our methods over the course of a full year.

The mean and standard deviation of the performance of each model over the 62 data sets are shown in Table 1. As expected, the DNN has a lower mean *WMAPE* than the linear regression and ANN forecasters, meaning that generally, the DNN performs better than the simpler models. Additionally, the large DNN marginally outperforms the small DNN in terms of *WMAPE*. Both results are shown to be statistically significant later in this section. In addition to the mean, the standard deviation of the performances of the two DNN architectures are smaller than that of the LR and ANN. This is an important result because it points to a more consistent performance across different areas as well as better performance overall.

Table 1. The mean and standard deviation of the performance of the four models on all 62 areas.

	LR *WMAPE*	ANN *WMAPE*	DNN *WMAPE*	Large DNN *WMAPE*
Mean	6.41	6.41	5.78	5.58
Standard Deviation	2.49	2.83	2.11	2.09

Simply stating the mean performance does not tell us much without looking at the differences in performance for each of the 62 areas individually, which is shown succinctly in Figures 6 and 7. Figure 6a,b and Figure 7 are histograms of the difference in performance on all 62 areas of two forecasting methods. By presenting the results this way, we can visualize the general difference in performance for each of the 62 operating areas. Additionally, *t*-tests can be performed on the histograms to determine the statistical significance of the difference. Right-tailed *t*-tests were performed on the distributions in Figure 6a,b. The resulting *p*-values are 1.2×10^{-7} and 6.4×10^{-4}, respectively, meaning that the DNN performed better, in general, than the ANN or LR, and that the difference in performance is statistically significant in both cases.

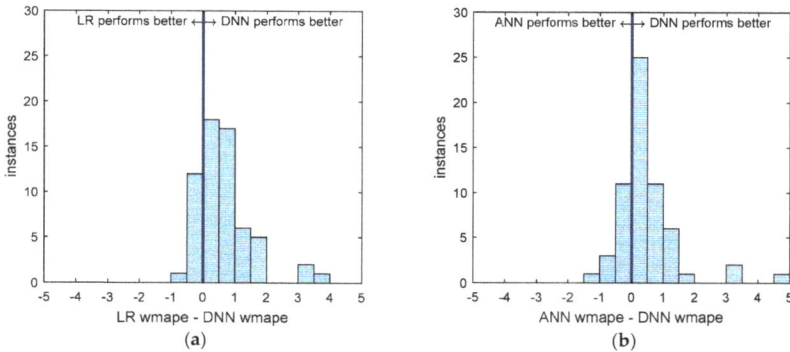

Figure 6. This figure shows two histograms: (**a**) A comparison of the performance of all 62 models between the DNN and the LR. Instances to the left of the center line are those for which the LR performed better, while those on the right are areas where the DNN performs better. The distance from the center line is the difference in *WMAPE*. (**b**) The same as (**a**) but comparing the ANN to the DNN. One instance (at 10.1) in (**b**) is cut off to maintain consistent axes.

It is also interesting to consider that in some areas, the LR and ANN forecasters perform better than the DNN. This implies that in some cases, the simpler model is the better forecaster. It is also important to point out that of the 13 areas where the LR outperforms the DNN, only two have LR *WMAPE*s greater than 5.5, which means that the simple LR models are performing very well when compared to industry standards for short-term load forecasting of natural gas on those areas.

Figure 7 compares the performance of the two DNNs. As with the two distributions in Figure 6, a left-tailed *t*-test was performed on the histogram in Figure 7 resulting in a *p*-value of 9.8×10^{-5}. This means that the Large DNN offers a statistically significant better performance over the 62 areas than the small DNN. However, much like in the comparison between the DNN and other models, the small DNN performs better in some areas, which supports the earlier claim that complex models do not necessarily outperform simpler ones.

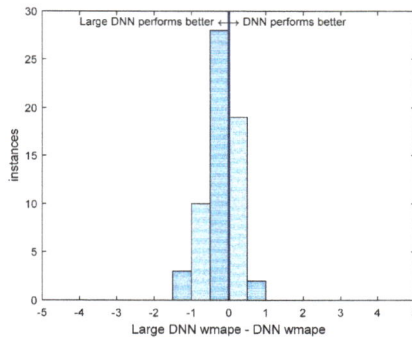

Figure 7. A comparison of the performance of all 62 models between the DNN and the Large DNN. Instances to the left of the center line are those for which the Large DNN performed better, while those on the right are areas where the DNN performs better. The distance from the center line is the difference in *WMAPE*.

8. Conclusions

We conclude that DNNs can be better short-term load forecasters than LR and ANNs. On average, over the 62 operating areas examined, a DNN outperformed an otherwise identical ANN at short-term load forecasting of natural gas, and a larger DNN offered even greater performance. However, these improvements to the performance are not present for all 62 operating areas. For some, even the much simpler linear regression model is shown to perform better than the DNN. For this reason, it is concluded that, although the DNN is a powerful option that in general will perform better than simpler forecasting techniques, it may not do so for every operating area. Therefore, DNNs can be used as a tool in short-term load forecasting of natural gas, but multiple other forecasting methods should be considered as well.

Author Contributions: G.D.M. and R.J.P. conceived and designed the experiments G.D.M. performed the experiments; G.D.M. and R.J.P. analyzed the data; R.H.B. contributed reagents/materials/analysis tools; G.D.M. and R.J.P. wrote the paper.

Funding: This research received no external funding.

Acknowledgments: The GasDay lab at Marquette University provided funding and data for this work.

Conflicts of Interest: The authors declare no conflicts of interest.

Nomenclature

b	bias term of a neural network node
c	the bias term of a restricted Boltzmann machine (RBM)
CDD	cooling degree days
DPT	dew point
Dth	dekatherm
h	vector of hidden nodes of a RBM
HDD	heating degree days
h_j	jth hidden node of a RBM
MAPE	mean absolute error
o	output of a neural network node
RMSE	root mean square error
s	natural gas demand
T	temperature in degrees Fahrenheit
T_{ref}	reference temperature for HDD and CDD
v	vector of visible nodes of a RBM
v_i	ith visible node of a RBM
W	weight matrix of a neural network
w_i	weight of the ith input of a neural network node
WMAPE	weighted mean absolute percentage error
x_i	ith input to the node of a neural network

References

1. Natural Gas Explained. Available online: https://www.eia.gov/energyexplained/index.php?page=natural_gas_home (accessed on 24 July 2018).
2. Merkel, G.D.; Povinelli, R.J.; Brown, R.H. Deep neural network regression for short-term load forecasting of natural gas. In Proceedings of the International Symposium on Forecasting, Cairns, Australia, 25–28 June 2015; p. 90.
3. Asbury, J.G.; Maslowski, C.; Mueller, R.O. *Solar Availability for Winter Space Heating: An Analysis of the Calendar Period, 1953–1975*; Argonne National Laboratory: Argonne, IL, USA, 1979.
4. Dahl, M.; Brun, A.; Kirsebom, O.; Andresen, G. Improving short-term heat load forecasts with calendar and holiday data. *Energies* **2018**, *11*, 1678. [CrossRef]
5. Vitullo, S.R.; Brown, R.H.; Corliss, G.F.; Marx, B.M. Mathematical models for natural gas forecasting. *Can. Appl. Math. Q.* **2009**, *17*, 807–827.
6. Ishola, B. Improving Gas Demand Forecast during Extreme Cold Events. Master's Thesis, Marquette University, Milwaukee, WI, USA, 2016.
7. Haida, T.; Muto, S. Regression based peak load forecasting using a transformation technique. *IEEE Trans. Power Syst.* **1994**, *9*, 1788–1794. [CrossRef]

8. Hornik, K.; Stinchcombe, M.; White, H. Multilayer feedforward networks are universal approximators. *Neural Netw.* **1989**, *2*, 359–366. [CrossRef]
9. Park, D.C.; El-Sharkawi, M.A.; Marks, R.J.; Atlas, L.E.; Damborg, M.J. Electric load forecasting using an artificial neural network. *IEEE Trans. Power Syst.* **1991**, *6*, 442–449. [CrossRef]
10. Längkvist, M.; Karlsson, L.; Loutfi, A. A review of unsupervised feature learning for time series modeling. *Pattern Recognit. Lett.* **2014**, *42*, 11–24. [CrossRef]
11. Szegedy, C.; Liu, W.; Jia, Y.; Sermanet, P.; Reed, S.; Anguelov, D.; Erhan, D.; Vanhoucke, V.; Rabinovich, A. Going deeper with convolutions. In Proceedings of the 2015 IEEE Conference on Computer Vision and Pattern Recognition (CVPR), Boston, MA, USA, 7–12 June 2015; pp. 1–9.
12. Hinton, G.E.; Hinton, G.E.; Osindero, S.; Osindero, S.; Teh, Y.W.; Teh, Y.W. A fast learning algorithm for deep belief nets. *Neural Comput.* **2006**, *18*, 1527–1554. [CrossRef] [PubMed]
13. Qiu, X.; Zhang, L.; Ren, Y.; Suganthan, P.N.; Amaratunga, G. Ensemble deep learning for regression and time series forecasting. In Proceedings of the 2014 IEEE Symposium on Computational Intelligence in Ensemble Learning, Orlando, FL, USA, 9–12 December 2014; pp. 1–6. [CrossRef]
14. Busseti, E.; Osband, I.; Wong, S. *Deep Learning for Time Series Modeling*; Stanford University: Stanford, CA, USA, 2012.
15. Dalto, M.; Matusko, J.; Vasak, M. Deep neural networks for time series prediction with applications in ultra-short-term wind forecasting. In Proceedings of the IEEE International Conference on Industrial Technology (ICIT), Seville, Spain, 17–19 March 2015; pp. 1657–1663.
16. Kuremoto, T.; Kimura, S.; Kobayashi, K.; Obayashi, M. Time series forecasting using a deep belief network with restricted Boltzmann machines. *Neurocomputing* **2014**, *137*, 47–56. [CrossRef]
17. Glorot, X.; Bengio, Y. Understanding the difficulty of training deep feedforward neural networks. In Proceedings of the Thirteenth International Conference on Artificial Intelligence and Statistics, Sardinia, Italy, 13–15 May 2010; Teh, Y.W., Titterington, M., Eds.; PMLR: London, UK, 2010; pp. 249–256.
18. Kennedy, J.; Eberhart, R. Particle swarm optimization. In Proceedings of the IEEE International Conference on Neural Networks, Perth, Australia, 27 November–1 December 1995; Volume 4, pp. 1942–1948.
19. Ryu, S.; Noh, J.; Kim, H. Deep neural network based demand side short term load forecasting. *Energies* **2017**, *10*, 3. [CrossRef]
20. Kuo, P.-H.; Huang, C.-J. A high precision artificial neural networks model for short-term energy load forecasting. *Energies* **2018**, *11*, 213. [CrossRef]
21. Li, C.; Ding, Z.; Yi, J.; Lv, Y.; Zhang, G. Deep belief network based hybrid model for building energy consumption prediction. *Energies* **2018**, *11*, 242. [CrossRef]
22. Chen, K.; Chen, K.; Wang, Q.; He, Z.; Hu, J.; He, J. Short-term load forecasting with deep residual networks. *IEEE Trans. Smart Grid* **2018**. [CrossRef]
23. Hosein, S.; Hosein, P. Load forecasting using deep neural networks. In Proceedings of the IEEE Power & Energy Society Innovative Smart Grid Technologies Conference (ISGT), Washington, DC, USA, 23–26 April 2017; pp. 1–5.
24. Lin, C.-T.; Lee, C.S.G. *Neural Fuzzy Systems—A Neuro-Fuzzy Synergism to Intelligent Systems*; Prentice-Hall: Upper Saddle River, NJ, USA, 1996.
25. Bengio, Y. Learning deep architectures for AI. *Found. Trends Mach. Learn.* **2009**, *2*, 1–127. [CrossRef]
26. Hinton, G. A practical guide to training restricted Boltzmann machines. In *Neural Networks: Tricks of the Trade*; Springer: Berlin/Heidelberg, Germany, 2012.
27. Hansen, L.K.; Salamon, P. Neural network ensembles. *IEEE Trans. Pattern Anal. Mach. Intell.* **1990**, *12*, 993–1001. [CrossRef]
28. Brown, R.H.; Vitullo, S.R.; Corliss, G.F.; Adya, M.; Kaefer, P.E.; Povinelli, R.J. Detrending daily natural gas consumption series to improve short-term forecasts. In Proceedings of the IEEE Power and Energy Society General Meeting, Denver, CO, USA, 26–30 July 2015.
29. Ruchti, T.L.; Brown, R.H.; Garside, J.J. Kalman based artificial neural network training algorithms for nonlinear system identification. In Proceedings of the IEEE International Symposium on Intelligent Control, Chicago, IL, USA, 25–27 August 1993; pp. 582–587.

![energies logo] *energies*

MDPI

Article

The Optimization of Hybrid Power Systems with Renewable Energy and Hydrogen Generation

Fu-Cheng Wang * [iD], Yi-Shao Hsiao and Yi-Zhe Yang

Department of Mechanical Engineering, National Taiwan University, Taipei 10617, Taiwan;
r03522831@ntu.edu.tw (Y.-S.H.); rogeryoun123@gmail.com (Y.-Z.Y.)
* Correspondence: fcw@ntu.edu.tw; Tel.: +886-2-3366-2680

Received: 26 June 2018; Accepted: 23 July 2018; Published: 26 July 2018

Abstract: This paper discusses the optimization of hybrid power systems, which consist of solar cells, wind turbines, fuel cells, hydrogen electrolysis, chemical hydrogen generation, and batteries. Because hybrid power systems have multiple energy sources and utilize different types of storage, we first developed a general hybrid power model using the Matlab/SimPowerSystem^{TM}, and then tuned model parameters based on the experimental results. This model was subsequently applied to predict the responses of four different hybrid power systems for three typical loads, without conducting individual experiments. Furthermore, cost and reliability indexes were defined to evaluate system performance and to derive optimal system layouts. Finally, the impacts of hydrogen costs on system optimization was discussed. In the future, the developed method could be applied to design customized hybrid power systems.

Keywords: hybrid power system; fuel cell; solar; wind; hydrogen; optimization; cost; reliability

---

1. Introduction

The development of alternative energy, such as solar, wind, geothermal, hydropower, ocean power, and hydrogen, has attracted much research attention because of the energy crisis and environmental pollution problems. Among these, solar, wind, and hydrogen are promising alternative energies. Solar cells and wind turbines (WTs) convert solar irradiation and wind power, respectively, into electrical power. Hydrogen energy can be converted into electricity via an electrochemical reaction of fuel cells. Each type of energy source has various strengths and weaknesses. For example, solar and wind energy are pollution free and relatively cheap to produce but lack stability because of their dependence on weather conditions. In contrast, hydrogen energy with fuel cells guarantees stable power supplies but is expensive at present. Therefore, hybrid systems that utilize multiple energy sources and storage methods are the best option for reducing system costs and increasing system reliability. Previously, in an Iranian study, Maleki and Askarzadeh [1] designed a hybrid power system containing photovoltaic (PV) arrays, a WT, a diesel generator, and a secondary battery. They showed that systems consisting of a WT, diesel generator, and a secondary battery satisfied the load demand at the lowest cost. Based on an analysis of weather data in Turkey, Devrim and Bilir [2] concluded that wind energy could compensate for solar (PV) energy in winter. Therefore, a hybrid system with a WT can achieve better performance than one without a WT. Martinez-Lucas et al. [3] studied the performance of a system based on WTs and pump storage hydropower on El Hierro Island in the Canary archipelago. This hybrid wind–hydropower plant showed improvements in system performance to different wind speeds and power demands.

The most important issues when designing hybrid power systems are the selection of the system components and the component sizes, according to load demands. Wang and Chen [4] considered a hybrid system consisting of PV arrays, a proton-exchange membrane fuel cell (PEMFC), and an Lithium iron (Li-Fe) battery. They showed that the integration of the PEMFC improved system reliability, and that tuning the PV and battery units greatly reduced the system cost. The present paper extends these ideas and discusses the impacts of WTs and a hydrogen electrolyzer on system performance.

A WT converts wind power into electricity. Many factors, such as wind speed, air density, the rotor swept area, and the power coefficient of the motor, affect the amount of power extracted from WTs. For example, Bonfiglio et al. [5] modeled WTs equipped with direct-drive permanent magnet synchronous generators. They used the model to examine the influences of active power loss on the effectiveness of wind generator control and applied Digsilent Power Factory to verify the results. Pedra et al. [6] built fixed-speed induction generator models using PSpice and PSCAD-EMTDC codes They compared single-cage and double-cage models, and showed that the latter was more suitable for fixed-speed WT simulation. Lee et al. [7] assessed large-scale application of solar and wind power in 143 urban areas. The proposed system was shown to lead to a balance of the building energy consumption. Maouedja et al. [8] constructed a small hybrid system in Adrar, Algeria, and concluded that wind energy can compliment solar energy. Al Ghaithi et al. [9] analyzed a hybrid energy system in Masirah Island in Oman. The simulation results showed that a hybrid system composed of PV, a WT, and an existing diesel power system is the most economically viable, and can significantly improve voltage profiles. Devrim and Bilir [2] also found that a hybrid system with a WT can perform better than one without a WT in Ankara, Turkey. However, Chen and Wang [10] reached the opposite conclusion in their analysis of a green building in Miao-Li county of Taiwan equipped with a hybrid power system consisting of PV arrays, a WT, a PEMFC, a hydrogen electrolyzer, and battery sets. They found that wind and solar energy had similar profiles, and concluded that a WT was unsuitable because it increased the cost of the system but did not significantly compensate the renewable energy of the PV array. Therefore, the inclusion of WTs in a hybrid system should depend on local weather conditions.

Hydrogen electrolyzation is a new method of energy storage, where redundant energy is used to produce hydrogen that can then be utilized by PEMFCs to produce electricity when the power supply is insufficient. For example, Chennouf et al. [11] utilized solar energy to produce hydrogen in Algeria. They demonstrated that hydrogen conversion efficiency was best under low voltage and high temperature conditions. Tribioli et al. [12] analyzed an off-grid hybrid power system with two energy storage methods: a lead-acid battery and reversible operation of a PEMFC. They combined the system with a diesel engine and showed that the consumption of fossil fuels can be greatly reduced by integrating a suitable renewable power plant to match the loads. Cozzolino et al. [13] applied the model to analyze a particular case: the TUNeIT (Tunisia and Italy) Project. The simulation demonstrated an almost self-sustaining renewable power plant that consisted of 1 MW WT, 1.1 MW PV, a 72 kWh battery, a 300 kW fuel cell, a 300 kW diesel engine to cope with power demand at a cost of 0.522 €/kWh. Aouali et al. [14] built a PV array and hydrogen electrolyzer model based on dynamic equations. They conducted small-scale experiments and showed that the experimental responses fitted the model responses. Rahimi et al. [15] analyzed the economic benefits of utilizing wind energy in hydrogen electrolysis in Manjil and Binaloud, Iran. They showed that a stand-alone application was more expensive than an on-grid one because the former required larger WTs. Bianchi et al. [16] analyzed a hybrid system that utilized two storage methods: a solar battery system and a solar battery–hydrogen electrolyzer fuel cell system. They found that the conversion efficiency of stored energy was about 90% with the use of battery, and about 20% with the electrolyzer and PEMFC. Bocklisch et al. [17] proposed a multistorage hybrid system that combined short-term storage by batteries and long-term storage by hydrogen. They converted excessive PV energy in summer into hydrogen and hydrogen into electricity and heat in winter. The power exchanged with the public grid was smaller and more predictable compared with that of a conventional PV battery–hybrid system. As weather conditions have a major influence on the performance of hybrid power systems, climate data must be incorporated into the design of any hybrid system. For instance, Ikhsan et al. [18] collected weather data to estimate the energy flow into hybrid systems and to resize the system components. Their results demonstrated an improvement in system costs after size adjustment. Chen and Wang [10] included irradiation and wind data in a hybrid system model to optimize system costs and reliability. The present paper will also utilize historic weather data and load conditions when analyzing the impacts of system configurations.

The paper is arranged as follows: Section 2 introduces a general hybrid power system that consists of solar cells, WTs, a fuel cell, hydrogen electrolysis, chemical hydrogen generation, and batteries. We extend a previous hybrid power model [4] by adding WT and hydrogen electrozation modules. Then, system cost and reliability functions are defined to evaluate system performance. Based on this general hybrid power model, we apply three standard load conditions (laboratory, office, and house) to four specified hybrid power systems to estimate the impact of system configuration on performance. Section 3 discusses the optimization of the four hybrid power models and shows that both system cost and reliability can be improved by tuning the system component sizes. Based on the results, the solar battery system is preferable because of high hydrogen costs at present. We also predict the system costs at which hydrogen energy could become feasible. Last, conclusions are drawn in Section 4.

2. Results

This section builds a general hybrid power model that consists of a PV array, a WT, a PEMFC, hydrogen electrolysis, chemical hydrogen generation, and batteries. We applied a Matlab/SimPowerSystem (r2014a, MathWorks, Inc., Natick, MA, USA) model to predict the performance of four different hybrid power systems under three typical loads. Furthermore, cost and reliability indexes were defined to quantify performance measures of the hybrid systems.

2.1. Hybrid Power Systems

Figure 1a shows a general hybrid power system, which consists of a 3 kW PEMFC, a chemical hydrogen production system with sodium borohydride ($NaBH_4$), a 410 W hydrogen electrolyzer, 1.32 kW PV arrays, a 0.2 kW WT, a 15 Ah Li-Fe battery set, and power electronic devices. The system specifications are illustrated in Table 1 [19–25]. The system has three energy sources (solar, wind, and a PEMFC) and two energy storage methods (battery and hydrogen electrolysis).

Regarding energy sources, solar power is connected directly to a DC bus. Wind power is transferred by a controller and connects to the DC bus. As both solar power and wind power are significantly influenced by the weather, a PEMFC is used to provide reliable energy when necessary. The PEMFC can transform hydrogen energy to electricity and can provide continuous power as long as the hydrogen supply is sufficient. Two hydrogen supply methods are considered: the chemical reaction of $NaBH_4$ and hydrogen electrolysis. The former can provide power with high-energy density using an auto-batching system developed previously [25,26]; the latter can be regarded as energy storage, because redundant renewable energy can be stored in the form of hydrogen [24].

For energy storage, a Li-Fe battery is used for short-term electricity storage [17] because the battery has high efficiency (about 90%), and can absorb power surges when the load changes rapidly. Hydrogen electrolysis is used for long-term storage, considering the self-discharging problems of batteries. A benefit of the electrolysis process is that it does not produce contaminants. However, the energy conversion efficiency is much lower than of the battery [16].

We developed the general hybrid power model using the Matlab/SimPowerSystem, as shown in Figure 1b, and analyzed the impacts of different energy sources and storage methods on the system. In a previous study [4], a SimPowerSystem model was built to include a PEMFC, an Li-Fe battery set, PV arrays, and a chemical hydrogen production system. The model parameters were tuned based on experimental data to enable the simulation model to predict the responses/behavior of the experimental system under various conditions. Currently, PEMFC, PV arrays, chemical hydrogen production, and battery sets are operated as follows [4,17,25,26]:

1. The PEMFC is switched on to provide a default current of 20 A with the highest energy efficiency [20] when the battery state-of-charge (SOC) is 30%. If the SOC continuously decreases to 25%, the PEMFC current output is increased by up to 50 A, according to load, until the SOC is

35%, where the PEMFC is set to provide a default current of 20 A. The PEMFC is switched off when the battery SOC is 40%.

2.   The PV array transfers irradiance into electricity as follows [4,20]:

$$P_{PV} = 0.69(Irr - 1.52)$$

where P_{PV} and Irr represent solar power and irradiance, respectively.

3.   The chemical hydrogen generation is switched on when the pressure of the hydrogen storage tank decreases to 3 bar [25,26]. Currently, hydrogen is generated from a $NaBH_4$ solution by a previously developed auto-batch process, with a maximum generation rate of 93.8 standard liters per min (SLPM) [25]. This can sustain the operation of a 3 kW PEMFC [25].

4.   The battery regulates the power supply and load demands as follows: it is charged (discharged) when the supply is greater (lower) than the demand. To avoid overcharging, battery charging is stopped when its SOC reaches 98%.

In this paper, we extend the previously developed model [4] by adding wind power and hydrogen electrolysis modules.

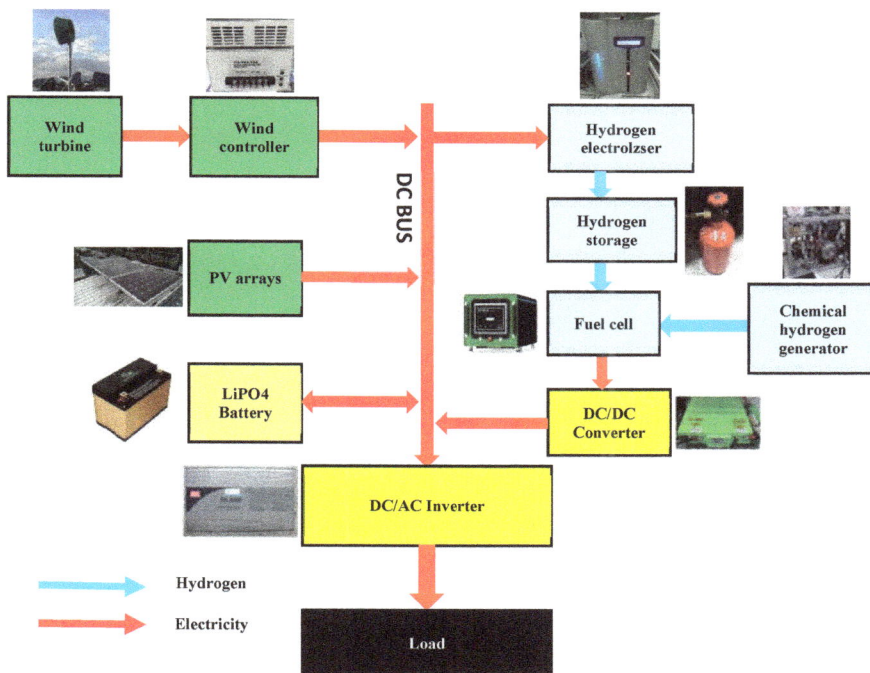

(a) System configuration.

Figure 1. *Cont.*

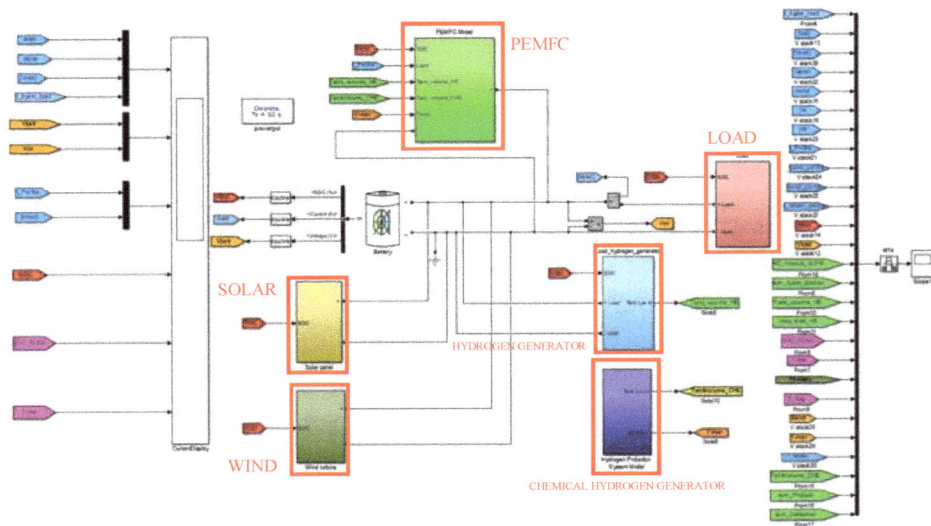

(**b**) SimPowerSystem model.

Figure 1. The general hybrid power system.

Table 1. Specifications of the hybrid system [19–25].

Component	Type	Specification
PEMFC Module	M-Field™ LPH8020	See Reference [19]
Solar Module [20]	ASEC-220G6S	Maximum Power: 220 W Open Circuit Voltage: 33.86 V Short Circuit Current: 8.61 A
Wind Turbine [21]	JPS-200	Rated Power: 200 W Voltage Output: DC 12 V Rotor Diameter: 0.68 m
LiFePO$_4$ Battery [22]	NA	Nominal Voltage: 52.8 V Nominal Capacity: 23 Ah
DC/DC Converter [23]	M-Field™ S/N:00051	Input Voltage: DC 44–85 V Output Voltage: DC 42–57 V Maximum Power: 3 kW
DC/AC Inverter [20]	MW™ TS-3000-148	Input Voltage: DC 42–60 V Output Voltage: AC 110 V Maximum Power: 4.5 kW
PEM Electrolyzer [24]	HGL-1000U	Gas Flow Rate: 1000 mL/min Power Consumption: <430 W Input Voltage: AC 100–240 V
Chemical Hydrogen Generation Module [25]	NA	Input Voltage: DC 24V Output: See Reference [25]

2.2. The Wind Power Model

The WT used in this paper was a commercial product, JPS-200, which is equipped with a permanent magnet synchronous generator that has a rating power of 200 W [21]. A wind power system and a theoretical model are developed to estimate wind power from wind speed based on experimental responses. The experiments were conducted using an industrial fan, which had a maximum wind speed of about 10 m/s. The wind power system structure is shown in Figure 2. We measured the AC current and voltage from a wind turbine and recorded the DC current and voltage from a wind controller.

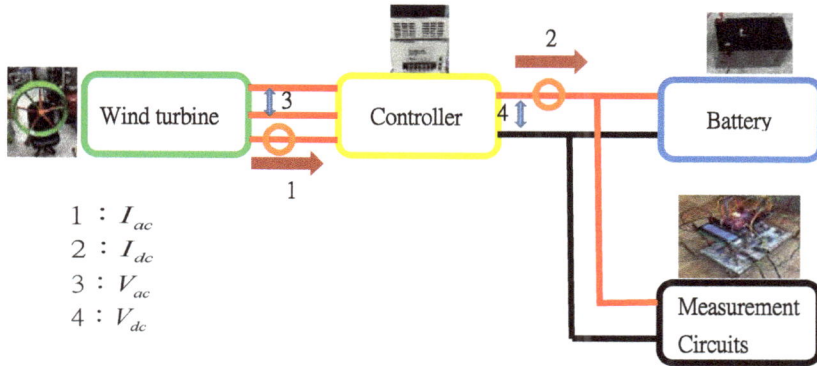

Figure 2. Measurement of the wind power.

The WT was tested under steady wind and varying wind conditions. The time responses are shown in Figure 3a, where the responses change slowly with steady wind, but quickly with varying wind. From the comparison of the wind speed and AC power, as illustrated in Figure 3b, the wind power can be theoretically described using the following equation:

$$P_{ac} = 0.11574 \, V_{wind}^3 \qquad (1)$$

where P_{ac} and V_{wind} represent the power and speed, respectively, of the wind. The experimental results show that the wind power can be predicted from the wind speed with maximum root mean square errors of 7.64 W and 17.32 W for steady and varying wind, respectively. The WT reached its maximum theoretical power of 200 W when the wind speed was greater than 12 m/s. We set the battery voltage at 12 V. The energy conversion relationship between AC and DC wind power is shown in Figure 3c, where the charging operation is divided into three zones according to the wind turbine voltage V_{ac}: (1) no charging (when $V_{ac} < 4.3V$), where the wind controller does not charge the battery; (2) linear charging (when $4.3V \leq V_{ac} < 8V$), where the DC charging voltage increases linearly; and (3) stable charging (when $V_{ac} \geq 8V$), where the DC charging voltage is 14.3 V. The conversion of AC and DC power can be described as follows:

$$P_{DC} = 0.70973 P_{ac} - 3.0958 = 0.0821 V_{wind}^3 - 3.0958 \qquad (2)$$

as illustrated in Figure 3c. Therefore, given wind speed data, the wind turbine DC power can be calculated by (1) and (2). Equations (1) and (2) can be applied to build the wind power module in Figure 2 for the simulation and optimization analyses.

the steady wind   the varying wind

(**a**) Time responses of the wind turbine.

the steady wind   the varying wind

(**b**) AC power v.s. wind speeds.

voltage   current   power

(**c**) Scatter diagrams of the DC power v.s. AC power

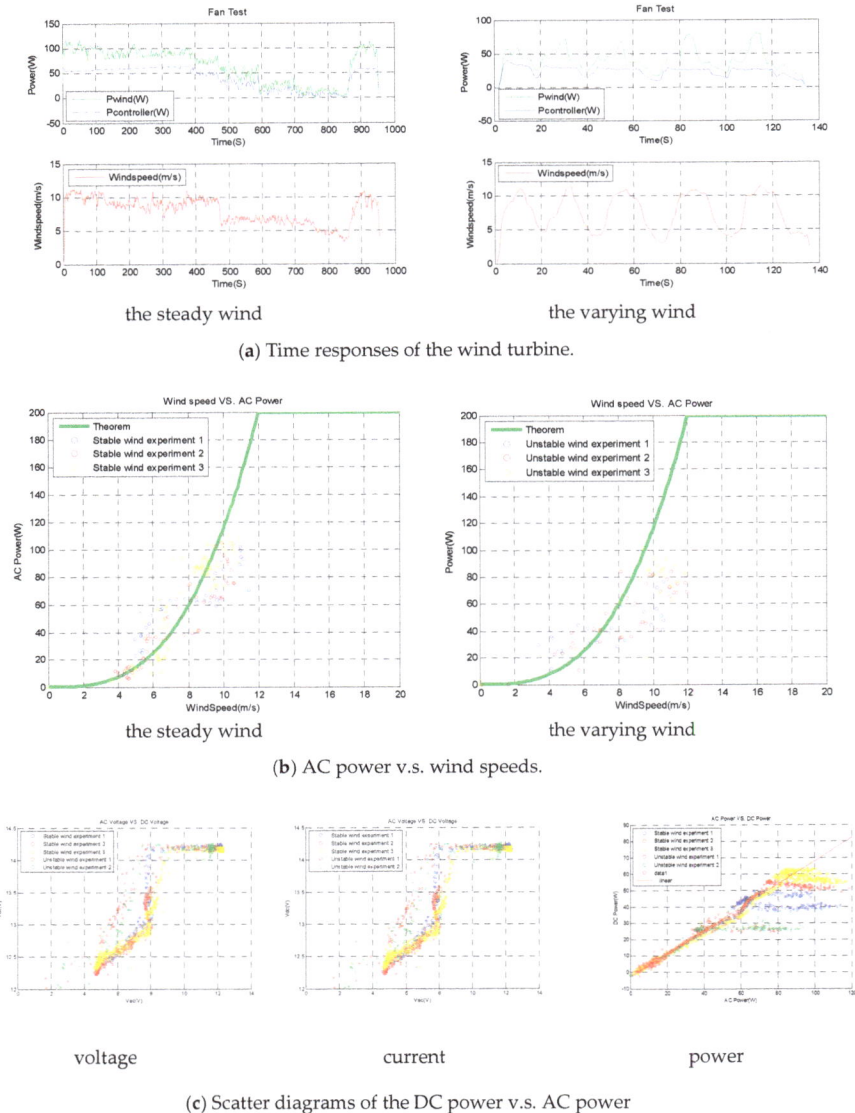

Figure 3. Experimental responses of the wind turbine.

2.3. The Hydrogen Electrolysis Model

The hydrogen electrolyzer transfers redundant energy, i.e., the extra-renewable energy when the battery SOC is near 100%, into hydrogen when the power supply is greater than the load. The stored hydrogen is then converted into electricity by a PEMFC when the load demand exceeds the power supply. Therefore, a theoretical model can be built to estimate hydrogen production based on redundant renewable energy. A hydrogen electrolyzer utilizes this redundant energy to produce hydrogen. The hydrogen electrolyzation system is shown in Figure 4. It consists of a commercial hydrogen electrolyzer, HGL-1000U, with a rating energy consumption of 400 W and hydrogen production of 1 L/min [24]. The electrolyzer has four operation modes: warm up, production, standby, and shut down. We developed a hydrogen electrolysis model using the Matlab/SimPowerSystem™ and applied the following management strategies (see Figure 4b):

1. Warm up: The extra-renewable energy is regarded as redundant energy when the battery SOC is greater than an upper limit of 95%.
2. Production: The electrolyzer is switched on after 10 min, when the integrated redundant renewable energy $\int_0^{10} (P_{renew} - P_{load})dt$ increases. P_{renew} and P_{load} represent the power sources from the renewable energy and power consumption of the loads, respectively.
3. Standby: The electrolyzer is switched off when the hydrogen tank is full (reaches the high-pressure limit).
4. Shut down: The hydrogen electrolyzer is switched off when the battery SOC falls to the lower limit of 85%.

To avoid frequent switching, the electrolyzer is allowed to produce hydrogen when the battery SOC is between 85% and 95%.

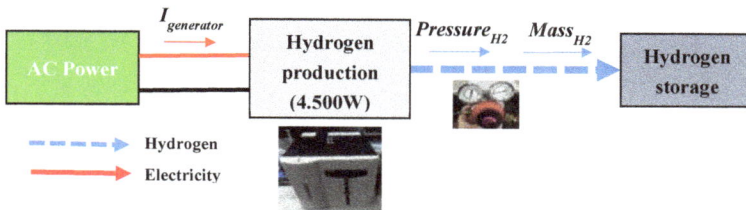

(a) System layout.

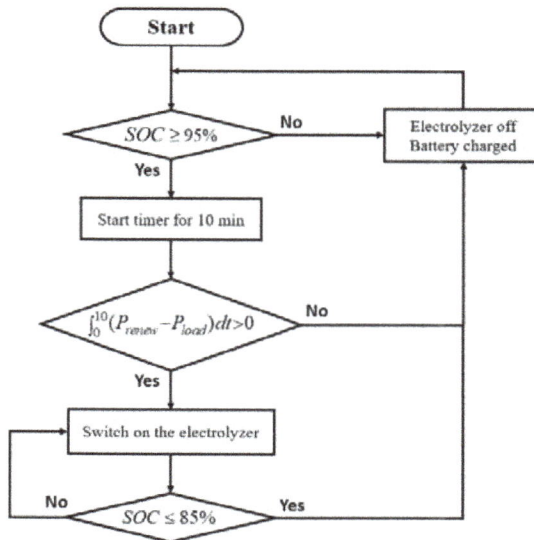

(b) Management strategy.

Figure 4. The hydrogen electrolyzation system.

A 3 L hydrogen cylinder was used to conduct the electrolyzation experiments. The results are shown in Figure 5, where the initial and final pressures of the cylinder are 8.6 bar and 10 bar, respectively. As a check valve is installed at the hydrogen outlet, hydrogen is produced only when the electrolyzer pressure exceeds the cylinder pressure, with a production rate of about 1.14 SLPM by consuming about 413 W. The hydrogen is purged every 350 s to prevent water flooding that could disturb the electrochemical reactions. The output hydrogen energy can be calculated using the following equation:

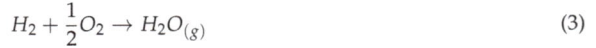

$$H_2 + \frac{1}{2}O_2 \rightarrow H_2O_{(g)} \tag{3}$$

The total enthalpy change of a reaction at 1 atm, 25 °C, referred to as the standard state, is the low heat value of hydrogen, which is equivalent to 241.32 kJ · mol^{-1} (or 120 MJ · kg^{-1}). Therefore, the hydrogen production efficiency can be defined as follows:

$$\eta_{LHV} = \frac{E_{fuel,production}}{E_{generator}} = \frac{H_2^{Exp} \cdot LHV}{E_{generator}} \tag{4}$$

where $E_{generator}$ and $E_{fuel,production}$ represent the ratio of input electric energy and the output hydrogen energy, respectively, and H_2^{Exp} is the produced hydrogen. For example, in one experiment, the input electric energy was $E_{generator}$ = 0.0372 kWh, and the output hydrogen volume was H_2^{Exp} = 6.233 L. The standard molecular weight and density of hydrogen was 2.0158 g/mole and 0.08228 g/L, respectively. Therefore, the output hydrogen energy can be calculated as follows:

$$E_{fuel,production} = \frac{6.233 \times 0.08228 \times 241.32}{2.0158 \times 3600} = 0.01705 \text{ (kWh)}$$

Hence, the hydrogen production efficiency was:

$$\eta_{LHV} = \frac{E_{fuel,production}}{E_{generator}} = \frac{0.01705}{0.0372} = 45.83\%$$

The hydrogen production efficiencies in all the experiments were about 45%. Based on the experimental results, the hydrogen production rate was set as follows to convert renewable energy to hydrogen storage:

$$H_2 = \frac{\eta_{LHV}}{LHV} \cdot E = 0.0465 \text{ (L/kJ)}$$

As the electrolyzer consumes an average power of 410 W during the production period, the hydrogen electrolyzer module was set to produce hydrogen at a rate of 1.14 L/min by consuming redundant renewable energy at a constant power of 410 W.

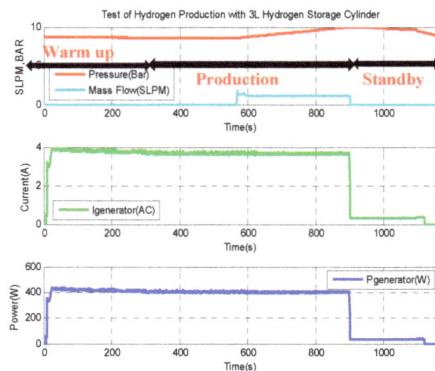

Figure 5. Experimental responses of the hydrogen electrolysis system.

2.4. Performance Indexes Hybrid Power Models

The hybrid power model of Figure 1 was applied to predict the system responses under different operation conditions based on the following management strategies (see Figure 6):

1. To avoid wasting renewable energy, the wind and solar power subsystems are operated as follows: when the battery SOC is greater than 98% and the input renewable power, including solar and wind power, is greater than the load, redundant renewable energy is dumped. Solar energy is reduced first because it is much more abundant than wind energy. When the battery SOC is less than 95%, all renewable energy is supplied to the system.
2. The PEMFC system is switched on when the battery SOC reaches a low bound of 30%. The PEMFC is then switched off when the battery SOC rises to a high limit of 40%. The PEMFC is controlled to provide a default current load of 20 A with the highest energy efficiency, and it is set to provide a load up to 50 A when the battery SOC continuously drops to 25% [20].
3. The chemical hydrogen generator system is switched on if the storage hydrogen level is lower than a safety limit [25,26]. We designed a batch procedure with suitable production rates to satisfy the system requirements. Each batch consumes 60 g of $NaBH_4$ and produces about 150 L of hydrogen [25]. Thus, the PEMFC can be continuously operated.

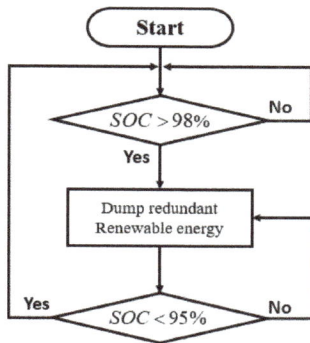

(a) Management strategy of the renewable energy.

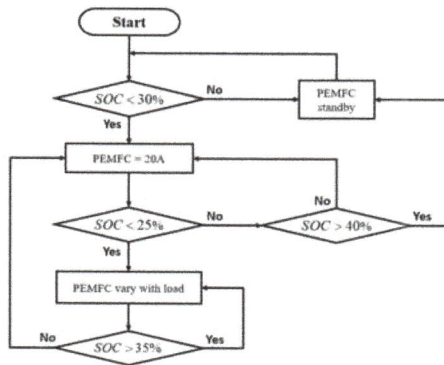

(b) Management strategy of the PEMFC.

Figure 6. Flow charts of the power management.

Hydrogen fuel can be obtained from two sources for PEMFC operation: chemical production and electrolysis. As the costs of different energy sources and storage are not the same, we utilized standard load profiles, irradiance, and wind data, as shown in Figure 7, for the simulation and optimization analyses.

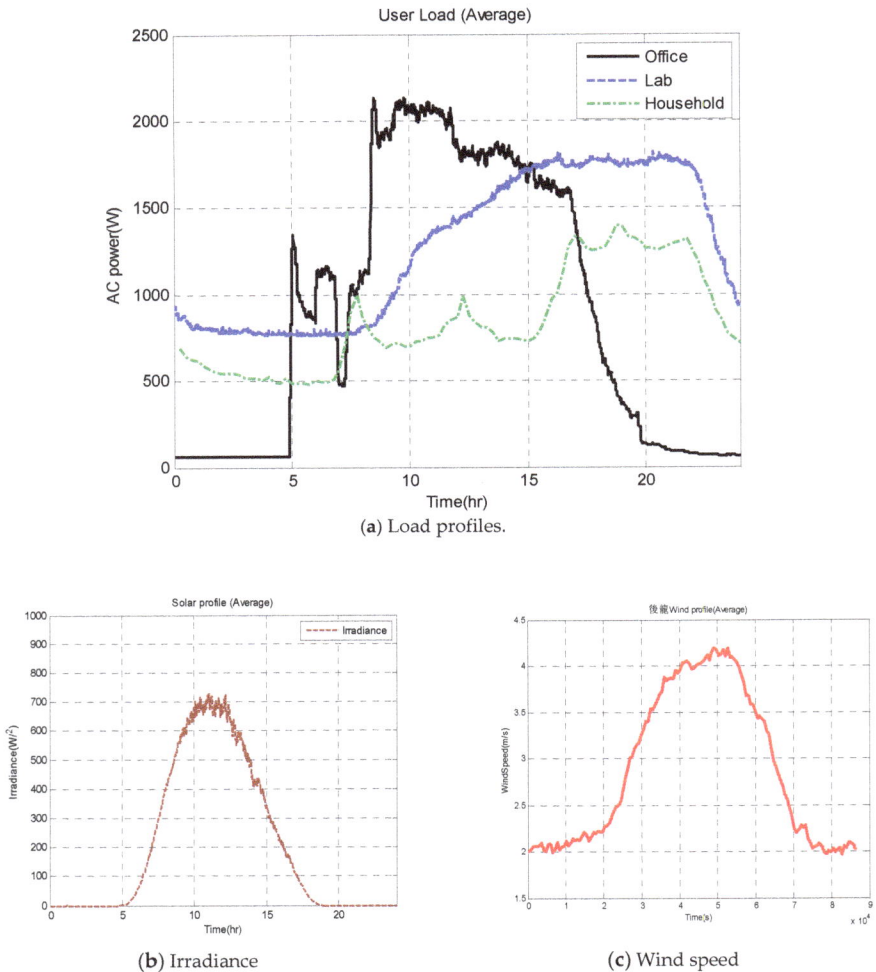

(**a**) Load profiles.

(**b**) Irradiance

(**c**) Wind speed

Figure 7. Daily average data.

The system responses can be applied to evaluate the system cost and reliability under different conditions.

The system cost $J_{(b,s,w)}$ was defined as follows [4]:

$$J_{(b,s,w)} = J_{i(b,s,w)} + J_{o(b,s,w)} \tag{5}$$

where J_i and J_o were the initial and operation costs, respectively, of the hybrid power system. In Equation (5), b, s, and w represent the numbers of the battery, PV array, and WT in units of

30 Ah, 660 W, and 200 W, respectively. For example, $(b, s, w) = (1, 2, 3)$ means that the system was equipped with one 30 Ah battery set, two 660 W PV arrays, and three 200 W WTs.

The initial cost J_i consists of several system components, as follows:

$$J_{i(b,s,w)} = \sum_{k=component} J^k_{i(b,s,w)} \tag{6}$$

where $k = PEMFC, DC, solar, WT, HE, CHG$, and $batt$ for the PEMFC, power electric devices, PV arrays, wind turbine, hydrogen electrolyzer, chemical hydrogen generator, and battery set, respectively. Similarly, the operation cost J_o includes two parts:

$$J_{o(b,s,w)} = \sum_{l=component} J^l_{o(b,s,w)} \tag{7}$$

where $l = NaBH_4, WT$, and *solar* for chemical hydrogen, WT maintenance, and PV maintenance, respectively.

The costs $J^k_{i(b,s,w)}$ and $J^l_{o(b,s,w)}$ can be calculated by the following equations:

$$J^k_{i(b,s,w)} = C_k \cdot n_k \cdot CRF_k \tag{8}$$

$$J^l_{o(b,s,w)} = C_l \cdot n_l \tag{9}$$

in which C is the component price per unit, and n is the component units. CRF represents the *capital recovery factor* and is defined as follows [10]:

$$CRF = \frac{ir(1+ir)^{ny}}{(1+ir)^{ny} - 1} \tag{10}$$

where ir is the inflation rate, and ny is the component life. The component life and cost are listed in Table 2. The inflation rate was set as 1.26% by referring to the average annual change of consumer price index of Taiwan [4].

Table 2. Simulation parameters.

Component	Lifetime	Price ($NT)
Hybrid system	15 (year)	NA
Fuel cell (3 kW)	8000 (h)	180,000
Power electronic devices (3 kW)	15 (year)	50,000
PV array (0.66 kW)	15 (year)	45,840
Wind turbine (0.2 kW)	15 (year)	19,333
Hydrogen electrolyzer (410 W)	8000 (h)	320,000
Chemical hydrogen generator	10 (year)	320,000
$NaBH_4$ (60 g/Batch, 150 L H_2)	NA	28

The system reliability is defined as the loss of power supply probability (LPSP), as follows [4]:

$$LPSP = \frac{\sum_1^T LPS(t)}{E_{load}(t)} \tag{11}$$

in which the numerator is the total loss of power supply during time interval T, and the denominator represents the required load demand during time interval T. The system is more reliable with a smaller LPSP.

2.5. Optimization of Four Hybrid Power Models

Based on the general hybrid power model, as shown in Figure 1, we considered the following four hybrid power systems with different combinations of energy sources and storage:

1. Solar_Wind (SW) system: The system contains two energy sources (a solar panel and WT) and one energy storage method (Li-Fe battery).
2. Solar_Wind_PEM_HE (SWPH) system: The system contains three energy sources (a solar panel, WT, and PEMFC) and two energy storage methods (a Li-Fe battery and a hydrogen electrolyzer).
3. Solar_Wind_PEM_CHG (SWPC) system: The system contains three energy sources (a solar panel, WT, and PEMFC) with a chemical hydrogen generator and one energy storage methods (an Li-Fe battery).
4. Solar_Wind_PEM_HE_CHG (SWPHC) system: The system contains three energy sources (a solar panel, WT, and PEMFC) with a chemical hydrogen generator and two energy storage methods (an Li-Fe battery and a hydrogen electrolyzer).

The corresponding SimPowerSystem models are illustrated in Figure 8.

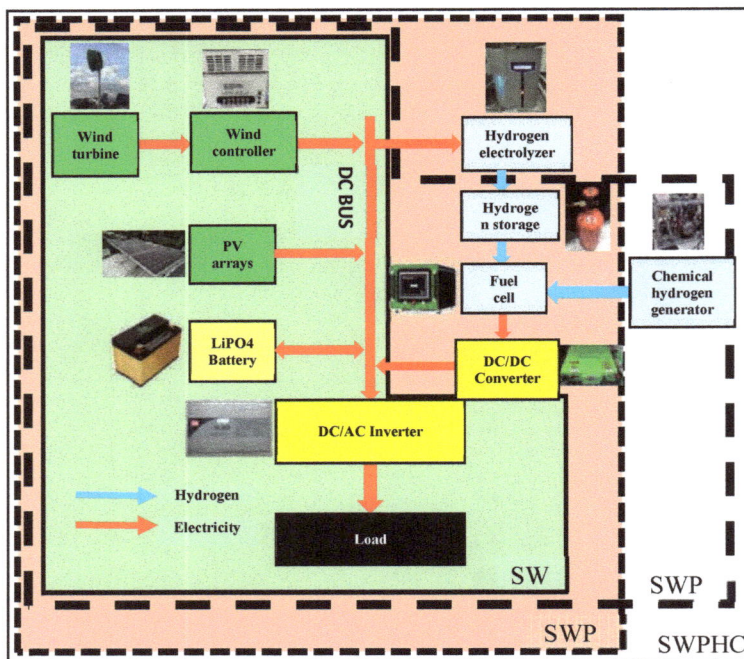

Figure 8. The four hybrid power models.

Three standard load conditions, as shown in Figure 7a, were applied to the four hybrid power models to predict systems responses. Then, we used Equations (5)–(11) to evaluate system cost and reliability using different component sizes. The resulting reference plots are shown in Figure 9, where the number of WTs was set to zero, because using a WT tended to increase the system costs. The optimal system costs of the four hybrid power systems are illustrated in Table 3.

Lab   Office   House

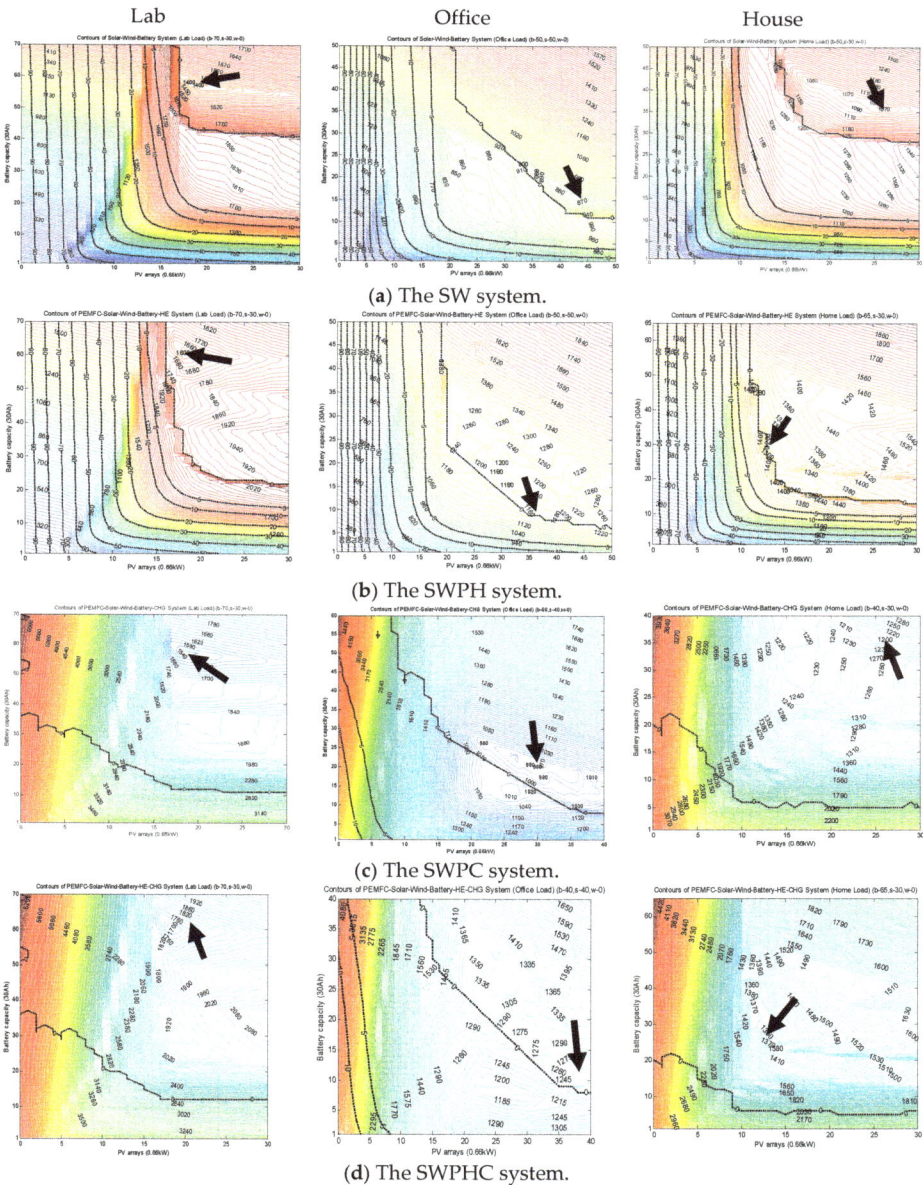

(a) The SW system.

(b) The SWPH system.

(c) The SWPC system.

(d) The SWPHC system.

Figure 9. Reference plots of four hybrid power models.

Table 3. Optimal system costs.

Daily Energy Consumption (kWh)	Lab	Office	House
	30.318	21.885	19.933
System Cost Per Day	-	-	-
SW	1399	865	1064
SWPH	1591	1148	1246
SWPC	1529	963	1194
SWPHC	1685	1241	1340
System Cost Per kWh	-	-	-
SW	46.144	39.525	53.379
SWPH	52.477	52.456	62.509
SWPC	50.432	44.003	59.901
SWPHC	55.578	56.706	67.225

3. Discussion

The analyses of the four hybrid power systems showed that system cost and reliability can be greatly improved by optimizing system sizes. For example, Figure 10 shows the reference plot of applying the SWPHC model to the laboratory load. If we use 10 units of battery (300 Ah), 10 units of solar (6.6 kW), and no WT, the system cost is estimated as NT$3208/day (or NT$106.17/kWh) with a possible power cut (LPSP = 0.33%). Based on Figure 10, the optimal system setting should be 61 units of battery (1830 Ah), 18 units of solar (11.88 kW), and no WT. Using these settings, the system cost is reduced to NT$1,685/day (or NT$55.6/kWh), and system reliability is improved to 100% (LPSP = 0).

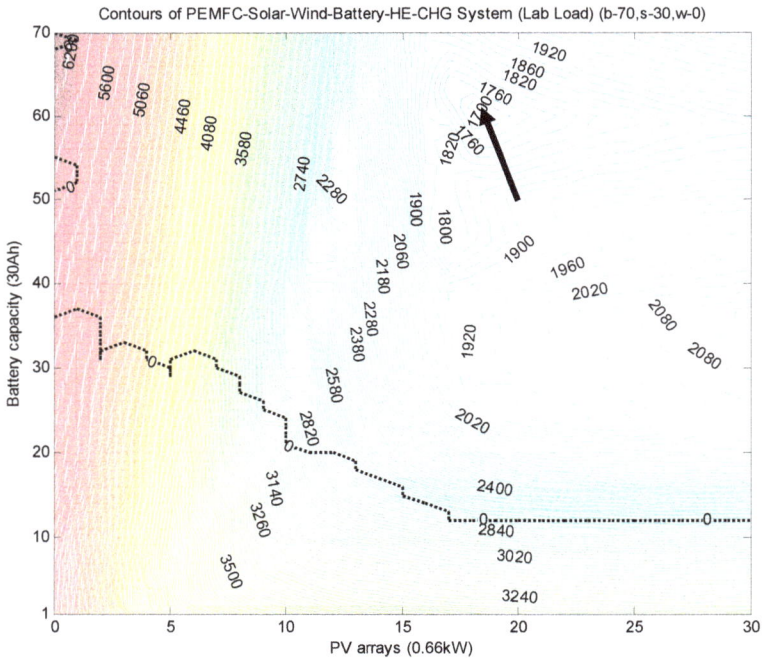

Figure 10. The reference plot of applying SWPHC to lab load.

The comparison of the four different hybrid power configurations shows that currently the SW system can achieve the cheapest system cost. For example, the daily cost for the office load is NT$865 using the SW system, but NT$963, 1,148, and 1,241 using the SWPC, SWPH, and SWPCH systems, respectively. However, the reliability (LPSP = 0) of the three systems is greater than the SW system (see Figure 9) (i.e., the reliability of the systems improved because the PEMFC can provide reliable energy when necessary). Under current conditions, the cost ranking is SW > SWPC > SWPH > SWPCH for all loads for the following reasons: (1) The cost of hydrogen is high at present; (2) energy storage efficiency by hydrogen electrolyzation is much lower than by Li-Fe batteries; (3) the extra hardware, such as the PEMFC and hydrogen electrolyzer, significantly increase systems costs.

The cost and energy distribution of applying the optimal SWPCH system to the laboratory load are shown in Table 4. First, due to system optimization, the PEMFC and Sodium borohydride tends not to be used, because the fuel cost is high (NT$28 per batch to produce 150 L of H_2, see Table 2). Therefore, the corresponding equipment (hydrogen electrolyzer, PEMFC, and chemical hydrogen production) can be saved to reduce the system cost by 13.39%. Second, the battery cost accounts for nearly 73% of the total system costs, whereas the PV panels to store the solar power constitute only 11.21% of the system cost. Thus, system optimization tends to use solar energy, although the system is equipped with three energy sources. Third, the system stores 4.62% energy as hydrogen; this was not used to produce electricity during the 61-day analyses because batteries are better for short-term storage. We further compare the cost and energy distribution of the twelve cases (four systems for three load conditions). For all four systems, the office load reaches the highest solar cost but the lowest battery cost, because the working hours are similar to the irradiation curve (see Figure 7). Contrarily, the lab load reaches the highest battery cost for the same reason (the working hours are different from the irradiation curve), so more batteries needs to be used for energy storage.

Table 4. The distribution of cost, energy sources, and loads.

SWPCH System to the Lab Load with (b, s, w) = (61, 18, 0)	
1. Cost Distribution (%)	
Li-Fe Battery	72.98% ($1229)
power electric devices	2.39% ($40)
WT	0% ($0)
Solar panels	9.87% ($166)
WT maintenance	0% ($0)
Solar maintenance	1.34% ($22)
Hydrogen electrolyzer	5.68% ($95)
PEMFC	2.15% ($36)
Chemical hydrogen production	5.56% ($93)
Sodium borohydride (NaBH$_4$)	0% ($0)
2. Energy Supply Distribution (%)	
Wind	0%
PEMFC	0%
Solar	99.32%
battery	0.679%
3. Load Distribution (%)	
Lab load	95.38%
Hydrogen electrolyzer	4.62%

The optimization of the hybrid systems demonstrates a preference for using the solar battery system because of the high cost of hydrogen production. Therefore, we investigated the impacts of hydrogen prices on the total system costs. Figure 11 shows the results of applying the SWPCH system to the laboratory load. First, the system costs begin to decrease when the hydrogen cost falls to about NT$10 per batch (60 g of NaBH$_4$ to produce about 150 L of hydrogen). When the cost of hydrogen declines from NT$28 to NT$9 (NT$1) per batch, the system cost drops from NT$1685 to NT$1662 (NT$1088) per day. Second, the energy supply ratio by the PEMFC increases to 19.6% (59.5%) when the hydrogen price is NT$9 or NT$1 per batch. Under this scenario, the system tends to use more hydrogen energy, as the cost is competitive with that of other renewable energies. Last, the stack price has little influence on the system cost because it is considered in the initial cost (from 10 k to 180 k). For example, when the hydrogen price is greater than NT$11/batch and the PEMFC stack cost drops from NT$180 k to NT$90 k, the system cost drops from NT$1685 to NT$1667 per day. When the hydrogen cost is NT$9 (NT$1) per batch and the PEMFC stack price drops from NT$180 k to NT$90 k, the system cost drops from NT$1662 (NT$1088) to NT$1615 (NT$999) per day.

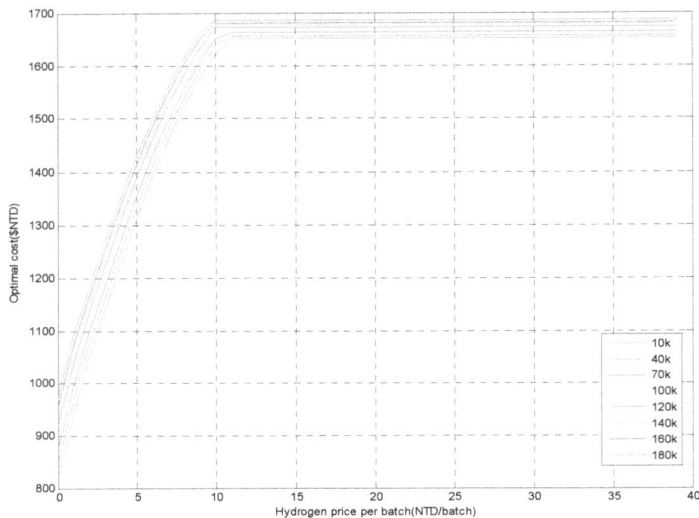

Figure 11. The influence of hydrogen prices on system costs.

4. Conclusions

This paper demonstrated the optimization of hybrid power systems. We developed a general hybrid power model that consisted of solar cells, a WT, a fuel cell, hydrogen electrolysis, chemical hydrogen generation, and batteries. The model parameters are tuned based on experimental data, so that system responses under different operation conditions can be predicted without conducting individual experiments. Then, the performance of four hybrid systems under three typical loads was evaluated by calculating system costs and reliability. The results showed that the costs and reliability of all the systems were effectively improved by optimizing the system sizes. The hybrid system with the solar panels and battery sets achieved the lowest costs, as wind and hydrogen energy are relatively expensive at present. Last, the impacts of stack and hydrogen prices on system costs was analyzed. The results indicated that hydrogen prices had a more substantial influence than the stack price on system costs, and that hydrogen energy would be competitive when its price fell to about one-third of the current price. In future research, the impact of cost of other components, such as the PV and WT, can be analyzed in a similar way.

Author Contributions: Conceptualization, F.-C.W.; Methodology, F.-C.W. and Y.-S.H.; Software, Y.-S.H.; Validation, F.-C.W., Y.-S.H. and Y.-Z.Y.; Formal Analysis, F.-C.W. and Y.-S.H.; Investigation, F.-C.W. and Y.-S.H.; Resources, F.-C.W. and Y.-S.H.; Data Curation, F.-C.W., Y.-S.H., and Y.-Z.Y.; Writing-Original Draft Preparation, Y.-S.H.; Writing-Review and Editing, F.-C.W., Y.-S.H., Y.-Z.Y.; Visualization, F.-C.W., Y.-S.H., Y.-Z.Y.; Supervision, F.-C.W.; Project Administration, F.-C.W.; Funding Acquisition, F.-C.W.

Energies **2018**, *11*, 1948

Funding: This research was funded by the Ministry of Science and Technology, R.O.C., in Taiwan under Grands MOST 105-2622-E-002-029 -CC3, MOST 106-2622-E-002-028 -CC3, and MOST 106-2221-E-002-165-. This research was also supported in part by the Ministry of Science and Technology of Taiwan (MOST 107-2634-F-002-018), National Taiwan University, Center for Artificial Intelligence & Advanced Robotics. The authors would like to thank M-Field™ for their collaboration and technical supports.

Acknowledgments: This work was financially supported in part by the Ministry of Science and Technology, R.O.C., in Taiwan under Grands MOST 104-2622-E-002-023 -CC3, MOST 104-2221-E-002-086-, and MOST 105-2622-E-002-029 -CC3. The authors would like to thank M-Field™ for their collaboration and technical supports.

Conflicts of Interest: The authors declare no conflict of interest.

References

1. Maleki, A.; Askarzadeh, A. Optimal sizing of a PV/wind/diesel system with battery storage for electrification to an off-grid remote region: A case study of Rafsanjan, Iran. *Sustain. Energy Technol. Assess.* **2014**, *7*, 147–153. [CrossRef]
2. Devrim, Y.; Bilir, L. Performance investigation of a wind turbine–solar photovoltaic panels–fuel cell hybrid system installed at İncek region–Ankara, Turkey. *Energy Convers. Manag.* **2016**, *126*, 759–766. [CrossRef]
3. Martinez-Lucas, G.; Sarasua, J.I.; Sanchez-Fernandez, J.A. Frequency Regulation of a Hybrid Wind-Hydro Power Plant in an Isolated Power System. *Energies* **2018**, *11*, 239. [CrossRef]
4. Wang, F.C.; Chen, H.C. The development and optimization of customized hybrid power systems. *Int. J. Hydrogen Energy* **2016**, *41*, 12261–12272. [CrossRef]
5. Bonfiglio, A.; Delfino, F.; Invernizzi, M.; Procopio, R. Modeling and Maximum Power Point Tracking Control of Wind Generating Units Equipped with Permanent Magnet Synchronous Generators in Presence of Losses. *Energies* **2017**, *10*, 102. [CrossRef]
6. Pedra, J.; Corcoles, F.; Monjo, L.; Bogarra, S.; Rolan, A. On Fixed-Speed WT Generator Modeling for Rotor Speed Stability Studies. *IEEE Trans. Power Syst.* **2012**, *27*, 397–406. [CrossRef]
7. Lee, J.; Park, J.; Jung, H.J.; Park, J. Renewable Energy Potential by the Application of a Building Integrated Photovoltaic and Wind Turbine System in Global Urban Areas. *Energies* **2017**, *10*, 2158. [CrossRef]
8. Maouedja, R.; Mammeri, A.; Draou, M.D.; Benyoucef, B. Performance evaluation of hybrid Photovoltaic-Wind power systems. *Energy Procedia* **2014**, *50*, 797–807. [CrossRef]
9. Al Ghaithi, H.M.; Fotis, G.P.; Vita, V. Techno-economic assessment of hybrid energy off-grid system—A case study for Masirah island in Oman. *Int. J. Power Energy Res.* **2017**, *1*, 103–116. [CrossRef]
10. Chen, P.J.; Wang, F.C. Design optimization for the hybrid power system of a green building. *Int. J. Hydrogen Energy* **2018**, *43*, 2381–2393. [CrossRef]
11. Chennouf, N.; Settou, N.; Negrou, B.; Bouziane, K.; Dokkar, B. Experimental Study of Solar Hydrogen Production Performance by Water Electrolysis in the South of Algeria. *Energy Procedia* **2012**, *18*, 1280–1288. [CrossRef]
12. Tribioli, L.; Cozzolino, R.; Evangelisti, L.; Bella, G. Energy management of an off-grid hybrid power plant with multiple energy storage systems. *Energies* **2016**, *9*, 661. [CrossRef]
13. Cozzolino, R.; Tribioli, L.; Bella, G. Power management of a hybrid renewable system for artificial islands: A case study. *Energy* **2016**, *106*, 774–789. [CrossRef]
14. Aouali, F.Z.; Becherif, M.; Tabanjat, A.; Emziane, M.; Mohammedi, K.; Krehi, S.; Khellaf, A. Modelling and experimental analysis of a PEM electrolyser powered by a solar photovoltaic panel. *Energy Procedia* **2014**, *62*, 714–722. [CrossRef]
15. Rahimi, S.; Meratizaman, M.; Monadizadeh, S.; Amidpour, M. Techno-economic analysis of wind turbine–pem (polymer electrolyte membrane) fuel cell hybrid system in standalone area. *Energy* **2014**, *67*, 381–396. [CrossRef]
16. Bianchi, M.; Branchini, L.; De Pascale, A.; Melino, F. Storage solutions for renewable production in household sector. *Energy Procedia* **2014**, *61*, 242–245. [CrossRef]
17. Bocklisch, T.; Böttiger, M.; Paulitschke, M. Multi-storage hybrid system approach and experimental investigations. *Energy Procedia* **2014**, *46*, 186–193. [CrossRef]
18. Ikhsan, M.; Purwadi, A.; Hariyanto, N.; Heryana, N.; Haroen, Y. Study of renewable energy sources capacity and loading using data logger for sizing of solar-wind hybrid power system. *Procedia Technol.* **2013**, *11*, 1048–1053. [CrossRef]

19. Wang, F.C.; Kuo, P.C. Control design and power management of a stationary PEMFC hybrid power system. *Int. J. Hydrogen Energy* **2013**, *38*, 5845–5856. [CrossRef]

20. Guo, Y.F.; Chen, H.C.; Wang, F.C. The development of a hybrid PEMFC power system. *Int. J. Hydrogen Energy* **2015**, *40*, 4630–4640. [CrossRef]

21. Jetpro Technology, INC. Available online: http://www.jetprotech.com.tw/products/jps200.htm (accessed on 25 June 2018).

22. A123 System. Available online: http://www.a123systems.com (accessed on 25 June 2018).

23. M-Field Energy Ltd.: Power Converter. Available online: http://www.m-field.com.tw/product_converter.php (accessed on 25 June 2018).

24. M-Field Energy Ltd.: Hydrogen Generator. Available online: http://www.general-optics.com/fuelcells/Fu_pdfs/English/HGL3610_EN.pdf (accessed on 25 June 2018).

25. Li, S.C.; Wang, F.C. The development of a sodium borohydride hydrogen generation system for proton exchange membrane fuel cell. *Int. J. Hydrogen Energy* **2016**, *41*, 3038–3051. [CrossRef]

26. Wang, F.C.; Fang, W.H. The Development of a PEMFC Hybrid Power Electric Vehicle with Automatic Sodium Borohydride Hydrogen Generation. *Int. J. Hydrogen Energy* **2017**, *42*, 10376–10389. [CrossRef]

energies

MDPI

Article

Uncertainty Analysis of Weather Forecast Data for Cooling Load Forecasting Based on the Monte Carlo Method

Jing Zhao * [iD], Yaoqi Duan [iD] and Xiaojuan Liu

Tianjin Key Lab of Indoor Air Environmental Quality Control, School of Environmental Science and Engineering, Tianjin University, Tianjin 300350, China; mgzcdyq@163.com (Y.D.); lxj15869159276@163.com (X.L.)
* Correspondence: zhaojing@tju.edu.cn; Tel.: +86-22-87402072

Received: 29 June 2018; Accepted: 17 July 2018; Published: 20 July 2018

Abstract: Recently, the cooling load forecasting for the short-term has received increasing attention in the field of heating, ventilation and air conditioning (HVAC), which is conducive to the HVAC system operation control. The load forecasting based on weather forecast data is an effective approach. The meteorological parameters are used as the key inputs of the prediction model, of which the accuracy has a great influence on the prediction loads. Obviously, there are errors between the weather forecast data and the actual weather data, but most of the existing studies ignored this issue. In order to deal with the uncertainty of weather forecast data scientifically, this paper proposes an effective approach based on the Monte Carlo Method (MCM) to process weather forecast data by using the 24-h-ahead Support Vector Machine (SVM) model for load prediction as an example. The data-preprocessing method based on MCM makes the forecasting results closer to the actual load than those without process, which reduces the Mean Absolute Percentage Error (MAPE) of load prediction from 11.54% to 10.92%. Furthermore, through sensitivity analysis, it was found that among the selected weather parameters, the factor that had the greatest impact on the prediction results was the 1-h-ahead temperature $T(h-1)$ at the prediction moment.

Keywords: uncertainty analysis; load forecasting; the Monte Carlo Method (MCM); the Support Vector Machine (SVM) model

1. Introduction

In recent years, heating, ventilation and air conditioning (HVAC) systems have become important elements in office buildings and are responsible for around 40% of the energy use in office buildings, which means a great energy-saving potential [1]. However, the operation management level of HVAC systems is generally low, and the refrigeration capacity of the equipment does not match with the actual demand, resulting in a large energy consumption. Precise load forecasting is the basis of the optimization of HVAC system operation, which is conducive to formulate an operation strategy according to the load change and can lay the theoretical foundation for enhancing the thermal comfort and reducing the energy consumption of office buildings. Among the influential factors, meteorological parameters play a very important role in the dynamic cooling load, which has a great influence on the actual energy consumption of a building.

In the relevant literature on building load forecasting, various prediction models are proposed for load forecasting and related research. Xia and Xiang et al. [2] proposed a prediction model based on a radial basis function (RBF) neural network to forecast a daily load, which mainly took some weather parameters into consideration including temperature, humidity, wind speed, atmospheric pressure and so on. The forecasting results illustrated that the model has better performance compared with the Back Propagation (BP) network. Ruzic et al. [3] put forward a regression-based adaptive weather-sensitive short-term load-forecasting algorithm. This algorithm was used for the load prediction of the Electric Power Utility of Serbia. Wi [4] presented a fuzzy polynomial regression method for holiday load prediction combined with the dominant weather feature, and pointed out that it was pivotal to select the previous data relevant to the given holiday for improving the accuracy of holiday load forecasting. Support Vector Machines (SVMs) have been widely applied in the field of pattern recognition, bioinformatics, and other artificial intelligence relevant areas to solve the classification and regression issues; these are called Support Vector Classification (SVC) and Support Vector Regression (SVR). Particularly, along with Vapnik's ε-insensitive loss function, the SVM also has been extended to solve nonlinear regression estimation problems by SVR. It has been widely used in many fields involving prediction problems, such as financial industry forecasting [5–7], engineering and software field forecasting [8], atmospheric science forecasting [9] and so on. Furthermore, the SVR model has also been successfully applied to predict the power load [10]. The selection of the three parameters (C, ε, and σ) in the SVR model influences the prediction accuracy significantly. Many studies have given recommendations on appropriate setting of SVR parameters [11]. But those methods do not comprehensively consider the interaction effects among the three parameters. Thus, the intelligent algorithms are adopted to determine appropriate parameter values. Barman et al. [12] proposed a regional hybrid STLF model utilizing SVM with a new technique to evaluate its suitable parameters and pointed out that the GOA-SVM model is targeted for forecasting the load under local climatic conditions. Li et al. [13] investigate the feasibility of using Least Squares Support vector regression (LS-SVR) to forecast building cooling load. The evaluation of the tests illustrated that the SVR model with the Particle Swarm Optimization (PSO) has a good generalization performance.

At present, the research on the inputs of the prediction model mainly involves the optimized selection of input parameters. Duanmu et al. [14] proposed a simplified prediction model of the cooling load based on the hourly cooling load coefficient method and analyzed the various influential factors of the cooling load. They pointed out that outdoor temperature is the key influential factor of the cooling load. Wang et al. [15] researched the influence of climate change on the heating and the cooling (H/C) energy requirements of residential houses, which is from cold to hot humid in five regional climates of Australia. They pointed out that the impacts of significant climate change on H/C energy requirements may occur during the lifecycle of existing housing stock. Jiang [16] considered that the accurate prediction of building thermal performance is dependent on meteorological data such as dry-bulb temperature, relative humidity, wind speed and solar radiation to a large extent. Chen et al. [17] selected different meteorological variables as inputs for different time scales, using building dynamics simulation to forecast the energy demand for cooling and heating of residential buildings. Petersen et al. [18] analyzed the impact of uncertainty on the indoor environment.

Indeed, only a few studies have formally dealt with the issue of uncertainty in load forecasting. For example, Sarjiya [19] adopted a decision analysis method to handle the uncertainty of the load forecast in power systems for the aim of optimization of the operating strategy. Domínguez-Muñoz [20] proposed a new approach based on stochastic simulation methods to research the impact of the uncertainty of the internal disturbance on the peak cooling load in the buildings. Douglas et al. [21] put forward a method to analyze the risk of short-term power system operational planning with the electrical load forecast uncertainty. MacDonald [22] focused on the problem of quantifying the effect of uncertainty on the predictions made by simulation tools. Two approaches including external and internal methods were used to quantify this effect. Domínguez-Muñoz et al. [23] quantified the uncertainty that can be expected in the thermal conductivity of insulation materials in the lack of specific experimental measurements. Sten et al. [24] analyzed the influence of the uncertainties of temperature stratification and pressure coefficients on buildings in term of natural ventilation through an expert review process.

Overviewing the previous research, few studies have paid attention to the influence of uncertainty of weather forecast data on the load forecasting. However, external disturbance factors such as meteorological parameters play a very important role in the dynamic cooling loads of a building, which have a great impact on the actual energy consumption of the building. It is effective to use the weather forecast data to predict the building load in advance and adjust the air conditioning units in time according to the forecast loads for the purposes of improvement of the indoor comfort and reduction of building energy consumption. If the uncertainty of weather forecast data is ignored, it may cause errors in model inputs, which reduces the accuracy of the forecast load. The paper fills a gap in terms of the correction of the uncertainty of weather forecast data.

This paper explored the impact of weather forecast uncertainty on load forecasting, and the Monte Carlo Method (MCM) was used to modify the input parameters of the model for load forecasting, which can increase the accuracy of the load forecasting before and after the correction. Furthermore, the sensitivity analysis was adopted to explore the factors that have a great impact on load forecasting results.

The contents of the paper are as follows. Section 2 presents a general overview of the principles of the MCM, the SVM and sensitivity analysis. Section 3 presents a case study, in which this case study is used to illustrate how the methodology can be applied to study the impact of uncertainty of the weather forecast data on load prediction, and the main factors contributing to the load prediction are identified through a sensitivity analysis. Section 4 presents a discussion of the results, as well as some proposals for future research. Section 5 summarizes two important conclusions in the research.

2. Methodology

In this paper, the MCM is used to analyze the uncertainties of weather forecasting parameters, and the model based on SVM is established to forecast the cooling load of an office building. In addition, the Standardized Regression Coefficient (SRC) method for sensitivity analysis is introduced comprehensively. The flowchart shown in Figure 1 depicts the main steps in developing the research, which facilitates the understanding of the proposed approach.

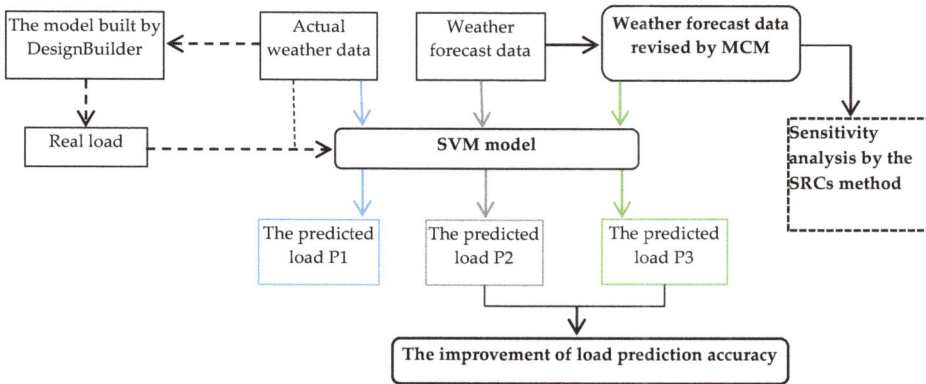

Figure 1. The framework of the research methods.

2.1. The MCM of Random Sampling in Processing Weather Forecast Data

The MCM, also called a statistical simulation method, is an important numerical calculation method guided by probability statistics theory due to the development of science technology and the invention of electronic computers in the mid–1940s. It is an effective way to use random numbers to solve many problems. The Monte Carlo simulation is a method of studying the distribution characteristics by setting up a stochastic process and calculating the estimates and statistics of parameters. Specifically, the reliability of the system is too complex, and it is difficult to establish an accurate mathematical model for reliability prediction. When the model is inconvenient to apply, the estimated value of the desired target can be approximated by the stochastic simulation method. As the number of simulations increases, the expected accuracy of the target is gradually increased.

2.1.1. The Principle of the MCM

The Theorem of Large Numbers and Central Limits in Probability Theory are the theoretical basis of the MCM [25]. The principle of the Monte Carlo simulation method is that when the problem or the object itself has a probability feature, a sampling result can be generated by a computer simulation method. The statistic or the value of the parameter can be calculated according to the sampling.

Based on these two theorems, the function can be expressed as follows.

Assuming the function [26]:

$$Y = f(X_1, X_2, \cdots, X_n),$$ (1)

where the probability distributions of the variables X_1, X_2, ... , X_n are known. The values $(x_1, x_2, \ldots, x_n)$ of each set of random variables $(X_1, X_2, \ldots, X_n)$ are obtained by direct or indirect sampling, then the value y_i of the function Y can be determined according to Formula (2) [26]:

$$y_i = f(x_{i1}, x_{i2}, \cdots, x_{in})$$ (2)

Sampling multiple times $(i = 1, 2, \ldots, m)$ repeatedly and independently, we can obtain a batch of sampling numbers $y_1, y_2, \ldots, y_n$ of the function Y, which are in accordance with the characteristics of the normal distribution.

For each output, m possible results are obtained [20]:

$$Y = \begin{bmatrix} y_1 \\ y_2 \\ \vdots \\ y_m \end{bmatrix} = \begin{bmatrix} f(x_{11}, x_{12}, \ldots, x_{1n}) \\ f(x_{21}, x_{22}, \ldots, x_{2n}) \\ \vdots \\ f(x_{m1}, x_{m2}, \ldots, x_{mn}) \end{bmatrix},$$ (3)

2.1.2. The Steps of the MCM

First, a statistical analysis tool, such as IBM SPSS Statistics 19.0 software (19.0, IBM, Armonk, NY, USA), is used to analyze the probability distribution of the errors of the weather forecast data and the real data. A statistical model related to the problem is determined, of which the solution is regarded as the probability distribution and mathematical expectation of the constructed model. Generally, an appropriate theoretical distribution (e.g., Uniform distribution, Normal distribution, Binomial distribution, Poisson distribution, Triangular distribution, etc.) is used to describe the empirical probability distribution of random variables. If there is no typical theoretical probability distribution that can be directly quoted, it is necessary to estimate an initial probability distribution of the research object based on historical statistics and subjective prediction.

Second, it is important to generate random numbers to simulate the random changes of variables. There are mainly two methods to generate random numbers. We can use an existing random numbers table, or they can be calculated by using a computer program. In this paper, the program of the MCM for the research was written into MATLAB to implement the Monte Carlo random sampling according to the probability distribution obtained by the previous step. After multiple sampling, we can get m possible results, such as Equation (2).

Finally, when the number of simulations is sufficiently large, the probability distribution of the function Y and the concerned digital feature information can be close to the actual situation. Stable conclusions could be obtained by averaging the statistics or estimates of the parameters.

$$\overline{Y} = \frac{\sum_{i=1}^{m} y_i}{m},$$ (4)

where y_i ($i = 1, 2, \ldots, m$) is the sampling result calculated by Equation (2). $\overline{Y}$ is the expected value of the results.

2.2. The Load Forecast Model Based on SVM

SVM is the technique first proposed by Vapnik to solve classification and regression problems [27]. SVR is a machine learning method based on statistical learning theory. It can effectively solve practical problems such as small samples and nonlinearities and has strong generalization ability. It mainly includes two regression models: ε-SVR and v-SVR. The ε-SVR model is used in this paper. By introducing a kernel function, the nonlinear problem of low-dimensional space is transformed into a linear problem in high-dimensional feature space using nonlinear mapping. After the transformation, the decision function [28] is:

$$f(x) = \omega^T \cdot \varphi(x) + b$$ (5)

In Formula (5), ω is a weight vector, b is a threshold, and $\varphi(x)$ is a sublinear mapping relationship from a low-dimensional space to a high-dimensional space.

The SVM uses the minimum structural risk to determine the parameters ω and b and introduces the insensitive loss function parameter ε, which translates the problem into the following optimization problems [28]:

$$\min \quad \frac{1}{2}\omega^T \omega + C \sum_{i=1}^{m} (\xi_i + \xi_i^*),$$ (6)

$$s.t. \quad y_i - \omega^T \cdot \phi(x_i) - b \le \varepsilon + \xi_i,$$ (7)

$$\omega^T \cdot \phi(x_i) + b - y_i \le \varepsilon + \xi_i^*,$$ (8)

$$\omega^T \cdot \phi(x_i) + b - y_i \le \varepsilon + \xi_i^*,$$ (9)

where $(x_1, y_1), \ldots, (x_m, y_m)$ are a pair of input and output vectors, m is the number of samples, ω is weight factor, b is the threshold value, C is error cost, input samples are mapped to higher dimensional space by using kernel function ϕ, ξ_i is the upper training error and the ξ_i^* is the lower training error subject to ε-insensitive tube $|y - (\omega^T \cdot \varphi(x) + b)| \le \varepsilon$.

The SVM includes two parameters: Intrinsic parameters of the support vector machine, including the penalty parameter 'C', the loss function parameter 'ε'; and parameters in the kernel function, such as the kernel width in the Gaussian kernel. The choices of these parameters are very important. The penalty parameter 'C' directly affects the complexity and stability of the model. It can make the model a tradeoff between complexity and training error. The loss function parameter 'ε' controls the simulation of SVR, which effects the number of support vectors and the generalization ability of the model. The width coefficient 'γ' in the kernel function that reflects the correlation between the vectors. The main types of kernel functions include linear kernel functions, polynomial kernel functions, Gaussian radial basis kernel functions, and sigmoid colony kernel functions. Among these functions, Gaussian kernel functions, suited to represent the complex nonlinear relationship between input and output [28,29], have the advantages of computational efficiency, simplicity, reliability, and ease of adaptation. Gaussian kernel functions [28] are as follows:

$$K(x_i, x_j) = exp\left(-\gamma \| x_i - x_j \|^2\right), \gamma > 0$$ (10)

where the γ is the kernel parameter. When training SVM models, two free parameters need to be identified, which are kernel parameter γ and regularization constant C.

Since the key parameters of the above support vector machine model directly affect the accuracy of the model, this paper uses the particle swarm optimization algorithm to determine the optimal combination of these parameters, and then substitutes the optimal combination parameters into the support vector machine model to obtain its regression model. The specific steps are as follows:

- Data normalization:

$$x_{ij}^* = \frac{x_{ij} - x_{jmin}}{x_{jmax} - x_{jmin}},\tag{11}$$

where x_{ij}, x_{ij}^* are data before and after normalization, respectively, and x_{jmin} and x_{jmax} are the respective minimum and maximum values of the column where x_{ij} is located. The normalization process of the dependent variable data is similar to the independent variable data, and will not be described here.
- Establishing the support vector machine objective function based on training samples.
- Using the particle swarm optimization algorithm to select the key parameters of the SVM to obtain the optimal combination of the key parameters of the SVM.
- Substituting the optimal combination parameters into the SVM model to obtain its regression model.
- Using the prediction sample and the model obtained above to forecast the energy consumption of the building.

2.3. The SRCs Method for Sensitivity Analysis

Sensitivity analysis is used to study the mapping relations of uncertainties of input parameters and outputs [30]. There are a lot of sensitivity analysis methods among previous studies [31]. Some methods directly research the input-output map generated by the Monte Carlo method without additional runs of the model. Other methods propagate specific samples are aimed at the sensitivity analysis, for example, the screening method of Morris [32]. The SRCs method has been adopted in this paper, of which the basis is to fit a linear multidimensional model [20] between model inputs and model outputs.

$$\hat{y}_i = \beta_0 + \sum_{j=1}^k \beta_j x_{ij}\tag{12}$$

The regression coefficients β_j are determined such that the sum of error squares

$$\sum_{i=1}^N (y_i - \hat{y}_i)^2 = \sum_{i=1}^N \left[y_i - \left(\beta_0 + \sum_{j=1}^k \beta_j x_{ij} \right) \right]^2\tag{13}$$

is minimized. The following ratio, called the coefficient of determination [20],

$$R^2 = \frac{\sum_{i=1}^N (\hat{y}_i - \bar{y}_i)^2}{\sum_{i=1}^N (y_i - \bar{y})^2}\tag{14}$$

is a measure of how well the model (12) matches the data. The closer to 1 the corresponding value of R^2, the greater the model matches the data, but considering the different units and orders of magnitude of parameters, these drawbacks are easily worked out reformulating Equation (12) [20] as

$$\frac{\hat{y} - \bar{y}}{\sigma_y} = \sum_{j=1}^k \frac{\beta_j \sigma_j}{\sigma_y} \frac{x_j - \bar{x}_j}{\sigma_j},\tag{15}$$

where $\bar{y}$ is the mean value and σ_y the variance of the output under the consideration

$$\sigma_y = \left[\sum_{i=1}^N \frac{(y_i - \bar{y})^2}{N - 1} \right]^{1/2},\tag{16}$$

and $\bar{x}_j$ is the mean value and σ_j the variance of the j input factor

$$\sigma_j = \left[\sum_{i=1}^{N} \frac{(x_{ij} - \bar{x}_j)^2}{N-1}\right]^{1/2}.$$ (17)

The Standardized Regression Coefficient [20] for the input factor j is defined as

$$SRC_j = \frac{\beta_j \sigma_j}{\sigma_y}.$$ (18)

Under the premise that the input variables are independent, the SRCs show the importance of each factor through moving each factor from its expected value by a fixed fraction of its standard deviation while keeping all other factors at their expected values [20]. Calculating the SRCs means to perform the regression analysis, with input and output parameters normalized to zero and standard deviation one. A positive sign indicates that the input is positively correlated with the output, while a negative sign indicates a negative correlation. The importance of these factors can be ranked according to the absolute value of the SRCs.

3. Case Study and Results

3.1. The Framework of the Case Study

The framework of the case study is shown in Figure 2. The real-time meteorological parameters that we collected were input into the DesignBuilder (DB), and the simulated cooling load was regarded as the real load of the office building. Next, the actual meteorological data and simulated load data are used as training samples to build a load forecasting model based on SVM. We use the actual weather data in July as test samples to perform load forecasting to obtain the predicted load P1. Then, the weather forecast data before and after processing are input into the SVM model to obtain predicted loads P2 and P3, which are used for comparing the prediction accuracy of P2 and P3. Sensitivity analysis was used to study the factors that have a significant impact on load forecasting in the input parameters.

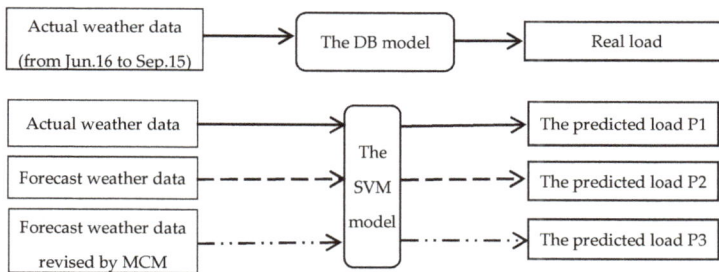

Figure 2. The framework of the case study.

In order to verify the validity of the MCM and the SVM model established, three evaluation indexes are used to compare the prediction results between P1, P2 and P3, which include the following: the Mean Absolute Percentage Error (MAPE) [33], the Mean Absolute Error (MAE) [34], the Root Mean Square Error (RMSE) [34]. Their definitions and the calculation results can then be shown as follows.

$$\text{MAPE} = \frac{1}{N} \sum_{i=1}^{N} \left| \frac{\hat{y}_i - y_i}{y_i} \right|$$ (19)

$$\text{MAE} = \frac{1}{N}|\hat{y}_i - y_i| \tag{20}$$

$$\text{RMSE} = \sqrt{\frac{1}{N}\sum_{i=1}^{N}(\hat{y}_i - y_i)^2} \tag{21}$$

where y_i is the real load, $\hat{y}_i$ is the forecasting load, and N is the number of samples.

MAPE not only considers the error between the predicted value and the true value but also the ratio between the error and the true value. It is a measure of the accuracy of the total prediction in the statistical field [32]. MAE and RMSE can amplify the value of the larger prediction bias, which can compare the stability of different prediction models.

3.2. Data Sources and Collection

The data used in the research work of this paper mainly include two parts: meteorological data and load data. For the load forecasting work, there are mainly two ways to obtain the energy consumption of the construction. First, it is obtained through testing. In addition, it is calculated using energy simulation software. For the first approach, due to the low level of operation and management techniques of the current HVAC system, it is usually regulated by the operating experience of workers. The adjustment of the HVAC system has a certain lag, which cannot immediately bring changes to the load even if it is adjusted according to actual conditions. In the case of fluctuations of indoor temperature, the load data obtained by this method do not reflect the impact of real-time changes in meteorological parameters. Therefore, this paper adopts the second method. We use DesignBuilder to simulate the cooling load of the building, analyze the relevant data and establish the model. The meteorological data used in this paper are composed of real-time weather data collected by a small weather station shown in Figure 3 and weather forecast data from the weather website (https://www.worldweatheronline.com). Table 1 shows the measurement information of the weather elements recorded by the station.

Figure 3. Meteorological station.

Table 1. Measurement information of the meteorological elements.

Meteorological Element	Measuring Range	Resolution Ratio	Accuracy
Dry-bulb temperature	$-50\sim+100\ °C$	$0.1\ °C$	$\pm0.2\ °C$
Relative humidity	$0\sim100\%$	0.1%	$\pm2\%\ (\leq80\%)\ \pm5\%\ (>80\%)$

This small weather station is located in Tianjin University, China, which consists of a PC-2-T solar radiation observer, a PC-4 meteorological monitoring recorder, transducers and the management software of the weather station monitoring system. It records the weather data every half hour by these devices and transfers data to the computer via wired cables. The weather data collected by the meteorological station mainly include dry-bulb temperature, relative humidity, wind speed, wind direction, sunshine hours, rainfall, solar radiation intensity. In addition, among the weather parameters from most existing weather websites, the prediction accuracies of the dry-bulb temperature and the relative humidity are relatively high, while the prediction accuracies of parameters such as wind speed, wind direction and solar radiation intensity are poor. Some cannot be predicted in advance, such as solar radiation intensity. Most of the previous literature selected temperature and relative humidity as inputs to establish the prediction model [14–16]. Therefore, in this paper, we mainly recorded the hourly weather forecast data from the weather website, including dry-bulb temperature and relative humidity, and discussed the influence of the uncertainty of forecast dry-bulb temperature and relative humidity on the cooling load forecast.

3.3. The DB Model of the Office Building

The case selected in this article is an office building in Tianjin City, located in Binhai New District, Tianjin, with a construction area of 10,723.16 square meters, building height of 22.80 m, 5 floors above ground, 1 floor underground and a roof set with skylights.

The final model created by the DesignBuilder software version 4.2.0.015 is shown in Figure 4. DB is the most comprehensive Graphical User Interface to the Energy Plus simulation engine which is widely used for modeling [35]. Parameters of the building structure are obtained through research, and other parameters refer to "Tianjin Public Building Energy Efficiency Design Standards" (DB 29-153-2014) for setting, such as personnel density, personnel per room rate, lighting density, running time. The heat source is supplied by the district heating pipe network in winter, and the terminal of the air conditioning system is the fan coil system, while it uses the split Variable Refrigerant Volume (VRV) air conditioning system for cooling in summer. It is difficult to obtain the hourly cooling load by measurement. In addition, the HVAC systems of the office building are normally used from Monday to Friday and are not used on weekends and holidays. Therefore, only the loads from 9 a.m. to 5 p.m. on weekdays are considered in the scope of the study of load forecasting. The error analysis of the simulated hourly heating load and the measured heating supply data from 9:00 to 17:00 for three working days is carried out to verify the simulation.

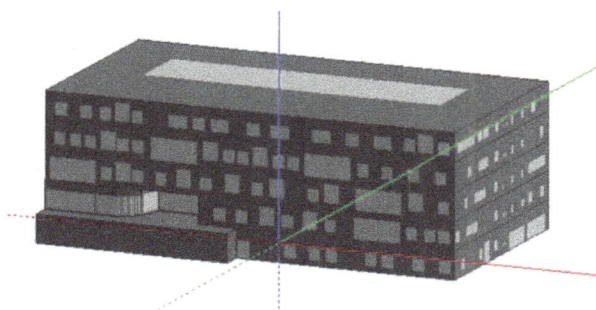

Figure 4. The office building model built by DesignBuilder.

The result is shown in Figure 5. The average relative error between the measured data and the simulated data was 16.1%, which is acceptable considering of the limitations of the on-site tests and measurement instruments. Therefore, the simulation load can be regarded as the real load to establish the database.

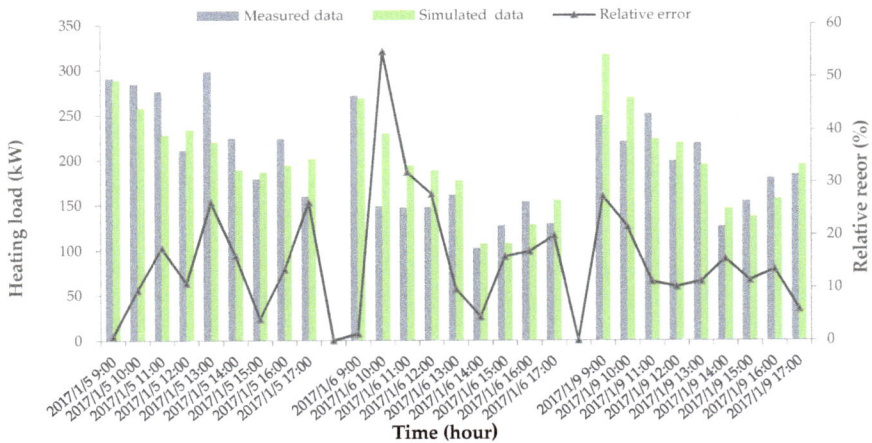

Figure 5. Comparison between the simulated heating load and the measured heating supply.

3.4. The SVM Model and Validation

The final input parameters of the dynamic cooling load forecasting model for the construction of this project still need to be determined in combination with the weather forecast and the actual situation. For the 24-h-ahead load forecasting model, it is difficult to obtain information about historical loads and solar radiation values for the 1-h to 3-h ahead, so we select weather data and load data from 16 June to 15 September as training samples (576 observation values) to establish a support vector machine model. Table 2 gives some examples of training samples.

Table 2. Some examples of training samples.

Time	Output		Inputs						
	$L(h)$	$L(d-1,h)$	$T(h)$	$T(h-1)$	$T(h-2)$	$T(h-3)$	$RH(h)$	$RH(h-1)$	$RH(h-2)$
7/3/9:00	−772.61	−584.99	27.4	26.0	25.2	24.7	78.9	84.2	87.8
7/3/10:00	−813.45	−649.59	28.3	27.4	26.0	25.2	75.1	78.9	84.2
7/3/11:00	−861.98	−700.47	28.3	28.3	27.4	26.0	74.3	75.1	78.9
7/3/12:00	−770.72	−660.08	28.5	28.3	28.3	27.4	71.8	74.3	75.1
7/3/13:00	−753.09	−723.44	28.8	28.5	28.3	28.3	71.3	71.8	74.3
7/3/14:00	−881.99	−876.29	29.3	28.8	28.5	28.3	68.2	71.3	71.8
7/3/15:00	−884.55	−844.72	29.2	29.3	28.8	28.5	68.6	68.2	71.3
7/3/16:00	−866.96	−824.41	28.8	29.2	29.3	28.8	69.2	68.6	68.2
7/3/17:00	−829.87	−815.52	28.5	28.8	29.2	29.3	70.2	69.2	68.6

$L(d-1, h)$ represents the historical load at the same moment we predict for the previous day. $T(h)$, $T(h-1)$, $T(h-2)$, $T(h-3)$ are the dry-bulb temperature at the moment we forecast and the time of the 1–h to 3–h ahead, respectively, $RH(h)$, $RH(h-1)$, $RH(h-2)$ are the relative humidity at the moment we forecast and the time of the 1–h to 2–h ahead, respectively.

The particle swarm optimization algorithm is used to optimize the parameters of the support vector machine, where we set $\varepsilon = 0.1$, $C\epsilon[0.1, 100]$, $g\epsilon[0.01, 100]$, where g is γ in Equation (10). The particle swarm optimization algorithm hyper-parameter optimization results are: Best C = 17.9873, Best g = 0.01, CVmse = 0.0068. Figure 6 shows the results of the fitness function of the particle swarm optimization algorithm.

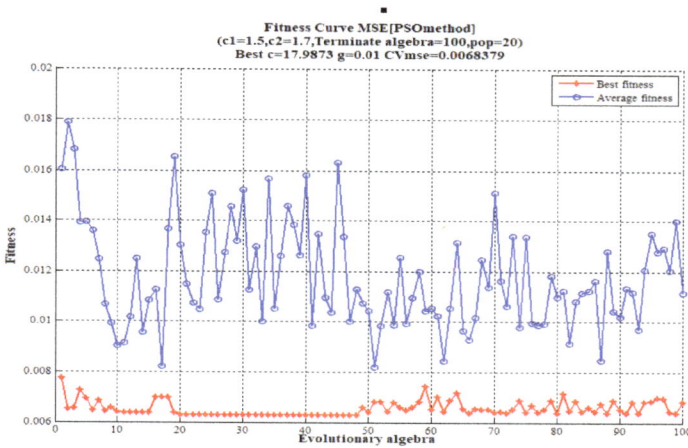

Figure 6. The fitness curve of the particle swarm optimization algorithm.

According to the above optimal parameter combination, the cooling load forecasting model based on SVM can be obtained. We use the actual meteorological parameters in July as test samples. Then, the forecasted data are anti-normalized to obtain the load forecast value, which is compared with the actual load simulated by DesignBuilder to verify the accuracy of the SVM model. The results of the comparison are shown in Figure 7.

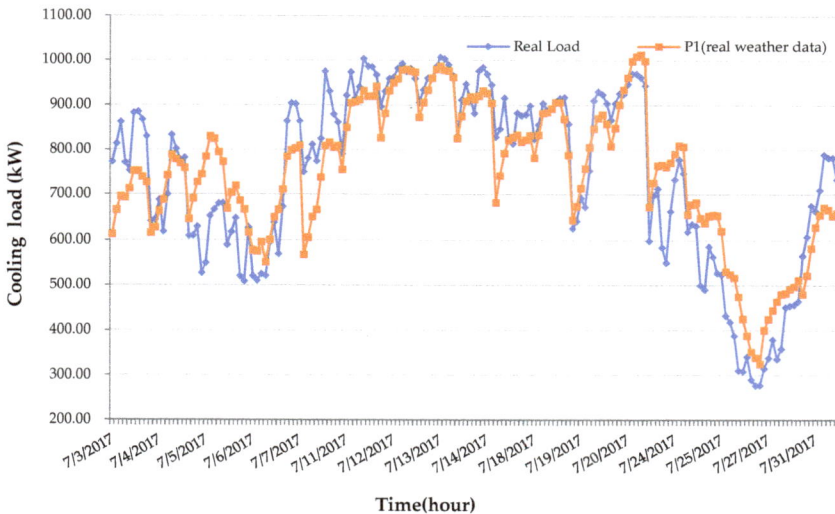

Figure 7. Comparison between the real load and forecast load P1. P1 is the forecast load adopting real weather data using the SVM model.

By calculation, the MAPE of the SVM model is 10.74% compared to the actual situation. Due to the 24-h-ahead load forecast model and the source limits of weather forecast data, we believe that the model basically meets the forecasting requirements. The next research can be done using this model.

3.5. Load Prediction with the Weather Forecast Data

3.5.1. Data Preprocessing Based on MCM

We recorded daily weather forecast data for July except for weekends with a total of 171 samples. Through analysis, it is found that the errors between weather forecast data and real weather data obey the normal distribution $N[\mu, \sigma^2]$. In order to study the influence of the uncertainty of the weather forecast data on the accuracy of load forecasting, we use the MCM to modify the input parameters of the SVM model, namely, the meteorological data. Then, the preprocessed weather forecast data are input into the model for load forecasting.

The IBM SPSS Statistics 19.0 software is used to analyze the error distribution of seven meteorological input parameters in turn. Figure 8 shows the error probability distribution between the weather forecast temperature $T(h-1)$ and the actual temperature one hour before the predicted time. The mean value is $\mu = 0.588$ and the standard deviation is $\sigma = 1.799$.

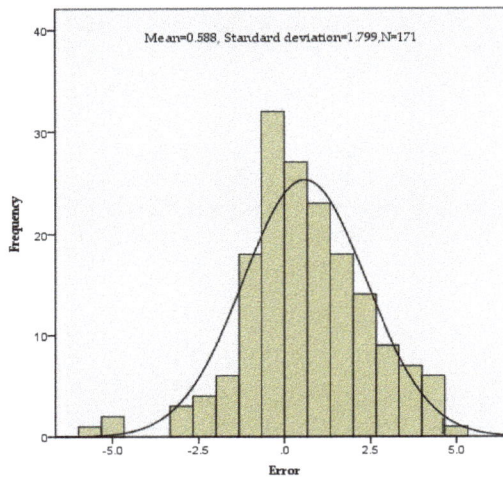

Figure 8. The error probability distribution of $T(h-1)$.

Next, we write the program of MCM into MATLAB to implement the Monte Carlo random sampling of its error Δw, setting the number of simulations M as 1000, and use a corresponding calculation program to obtain a set of revised weather forecast data $T(h-1)^*$. The formula is as follows:

$$T(h-1)^* = T(h-1) + \Delta w \tag{22}$$

For example, for forecasting the load at 9 o'clock on 3 July, it is known that the 1-h-ahead weather forecast dry bulb temperature of the prediction moment $T(h-1)$ is 27.2 °C. The result of the random sampling for $T(h-1)$ using the MCM based on 1000 runs of the model is shown in Figure 9. The most frequent values of $T(h-1)$ in the results of random sampling simulation are near 26.7 °C. Actually, the error Δw obtained by random sampling is −0.6 through calculation, and the revised weather forecast data $T(h-1)^*$ is 26.6 °C, which means that the expected value of $T(h-1)$ is 26.6 °C and is closer to the real weather data, i.e., 26.0 °C.

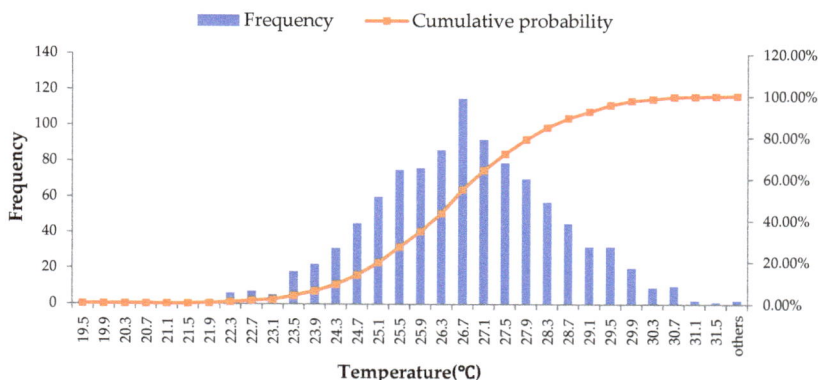

Figure 9. Histogram and the cumulative probability distribution of the T(h−1)*.

Table 3 lists the example of the corrected weather forecast values using MCM and actual values to predict the cooling load at 9 o'clock on 3 July. It can be seen that the weather forecast data corrected by MCM are closer to the actual data. The other samples have the same effects and will not be repeated here due to space limitations.

Table 3. The input parameters of the load forecasting model.

Data Type	L(d−1,h) kw	T(h) °C	T(h−1) °C	T(h−2) °C	T(h−3) °C	RH(h) %	RH(h−1) %	RH(h−2) %
Actual data	−584.99	27.4	26.0	25.2	24.7	78.9	84.0	87.8
Forecast data	−584.99	28.9	27.2	25.6	24.4	58.0	66.0	78.0
Revised data	−584.99	28.2	26.6	25.1	24.1	68.0	75.1	86.0

Table 4 shows the probability distribution functions obeyed by the errors of input parameters obtained from the statistical analysis by SPSS. The procedures for the correction of other parameters are similar to that of the parameter T(h−1). We no longer describe more details here.

Table 4. The probability distribution of each input parameter.

Factor	Parameter	Probability Distribution
X2	T(h)	$N[0.677, 1.889^2]$
X3	T(h−1)	$N[0.588, 1.799^2]$
X4	T(h−2)	$N[0.494, 1.676^2]$
X5	T(h−3)	$N[0.358, 1.539^2]$
X6	RH(h)	$N[-9.424, 10.379^2]$
X7	RH(h−1)	$N[-8.882, 9.822^2]$
X8	RH(h−2)	$N[-8.038, 9.178^2]$

We imported the forecast data for July before and after the correction into the SVM model for load forecasting. Both of the results are compared with the real cooling load simulated by the DesignBuilder, which are shown in Figure 10.

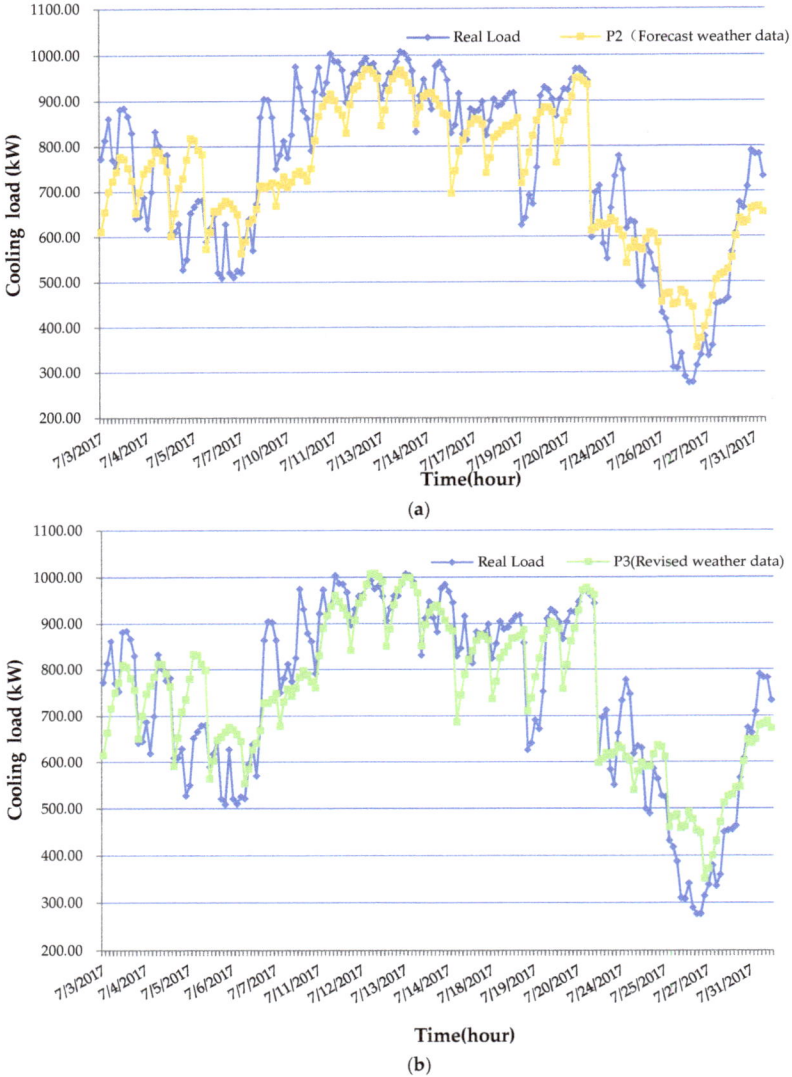

Figure 10. In (**a**), P2 is the forecasting load adopting weather forecast data, and in (**b**), P3 is the forecasting load adopting the weather forecast data dealt with MCM.

As can be seen from the figures above, the prediction results of P2 and P3 are still different from the actual load at some points, but P3 is closer to the real load from the overall level than P2, especially from 10th to 24th July. This proves that the uncertainty analysis of weather forecast data for cooling load forecasting based on MCM is beneficial to improve the accuracy of load prediction.

According to the results of load forecasting under the two scenarios, the evaluation of the prediction performance is shown in Table 5. As can be seen from the table, the 24-*h*-ahead load forecasting model based on SVM has good prediction accuracy. We established the SVM model and used actual meteorological data for load forecasting. The MAPE of P1 compared with the actual load is 10.74%, which includes the uncertainty of the model itself. Then, using the weather forecast data before and after processing with MCM to forecast load separately, we obtain the forecast results P2 and P3. The accuracy of load forecasting using the meteorological forecast data directly and that of the data processed by the MCM are 11.54%, 10.92%, respectively. In terms of MAE and RMSE, the values of P2 are 74.3807 kW and 90.8474 kW, respectively, while the values of P3 are 67.0291 kW and 85.4057 kW. It is clear that the accuracy of P3 is better than that of P2.

Table 5. Comparison between P1, P2, and P3.

Prediction Load	MAPE (%)	MAE (kW)	RMSE (kW)
P1	10.74%	67.8305	84.4138
P2	11.54%	74.3807	90.8474
P3	10.92%	67.0291	85.4057

3.5.2. Sensitivity Analysis

The SRCs for the case study are shown in Figure 11. The uncertainty in the previous seven factors explains most of the variance in the cooling load forecasting that is observed in Figure 10b. The remaining one uncertain factor RH(h–2) has little or no effect on the load forecasting.

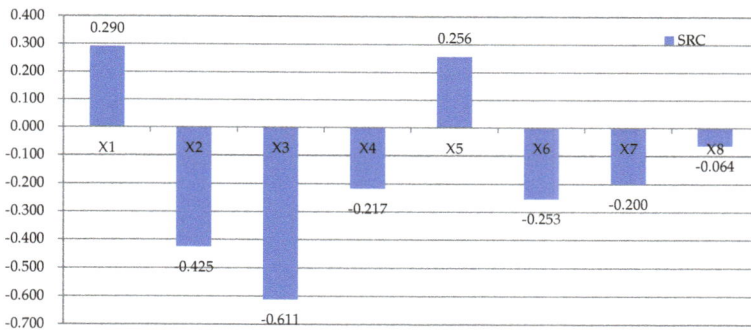

Figure 11. The SRCs of input parameters of the case study. X1 = L(d–1, h); X2 = T(h);X3 = T(h–1); X4 = T(h–2); X5 = T(h–3); X6 = RH(h); X7 = RH(h–1); X8 = RH(h–2).

It can be found that from the input parameters we selected, the factors that have the greatest impact on the load forecast are T(h–1) > T(h) > L(d–1, h) > T(h–3) > RH(h) > T(h–2) > RH(h–1) > RH(h–2) in turn. Obviously, the load at the predicted moment is mostly affected by the outdoor temperature at the previous moment. Due to the thermal inertia of the enclosure, the disturbance caused by the change of the outdoor temperature will not immediately affect the indoor temperature. Heat is transferred between the envelopes with detention and attenuation, which affects the load of the building in the next moments.

4. Discussion

Overviewing the previous research in Section 1, most focused on the optimization of the load prediction model itself to improve the accuracy of load prediction, and few studies have paid attention to the influence of uncertainty of the weather forecast data on the load forecasting. This paper fills a gap in this aspect by uncertainty analysis of weather forecast data for cooling load forecasting based on MCM. Three evaluation indexes are used to compare the prediction results between P1, P2 and P3 in Section 3.4. The results illustrate that the evaluation of the load forecasting with the data processed by the MCM is better than that of the load forecasting using the meteorological forecast data directly according to Figure 10 and Table 5. Furthermore, through sensitivity analysis, it was found that among the selected weather parameters shown in Figure 11, the factor that had the greatest impact on the prediction results was the 1-h-ahead temperature $T(h-1)$ at the prediction moment.

It is worth noting that both of the results P2 and P3 are obtained by the SVM model, and their uncertainties include two parts: one is the uncertainty from the SVM model itself, and the other is the uncertainty from the weather forecast data. This paper mainly focuses on adopting the MCM to process weather forecast data and explore the impact of the uncertainty of weather forecast on load forecasting, not the load forecasting model itself. The research just selects the SVM model as an example, because it is accepted by most researchers due to good performance. With the development of artificial intelligence algorithm technology, the optimal combination of various algorithms is used for load prediction [35], which indicates that the accuracy of load forecasting model itself can be improved to some extent.

The data-preprocessing method based on MCM makes the forecasting results closer to the actual load compared to those without processing, which is suitable for not only office buildings but also other types of buildings. The precise load forecasting results are conducive to the HVAC system operation control. Moreover, the MCM method is convenient for application. Historical weather forecast data and real-time meteorological data are obtained from reliable weather forecasting agencies. In addition, SPSS is used to analyze and obtain the probability distributions of the errors between weather forecast data and real-time meteorological data. The revised weather data are obtained by MATLAB with the relevant programs according to the probability distribution of the errors. Both of the tools are free for application.

When using the MCM to process weather forecast data, it is necessary to analyze the probability distribution characteristics that the error between the weather forecast data and actual weather data obey. The current work is limited by the sources of historical weather forecast data. The larger the historical samples size we collect from the weather forecast websites, the more accurate the probability distribution function of the errors, and then the closer the modified weather forecast data to the actual weather data. In the future, under the condition that the meteorological forecast data sources are more widely available and reliable, the 1-*h*-ahead load forecasting model can be established to predict the load combined with the MCM for data processing. It seems that more precise results of load predictions will be obtained. With the completion of follow-up work, software of the data-preprocessing method based on MCM will be developed.

5. Conclusions

This paper investigated the influence of the uncertainty of weather forecast data on the cooling load forecast. Here, taking the 24-*h*-ahead SVM model as an example, the MCM was adopted to preprocess meteorological forecast data to improve the accuracy of load forecasting. It was indicated that the accuracy of the load forecasting with the data processed by the MCM is better than that of the load forecasting using the meteorological forecast data directly, which is closer to the real load.

Among the selected input parameters, the factors that have the greatest impact on the load forecast are $T(h-1) > T(h) > L(d-1, h) > T(h-3) > RH(h) > T(h-2) > RH(h-1) > RH(h-2)$ in turn. Therefore, we must improve the accuracy of model input parameters to reduce the influence of uncertainty deriving from input parameters on load forecasting, especially those influential input parameters.

Author Contributions: Conceptualization, J.Z. and Y.D.; Methodology, J.Z. and Y.D.; Software, Y.D. and X.L.; Validation, Y.D. and X.L.; Formal Analysis, Y.D.; Writing—Original Draft Preparation, Y.D. and X.L.; Writing—Review & Editing, J.Z. and Y.D.

Funding: This research was funded by the Natural Science Foundation of China, grant number [51508380].

Conflicts of Interest: The authors declare no conflict of interest.

References

1. Verhelst, J.; Van, H.G.; Saelens, D. Model selection for continuous commissioning of HVAC-systems in office buildings: A review. *Renew. Sustain. Energy Rev.* **2017**, *76*, 673–686. [CrossRef]
2. Xia, C.H.; Xiang, X.J.; Hu, X.Y. Design of virtual instrument for load forecasting based on RBF neural network and weather data. *Relay* **2007**, *35*, 29–32.
3. Ruzic, S.; Vuckovic, A.; Nikolic, N. Weather sensitive method for short term load forecasting in Electric Power Utility of Serbia. *IEEE Trans. Power Syst.* **2003**, *18*, 1581–1586. [CrossRef]
4. Wi, Y.M.; Joo, S.K.; Songk, K.B. Holiday load forecasting using fuzzy polynomial regression with weather feature selection and adjustment. *IEEE Trans. Power Syst.* **2012**, *27*, 596–603. [CrossRef]
5. Cao, L. Support vector machines experts for time series forecasting. *Neurocomputing* **2003**, *51*, 321–339. [CrossRef]
6. Huang, W.; Nakamori, Y.; Wang, S.Y. Forecasting stock market movement direction with support vector machine. *Comput. Oper. Res.* **2005**, *32*, 2513–2522. [CrossRef]
7. Pai, P.F.; Lin, C.S. A hybrid ARIMA and support vector machines model in stock price forecasting. *Omega* **2005**, *33*, 497–505. [CrossRef]
8. Pai, P.F.; Hong, W.C. Software reliability forecasting by support vector machines with simulated annealing algorithms. *J. Syst. Softw.* **2006**, *79*, 747–755. [CrossRef]
9. Mohandes, M.A.; Halawani, T.O.; Rehman, S.; Hussain, A.A. Support vector machines for wind speed prediction. *Renew. Energy* **2004**, *29*, 939–947. [CrossRef]
10. Amari, S.; Wu, S. Improving support vector machine classiers by modifying kernel functions. *Neural Netw.* **1999**, *12*, 783–789. [CrossRef]
11. Cherkassky, V.; Ma, Y. Practical selection of SVM parameters and noise estimation for SVM regression. *Neural Netw.* **2004**, *17*, 113–126. [CrossRef]
12. Barman, M.; Choudhury, N.B.D.; Sutradhar, S. A regional hybrid GOA-SVM model based on similar day approach for short-term load forecasting in Assam, India. *Energy* **2018**, *145*, 710–720. [CrossRef]
13. Li, X.; Shao, M.; Ding, L.; Xu, G. Particle swarm optimization-based LS-SVM for building cooling load prediction. *J. Comput.* **2010**, *5*, 614–621. [CrossRef]
14. Duanmu, L.; Wang, Z.; Zhai, Z.J. A simplified method to predict hourly building cooling load for urban energy planning. *Energy Build.* **2013**, *58*, 281–291. [CrossRef]
15. Wang, H.; Chen, Q.; Ren, Z. Impact of climate change heating and cooling energy use in buildings in the United States. *Energy Build.* **2014**, *82*, 428–436. [CrossRef]
16. Jiang, Y. Generation of typical meteorological year for different climates of China. *Energy* **2010**, *35*, 1946–1953. [CrossRef]
17. Chen, D.; Wang, X.; Ren, Z. Selection of climatic variables and time scales for future weather preparation in building heating and cooling energy predictions. *Energy Build.* **2012**, *51*, 223–233. [CrossRef]
18. Petersen, S.; Bundgaard, K.W. The effect of weather forecast uncertainty on a predictive control concept for building systems operation. *Appl. Energy* **2014**, *116*, 311–321. [CrossRef]
19. Eua-Arporn, B.; Yokoyama, A. Short-term operating strategy with consideration of load forecast and generating unit uncertainty. *IEEJ Trans. Power Energy* **2007**, *127*, 1159–1167. [CrossRef]
20. Domínguez-Muñoz, F.; Cejudo-López, J.M.; Carrillo-Andrés, A. Uncertainty in peak cooling load calculations. *Energy Build.* **2010**, *42*, 1010–1018. [CrossRef]
21. Douglas, A.P.; Breipohl, A.M.; Lee, F.N.; Adapa, R. The impacts of temperature forecast uncertainty on Bayesian load forecasting. *IEEE Trans. Power Syst.* **1998**, *13*, 1507–1513. [CrossRef]
22. MacDonald, I. Quantifying the Effects of Uncertainty in Building Simulation. Ph.D Thesis, Department of Mechanical Engineering, University of Strathclyde, Glasgow, UK, 2002.
23. Domínguez-Muñoz, F.; Anderson, B.; Cejudo-López, J.M.; Carrillo-Andrés, A. Uncertainty in the thermal conductivity of insulation materials. In Proceedings of the Eleventh International IBPSA Conference, Glasgow, UK, 27–30 July 2009.
24. Sten, D.W.; Augenbroe, G. Analysis of uncertainty in building design evaluations and its implications. *Energy Build.* **2002**, *34*, 951–958.
25. Reiter, D. The Monte Carlo Method, an Introduction. *Lect. Notes Phys.* **2008**, *50*, 63–78.

26. Papadopoulos, C.E.; Yeung, H. Uncertainty estimation and Monte Carlo simulation method. *Flow Meas. Instrum.* **2002**, *12*, 291–298. [CrossRef]

27. Selakov, A.; Cvijetinović, D.; Milović, L. Hybrid PSO-SVM method for short-term load forecasting during periods with significant temperature variations in city of Burbank. *Appl. Soft Comput.* **2014**, *16*, 80–88. [CrossRef]

28. Fu, Y.; Li, Z.; Zhang, H. Using Support vector machine to predict next-day electricity load of public buildings with sub-metering devices. *Procedia Eng.* **2015**, *121*, 1016–1022. [CrossRef]

29. Ebtehaj, I.; Bonakdari, H.; Shamshirband, S. A combined support vector machine-wavelet transform model for prediction of sediment transport in sewer. *Flow Meas. Instrum.* **2016**, *47*, 19–27. [CrossRef]

30. Saltelli, A.; Tarantola, S.; Campolongo, F.; Ratto, M. Sensitivity Analysis in Practice. *J. Am. Stat. Assoc.* **1989**, *101*, 398–399.

31. Saltelli, A.; Chan, K.; Scott, E.M. *Sensitivity Analysis: Gauging the Worth of Scientific Models*; John Wiley and Sons: Chichester, UK, 2000.

32. Morris, M.D. Factorial sampling plans for preliminary computational experiments. *Technometrics* **1991**, *33*, 161–174. [CrossRef]

33. Tayman, J.; Swanson, D.A. On the validity of MAPE as a measure of population forecast accuracy. *Popul. Res. Policy Rev.* **1999**, *18*, 299–322. [CrossRef]

34. Zhao, J.; Liu, X.J. A hybrid method of dynamic cooling and heating load forecasting for office buildings based on artificial intelligence and regression analysis. *Energy Build.* **2018**. [CrossRef]

35. Iman, W.; Mohd, W.; Royapoor, M.; Wang, Y.; Roskilly, A.P. Office building cooling load reduction using thermal analysis method-A case study. *Appl. Energy* **2017**, *185*, 1574–1584.

energies

MDPI

Article

Scalable Clustering of Individual Electrical Curves for Profiling and Bottom-Up Forecasting

Benjamin Auder [1], Jairo Cugliari [2,*] [iD], Yannig Goude [3] and Jean-Michel Poggi [4]

[1] LMO, University Paris-Sud, 91405 Orsay, France; benjamin.auder@math.u-psud.fr
[2] ERIC EA 3083, University de Lyon, Lyon 2, 69676 Bron, France
[3] EDF R & D, LMO, Univ Paris-Sud, 91405 Orsay, France; yannig.goude@edf.fr
[4] University Paris Descartes & LMO, Univ. Paris-Sud, 91405 Orsay, France; jean-michel.poggi@math.u-psud.fr
[*] Correspondence:jairo.cugliari@univ-lyon2.fr; Tel.: +33-4-7877-3155

Received: 29 June 2018; Accepted: 16 July 2018; Published: 20 July 2018

Abstract: Smart grids require flexible data driven forecasting methods. We propose clustering tools for bottom-up short-term load forecasting. We focus on individual consumption data analysis which plays a major role for energy management and electricity load forecasting. The first section is dedicated to the industrial context and a review of individual electrical data analysis. Then, we focus on hierarchical time-series for bottom-up forecasting. The idea is to decompose the global signal and obtain disaggregated forecasts in such a way that their sum enhances the prediction. This is done in three steps: identify a rather large number of super-consumers by clustering their energy profiles, generate a hierarchy of nested partitions and choose the one that minimize a prediction criterion. Using a nonparametric model to handle forecasting, and wavelets to define various notions of similarity between load curves, this disaggregation strategy gives a 16% improvement in forecasting accuracy when applied to French individual consumers. Then, this strategy is implemented using R—the free software environment for statistical computing—so that it can scale when dealing with massive datasets. The proposed solution is to make the algorithm scalable combine data storage, parallel computing and double clustering step to define the super-consumers. The resulting software is openly available.

Keywords: clustering; forecasting; hierarchical time-series; individual electrical consumers; scalable; short term; smart meters; wavelets

---

1. Introduction

1.1. Industrial Context

Energy systems are facing a revolution and many challenges. On the one hand, electricity production is moving to more intermittency and complexity with the increase of renewable energy and the development of small distributed production units such as photovoltaic panels or wind farms. On the other hand, consumption is also changing with plug-in (hybrid) electric vehicles, heat pumps, the development of new technologies such as smart phones, computers, robots that often come with batteries. To maintain the electricity quality, energy stakeholders are developing smart grids (see [1,2]), the next generation power grid including advance communication networks and associated optimisation and forecasting tools. A key component of the smart grids are smart meters. They allow two-sided communication with the customers, real time measurement of consumption and a large scope of demand side management services. A lot of countries have deployed smart meters, as stated in [3], the UK, the US and China have respectively deployed 2.9, 70 and 96 million of such equipments in 2016. In France, 35 million will be deployed before 2021 for a global cost of 5 billion (see e.g., [4]).

Ref. [5] mentions that Sweden and Italy have achieved full deployment and [6] that Italian distribution system operators are planning the second wave of roll-outs.

This results in new opportunities such as local optimisation of the grid, demand side management and smart control of storage devices. Exploiting the smart grid efficiently requires advanced data analytics and optimisation techniques to improve forecasting, unit commitment, and load planning at different geographical scales. Massive data sets are and will be produced as explained in [7]: data from energy consumption measured by smart meters at a high frequency (every half minute instead of every 6 months); data from the grid management (e.g., Phasor Measurement Units); data from energy markets (prices and bidding, transmission and distribution system operators data, such as balancing and capacity); data from production units and equipments for their maintenance and control (sensors, periodic measures...). A lot of efforts are made by utilities to develop datalakes and IT structures to gather and make these data available for their business units in real time. Designing new algorithms to analyse and process these data at scale is a key activity and a real competitive advantage.

We will focus on individual consumption data analysis which plays a major role for energy management and electricity load forecasting, designing marketing offers and commercial strategies, proposing new services as energy diagnostics and recommendations, detect and prevent non-technical losses.

1.2. Individual Electrical Consumption Data: A State-of-the-Art

Individual consumption data analysis is, according to the development of smart meters, a popular and growing field of research. Composing an exhaustive survey of recent realizations is then a difficult challenge not addressed here. As detailed in [3], individual consumption data analytics covers various fields of statistics and machine learning: time series, clustering, outlier detection, deep learning, matrix completion, online learning among others.

Given a data set of individual consumptions, a first natural step is exploratory: clustering, which is the most popular unsupervised learning approach. The purpose of clustering is to partition a dataset into homogeneous subsets called clusters (see [8]). Homogeneity is measured according to various criteria such as intra and inter class variances, or distance/dissimilarity measures. The elements of a given cluster are then more similar to those of the same cluster than the elements of the other clusters. Time series clustering is an active subfield where each individual is not characterised by a set of scalar variables but are described by time series, signals or functions, considered as a whole, opening the way for signal processing techniques or functional data analysis methods (see [9,10] for general surveys).

Clustering methods for electricity load data have been widely applied for profiling or demand response management. Refs. [11,12] give an overview of the clustering techniques for customer grouping, finding patterns into electricity load data or detecting outliers and apply it to 400 non-residential medium voltage customers. Clustering can be seen as longitudinal when the objective is to cluster temporal patterns (e.g., daily load curves) from a single individual or transversal when the goal is to build clusters of customers according to their load consumption profile and/or side information. The main application of clustering is load profiling which is essential for energy management, grid management and demand response (see [13]). For example, in [14] data mining techniques are applied to extract load profiles from individual load data of a set of low voltage Portuguese customers, and then supervised classification methods are used to allocate customers to the different classes. In [15], load profiles are obtained by iterative self-organizing data analysis on metered data and demonstrated on a set of 660 hourly metered customers in Finland. Ref. [16] proposes an unsupervised clustering approach based on k-means on features obtained by average seasonal curves using minute metered data from 103 homes in Austin, TX. Correspondence between clusters, their associated profiles and survey data are also studied. Authors of [17] suggest a k-means clustering to derive daily profiles from 220,000 homes and a total of 66 millions daily curves in California. Other approaches based on mixture models are presented in [18] for customers categorization and load profiling on a data set of 2613 smart metered household from London.

Another interest of clustering is forecasting, more precisely bottom-up forecasting which means forecasting the total consumption of a set of customers using individual metered data. Forecasting is an obvious need for optimisation of the grid. As pointed previously, it becomes more and more challenging but also crucial to forecast electricity consumption at different "spatial" scale (a district, a city, a region but also a segment of customers). Bottom up methods are a natural approach consisting of building clusters, forecasting models in each cluster and then aggregating them. In [19], clustering algorithms are compared according to their forecasting accuracy on a data set consisting of 6000 residential customers and SME in Ireland. Good performances are reached but the proposed clustering methods are defined quite independently of the model used for forecasting. On the same data set, Ref. [20] associate a longitudinal clustering and a functional forecasting model similar to KWF (see [21]) for forecasting individual load curves.

A clustering method supervised by forecasting accuracy is proposed in [22] to improve the forecast of the total consumption of a French industrial subset obtaining a substantial gain but suffering from high computational time. In [23], a k-means procedure is applied on features consisting of mean consumption for 5 well chosen periods of day, mean consumption per day of a week and peak position into the year. In each cluster a deep learning algorithm is used for forecasting and then the bottom up forecast is the simple sum of clusters forecasts. Results showing a minimum gain of 11% in forecast accuracy are provided on the Irish data set and smart meter data from New-York. On the Irish data again, Ref. [3] propose to build an ensemble of forecasts from a hierarchical clustering on individual average weekly profiles, coupled with a deep learning model for forecasting in each cluster. Different forecasts corresponding to different sizes of the partition are at the end aggregated using linear regression.

We propose here a new approach, following the previous work of [24], to build clusters and forecasting models that are performant for the bottom-up forecasting problem as well as from the computational point of view.

The paper is organized as follows. After this first section introducing the industrial context and a state-of-the-art review of individual electrical data analysis, Section 2 provides the big picture of our proposal for bottom-up forecasting from smart meter data, without technical details. The next three sections focus on the main tools: wavelets (Section 3) to represent functions and to define similarities between curves, the nonparametric forecasting method KWF (Section 4) and the wavelet-based clustering tools to deal with electrical load curves (Section 5). Section 6 is specifically devoted to the upscaling issue and strategy. Section 7 describes an application for forecasting a French electricity dataset. Finally, Section 8 collects some elements of discussion. It should be noted that we tried to write the paper in such a way that each section could be read independently of each other. Conversely, some sections could be skipped by some readers, without altering the local consistency of the others.

2. Bottom-Up Forecasting from Smart Meter Data: Big Picture

We present here our procedure to get a hierarchical partition of individual customers, schematically represented in Figure 1. On the bottom line, there are N individual customers, say $I_1, \ldots, I_N$. Each of them has an individual demand coded into an electrical load curve. At the top of the schema, there is one single global demand G obtained by the simple aggregation of the individual ones at each time step, i.e., $G = \sum_n I_n$. We look for the construction of a set of K medium level aggregates, $A_1, \ldots, A_K$ such that they form a partition of the individuals. Each of the considered entities (individuals, medium level aggregates and global demand) can be considered as time series since they carry important time dependent information.

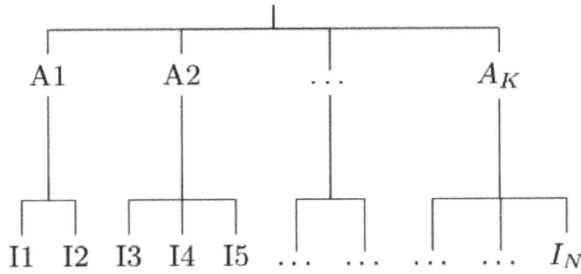

Figure 1. Schematic representation of a hierarchy of customers.

Seasonal univariate time series can naturally be partitioned with regards to time. For example, electrical consumption could be viewed as a sequence of consecutive daily curves which exhibit rich information related to calendar, weather conditions or tariff options. Functional data analysis and forecasting is then a very elegant method to consider. Ref. [25] propose a non-parametric functional method called KWF (Kernel + Wavelet + Functional) to cope with nonstationary time series. Briefly, the main idea is to see the forecasts as a weighted mean of futures of past situations. The weights are obtained according to a similarity index between past curves and actual ones.

Pattern research-based methods repose on a fully non parametric and thus more general frame than prediction approaches more adapted to electricity load demand. This point can be seen as both a weakness and a strength. Specific models can better express the known dependences of electricity demand to long-term trend, seasonal components (due to the interaction of economic and social activities) and climate. However, they usually need more human time to be calibrated. The arrival of new measurement technologies structure of intelligent networks, with more local and high resolution information, unveils forecasting electricity consumption at local scale.

Several arguments can be given to prefer bottom-up approaches with respect to some descending alternatives. Let us briefly mention two of them. The first is related to electrical individual signals themselves which need to be smoothed and the most natural and interpretable way is to define small aggregates of individuals leading to more stable signals, easier to analyse and to forecast (see [17]). The second reason is more statistically related and refers to descending clustering strategies which involve supervised strategies which appear to be especially time consuming (see [22]).

Bottom-up forecasting methods are composed of two successive steps: clustering and forecasting. In the clustering step, the objective is to infer classes from a population such that each class could be accurately forecast. Typically, each class corresponds to customers with specific daily/weekly profile, different relationships to temperature, tariff options or prices (see e.g., [26] regarding demand response). The second step consists in aggregating forecasts to predict the total or any subtotal. In the context of demand response and distribution grid management it could be forecasting the consumption of a district, a town or a substation on the distribution grid.

Recently, Ref. [24] suggested a clustering method achieving both clustering and forecasting of a population of individual customers. They decompose the total consumption such that the sum of disaggregated forecasts improves significantly the forecast of the total. The strategy includes three steps: in the first one super-consumers are designed with a greedy but computationally efficient clustering, then a hierarchical partitioning is done and among which the best partition is chosen according to a disaggregated forecast criterion. The predictions are made with the KWF model which allows one to use it as a off-the-shelve tool.

In concrete, data for each customer is a set of P time dependent (potentially noisy) records evenly sampled at a relatively high frequency (e.g., 1/4, 1/2 or hourly records). Then, we consider the data for each individual as a time series that we treat as a function of time. Wavelets are used to code the information about the shape of the curves. Thanks to nice mathematical properties of wavelets, we compress the information of each curve into a handy number of coefficients (in total $J = [\log_2(P)]$) that are called relative energetic contributions. The compression is such that discriminative power is kept even if information is lost. Data are so tabulated in a matrix where lines correspond to observations and columns to variables (see Figure 2).

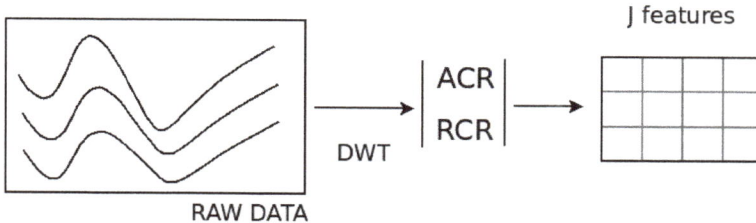

Figure 2. From curves to matrices.

3. Wavelets

A wavelet ψ is a sufficiently regular and well localized function verifying a simple admissibility condition. During a certain time a wavelet oscillates like a wave and is then localized in time due to a damping. Figure 3 represents the Daubechies least-asymmetric wavelet of order 6. From this single function ψ, using translation and dilation a family of functions that form the basic atoms of the Continuous Wavelet Transform (CWT) is derived From this single function ψ, a family of functions is derived using translation and dilation that form the basic atoms of the Continuous Wavelet Transform (CWT)

$$\psi_{a,b}(t) = \frac{1}{\sqrt{a}} \psi\left(\frac{t-b}{a}\right), a \in \mathbf{R}_*^+, b \in \mathbf{R}.$$

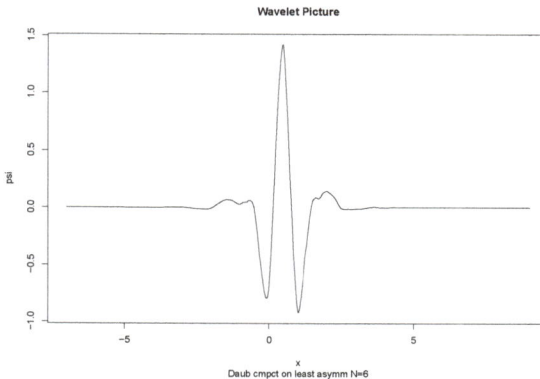

Figure 3. Daubechies least-asymmetric wavelet with filter size 6.

For a function $z(t)$ of finite energy we define its CWT by the function C_z of two real-valued variables:

$$C_z(a,b) = \int_{-\infty}^{\infty} z(t)\psi_{a,b}(t)\,dt$$

Each single coefficient measures the fluctuations of function f at scale a, around the position b. Figure 4 gives a visual representation of $|C_z(a,b)|^2$, also known as wavelet spectrum, for a 10 day period of load demand sampled at 30 min. The waves that one can visually find on the image indicate the highest zone of fluctuations which corresponds to the days. CWT is then extremely redundant but it is useful for example, to characterize the Holderian regularity of functions or to detect transient phenomena or change-points. A more compact wavelet transform can also be defined.

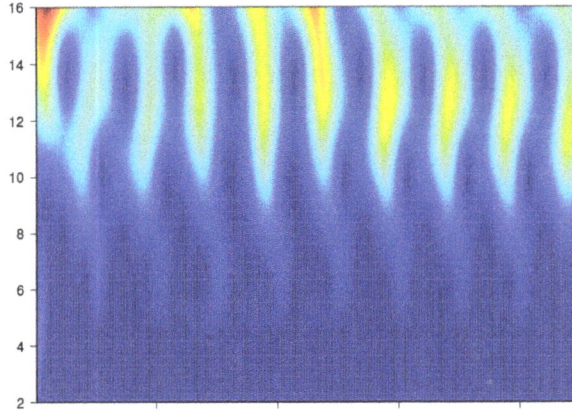

Figure 4. Wavelet spectrum of a week of electrical load demand.

The Discrete Wavelet Transform is a technique of hierarchical decomposition of the finite energy signals. It allows representing a signal in the time-scale domain, where the scale plays a role analogous to that of the frequency in the Fourier analysis ([27]). It allows to describe a real-valued function through two objects: an approximation of this function and a set of details. The approximation part summarizes the global trend of the function, while the localized changes (in time and frequency) are captured in the detail components at different resolutions. The analysis of signals is carried out by wavelets obtained as before from simple transformations of a single well-localized (both in time and frequency) mother wavelet. A compactly supported wavelet transform provides with an orthonormal basis of waveforms derived from scaling (i.e., dilating or compressing) and translating a compactly supported scaling function $\widetilde{\phi}$ and a compactly supported mother wavelet $\widetilde{\psi}$. If one works over the interval $[0, 1]$, periodized wavelets are useful denoting by

$$\phi(t) = \sum_{l \in \mathbb{Z}} \widetilde{\phi}(t - l) \quad \text{and} \quad \psi(t) = \sum_{l \in \mathbb{Z}} \widetilde{\phi}(t - l), \quad \text{for } t \in [0, 1],$$

the periodized scaling function and wavelet, that we dilate or stretch and translate

$$\phi_{j,k}(t) = 2^{j/2} \phi(2^j t - k), \quad \psi_{j,k}(t) = 2^{j/2} \phi(2^j t - k).$$

For any $j_0 \geq 0$, the collection

$$\{\phi_{j_0,k}, k = 0, 1, \ldots, 2^{j_0} - 1; \psi_{j,k}, j \geq j_0, k = 0, 1, \ldots, 2^j - 1\},$$

is an orthonormal basis of $\mathcal{H}$. Then, for any function $z \in \mathcal{H}$, the orthogonal basis allows one to write the development

$$z(t) = \sum_{k=0}^{2^{j_0}-1} c_{j_0,k} \phi_{j_0,k}(t) + \sum_{j=j_0}^{\infty} \sum_{k=0}^{2^j-1} d_{j,k} \psi_{j,k}(t), \tag{1}$$

where $c_{j,k}$ and $d_{j,k}$ are called respectively the scale and the wavelet coefficients of z at the position k of the scale j defined as

$$c_{j,k} = \langle z, \phi_{j,k} \rangle_{\mathcal{H}} \quad d_{j,k} = \langle z, \psi_{j,k} \rangle_{\mathcal{H}}.$$

The wavelet transform can be efficiently computed using the notion of mutiresolution analysis of $\mathcal{H}$ (MRA), introduced by Mallat, who also designed a family of fast algorithms (see [27]). Using MRA, the first term at the right hand side of (1) describe a smooth approximation of the function z at a resolution level j_0 while the second term is the approximation error. It is expressed as the aggregation of the details at scales $j \geq j_0$. If one wants to focus on the finer details then only the information at the scales $\{j : j \geq j_0\}$ is to be looked.

Figure 5 is the multiresolution analysis of a daily load curve. The original curve is represented on the top leftmost panel. The bottom rightmost panel contains the approximation part at the coarsest scale $j_0 = 0$, that is, a constant level function. The set of details are plotted by scale which can be connected to frequencies. With this, the detail functions clearly show the different patterns ranging between low and high frequencies. The structure of the signal is centred on the highest scales (lowest frequencies), while the lowest scale (highest frequencies) keep the noise of the signal.

MRA

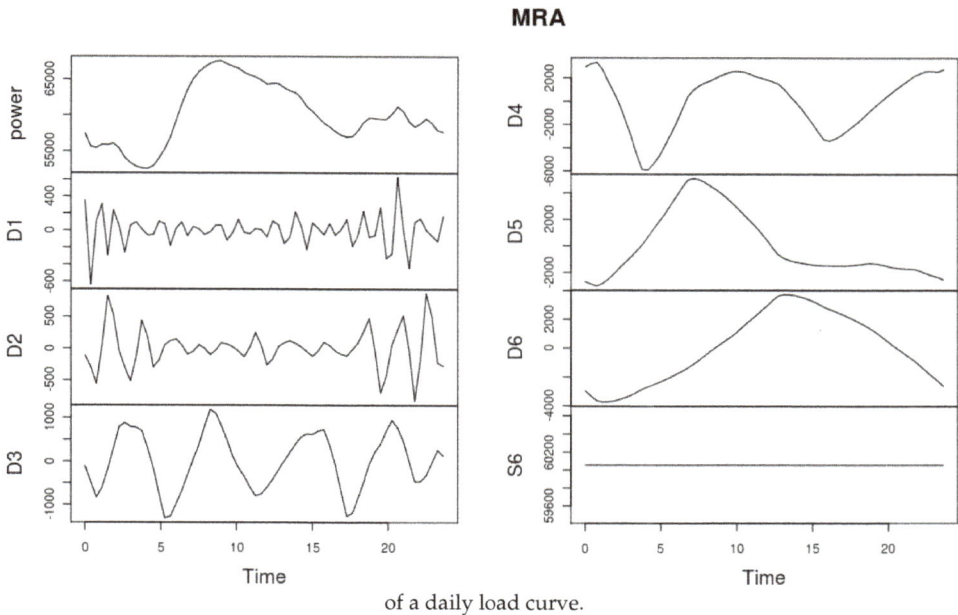

of a daily load curve.

Figure 5. Multiresolution analysis

From a practical point of view, let us suppose for simplicity that each function is observed on a fine time sampling grid of size $N = 2^J$ (if not, one may interpolate data to the next power of two). In this context we use a highly efficient pyramidal algorithm ([28]) to obtain the coefficients of the Discrete Wavelet Transform (DWT). Denote by $\mathbf{z} = \{z(t_l) : l = 0, \ldots, N_i - 1\}$ the finite dimensional sample of the function z. For the particular level of granularity given by the size N of the sampling grid, one rewrites (1) using the truncation imposed by the 2^J points and the coarser approximation level $j_0 = 0$, as:

$$\tilde{z}_J(t) = c_0 \phi_{0,0}(t) + \sum_{j=0}^{J-1} \sum_{k=0}^{2^j-1} d_{j,k} \psi_{j,k}(t). \tag{2}$$

Hence, for a given wavelet ψ and a coarse resolution $j_0 = 0$, one may define the DWT operator:

$$W_\psi : \mathbb{R}^N \to \mathbb{R}^N, \quad \mathbf{z} \mapsto (\mathbf{d}_0, \ldots, \mathbf{d}_{J-1}, c_0 f)$$

with $\mathbf{d}_j = \{d_{j,0}, \ldots, d_{j,2^j-1}\}$. Since the DWT operator is based on an L_2-orthonormal basis decomposition, the energy of a square integrable signal is preserved:

$$\|\mathbf{z}\|_2^2 = c_0^2 + \sum_{j=0}^{J-1} \sum_{k=0}^{2^j-1} d_{j,k}^2 = c_0^2 + \sum_{j=0}^{J-1} \|\mathbf{d}_j\|_2^2. \tag{3}$$

Hence, the global energy $\|\mathbf{z}\|_2^2$ of $\mathbf{z}$ is distributed over some energetic components. The key fact that we are going to exploit is how these energies are distributed and how they contribute to the global energy of a signal. Then we can generate a handy number of features that are going to be used for clustering.

4. KWF

4.1. From Discrete to Functional Time Series

Theoretical developments and practical applications associated with functional data analysis were mainly guided by the case of independent observations. However, there is a wide range of applications in which this hypothesis is not reasonable. In particular, when we consider records on a finer grid of time assuming that the measures come from a sampling of an underlying unknown continuous-time signal.

Formally, the problem can be written by considering a continuous stochastic process $X = (X(t), t \in \mathbb{R})$. So the information contained in a trajectory of X observed on the interval $[0, T]$, $T > 0$ is also represented by a discrete-time process $Z = (Z_k(t), k = 0, \ldots, n; t \in [0, \delta])$ where $Z_k(t) = X((\delta - 1)k + t)$ comes from the segmentation of the trajectory X in n blocks of size $\delta = T/n$ ([29]). Then, the process Z is a time series of functions. For example, we can forecast $Z_{n+1}(t)$ from the data $Z_1, \ldots, Z_n$. This is equivalent to predicting the future behaviour of the X process over the entire interval $[T, T + \delta]$ by having observed X on $[0, T]$. Please note that by construction, the $Z_1, \ldots, Z_n$ are usually dependent functional random variables.

This framework is of particular interest in the study of electricity consumption. Indeed, the discrete consumption measurements can naturally be considered as a sampling of the load curve of an electrical system. The usual segment size, $\delta = 1$ day, takes into account the daily cycle of consumption.

In [21], the authors proposed a prediction model for functional time series in the presence of non stationary patterns. This model has been applied to the electricity demand of *Electricité de France* (EDF). The general principle of the forecasting model is to find in the past, situations similar to the present and linearly combine their futures to build the forecast. The concept of similarity is based on wavelets and several strategies are implemented to take into account the various non stationary sources. Ref. [30] proposes for the same problem to use a predictor of a similar nature but applied to a multivariate process. Next, [31] provide an appropriate framework for stationary functional processes using the wavelet transform. The latter model is adapted and extended to the case of non-stationary functional processes ([32]).

Thus, a forecast quality of the same order of magnitude as other models used by EDF is obtained for the national curve (highly aggregated) even though our model can represent the series in a simple and parsimonious way. This avoids explicitly modeling the link between consumption and weather covariates, which are known to be important in modeling and often considered essential to take into account. Another advantage of the functional model is its ability to provide multi-horizon forecasts simultaneously by relying on a whole portion of the trajectory of the recent past, rather than on certain points as univariate models do.

4.2. Functional Model KWF

4.2.1. Stationary Case

We consider a stochastic process $Z = (Z_i)_{i \in \mathbb{Z}}$ assumed for the moment, to be stationary, with values in a functional space H (for example $H = L_2([0,1])$). We have a sample of n curves $Z_1, \ldots, Z_n$ and the goal is to forecast Z_{n+1}. The forecasting method is divided in two steps. First, find among the blocks of the past those that are most similar to the last observed block. Then build a weight vector $w_{n,i}, i = 1, \ldots, n-1$ and obtain the desired forecast by averaging the future blocks corresponding to the indices $2, \ldots, n$ respectively.

First step.

To take into account in the dissimilarity the infinite dimension of the objects to be compared, the KWF model represents each segment $Z_i, i = 1, \ldots, n$, by its development on a wavelet basis truncated to a scale $J > j_0$. Thus, each observation Z_i is described by a truncated version of its development obtained by the discrete wavelet transform (DWT):

$$Z_{i,J}(t) = \sum_{k=0}^{2^{j_0}-1} c_{j_0,k}^{(i)} \phi_{j_0,k}(t) + \sum_{j=j_0+1}^{J} \sum_{k=0}^{2^j-1} d_{j,k}^{(i)} \psi_{j,k}(t), \qquad t \in [0,1].$$

The first term of the equation is a smooth approximation to the resolution j_0 of the global behaviour of the trajectory. It contains non-stationary components associated with low frequencies or a trend. The second term contains the information of the local structure of the function. For two observed segments $Z_i(t)$ and $Z_{i'}(t)$, we use the dissimilarity D defined as follows:

$$D(Z_i, Z_{i'}) = \sum_{j=j_0+1}^{J} 2^{-j} \sum_{k=0}^{2^j-1} (d_{j,k}^{(i)} - d_{j,k}^{(i')})^2. \tag{4}$$

Since the Z process is assumed to be stationary here, the approximation coefficients do not contain useful information for the forecast since they provide local averages. As a result, they are not taken into account in the proposed distance. In other words, the dissimilarity D makes it possible to find similar patterns between curves even if they have different approximations.

Second step.

Denote $\Xi_i = \{c_{j,k}^{(i)} : k = 0, 1, \ldots, 2^J - 1\}$ the set of scaling coefficients of the i-th segment Z_i at the finer resolution J. The prediction of scaling coefficients (at the scale J) $\widehat{\Xi_{n+1}}$ of Z_{n+1} is given by:

$$\widehat{\Xi_{n+1}} = \frac{\sum_{m=1}^{n-1} K_{h_n}(D(Z_{n,J}, Z_{m,J})) \Xi_{m+1}}{1/n + \sum_{m=1}^{n-1} K_{h_n}(D(Z_{n,J}, Z_{m,J}))},$$

where K is a probability kernel. Finally, we can apply the inverse transform of the DWT to $\widehat{\Xi_{n+1}}$ to obtain the forecast of the Z_{n+1} curve in the time domain. If we note

$$w_{n,m} = \frac{K_{h_n}(D(Z_{n,J}, Z_{m,J}))}{\sum_{m=1}^{n-1} K_{h_n}(D(Z_{n,J}, Z_{m,J}))}, \tag{5}$$

these weights allow to rewrite the predictor as a barycentre of future segments of the past:

$$\widehat{Z_{n+1}}(t) = \sum_{m=1}^{n-1} w_{n,m} Z_{m+1}(t). \tag{6}$$

4.2.2. Beyond the Stationary Case

In the case where Z is not a stationary functional process, some adaptations in the predictor (6) must be made to account for nonstationarity. In Antoniadis et al, (2012) corrections are proposed and their efficiency is studied for two types of non-stationarities: the presence of an evolution of the mean level of the approximations of the series and the existence of classes segments. Let us now be more precise.

It is convenient to express each curve Z_i according to two terms $S_i(t)$ and $D_i(t)$ describing respectively the approximation and the sum of the details,

$$Z_i(t) = \sum_k c_{j_0,k}^{(i)} \phi_{j_0,k}(t) + \sum_{j \geq j_0} \sum_k d_{j,k}^{(i)} \psi_{j,k}(t)$$

$$= S_i(t) + D_i(t).$$

When the curves Z_{m+1} have very different average levels, the first problem appears. In this case, it is useful to centre the curves before calculating the (centred) prediction, and then update the forecast in the second phase. Then, the forecast for the segment $n+1$ is $\widehat{Z_{n+1}}(t) = \widehat{S_{n+1}}(t) + \widehat{D_{n+1}}(t)$. Since the functional process $D_{n+1}(t)$ is centred, we can use the basic method to obtain its prediction

$$\widehat{D_{n+1}}(t) = \sum_{m=1}^{n-1} w_{m,n} D_{n+1}(t), \tag{7}$$

where the weights $w_{m,n}$ are given by (5). Then, to forecast $S_{n+1}(t)$ we use

$$\widehat{S_{n+1}}(t) = S_n(t) + \sum_{m=1}^{n-1} w_{m,n} \Delta(S_n)(t). \tag{8}$$

To solve the second problem, we incorporate the information of the groups in the prediction stage by redefining the weights $w_{m,n}$ according to the belonging of the functions m and n to the same group:

$$\widetilde{w}_{m,n} = \frac{w_{m,n} \mathbf{1}_{\{gr(m)=gr(n)\}}}{\sum_{m=1}^{n} w_{m,n} \mathbf{1}_{\{gr(m)=gr(n)\}}}, \tag{9}$$

where $\mathbf{1}_{\{gr(m)=gr(n)\}}$ is equal to 1 if the groups $gr(n)$ of the n-th segment is equal to the group of the m-th segment and zero elsewhere. If the groups are unknown, they can be determined from an unsupervised classification method.

The weight vector can give an interesting insight into the prediction power carried out by the shape of the curves. Figure 6 represents the computed weights obtained for the prediction of a day during Spring 2007. When plotted against time, it is clear that the only days found similar to the current one are located in a remarkably narrow position of each year in the past. Moreover, the weights seem to decrease with time giving more relevance to those days closer to the prediction past. A closer look at the weight vector (not shown here) reveals that only days in Spring are used. Please note that no information about the position of the year was used to compute the weights. Only the information coded in the shape of the curve is necessary to locate the load curve at its effective position inside the year.

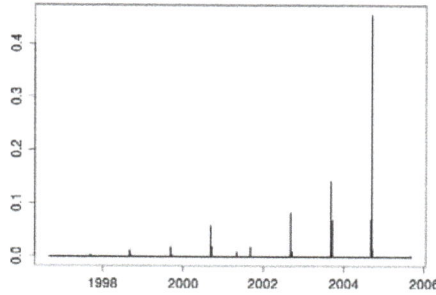

Figure 6. Vector of weights (sorted chronologically) obtained for the prediction of a day during Spring.

Figure 7 is also of interest to understand how the prediction works. There, the plot on the left contains all the days of the dataset against which the similarity was computed with respect to the curve in blue. A transparency scale which makes visible only those curves with a relatively high similarity index. The plot on the right contains the futures of the past days on the left. These are also plot on the transparent scale with the curve in orange which is the prediction given by the weighted average.

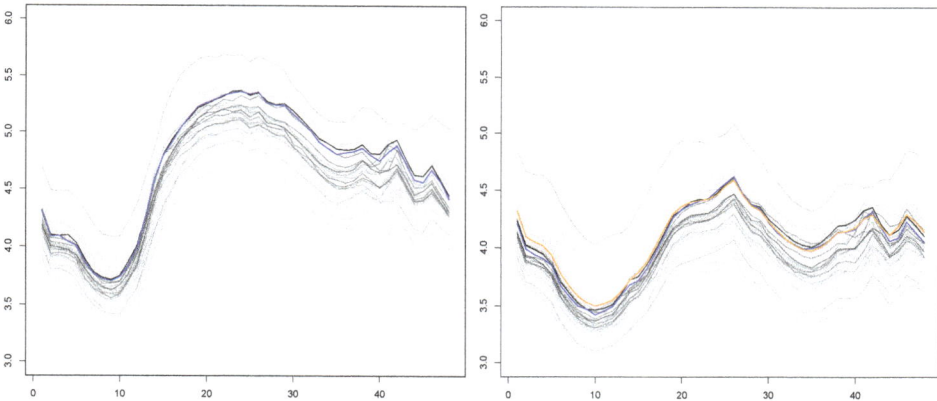

Figure 7. Past and future segments involved in the construction of the prediction by KWF. On each panel, all the days are represented with a transparent colour making visible only the most relevant days for the construction of the predictor.

5. Clustering Electrical Load Curves

Individual load prediction is a difficult task as individual signals have a high volatility. The variability of each individual demand is such that the ratio signal to noise decreases dramatically when passing from aggregate to individual data. With almost no hope of predicting individual data, an alternative strategy is to use these data to improve the prediction of the aggregate signal. For this, one may rely on clustering strategies where customers of similar consumption structure will be put into classes in order to form groups of heterogeneous clients. If the clients are similar enough, the signal of the aggregate will gain in regularity and thus in predictability.

Many clustering methods exist in the specialized literature. We adopt the point of view of [33] where two strategies for clustering functional data using wavelets are presented. While the first one allows to rapidly create groups using a dimension reduction approach, the second one permits to better exploit the time-frequency information at the price of some computational burden.

5.1. Clustering by Feature Extraction

From Equation (3), we can see that the global energy of the curve is approximately decomposed into energy components associated with the smooth approximation of the curve (c_0^2) plus a set of components related to each detail level. In [33] these detail levels were called the absolute contributions $AC_j = \|d_j\|_2^2, j = 0, \ldots, J - 1$ of each scale to the global energy of the curve. Notice that the approximation part is not of primary interest since the underlying process of electrical demand may be highly non stationary. With this, we focus only on the shape of the curves and on its frequency content in order to unveil the structure of similar individual consumers to construct clusters. A normalized version of absolute contributions can be considered, which is called relative contributions and is defined as $RC_j = AC_j / \sum_j AC_j$. After this representation step, the result is depicted by the schema in Figure 2, where the original curves are now embedded into a multi dimensional space of dimension J. Moreover if relative contributions are used, the points are in the simplex of $\mathbb{R}^J$.

Let us describe now the clustering step more precisely. For this any clustering algorithm on multivariate data can be used. Since the time complexity of this step is not only dependant on the sample size N but also on the number of variables P, we choose to detect and remove irrelevant features using a variable selection algorithm for unsupervised learning introduced in [34]. Besides the reduction of the computation time, feature selection allows also to gain in interpretability of the clustering since it highly depends on the data.

The aim of this first clustering step is to produce first a coarse clustering with a rather large quantity K' of aggregated customers, each of them called super-customer (SC). The synchrone demand—that is to say, the sum of all individual demand curves in a specific group—is computed then in all clusters. A parallel can be drawn with the primary situation: we now obtain K' coarsely aggregated demands over P features, that can be interpreted as a discrete noisy sampling of a curve.

Figure 8 shows this first clustering round on the first row of the schema.

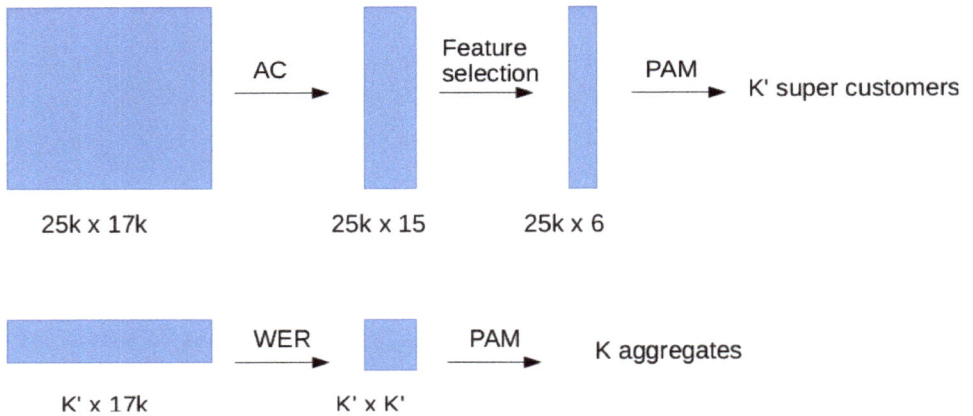

Figure 8. Two step clustering.

5.2. Clustering Using a Dissimilarity Measure

The second clustering stage consists in grouping the SC into a small quantity K of (super-)aggregates, and building a hierarchy of partitions—as seen before. We consider the samples as functional objects and thus define a dissimilarity between curves, to obtain a dissimilarity matrix between the SC. The projection alternative (working with coefficients) was discarded because of the loss of information.

Following [33] we use the wavelet extended R^2 based distance (WER) which is constructed on top of the wavelet coherence. If $x(t)$ and $z(t)$ are two signals, the wavelet coherence between them is defined as

$$R_{x,z}(a,b) = \frac{|S(C_{x,z}(a,b))|}{|S(C_{x,x}(a,b))|^{1/2}|S(C_{z,z}(a,b))|^{1/2}},$$

where $C_{x,z}(a,b)) = C_x(a,b)C_z^*(a,b))$ is the cross-wavelet transform, and S is a smooth operator. Then, the wavelet coherence can be seen as a linear correlation coefficient computed in the wavelet domain and so localized both in time and scale. Notice that smoothing is a mandatory step in order to avoid a trivial constant $R_{x,z}(a,b) = 1$ for all a,b.

The wavelet coherence is then a two dimensional map that quantifies for each time-scale location the strength of the association between the two signals. To produce a single measure of this map, some kind of aggregation must be done. Following the construction of the extended determination coefficient R^2, Ref. [33] propose to use the wavelet extended R^2 which can be computed using

$$WER_{x,z}^2 = \frac{\sum_{j=1}^{J}\left(\sum_{k=1}^{N}|S(C_{x,z}(j,k))|\right)^2}{\sum_{j=1}^{J}\left(\sum_{k=1}^{N}|S(C_{x,x}(j,k))|\sum_{k=1}^{N}|S(C_{z,z}(j,k))|\right)}.$$

Notice that $WER_{x,z}^2$ is a similarity measure and it can easily be transformed into a dissimilarity one by

$$D(x,z) = \sqrt{JN(1 - WER_{x,z}^2)},$$

where the computations are done over the grids $\{1,\dots,N\}$ for the locations b and $\{a_j, j = 1,\dots J\}$ for the scales a.

The boundary scales (smallest and greatest) are generally taken as a power of two which depend respectively on the minimum detail resolution and the size of the time grid. The other values correspond usually to a linear interpolation over a base 2 logarithmic scale.

While the measure is justified by the power of the wavelet analysis, in practice this distance implies heavy computations involving complex numbers and so requires of a larger memory space. This is one of the two reason that renders its use on the original dataset intractable. The second reason is related to the size of the dissimilarity matrix that results from its applications and that grows with the square of the number of time series. Indeed, such a matrix obtained from the SC is largely tractable for a moderate number of super customers of about some hundreds, but it is not if applied on the whole dataset of some tens of millions of individual customers. The trade off between computation time and precision is resolved by a first clustering step that dramatically reduces the number of time series using the RC features; and a second step that introduces the finer but computationally heavier dissimilarity measure on the SC aggregates.

Since K' (the number of SC) is sufficiently small, a dissimilarity matrix between the SC can be computed in a reasonable amount of time. This matrix is then used as the input of the classical Hierarchical Agglomerative Clustering (HAC) algorithm, used here with the Ward link. Its output corresponds to the desired hierarchy of (super-)customers.

Otherwise, one may use other clustering algorithms that use a dissimilarity matrix as input (for instance Partitioning Around Mediods, PAM) to get an optimal partitioning for a fixed number of clusters. The second row of the scheme in Figure 8 represents this second step clustering.

6. Upscaling

We discuss in this section the ideas we develop to upscale the problem. Our final target is to work over twenty million time-series. For this, we run many independent clustering tasks in parallel, before merging the results to obtain an approximation of the direct clustering. We give proposed solutions that were tested in order to improve the code performance. Some of our ideas proved to be useful for moderate sample sizes (say tens of thousands) but turned to be counter-productive for larger sizes (tens of millions). Of course all these considerations depend heavily on the specific material and technology. We recall that our interest is on relatively standard scientific workstations. The algorithm we use on the first step of the clustering is described below. We then show the results of the profiling of our whole strategy to highlight where are the bottlenecks when one wishes to up-scale the method. We end this section discussing the solutions we proposed.

6.1. Algorithm Description

The massive dataset clustering algorithm is as follows:

1. *Data serialization.* Time series are given in a verbose by-column format. We re-code all of them in a binary file (if suitable), or a database.
2. *Dimensionality reduction.* Each series of length N is replaced by the $\log_2(N)$ energetic coefficients defined using a wavelet basis. Eventually a feature selection step can be performed to further reduction on the number of features.
3. *Chunking.* Data is chunked into groups of size at most n_c, where n_c is a user parameter (we use $n_c = 5000$ in the next section experiments).
4. *Clustering.* Within each group, the PAM clustering algorithm is run to obtain K_0 clusters.
5. *Gathering.* A final run of PAM is performed to obtain K' mediods, $K' \ll n$ out of the $n_c \times K_0$ mediods obtained on the chunks..

From these K' medoids the synchrone curves are computed (i.e., the sum of all curves within each group for each time step), and given on output for the prediction step.

6.2. Code Profiling

Figure 9 gives some timings obtained by profiling the runs of our initial (C) code. To give a clearer insight, we also report the size of the objects we deal with. The starting point is the ensemble of individual records of electricity demand for a whole year. Here, we treat over 25,000 clients sampled half-hourly during a year. The tabulation of these data to obtain a matrix representation suitable to fit in memory take about 7 min. and requires over 30 Gb of memory.

Task	Time	Memory	Disk
Raw (15 Gb) to matrix	7 min	30 Gb	2.7 Gb
Compute contributions	7 min	<1 Gb	7 Mb
1st stage clustering	3 min	<1 Gb	–
Aggregation	1 min	6 Gb	30 Mb
Wer distance matrix	40 min	64 Gb	150 Kb
Forecasts	10 min	<1 Gb	–

Figure 9. Code profiling by tasks.

6.3. Proposed Solutions

Two main solutions are to be discussed, concerning the internal data storage strategy and the use of a simple parallelization scheme. The former looks for reducing the communication time of internal operations using serialization. The latter attacks the major bottleneck of our clustering approach, that is the construction of the WER dissimilarity matrix.

The initial format (verbose, by-column) is clearly inappropriate for efficient data processing. There are several options starting from this data format, they imply having all series stored as

- an ASCII file, one sample per line; very fast, but data retrieval will depend on line number;
- a binary format (3 or 4 octets per value); compression is unadvised since it would increase both preprocessing time and (by a large amount) reading times;

- a database (this is the slowest option), so that retrieval can be very quick.

Since we plan to deal with millions of series of thousands time steps, binary files seemed like a good compromise because they can easily fit on disk—and often also in memory. Our R package uses this format internally, although it allows to input data in any of these three shapes. If we were speaking of billions of series of a million time steps or more, then distributed databases would be required. In this case one would only has to fill the database and tell the R package how to access time-series.

The current version is mostly written in R using the `parallel` package for efficiency, [35]. A partial version written fully in C was slightly faster, but not enough compared to the loss of code clarity. The current R version can handle the 25 millions samples on an overnight computation over a standard desktop workstation—assuming the curves can be stored and accessed quickly. Our implementation is called `iecclust` is available as open source software.

7. Forecasting French Electricity Dataset

7.1. Data Presentation

We work on the data provided by EDF also used in [24] which is composed of big customers equipped with smart meters. Unfortunately, this dataset is confidential and cannot be shared. Nevertheless we suggest to the reader interested in bottom up electricity consumption forecasting problems to refer to the open data sets listed in [3].

The dataset consists in approximately 25000 half-hourly load consumption series over two years (2009–2010). The first year is used for partitioning and the calibration of our forecasting algorithm, then the second year is used as a test set to simulate a real forecasting use-case.

The initial dataset contains over 25,000 individual load curves. To test the up-scaling ability of our implementation, we create three datasets of sizes 250,000; 2,500,000 and 25,000,000. In other words, we progressively increase the sample sizes by a factor of 10, 100 and 1000 respectively. The creation follows a simple scheme where each individual curve is multiplied by the realization of independent variables uniformly distributed on $[0.95, 1.05]$ at each time step. Each curve is then replicated using this scheme by several times equal to the up-scaling factor.

7.2. Numerical Experiments

The first task clustering is crucial for reducing the dimension of the dataset. We give some timings in order to illustrate how our approach can deal with tens of thousands of time series. Of course, the total computation time depends on the technical specification of the structure used to perform the computation. In our case, we restrict ourselves to a standard scientific workstation with 8 physical cores and 70 Gigabits of live memory. We use all the available cores to cluster chunks of 5000 observations following the algorithm described in Section 6 for both the first and second clustering task.

A very simple pretreatment is done in order to eliminate load curves with eventual errors. For this, we measure the standard deviation of the contributions of each curve to keep only the 99% central observations eliminating the extremest ones. With this, too flat curves (maybe constant) consumptions or very wiggle ones are considered to be abnormal.

Table 1 gives mean average running times over 5 replicates for each of the different sample sizes. These figures show that our strategy yields on a linear increment on the computation time with respect to the number of time series. The maximum number of series we treat, that is 25 millions of individual curves, needs about 12 h to achieve the first task clustering.

Table 1. Mean average running times (over 5 replicates) for different sample sizes (in log).

Sample Size	Time (In Seconds)
25×10^3	67
25×10^4	513
25×10^5	4420
25×10^6	43,893

The result of this first clustering task is the load curves of 200 super consumers (SC). We now explore how much time series contains the super consumers. For this, we plot (in Figure 10) the relative frequency of each SC cluster (i.e., proportion of observations in the cluster) against its size rank (in logarithmic scale). With this, the leftmost point of the curve represents the largest cluster, while the following ones are sorted in decreasing order of size. To compare between sample sizes four curves are plotted, one for each sample size. A common decreasing trend of the curves appears producing several relatively small clusters. This is not a desired behaviour for a final clustering task. However, we are in an intermediate step which aims at reducing the number of curves n to a certain number K' of super customers, here $K' = 200$. The isolated super customers may merge together in the following step, producing meaningful aggregates.

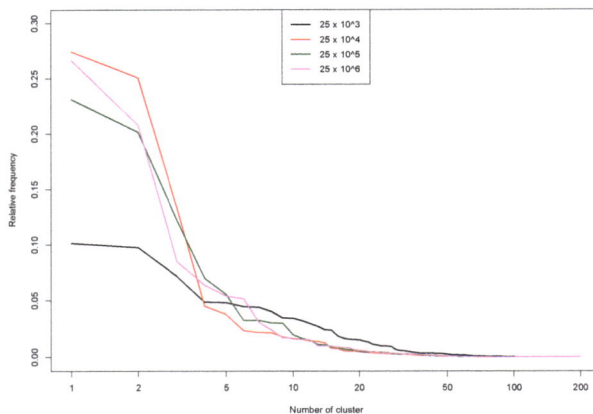

Figure 10. Relative frequency of observations by cluster, in decreasing order, for different sample sizes, $n = 10^3, 10^4, 10^5, 10^6$.

In what follows we focus on the results for the largest dataset, which is the one with over 25 millions of load curves. The resulting 200 super consumers are used to construct the WER dissimilarity matrix, which contains rich information about the clustering structure. One may use for instance a hierarchical clustering algorithm to obtain a hierarchy of SC. A graphical result of this structure in the object of Figure 11, which corresponds to the dendrogram obtained by agglomerative hierarchical clustering using the Ward link function. Then, one may get a partitioning of the ensemble of SC by setting some threshold (a value of height in the figure). However, we will not follow this idea to concentrate on the bottom-up prediction task.

The WER dissimilarity matrix encodes rich information about the pairwise closeness between the 200 super consumers. A way to visualize this matrix is to obtain a multidimensional scaling, that is to construct a setting of low dimension coordinates that best represent the dissimilarities between the curves. Figure 12 contains the matrix scatter plot of the first 4 dimensions of such a setting. For each bivariate scatter plot, the points are drawn with a discrete scale of 15 colours, each one representing a different cluster. This low dimensional representation succeed to represent the clustering structure since points with the same colour are closer forming compact groups.

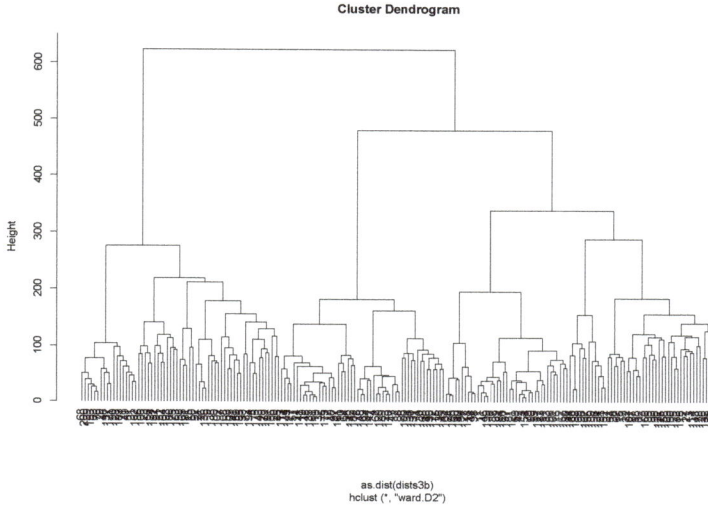

Figure 11. Dendrogram obtained from the WER dissimilarity matrix.

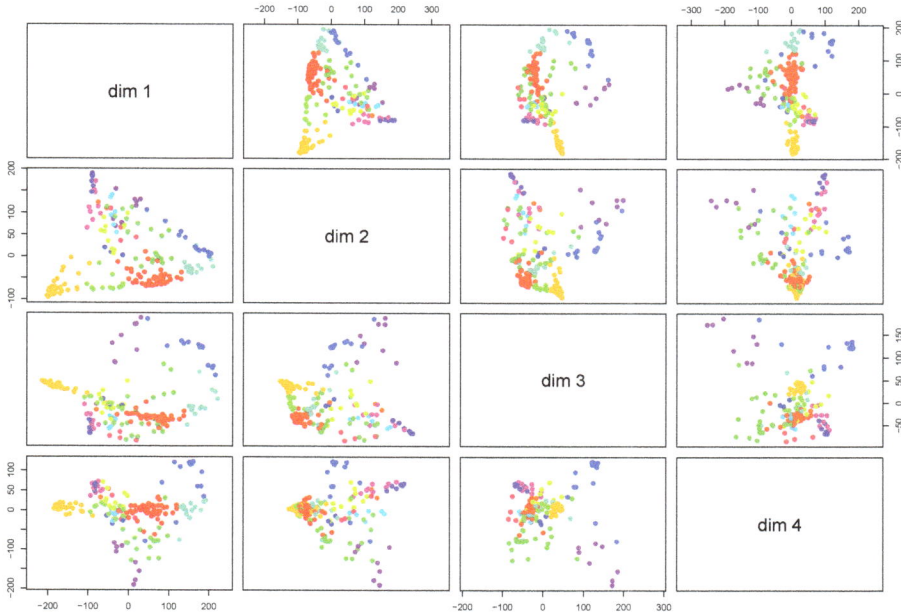

Figure 12. Multidimensional scaling of the WER dissimilarity matrix for the 200 super consumers.

Bottom-up forecast is the leading argument of using the individual load curve clustering. We test the appropriateness of our proposition by getting for a final number of clusters ranging from 2 to 20, 50, 100 and 200 the respective aggregates in terms of load demand. Then, we use KWF as an automatic prediction model for both strategies: prediction of the global demand using the global demand, and the one based on the bottom-up approach.

We use the second year on the dataset to measure the quality of the daily prediction using a rolling basis. Figure 13 reports the prediction error using the MAPE for both the two forecasting strategies. The full horizontal line indicates the annual mean MAPE using the direct method and so it is independent of the number of clusters. For different choices of the number of clusters, the dashed line represent the associated MAPE. All possible clusterings produce then bottom-up forecasts that are better than the one obtained from direct global forecasting.

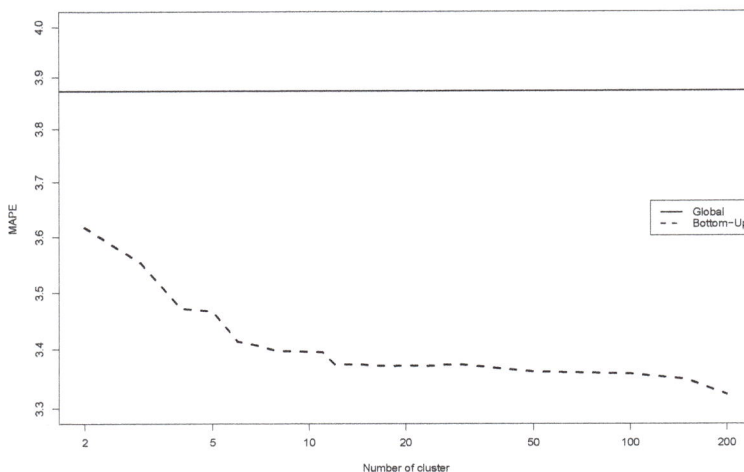

Figure 13. MAPE for the aggregate demand by number of classes for the two strategies: direct global demand forecast (full) and bottom-up forecast (dashed).

8. Discussion

In this final section, we discuss the various choices made as well as some possible extension to cope with multiscale model point of view and how to handle non stationarity.

8.1. Choice of Methods

The three main tools are:

- the wavelet decomposition to represent functions and compute dissimilarities. Of course, several other choices could be interesting, such as splines for bases of functions which are independent of the data or even some data-driven bases like those coming from functional principal component analysis. With respect to these two classical alternatives, (more or less related to a monoscale strategy) the choice of wavelets allows simultaneously a parsimonious representation capturing local features of the data as well as redundant one delivering a more accurate multiscale representation. In addition, from a computational viewpoint, DWT is a very fast: of linear complexity. So to design the super-customers the discrete transform is good enough, for the final clusters, the continuous transform leads to better results. Let us remark that combining wavelets and clustering has recently been considered in [36] from a different viewpoint: details and approximations of the daily load curves are clustered separately leading to two different partitions which are then fused.

- the PAM algorithm and the hierarchical clustering to build the clusters are of very common use and well adapted to their specific role in the whole strategy. It should be noted that the use of PAM to construct the super customers must necessarily be biased towards a large number of clusters (defining the super customers) so it is useless to include sophisticated model-selection rules to choose an optimal number of clusters since the strategy is used only to define a sufficiently large number of clusters.

- the Kernel-Wavelet-Functional (KWF) method to forecast time-series. The global forecasting scheme is clearly fully modular and then, KWF could be replaced by any other time-series model forecasting. The model must be flexible and easy to automatically be tuned because the modeling and forecasting must be performed in each cluster in a rather blind way. The main difficulty with KWF is to introduce exogenous variables. We could imagine to include a single one quite easily but not a richer family in full generality. Nevertheless, it is precisely when dealing with models corresponding to some specific clusters that it could be of interest to use exogenous variables especially informative, for example describing meteo at a local level or some specific market segment. Therefore, some alternatives could be considered, such as generalized additive models (see [37] for a strategy which could be plugged into our scheme).

8.2. Multiscale Modeling and Forecasting

In fact, such a forecasting strategy combining clustering in individuals and forecasting of the total consumption of each cluster can be also viewed as a multiscale modeling. Indeed a by-product is a forecasting at different levels of aggregation from the super customers to the total population. Therefore, instead of restricting our attention on the forecasting of the global signal for a given partition we could imagine to combine in time the different predictions given by each piece of the different partitions in such a way that all the levels could contribute to the final forecasting. The way to weight the different predictions could be fixed for all the instants (see [38] for a large choice of proposals) or, on the contrary, time-dependent according to a convenient choice of the updating policy (see the sequential learning strategies already used in the electrical context in [39]). An attempt in this direction can be found in [40].

Another related topic is individual forecasting or prediction. It must be mentioned since it is interesting to have some ideas about the kind of statistical models or strategies used in this especially hard context, due to extreme volatility and wild nonstationarity. Ref. [41] examine the short-term (one hour) forecasting of individual consumptions using a sparse autoregressive model which is compared against well-known alternatives (support vector machine, principal component regression, and random forests). In general, exogenous variables are used to forecast electricity consumptions, but some authors focus on the reverse. Ref. [42,43] are interested in determining household characteristics or customers information based on temporal load profiles of household electricity demand. They use sophisticated deep learning algorithm for the first one and more classical tools for the second one. In the context of customers surveys, Ref. [44] use smart meter data analytics for optimal customer selection in demand response programs.

8.3. How to Handle Non Stationarity?

Even if the model KWF is well suited to handle non stationarities in the time-domain, it remains that the clusters of customers are also subjected to some dynamics which could be of interest to model in order to control these changes. A first naive possibility is to periodically recompute the entire process including a new calculation of the super-customers and decide, at some stage if the change is significant to be taken into account. A second possibility could be to directly model the evolution of the clusters. For example, in [45] a time-varying extension of the K-means algorithm is proposed. A multivariate vector autoregressive model is used to model the evolution of clusters' centroids over time. This could help to model the changes of clusters along time but we have to think about a penalty mechanism allowing to make changes in the cluster only when it is useful.

Author Contributions: B. Auder, J. Cugliari, Y. Goude and J.-M. Poggi equally contributed to this work.

Acknowledgments: This research benefited from the support of the FMJH 'Program Gaspard Monge for optimization and operations research and their interactions with data science', and from the support from EDF and Thales.

Conflicts of Interest: The authors declare no conflict of interest.

References

1. Yan, Y.; Qian, Y.; Sharif, H.; Tipper, D. A Survey on Smart Grid Communication Infrastructures: Motivations, Requirements and Challenges. *IEEE Commun. Surv. Tutor.* **2013**, *15*, 5–20. [CrossRef]
2. Mallet, P.; Granstrom, P.O.; Hallberg, P.; Lorenz, G.; Mandatova, P. Power to the People!: European Perspectives on the Future of Electric Distribution. *IEEE Power Energy Mag.* **2014**, *12*, 51–64. [CrossRef]
3. Wang, Y.; Chen, Q.; Hong, T.; Kang, C. Review of Smart Meter Data Analytics: Applications, Methodologies, and Challenges. *IEEE Trans. Smart Grid* **2018**. [CrossRef]
4. Jamme, D. Le compteur Linky: Brique essentielle des réseaux intelligents français. In Proceedings of the Office Franco-Allemand Pour La Transition énergétique, Berlin, Germany, 11 May 2017.
5. Alahakoon, D.; Yu, X. Smart Electricity Meter Data Intelligence for Future Energy Systems: A Survey. *IEEE Trans. Ind. Inform.* **2016**, *12*, 425–436. [CrossRef]
6. Ryberg, T. The Second Wave of Smart Meter Rollouts Begin in Italy and Sweden. 2017. Available online: https://www.metering.com/regional-news/europe-uk/second-wave-smart-meter-rollouts-begins-italy-sweden/ (accessed on 1 June 2018).
7. Jiang, H.; Wang, K.; Wang, Y.; Gao, M.; Zhang, Y. Energy big data: A survey. *IEEE Access* **2016**, *4*, 3844–3861. [CrossRef]
8. Kaufman, L.; Rousseeuw, P. *Finding Groups in Data: An Introduction to Cluster Analysis*; Wiley: Hoboken, NJ, USA, 1990.
9. Liao, T.W. Clustering of time series data a survey. *Pattern Recognit.* **2005**, *38*, 1857–1874. [CrossRef]
10. Jacques, J.; Preda, C. Functional Data Clustering: A Survey. *Adv. Data Anal. Classif.* **2014**, *8*, 231–255. [CrossRef]
11. Chicco, G. Overview and performance assessment of the clustering methods for electrical load pattern grouping. *Energy* **2012**, *42*, 68–80. [CrossRef]
12. Zhou, K.; Yang, S.; Shen, C. A review of electric load classification in smart grid environment. *Renew. Sustain. Energy Rev.* **2013**, *24*, 103–110. [CrossRef]
13. Wang, Y.; Chen, Q.; Kang, C.; Zhang, M.; Wang, K.; Zhao, Y. Load profiling and its application to demand response: A review. *Tsinghua Sci. Technol.* **2015**, *20*, 117–129. [CrossRef]
14. Figueiredo, V.; Rodrigues, F.; Vale, Z.; Gouveia, J.B. An electric energy consumer characterization framework based on data mining techniques. *IEEE Trans. Power Syst.* **2005**, *20*, 596–602. [CrossRef]
15. Mutanen, A.; Ruska, M.; Repo, S.; Jarventausta, P. Customer Classification and Load Profiling Method for Distribution Systems. *IEEE Trans. Power Deliv.* **2011**, *26*, 1755–1763. [CrossRef]
16. Rhodes, J.D.; Cole, W.J.; Upshaw, C.R.; Edgar, T.F.; Webber, M.E. Clustering analysis of residential electricity demand profiles. *Appl. Energy* **2014**, *135*, 461–471. [CrossRef]
17. Kwac, J.; Flora, J.; Rajagopal, R. Household Energy Consumption Segmentation Using Hourly Data Smart Grid. *IEEE Trans.* **2014**, *5*, 420–430.
18. Sun, M.; Konstantelos, I.; Strbac, G. C-Vine copula mixture model for clustering of residential electrical load pattern data. In Proceedings of the 2017 IEEE Power Energy Society General Meeting, Chicago, IL, USA, 16–20 July 2017. [CrossRef].
19. Alzate, C.; Sinn, M. Improved electricity load forecasting via kernel spectral clustering of smartmeter. In Proceedings of the International Conference on Data Mining, Dallas, TX, USA, 7–10 December 2013; pp. 943–948.
20. Chaouch, M. Clustering-Based Improvement of Nonparametric Functional Time Series Forecasting: Application to Intra-Day Household-Level Load Curves. *IEEE Trans. Smart Grid* **2014**, *5*, 411–419. [CrossRef]

21. Antoniadis, A.; Brossat, X.; Cugliari, J.; Poggi, J.M. Prévision d'un processus à valeurs fonctionnelles en présence de non stationnarités. Application à la consommation d'électricité. *J. Soc. Française Stat.* **2012**, *153*, 52–78.

22. Misiti, M.; Misiti, Y.; Oppenheim, G.; Poggi, J.M. Optimized Clusters for Disaggregated Electricity Load Forecasting. *Rev. Stat. J.* **2010**, *8*, 105–124.

23. Quilumba, F.L.; Lee, W.J.; Huang, H.; Wang, D.Y.; Szabados, R.L. Using Smart Meter Data to Improve the Accuracy of Intraday Load Forecasting Considering Customer Behavior Similarities. *IEEE Trans. Smart Grid* **2015**, *6*, 911–918. [CrossRef]

24. Cugliari, J.; Goude, Y.; Poggi, J.M. Disaggregated Electricity Forecasting using Wavelet-Based Clustering of Individual Consumers. In Proceedings of the 2016 IEEE International Energy Conference (ENERGYCON), Leuven, Belgium, 4–8 April 2016.

25. Antoniadis, A.; Brossat, X.; Cugliari, J.; Poggi, J.M. Une approche fonctionnelle pour la prévision non-paramétrique de la consommation d'électricité. *J. Soc. Française Stat.* **2014**, *155*, 202–219.

26. Labeeuw, W.; Stragier, J.; Deconinck, G. Potential of active demand reduction with residential wet appliances: A case study for Belgium. *Smart Grid IEEE Trans.* **2015**, *6*, 315–323. [CrossRef]

27. Mallat, S. *A Wavelet Tour of Signal Processing*; Academic Press: Cambridge, MA, USA, 1999.

28. Mallat, S. A theory for multiresolution signal decomposition: The wavelet representation. *IEEE Trans. Pattern Anal. Mach. Intell.* **1989**, *11*, 674–693. [CrossRef]

29. Bosq, D. Modelization, nonparametric estimation and prediction for continuous time processes. In *Nonparametric Functional Estimation and Related Topics*; Roussas, G., Ed.; NATO ASI Series, (Series C: Mathematical and Physical Sciences); Springer: Dordrecht, The Netherland; 1991; Volume 335, pp. 509–529.

30. Poggi, J.M. Prévision non paramétrique de la consommation électrique. *Rev. Stat. Appl.* **1994**, *4*, 93–98.

31. Antoniadis, A.; Paparoditis, E.; Sapatinas, T. A functional wavelet-kernel approach for time series prediction. *J. R. Stat. Soc. Ser. B Stat. Meth.* **2006**, *68*, 837. [CrossRef]

32. Cugliari, J. Prévision Non Paramétrique De Processus à Valeurs Fonctionnelles. Application à la Consommation D'électricité. Ph.D. Thesis, Université Paris Sud, Orsay, France, 2011.

33. Antoniadis, A.; Brossat, X.; Cugliari, J.; Poggi, J.M. Clustering functional data using wavelets. *Int. J. Wave. Multiresolut. Inform. Proc.* **2013**, *11*. [CrossRef]

34. Steinley, D.; Brusco, A.M. new variable weighting and selection procedure for k-means cluster analysis. *Multivar. Behav. Res.* **2008**, *43*, 32. [CrossRef] [PubMed]

35. R Core Team. *R: A Language and Environment for Statistical Computing*; R Foundation for Statistical Computing: Vienna, Austria, 2018.

36. Jiang, Z.; Lin, R.; Yang, F.; Budan, W. A Fused Load Curve Clustering Algorithm based on Wavelet Transform. *IEEE Trans. Ind. Inform.* **2017**. [CrossRef]

37. Thouvenot, V.; Pichavant, A.; Goude, Y.; Antoniadis, A.; Poggi, J.M. Electricity forecasting using multi-stage estimators of nonlinear additive models. *IEEE Trans. Power Syst.* **2016**, *31*, 3665–3673. [CrossRef]

38. Polikar, R. Ensemble learning. In *Ensemble Machine Learning*; Springer: Berlin, Germany, 2012; pp. 1–34.

39. Gaillard, P.; Goude, Y. Forecasting electricity consumption by aggregating experts; how to design a good set of experts. In *Modeling and Stochastic Learning for Forecasting in High Dimensions*; Springer: Berlin, Germany, 2015; pp. 95–115.

40. Goehry, B.; Goude, Y.; Massart, P.; Poggi, J.M. Forêts aléatoires pour la prévision à plusieurs échelles de consommations électriques. In Proceedings of the 50 èmes Journées de Statistique, Paris Saclay, France, 28 May–1 June 2018; talk 112.

41. Li, P.; Zhang, B.; Weng, Y.; Rajagopal, R. A sparse linear model and significance test for individual consumption prediction. *IEEE Trans. Power Syst.* **2017**, *32*, 4489–4500. [CrossRef]

42. Wang, Y.; Chen, Q.; Gan, D.; Yang, J.; Kirschen, D.S.; Kang, C. Deep Learning-Based Socio-demographic Information Identification from Smart Meter Data. *IEEE Trans. Smart Grid* **2018**. [CrossRef]

43. Anderson, B.; Lin, S.; Newing, A.; Bahaj, A.; James, P. Electricity consumption and household characteristics: Implications for census-taking in a smart metered future. *Comput. Environ. Urban Syst.* **2017**, *63*, 58–67. [CrossRef]

Energies **2018**, *11*, 1893

44. Martinez-Pabon, M.; Eveleigh, T.; Tanju, B. Smart meter data analytics for optimal customer selection in demand response programs. *Energy Proc.* **2017**, *107*, 49–59. [CrossRef]
45. Maruotti, A.; Vichi, M. Time-varying clustering of multivariate longitudinal observations. *Commun. Stat. Theory Meth.* **2016**, *45*, 430–443. [CrossRef]

![energies logo] *energies*

MDPI

Article

Improving Short-Term Heat Load Forecasts with Calendar and Holiday Data

Magnus Dahl [1,2,*] ⓘ, **Adam Brun** [2], **Oliver S. Kirsebom** [3] ⓘ **and Gorm B. Andresen** [1] ⓘ

[1] Department of Engineering, Aarhus University, Inge Lehmanns Gade 10, 8000 Aarhus, Denmark; gba@eng.au.dk
[2] AffaldVarme Aarhus, Municipality of Aarhus, Bautavej 1, 8210 Aarhus, Denmark; adbr@aarhus.dk
[3] Department of Physics and Astronomy, Aarhus University, Ny Munkegade 120, 8000 Aarhus, Denmark; oliskir@phys.au.dk
* Correspondence: magnus.dahl42@gmail.com

Received: 28 May 2018; Accepted: 25 June 2018; Published: 27 June 2018

Abstract: The heat load in district heating systems is affected by the weather and by human behavior, and special consumption patterns are observed around holidays. This study employs a top-down approach to heat load forecasting using meteorological data and new untraditional data types such as school holidays. Three different machine learning models are benchmarked for forecasting the aggregated heat load of the large district heating system of Aarhus, Denmark. The models are trained on six years of measured hourly heat load data and a blind year of test data is withheld until the final testing of the forecasting capabilities of the models. In this final test, weather forecasts from the Danish Meteorological Institute are used to measure the performance of the heat load forecasts under realistic operational conditions. We demonstrate models with forecasting performance that can match state-of-the-art commercial software and explore the benefit of including local holiday data to improve forecasting accuracy. The best forecasting performance is achieved with a support vector regression on weather, calendar, and holiday data, yielding a mean absolute percentage error of 6.4% on the 15–38 h horizon. On average, the forecasts could be improved slightly by including local holiday data. On holidays, this performance improvement was more significant.

Keywords: district heating; load forecasting; machine learning; weather data; consumer behavior; neural networks; support vector machines

---

1. Introduction

Energy systems are changing throughout the world, and heat load forecasting is gaining importance in modern district heating systems [1]. The growing penetration of renewable energy sources makes energy production fluctuate beyond human control and increases the volatility in electricity markets. Stronger coupling between the heating and electricity sectors means that production planners in systems with combined heat and power generation need accurate heat load forecasts in order to optimize the production.

It is not trivial to forecast district heating demand on time scales that are relevant for trading on the day-ahead electricity market. The total heat load in a district heating system is influenced by several factors—most importantly, the weather, the building mass of the city, and the behavior of the heat consumers. Cold and windy weather increases the heat demand, and warm and sunny weather decreases it. The constitution of the building mass influences how the heat load responds to changes in the weather [2]. Human behavior is an often overlooked factor, and, especially in summer, the heat demand is dominated by hot water consumption rather than space heating. Consumer behavior can vary considerably from day to day, and the heat load on special occasions, e.g., New Year's Eve, is notoriously difficult to forecast accurately.

Heat load forecasting has been studied extensively in the scientific literature. The successful application of simple linear models in [3,4] has inspired us to use an ordinary least squares (OLS) model as a simple benchmarking model. Statistical time-series models, such as SARIMA (seasonal autoregressive integrated moving average) models [4,5] and grey-box models combining physical insight with statistical modeling [6], are natural ways of handling the temporal nature of load forecasting. These models are usually linear and struggle with multiple seasonality. In [7], the authors compared a number of machine learning algorithms, including a simple feed forward neural network, support vector regression (SVR), and OLS. They concluded that the SVR model performs best. The strong forecasting capabilities of SVR models have also been demonstrated in [8], where heat demand was forecasted based on natural gas consumption. Neural networks have been widely applied in load forecasting. Several studies apply simple feed forward networks with one hidden layer such as the multilayer perceptron (MLP) [7,9,10]. A recurrent neural network is used in [11] to better handle non-stationarities in the heat load. A comprehensive review of load forecasting in districts can be found in [1].

In the present study, we chose to compare three different machine learning models: OLS, MLP, and SVR, as they have all proven effective for heat load forecasting. Some studies attempt to include the different consumer behavior on weekdays and weekends. Working days and non-working days are modeled with distinct profiles in [12], and in [4] mid-week holidays were treated as Saturdays or Sundays. In [13], the correlation between electric load and weather variables was exploited to forecast the aggregated load using MLP models, and the authors explored the different autocorrelations of the load on weekdays and weekends. In this study, we include generic calendar data such as the day of the week, as well as local holiday data to account for observances, national holidays, and city-specific holidays, i.e., school holidays.

School holidays are often planned locally, and some religious holidays, e.g., Easter, fall on different dates each year. Therefore, generic calendar data is insufficient for modeling events that depend on local holidays. Heat consumers behave differently on holidays and change the pattern of consumption, so including local holiday data in heat load forecast models has the potential to improve forecast accuracy.

The novelty of this work lies in the application of new data sources, specifically local holiday data, to create heat load forecasting models that more accurately capture consumer behavior. To the best of our knowledge, school holiday data has not previously been used for heat load forecasting. We isolate the effect of using local holiday data by employing machine learning models that have proven effective for heat load forecasting in the past. Moreover, we base our modeling on a very large amount of data. Seven years of hourly heat load and weather data supplemented with data about national holidays, observances, and school holidays help the forecast models capture rare load events.

The remainder of the paper is structured as follows. The Methodology section describes the data foundation, the machine learning models, and the validation and testing procedure. In the Results section, the forecasting models are benchmarked and compared, and the potential of using new data sources is evaluated. The paper is wrapped up in the Conclusion section.

2. Methodology

In this section, we describe the data foundation and how the heat load forecasting models were built, validated and tested.

The focus of this paper is to create heat load forecasts that are relevant on the time horizon of the day-ahead electricity market. Therefore, a forecast must be produced each morning at 10:00 for each hour of the following day. This timeline, illustrated in Figure 1, allows time for communication between different actors in a production system and for planning of the following day's heat production in accordance with the bids in the day-ahead electricity market.

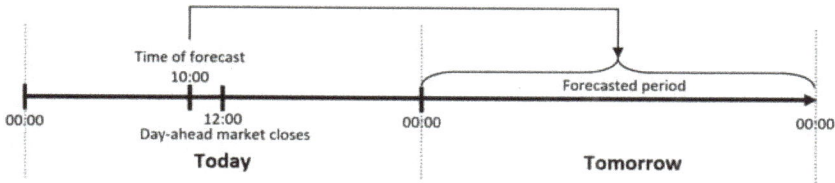

Figure 1. Timeline for the heat load forecast that is relevant for the trading decisions in the day-ahead electricity market. Every day at 10:00 a forecast is produced for each hour of the following day.

The analysis in this paper is based on seven years of data for the total hourly heat load of Aarhus, Denmark. The years 2009, 2010, 2012, 2013, 2014, 2015, and 2016 were used. Unfortunately, heat load data from 2011 has not been available to us. We denote the heat load in hour t by P_t.

2.1. Weather Data

The heat demand depends strongly on the weather. Hourly outdoor temperature, wind speed, and solar irradiation for the seven years were obtained from [14]. Weather data from the geographical point N 56°2′42.24″, E 9°59′59.95″ in the southern part of Aarhus was used. Weather forecasts of the outdoor temperature, wind speed, and solar irradiation were provided by the Danish Meteorological Institute (DMI) and used to test the performance of the heat load forecasts as realistically as possible. These weather forecasts were based on the HIRLAM (High Resolution Limited Area Model), a numerical weather prediction system developed by a consortium of European meteorological institutes with the purpose of providing state-of-the-art short-range weather predictions [15], for numerical weather prediction, had a forecast horizon of up to 54 h, and were disseminated four times a day [15]. We denote the outdoor temperature, wind speed, and solar irradiance by T_t^{out}, v_t^{wind} and I_t^{sun}, respectively.

2.2. Calendar Data

The heat demand has a strong social component that depends on human behavior. The social component is part of the reason for the daily and weekly patterns in the heat load. Different load profiles on weekdays and weekends can also be explained by consumer behavior. In order to allow the forecast models to account for load variations that are tied to specific days, seasons, and times of day, certain calendar data were included as input variables. Specifically, the hour of the day, the day of the week, the weekend, and the month of the year were used as input. How the calendar data was encoded and included in the models is described in Section 2.4.1.

2.3. Holiday Data

In addition to generic calendar data, we also used more specific local data about special days that may influence the heat consumption pattern. The district heating system of Aarhus, Denmark, served as our case study. Therefore, we used data about Danish national holidays, observances, and local school holidays. National holidays and observances were sourced from [16]. National holidays include New Year's Day, Christmas Day, Easter Day, etc. and constitute 11 days per year. Observances include, e.g., Christmas Eve and Constitution Day and amount to six days per year. Information about the municipal school holidays was collected from local schools in the Aarhus area and amounts to 96 days per year on average. Note that all national holidays are also school holidays. It is clear that this kind of information is highly local and that gathering such data, compared to the generic calendar data, is more difficult. The following analysis will illuminate whether including this data significantly improves heat load forecasts, or if more easily available data types are sufficient.

2.4. Data Exploration

Figure 2 shows the average hourly heat load for the example year of 2010. Notice how much the heat load varies over the year both in magnitude and in variance. The zoomed inset in the plot shows the heat load variations over a week in March. A clear daily pattern can be observed, with a sharp morning peak between 7:00 and 8:00 on weekday mornings. The morning peak is a well-known phenomenon in the district heating community and is caused by many people showering around the same time every morning. On weekends, morning peaks can be observed later in the morning and tend to be less sharp compared to weekdays. From the inset, it is clear that the daily load pattern varies substantially within just one week.

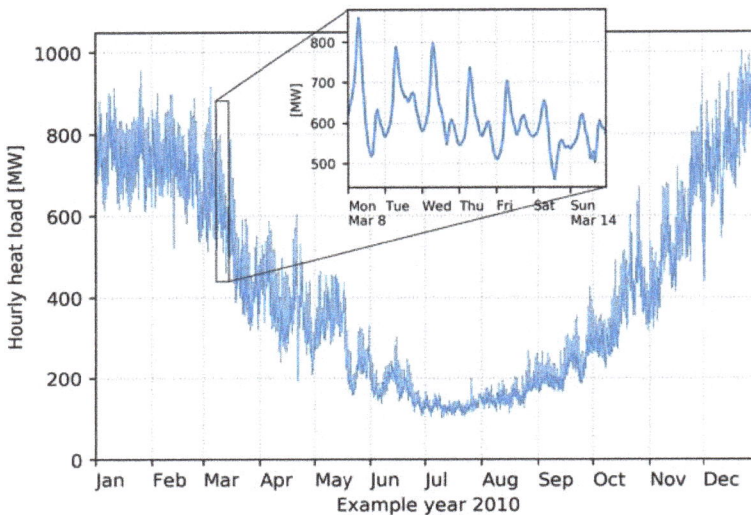

Figure 2. Time series for the hourly heat load in the year 2010. The inset shows a zoom of a week in March.

The heat demand has peaks in its autocorrelation function at 24 h, 48 h, 72 h, and so on. This is due to the strong daily pattern. There is also a notable peak at 168 h (one week). In order to capture this behavior, lagged values of the heat load were used as input variables in the modeling. Specifically, we included the heat load lagged with 24 h, 48 h, and 168 h. Looking at Figure 1, we see that the forecast horizon varies between 15 h and 38 h. The heat load in the first hour of the day can be forecasted with the shortest horizon, and the last hour of each day is forecasted with the longest horizon. When forecasting hours with a forecast horizon of 24 h or less, the heat load lagged 24 h can be used. When forecasting hours with a longer horizon than 24 h, the heat load lagged with 48 h must be used instead. A power spectrum analysis confirmed strong peaks at frequencies $1/12\,\text{h}^{-1}$ and $1/24\,\text{h}^{-1}$, but 12 h is shorter than the shortest forecast horizon and was thus discarded. The two lags that best captured the daily and weekly pattern of the heat load were included. We denote the lagged heat load by P_{t-24}, P_{t-48}, and P_{t-168}, respectively.

The most important weather variable when modeling district heating loads is the outdoor temperature, because there is a strong negative correlation between the heat demand and the outdoor temperature. Depending on the specific district heating system, solar irradiation and wind speed can also be significant predictors for the heat load [2]. Due to the thermal mass of the buildings in a district heating system, there is a certain inertia in the heat load when changes in the weather occur. On the individual building level, this inertia is handled in great detail in the civil engineering literature [17]. Since we were forecasting the heat load of an entire city, we took a more simplified approach. In the Aarhus district heating system, the heat load is most strongly correlated with the outdoor temperature lagged by 4 h, compared to other time lags of the temperature. The heat load also correlates most strongly with the solar irradiation lagged by 4 h. There seemed to be no benefit in lagging the wind speed. Only including two specific lags, is of course a simplification of the dynamics of the system, but the results of including them were significantly better than just using simultaneous (lag 0 h) weather variables. Summing up, the following five weather variables were included in the modeling: T_t^{out}, v_t^{wind}, I_t^{sun}, T_{t-4}^{out}, and I_{t-4}^{sun}.

Outdoor temperature and, as a consequence, the heat load varies substantially from year to year [18]. The mean annual temperatures in our dataset spanned a range of 2.5 °C. Compared to the mean load of the whole dataset (excluding 2016), the annual mean heat load was 15% higher in the coldest year and 11% lower in the warmest year.

2.4.1. Data Scenarios and Pre-Processing

In order to evaluate the effect of including the various types of input data for forecasting heat load, three different data scenarios have been defined. We call these scenarios: "Only Weather Data," "Weather and Calendar," and "Weather, Calendar, and Holidays". Table 1 details the input data used in each scenario.

Table 1. Input variables used in the three data scenarios (in bold).

		Only Weather Data	Weather and Calendar	Weather, Calendar and Holidays
Lagged heat load	P_{t-24} or P_{t-48}	✓	✓	✓
	P_{t-168}	✓	✓	✓
Weather data	T_t^{out}	✓	✓	✓
	v_t^{wind}	✓	✓	✓
	I_t^{sun}	✓	✓	✓
	T_{t-4}^{out}	✓	✓	✓
	I_{t-4}^{sun}	✓	✓	✓
Calendar data	Hour of day		✓	✓
	Day of week		✓	✓
	Weekend		✓	✓
	Month of year		✓	✓
Holiday data	National holiday			✓
	Observance			✓
	School holiday			✓

To achieve the best performance of the models, the input data were scaled and encoded as follows. All the continuous variables (lagged heat load and weather) were standardized to have mean 0 and standard deviation 1. The calendar data and holiday data were included as so-called dummy variables. Dummy variables are a way to represent categorical variables as binary variables. For instance, whether or not a given hour falls on a school holiday can be encoded as a binary variable (0 or 1). The day of week can be encoded as six binary variables: one variable indicating if it is Monday, one indicating if it is Tuesday, etc. Only six variables are needed to encode seven days, because if it is not any of the days from Monday to Saturday, then it must be Sunday. Using similar dummy variables all the calendar and holiday data was included. Encoding categorical data as dummy variables is a standard machine learning method [19].

2.5. Machine Learning Models

We benchmarked and compared three different machine learning models that have all previously been proven adequate for heating load forecasting [7,8]: ordinary least squares regression, multilayer perceptron, and support vector regression.

2.5.1. OLS—Ordinary Least Squares Regression

Ordinary least squares regression is a simple model type in which the output is modeled with the hyperplane that minimizes the squared residuals between the target and the output of the model. Sometimes referred to as multiple linear regression, it is a popular model due to its simplicity, computational speed, and the fact that results can be easily interpreted. Because of its linear structure, the OLS model underperforms when modeling nonlinear input–output relationships.

2.5.2. MLP—Multilayer Perceptron

A multilayer perceptron is a simple kind of neural network. Neural networks have been applied to problems in many fields, including heat load forecasting, due to their ability to capture complicated relationships between input and output [7,9,11]. A multilayer perceptron has at least one hidden layer between the input and output layers of the model and a nonlinear activation function allows the model to capture nonlinear relationships between input and output. A good coverage of neural network models and the multilayer perceptron can be found in [20]. We used a multilayer perceptron with one hidden layer and the rectifier activation function: $f(x) = \max(0, x)$. We have experimented with adding more hidden layers, but the increase in the model accuracy was not large enough to justify the growth in model complexity and the risk of overfitting.

Besides the simple multilayer perceptron, we have experimented with a more advanced type of neural network. Recurrent neural networks with long short-term memory (LSTM) units [21] were implemented in an attempt to simplify the feature selection and leave it to the model to discover relevant lags of heat load and weather data. Our initial LSTM networks yielded results comparable to the simpler models included in this work, but their performance tended to be inconsistent. The benefit of simplified feature selection may also be outweighed by a more complicated model selection and training procedure. The LSTM modeling for heat load forecasting requires more work and will be left for future work.

2.5.3. SVR—Support Vector Regression

Support vector regression is the application of support vector machine models to regression problems and was first introduced in [22]. Support vector regression has a computational advantage in very high dimensional feature spaces. The model only depends on a subset of input data, because it minimizes a cost function that is insensitive to points within a certain distance from the prediction. The cost function is less sensitive to small errors and less sensitive to very large errors and outliers, compared to the quadratic cost function minimized in the ordinary least squares regression. To avoid overfitting, the model is governed by a regularization parameter C, that ensures that the parameters of the model do not grow uncontrollably. The smaller the value of C, the harder large model parameters are penalized. Support vector regression is explained in great detail in [19,20]. By employing the so-called "kernel trick", support vector regression can handle nonlinear input–output relationships. A very popular kernel function is the radial basis function kernel (RBF), which has been proven effective in this application as well. The RBF kernel is governed by a kernel parameter γ. The greater the value of γ, the more prone the model is to overfitting, but if γ is chosen too small, the model may be underfitting and fail to capture actual input–output relationships.

Summing up, the three machine learning models OLS, MLP, and SVR were chosen because they have all been successfully applied to heat load forecasting in the past. Using well-established algorithms allows us to focus on the main research question: whether conventional heat load forecasts can be improved by adding new types of data. Each of the models have advantages and disadvantages. The advantage of the OLS model is that it is computationally cheap, and its estimated parameters carry a physical interpretation. The disadvantage is that the model is linear and fails to capture nonlinearities in input–output relationships. The advantage of the MLP model is that it is capable of capturing very complex relationships between input and output. A disadvantage of neural network models, such as

MLP, is the risk of overfitting and that they require careful tuning of several hyperparameters. Finally, the SVR model has the advantage of being robust to outliers and that the final model depends only on a subset of the training data. The SVR model, however, is sensitive to the scaling of the input data and the correct tuning of regularization and kernel parameters.

2.6. Model Selection and Testing

A good forecast model is one that performs well on previously unseen data. This is the generalization ability of the model. In order to accurately measure the generalization performance of the models, we divided the full dataset (seven years of hourly data) into a training and validation set and a test set. All model selection and training was performed on the years from 2009 to 2015 (2011 not included). This is the training and validation set. The entire year of 2016 was used as a blind test set to estimate the generalization performance of the forecasts.

The three models were chosen and their hyperparameters tuned based on sixfold cross-validation on the years 2009, 2010, 2012, 2013, 2014, and 2015. Using six folds ensured that each fold contained an entire year and thus represented the full annual variation of the heat load. In the cross-validation, the different models and data scenarios were scored according to the hourly root mean square error (RMSE)

$$\text{RMSE} = \sqrt{\frac{1}{N}\sum_t (\hat{P}_t - P_t)^2} \tag{1}$$

where $\hat{P}_t$ is the forecasted heat load for hour t, and N is the number of hours.

The OLS model does not have any hyperparameters to tune, but a model with a nonzero constant term was chosen. In the MLP model, we tuned the number of neurons in the hidden layer using a grid search on the cross-validation scores. A MLP model with one hidden layer consisting of 110 hidden neurons was chosen, and the L2 regularization parameter α was set to 0.1. In the SVR model, the best choices for the regularization parameter and the kernel parameter were found to be $C = 4.3$ and $\gamma = 0.2$. All modeling has been performed in Python 2.7 using the scikit-learn framework (version 0.19.0) [23].

All results presented in the following section were produced using the blind test year 2016. This year was not used for any of the training, data exploration, or model selection. In the Results section, we employ two other forecast error metrics, in addition to the RMSE. The mean absolute error (MAE) is also an absolute error metric (here in units of MW), but it is less sensitive to large errors, compared to the RMSE. The MAE is defined as

$$\text{MAE} = \frac{1}{N}\sum_t |\hat{P}_t - P_t| \ . \tag{2}$$

Finally, we use the relative error metric mean absolute percentage error (MAPE) to facilitate easier comparison between different district heating systems. The MAPE is defined as

$$\text{MAPE} = \frac{1}{N}\sum_t \left| \frac{\hat{P}_t - P_t}{P_t} \right| \ . \tag{3}$$

3. Results

The heat load in a district heating system has been forecasted using three different machine learning models, described in the previous section: OLS, MLP, and SVR. The performance of these models have been tested by letting them produce a forecast for the following day using the input data available each day at 10:00 a.m. The models have been trained exclusively on data prior to the test year 2016 to be able to accurately gauge their generalization performance. Figure 3 shows an example of the forecasts produced for 4 May. Only the heat load up to the time of the forecast was used as input to produce the forecast. Real weather forecasts were used as weather inputs for 4 May, as opposed to the historical weather data used for training. It is clear how the three forecast models produce similar, yet distinct forecasts. On 4 May, the MLP model appears to produce the best forecast, especially in the morning.

Figure 4 summarizes the performance of the three models in the three different data scenarios. The top panel shows the forecast performance that could be achieved if weather forecasts were 100% accurate, simulated by using historical weather data. The bottom panel shows the performance using real weather forecasts. Comparing the three data scenarios, we see the benefit of including different data types in the modeling. In the first scenario, only lagged heat load and weather data are used as input. In the second scenario, generic calendar data is included as well, and in the third scenario, local observances, national holidays, and school holidays are also included as inputs to the model. Including calendar data significantly improves performance, compared to only using weather data. Extending the input data with holiday data as well results in an additional, but small improvement compared to using generic calendar data only. Obtaining the local holiday data can be laborious or impossible, so it is positive to see generic calendar data yielding comparable results. It is much easier to apply these models to a wide range of district heating systems around the world if it can be done without collecting local holiday data.

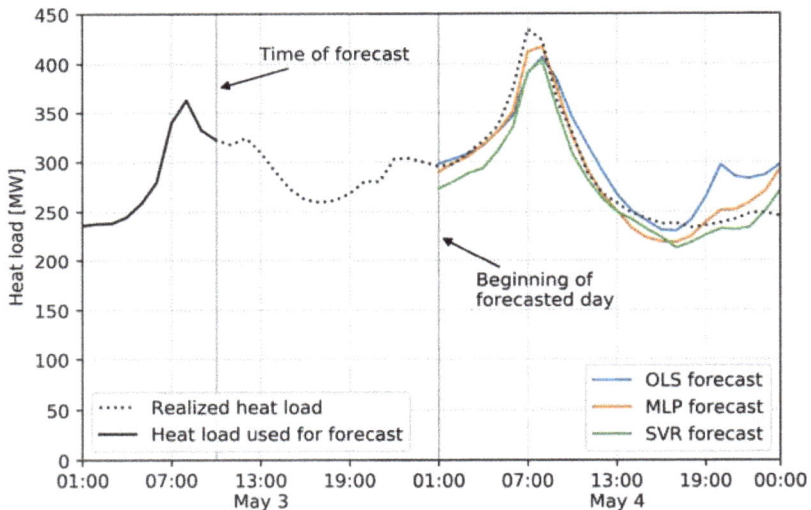

Figure 3. Example forecasts for 4 May 2016. The forecasts were produced on 3 May at 10:00 and based on real weather forecasts, calendar, and holiday data.

Figure 4 allows for comparison of the performance of the three machine learning models as well. The OLS model stands out by performing significantly worse than the other two models in all scenarios. The OLS model has a root mean square error of 38.9 MW, compared to 31.1 MW and 29.3 MW for the other two models when using real weather forecasts, calendar, and holiday data (bottom panel). The poor performance of the OLS model can be attributed to its linear structure. The relationship between the outdoor temperature and the heat load in a temperate climate is nonlinear. This causes the linear model to perform poorly during summer by undershooting the heat load and overestimating its variance. The two nonlinear models, MLP and SVR, perform similarly in these scenarios. The SVR model has the smallest error, and the focus in the rest of this paper will be on the SVR model using weather, calendar, and holiday data.

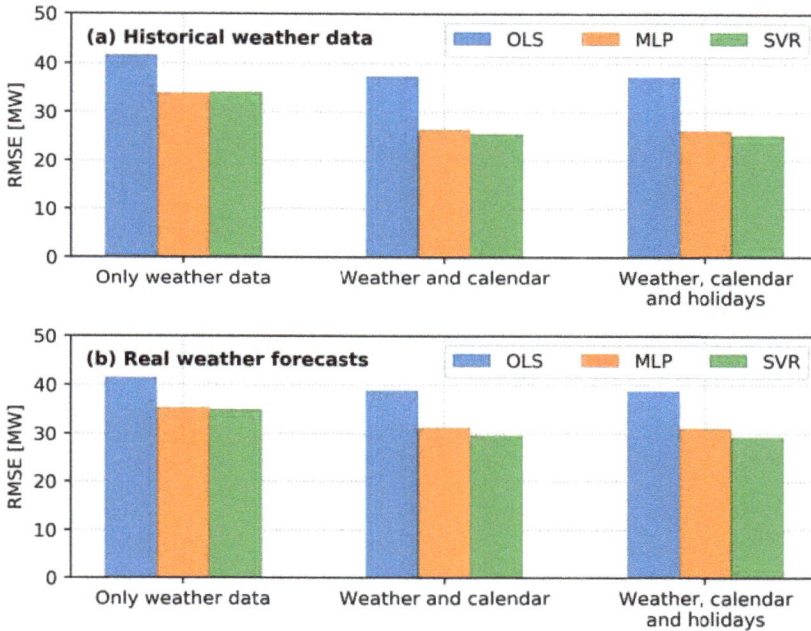

Figure 4. Root mean square error of the three forecast models OLS, MLP, and SVR on the year 2016. The top panel (**a**) shows the error using historical weather data to simulate 100% accurate weather forecasts. The bottom panel (**b**) shows the error using real weather forecasts.

3.1. The Value of Improving Weather Forecasts

Figure 4 has two panels. The top panel shows the forecast errors that could be achieved if weather forecasts predicted the measured weather completely accurately. This has been simulated by allowing the models to use actual measured weather data, instead of weather forecasts as input when producing the load forecast. The top panel reflects the scenario in the which future weather is known. The bottom panel shows the results in the case where real forecast data has been used instead. This is the actual forecast performance that can be achieved in an operational situation, given the current quality of weather forecasts. Having access to weather forecasts without prediction errors could, in a perfect world, reduce the error from 29.3 MW to 25.2 MW in the forecasts from the best model. While an error reduction of 4.1 MW is a start, perfecting the weather forecast only shaves 14% off the error. The remainder of the load forecast error has other causes than weather forecast errors, a result that was also found in [24], where ensemble weather predictions were used to quantify heat load forecasting uncertainty.

The OLS model using only weather data and lagged heat load does not perform notably different on historical weather data compared to real weather forecasts. This can be explained by the OLS model attributing greater weight to the lagged heat load compared to the weather, because the relationship between the heat load and the weather is nonlinear.

The forecast performance, shown in the top of Figure 4, is similar to the performance that was achieved during training and cross-validation. This indicates that the models have not been overfitted and generalize well to out-of-sample predictions.

It is worth pointing out that the performance of all these models, even the OLS model using only weather data, exceeds the performance of the commercial forecasting system that is currently in operation in the Aarhus district heating system. This commercial forecasting system had an RMSE of 41.9 MW in year of 2016 on the same forecast horizons. In relative terms, the SVR model has a MAPE of 6.4% versus 8.3% for the commercial system. The models presented here perform better than all other forecast models that have been used in the Aarhus district heating system.

3.2. Seasonal Performance Variations

The heat load varies significantly over the year, both in magnitude and in variance, as exemplified in Figure 2. It can be a challenge for a single model to adequately forecast both winter and summer heat loads. Therefore, it is relevant to further investigate the model performance throughout the year. The forecast error of the best model, SVR using weather, calendar, and holiday data, is illustrated in Figure 5. Three different error metrics are shown: on the left axes, the RMSE (blue) and MAE (yellow) are shown in MW; on the right axes, the MAPE (red) is shown in percent. The horizontal axes show the hour of day for the forecasted hour, and each subplot depicts a month in the year. This makes it possible to see if it is harder to forecast the morning peak and if the forecast horizon impacts the accuracy. Keep in mind that Hour 1 has the shortest forecast horizon (15 h), and Hour 24 has the longest horizon (38 h), since the forecasts are produced at 10:00 a.m. the previous day.

Inspecting Figure 5, it is clear that the absolute error measures RMSE and MAE are largest in winter and smallest in summer. This is a reflection of the annual heat load profile and the large load with large variance during winter. In late fall and winter, the RMSE can be above 50 MW in some hours, whereas it can be below 10 MW in some hours in July. The relative error metric MAPE behaves in the opposite way. The relative error is smaller in the winter months and larger in summer months, but it stays between 2.5% and 10.5%. This is a consequence of the annual load variations being larger than the annual variations in the absolute error.

There is no clear pattern in the way the error changes during the day. The model does not seem to perform worse between 7:00 and 8:00 in the morning, where the morning peak falls. November and May are exceptions to this rule. In many applications, the error of a forecast model increases with the forecast horizon (here the hour of day). We do not observe a general increasing trend in the error with the hour of day. This indicates that the weather forecasts that are used as inputs to create the forecast are not significantly worse at the longest horizon compared to the shortest horizon. It may also be due to weather forecasting accuracy having a minor impact on the heat load forecasting error, as we saw from Figure 4. If we were to increase the forecast horizon further, the forecast error would most likely increase.

The forecast error varies significantly over the year, but aggregated error metrics such as RMSE, MAE, or MAPE do not tell the full story. Maximum errors can be relevant for unit commitment in the production planning and for evaluating risk regarding trading in the electricity market. Figure 6 shows histograms for the hourly error for each month of the blind test year 2016. The 10% and 90% quantiles have been indicated in each plot. It is clear that the width of the error distribution varies substantially from month to month. During the summer, the forecast error is quite confined, but the distribution widens in late fall and becomes widest in December.

In Table 2, a summary of the error distribution is shown. The 99% and 1% quantiles of the error distribution are indications of the maximum errors that can be expected. Ninety-eight percent of the forecasted hours have forecast errors between the 1% and the 99% quantile. The best month is July with 98% of the errors falling between −16.0 and 25.8 MW. The worst month is December, where there is a 1% risk of the forecast overshooting by more than 115.0 MW and a 1% risk of the forecast being more than 96.7 MW too low. These extreme errors can approach 20% of the mean heat load in December.

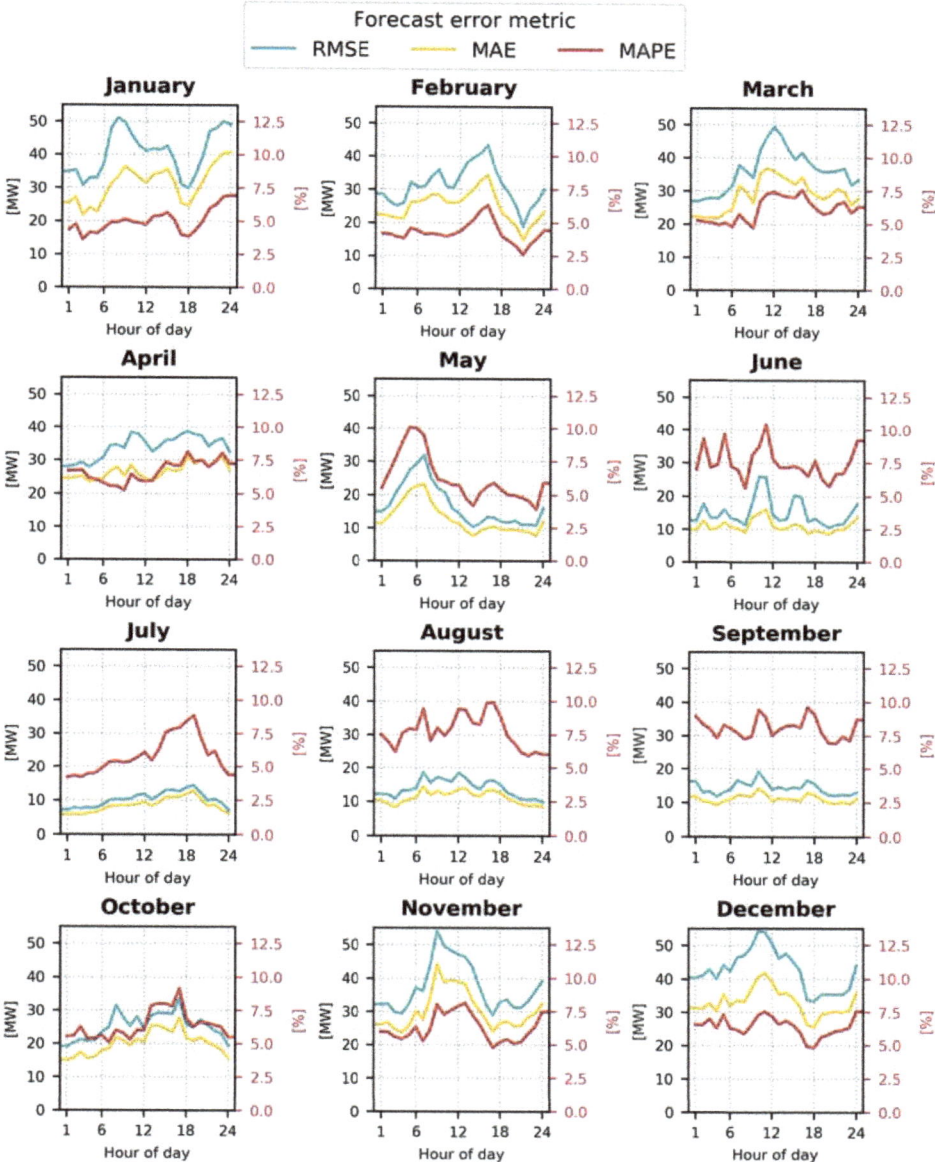

Figure 5. Performance of the SVR model on the year 2016, using real weather forecasts, calendar, and holiday data. Three different error metrics are shown for each month of the year. The forecast error varies with the time of day, shown on the horizontal axes. RMSE (blue) and MAE (yellow) are shown units of MW on the left axes. MAPE (red) is shown in percent on the right axes.

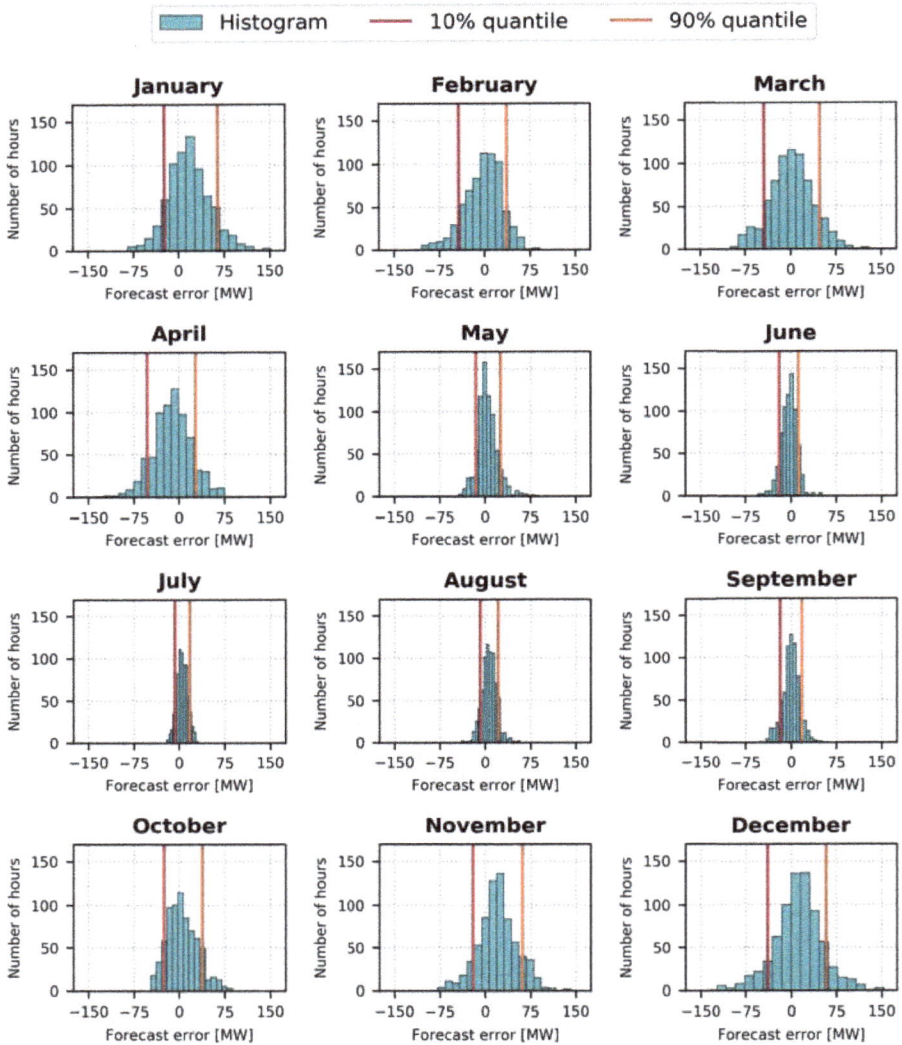

Figure 6. Histograms for the forecast error of the SVR model on the year 2016 using real weather forecasts, calendar, and holiday data. The distribution of the forecast error is depicted for each month in the year along with the 10% and 90% quantiles. The number of bins was chosen using Scott's rule [25] within each month. A positive error indicates that the forecast was too high, a negative error that it was too low.

From the histograms in Figure 6, it is also clear that the error distributions are not completely symmetric around 0. In January, for instance, the distribution is shifted slightly to the positive, and in April it is shifted to the negative side. The forecast appears to be biased differently in different months. The mean error for each month (ME) is shown in Table 2. The bias can be as large as 20.5 MW, with November being the worst month. September performs the best with a mean error of merely 0.3 MW. Varying monthly biases could be remedied by training separate models for each month. However, the focus in this paper is to investigate the effects of including holiday data to model human behavior, and training monthly forecast models would obscure the effects of using holiday data. There is also the possibility that the weather forecasts perform differently at different times of year.

Table 2. Summary of the hourly forecast error for each month for the SVR model using real weather forecasts, calendar, and holiday data. Histograms of the forecast error appear in Figure 6. The months with the worst performance are indicated in red, the best in green. The quantiles are evaluated in pairs, so the widest symmetric quantile interval is considered the worst.

	RMSE (MW)	ME (MW)	Error Quantiles (MW)			
			10%	90%	1%	99%
January	41.2	18.7	−24.0	64.1	−66.8	117.6
February	31.8	−2.2	−42.8	36.6	−91.6	61.3
March	36.9	2.0	−43.7	48.8	−80.4	89.3
April	34.2	−11.7	−52.9	27.1	−93.6	64.8
May	18.2	4.3	−14.7	25.6	−33.6	64.3
June	15.8	−3.3	−19.3	12.8	−45.3	34.0
July	10.6	5.0	−7.0	17.2	−16.0	25.8
August	14.2	6.9	−8.2	21.6	−20.1	41.6
September	14.3	0.3	−17.9	17.6	−35.0	33.3
October	25.6	4.8	−26.0	37.8	−43.5	70.0
November	38.1	20.5	−20.0	61.5	−59.1	98.1
December	43.1	12.5	−38.4	58.2	−96.7	115.0

In conclusion, there are significant seasonal variations in the performance of the best heat load forecast. The absolute errors are largest in winter and smallest in summer, with December being the hardest month to forecast and July being the easiest.

3.3. The Value of Calendar and Holiday Data

The goal of this analysis is to gauge the potential of including local holiday data in heat load forecasts in order to better capture the consumer behavior. The reduction in the annual error was very small when comparing models with only generic calendar data to models including local holiday data. This was clear from Figure 4b. It is well known among district heating operators that heat load forecasts tend to perform poorly on special occasions, such as Christmas or New Year's Eve. These special days are rare, so the performance on those specific days has little impact on the average annual performance (Figure 4b). Improved performance on special days is valuable to production planners, and whether including local holiday data can improve forecast performance on specific days is worth investigating in more detail.

Figure 7 shows the performance of the SVR model in the three data scenarios on different sets of days during the year. "Holidays" refer to all days that are observances, national holidays, or school holidays. "Weekdays" include all weekdays that are not also in holidays, and "weekends" include all weekend days not included in holidays. In 2016, there were 201 weekdays, 65 weekend days, and 100 holidays.

There is significant benefit in including generic calendar data in the forecast models for all day types. On weekdays, there is no performance improvement to gain by including local holiday data. The forecast error on weekends can be reduced by 0.5 MW. Not surprisingly, the greatest performance increase can be observed on holidays. The holiday error decreases by 1.3 MW when augmenting the modeling with local holiday data. The holiday error is generally smaller than the error for the other day types. This is due to the holidays being dominated by the schools' summer holidays, and the error is generally smaller during the summer. Summing up, including local holiday data only improves the forecasts slightly on average. The largest improvement is seen on holidays where the error can be reduced by 5%, compared to only using generic calendar data.

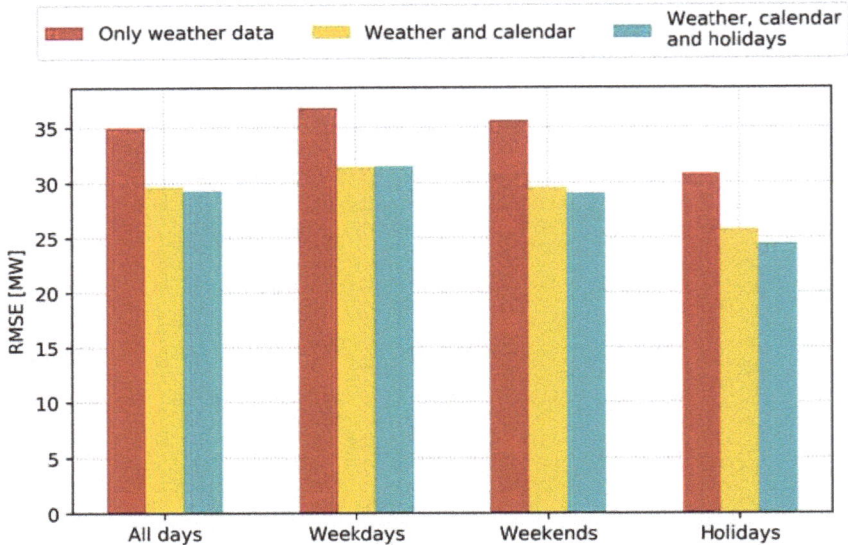

Figure 7. Forecast performance of the SVR model on the year 2016 using real weather forecasts, calendar, and holiday data. The second and third group of bins refer to weekdays and weekends that are not also included in holidays. Holidays refer to all days that are observances, national holidays, or school holidays (see Table 1).

4. Conclusions

We have tested heat load forecasts with horizons from 15 h to 38 h, relevant for district heating production planning considering the day-ahead electricity market. The work was based on seven years of heat load and weather data for the large district heating system of Aarhus, Denmark. In order to measure the forecast performance that can realistically be experienced in actual operation, we used blind testing on a whole year with real weather forecasts.

Three machine learning models have been tested: an ordinary least squares model, a multilayer perceptron, and a support vector regression model. The SVR model performed best, beating the OLS model by a large margin and the MLP model by a small margin. All the models were trained on lagged heat load data and weather data. The forecast performance could be significantly improved by including generic calendar data, such as month, weekday, and hour of day. A smaller improvement of the forecasts could be gained by supplying the models with local holiday data including observances, national holidays, and school holidays. This improvement was most significant on holidays and weekends. Local holiday data can be difficult and time-consuming to obtain, but merely including lagged heat load, weather, and generic calendar data can provide a good overall forecast performance.

The SVR model using weather, calendar, and holiday data had the best performance. The root mean square error was 29.3 MW, and the mean absolute percentage error was 6.4%. This forecast model beat all other models that we have seen for the Aarhus system. The commercial forecast system, currently in operation in the Aarhus district heating system, had an RMSE of 41.9 MW, and a MAPE of 8.3% on the test year.

Including local holiday data showed only minor overall improvements in forecast performance, and including new data types in forecast models requires a careful evaluation of the trade-off between

Energies **2018**, *11*, 1678

forecast accuracy and reliability of the data source. In live operational forecast systems, reliability is valued highly, and inputting data into a simpler model may work to make a more robust system. More features are thus not always an advantage, if the improvement in accuracy is insufficient to justify the added implementation and maintenance cost.

Initial experiments using long short-term memory networks have not shown notable improvement over the results attainable with the SVR model. However, future works should explore this type of model further, as it has the potential to simplify the feature selection procedure and make it easier to transfer these results to a wide range of district heating systems around the world.

Author Contributions: Conceptualization, G.B.A. and M.D.; Methodology, G.B.A. and M.D.; Formal Analysis, M.D.; Investigation, M.D. and O.S.K.; Data Curation, M.D.; Writing—Original Draft Preparation, M.D.; Writing—Review & Editing, G.B.A. and M.D.; Visualization, M.D.; Supervision, G.B.A. and A.B.; Project Administration, A.B.; Funding Acquisition, G.B.A. and A.B.

Funding: This research has received funding from the European Union's Seventh Framework Programme for research, technological development and demonstration under grant agreement no ENER/FP7/609127/READY.

Acknowledgments: We would like to thank the Danish Meteorological Institute for providing the weather forecast data. We also thank AffaldVarme Aarhus for providing data about the heat load and production system in Aarhus.

Conflicts of Interest: The authors declare no conflict of interest.

Abbreviations

The following abbreviations are used in this manuscript:

P_t	Heat load in hour t (MW)
P_{t-l}	Heat load lagged by l hours (MW)
T_t^{out}	Outdoor temperature in hour t (°C)
v_t^{wind}	Wind speed in hour t (m/s)
I_t^{sun}	Solar irradiation in hour t (W/m^2)
T_{t-l}^{out}	Outdoor temperature lagged by l hours (°C)
I_{t-l}^{sun}	Solar irradiation lagged by l hours (W/m^2)
$\hat{P}_t$	Heat load forecasted for hour t (MW)
α	L2 regularization parameter of the MLP model
C	Regularization parameter of the SVR model
γ	RBF kernel parameter of the SVR model
RMSE	Root mean square error (MW)
MAE	Mean absolute error (MW)
MAPE	Mean absolute percentage error (%)
ME	Mean error (MW)
OLS	Ordinary least squares regression model
MLP	Multilayer perceptron model
SVR	Support vector regression model
RBF	Radial basis function kernel
LSTM	Long short-term memory network model

References

1. Ma, W.; Fang, S.; Liu, G.; Zhou, R. Modeling of district load forecasting for distributed energy system. *Appl. Energy* **2017**, *204*, 181–205, doi:10.1016/j.apenergy.2017.07.009. [CrossRef]
2. Frederiksen, S.; Werner, S. *District Heating and Cooling*; Studentlitteratur: Lund, Sweden, 2013.
3. Dotzauer, E. Simple model for prediction of loads in district-heating systems. *Appl. Energy* **2002**, *73*, 277–284, doi:10.1016/s0306-2619(02)00078-8. [CrossRef]
4. Fang, T.; Lahdelma, R. Evaluation of a multiple linear regression model and SARIMA model in forecasting heat demand for district heating system. *Appl. Energy* **2016**, *179*, 544–552, doi:10.1016/j.apenergy.2016.06.133. [CrossRef]

5.   Grosswindhager, S.; Voigt, A.; Kozek, M. Online Short-Term Forecast of System Heat Load in District Heating Networks. In Proceedings of the 31st International Symposium on forecasting, Prague, Czech Republic, 26–29 June 2011.

6.   Nielsen, H.A.; Madsen, H. Modelling the heat consumption in district heating systems using a grey-box approach. *Energy Build.* **2006**, *38*, 63–71, doi:10.1016/j.enbuild.2005.05.002. [CrossRef]

7.   Idowu, S.; Saguna, S.; Åhlund, C.; Schelén, O. Forecasting heat load for smart district heating systems: A machine learning approach. In Proceedings of the 2014 IEEE International Conference on Smart Grid Communications (SmartGridComm), Venice, Italy, 3–6 November 2014.

8.   Izadyar, N.; Ghadamian, H.; Ong, H.C.; Moghadam, Z.; Tong, C.W.; Shamshirband, S. Appraisal of the support vector machine to forecast residential heating demand for the District Heating System based on the monthly overall natural gas consumption. *Energy* **2015**, *93*, 1558–1567, doi:10.1016/j.energy.2015.10.015. [CrossRef]

9.   Kusiak, A.; Li, M.; Zhang, Z. A data-driven approach for steam load prediction in buildings. *Appl. Energy* **2010**, *87*, 925–933, doi:10.1016/j.apenergy.2009.09.004. [CrossRef]

10.   Powell, K.M.; Sriprasad, A.; Cole, W.J.; Edgar, T.F. Heating, cooling, and electrical load forecasting for a large-scale district energy system. *Energy* **2014**, *74*, 877–885, doi:10.1016/j.energy.2014.07.064. [CrossRef]

11.   Kato, K.; Sakawa, M.; Ishimaru, K.; Ushiro, S.; Shibano, T. Heat load prediction through recurrent neural network in district heating and cooling systems. In Proceedings of the 2008 IEEE International Conference on Systems, Man and Cybernetics, Singapore, 15–16 May 2008; pp. 1401–1406.

12.   Nielsen, T.S.; Madsen, H. Control of Supply Temperature in District Heating Systems. In Proceedings of the 8th International Symposium on District Heating and Cooling, Trondheim, Norway, 14–16 August 2002.

13.   Hernández, L.; Baladrón, C.; Aguiar, J.M.; Calavia, L.; Carro, B.; Sánchez-Esguevillas, A.; García, P.; Lloret, J. Experimental analysis of the input variables' relevance to forecast next day's aggregated electric demand using neural networks. *Energies* **2013**, *6*, 2927–2948. [CrossRef]

14.   Saha, S.; Moorthi, S.; Pan, H.L.; Wu, X.; Wang, J.; Nadiga, S.; Tripp, P.; Kistler, R.; Woollen, J.; Behringer, D.; et al. The NCEP Climate Forecast System Reanalysis. *Bull. Am. Meteorol. Soc.* **2010**, *91*, 1015–1057, doi:10.1175/2010BAMS3001.1. [CrossRef]

15.   Unden, P.; Rontu, L.; Järvinen, H.; Lynch, P.; Calvo, J.; Cats, G.; Cuxart, J.; Eerola, K.; Fortelius, C.; Garcia-Moya, J.A.; et al. *HIRLAM-5 Scientific Documentation*; Technical Report; Swedish Meteorological and Hydrological Institute: Norrkoping, Sweden, 2002.

16.   Holidays in Denmark. Available online: www.timeanddate.com/holidays/denmark/ (accessed on 13 June 2017).

17.   Crawley, D.B.; Hand, J.W.; Kummert, M.; Griffith, B.T. Contrasting the capabilities of building energy performance simulation programs. *Build. Environ.* **2008**, *43*, 661–673, doi:10.1016/j.buildenv.2006.10.027. [CrossRef]

18.   Dahl, M.; Brun, A.; Andresen, G.B. Decision rules for economic summer-shutdown of production units in large district heating systems. *Appl. Energy* **2017**, *208C*, 1128–1138, doi:10.1016/j.apenergy.2017.09.040. [CrossRef]

19.   Alpaydin, E. *Introduction to Machine Learning*; MIT Press: Cambridge, MA, USA, 2014.

20.   Bishop, C.M. *Pattern Recognition and Machine Learning*; Springer: New York, NY, USA, 2006.

21.   Hochreiter, S.; Schmidhuber, J. Long short-term memory. *Neural Comput.* **1997**, *9*, 1735–1780. [CrossRef] [PubMed]

22.   Drucker, H.; Burges, C.J.; Kaufman, L.; Smola, A.J.; Vapnik, V. Support vector regression machines. In *Advances in Neural Information Processing Systems*; MIT Press: Cambridge, MA, USA, 1997; pp. 155–161.

23.   Pedregosa, F.; Varoquaux, G.; Gramfort, A.; Michel, V.; Thirion, B.; Grisel, O.; Blondel, M.; Prettenhofer, P.; Weiss, R.; Dubourg, V.; et al. Scikit-learn: Machine learning in Python. *J. Mach. Learn. Res.* **2011**, *12*, 2825–2830.

24.   Dahl, M.; Brun, A.; Andresen, G.B. Using ensemble weather predictions in district heating operation and load forecasting. *Appl. Energy* **2017**, *193*, 455–465, doi:10.1016/j.apenergy.2017.02.066. [CrossRef]

25.   Scott, D.W. On optimal and data-based histograms. *Biometrika* **1979**, *66*, 605–610. [CrossRef]

energies

MDPI

Article

Short-Term Forecasting for Energy Consumption through Stacking Heterogeneous Ensemble Learning Model

Mergani A. Khairalla [1,2,*] (iD), Xu Ning [1], Nashat T. AL-Jallad [1] and Musaab O. El-Faroug [3] (iD)

[1] School of Computer Science and Technology, Wuhan University of Technology, Wuhan 430070, China; xuning@whut.edu.cn (X.N.); jallad@whut.edu.cn (N.T.A.-J.)
[2] School of Science and Technology, Nile Valley University, Atbara 346, Sudan
[3] Faculty of Engineering, Elimam Elmahdi University, Kosti 11588, Sudan; musaabgaffar@mahdi.edu.sd
[*] Correspondence: mergani@whut.edu.cn; Tel.: +86-188-7609-0760

Received: 20 May 2018; Accepted: 13 June 2018; Published: 19 June 2018

Abstract: In the real-life, time-series data comprise a complicated pattern, hence it may be challenging to increase prediction accuracy rates by using machine learning and conventional statistical methods as single learners. This research outlines and investigates the Stacking Multi-Learning Ensemble (SMLE) model for time series prediction problem over various horizons with a focus on the forecasts accuracy, directions hit-rate, and the average growth rate of total oil demand. This investigation presents a flexible ensemble framework in light of blend heterogeneous models for demonstrating and forecasting nonlinear time series. The proposed SMLE model combines support vector regression (SVR), backpropagation neural network (BPNN), and linear regression (LR) learners, the ensemble architecture consists of four phases: generation, pruning, integration, and ensemble prediction task. We have conducted an empirical study to evaluate and compare the performance of SMLE using Global Oil Consumption (GOC). Thus, the assessment of the proposed model was conducted at single and multistep horizon prediction using unique benchmark techniques. The final results reveal that the proposed SMLE model outperforms all the other benchmark methods listed in this study at various levels such as error rate, similarity, and directional accuracy by 0.74%, 0.020%, and 91.24%, respectively. Therefore, this study demonstrates that the ensemble model is an extremely encouraging methodology for complex time series forecasting.

Keywords: time series forecasting; ensemble learning; heterogeneous models; SMLE; oil consumption

1. Introduction

In Machine Learning (ML), ensemble methods combine various learners to calculate prediction based on constituent learning algorithms [1]. The standard Ensemble Learning (EL) methods include bootstrap aggregating (or bagging) and boosting. Random Forest (RF) [2]; for instance, bagging combines random decision trees and can be used for classification, regression, and other tasks. The effectiveness of RF for regression has been investigated and analyzed in [3]. The boosting method, which builds an ensemble by adding new instances to emphasize misclassified cases, yields competitive performance for time series forecasting [4]. As the most generally utilized usage of boosting, Ada-Boost [5] has been compared with other ML algorithms such as support vector machines (SVM) [6] and furthermore combined with this algorithm to additionally enhance the forecasting performance [7]. Also, stacking [8] is an instance of EL multiple algorithms. It combines the yield which is produced by various base learners in the first level. In addition, by utilizing a meta-learner, it tries to combine the outcomes from these base learners in an ideal method to augment the generalization ability [9].

Although multistep predictions are desired in various applications, they are more difficult tasks than the one-step, due to lack of information and accumulation of errors. In some universal forecasting rivalries held lately, different forecasting methods were proposed to solve some genuine issues. In numerous studies, authors compared the performance of hybrid model on long-term forecasting, for instance, in [10], comparison results demonstrated that an ensemble of neural networks, such as multilayer perceptron (MLP), performed well in these competitions [10]. Also, Ardakani et al. [11] proposed optimal artificial neural networks (ANN) models based on improved particle swarm optimization for long-term electrical energy consumption. Regarding the same aspect this study, [12] introduced a model named the hybrid-connected complex neural network (HCNN), which is able to capture the dynamics embedded in chaotic time series and to predict long horizons of such series. In [13], researchers combined models with self-organizing maps for long-term forecasting of chaotic time series.

On the other hand, in short-term forecasting models, such as ANN and SVM, provide excellent performance for one-step forecasting task [14,15]. However, these models perform poorly or suffer severe degradation when applied to the general multistep problems. As well, the long-term forecasting models are designed for long time prediction tasks (for instance monthly or weekly time series prediction). That means they may perform better in multistep forecasting, while worse in one-step ahead than other methods. In general, the performance of combined forecasting models (e.g., mixing short-term and long-term approaches) is better when compared to single models [16]. Therefore, a forecasting combination can be benefit from performance advantages of short-term and long-term models, while avoiding their disadvantages. Furthermore, major static combination approaches [17–19] depend on assign a fixed weight for each model such as (average, inverse mean), while dynamic combinations methods such as bagging and boosting investigated to combine the results of complementary and diverse models generated by actively perturbing, reweighting, and resampling training data [20,21]. Therefore horizon dependent weights used to avoid the shortcoming of a static and dynamic combination for short- and long-term forecasts [14].

Oil Consumption (OC) is a significant factor for economic development, while the accuracy of demand forecasts is an essential factor leading to the accomplishment of proficiency arranging. Due to this reason, energy analysts are concerned with how to pick the most suitable forecasting methods to provide accurate forecasts of OC trends [22]. However, numerous techniques contribute to estimating the oil demand in future. The field of energy production, consumption, and price forecasting have been gaining significance as a current research theme in the entire energy sectors. For instance, numerous studies investigated foe electricity price forecasting such as Rafał [23], this review article aims to explain and partition the primary methods of electricity price forecasting. Furthermore, Silvano et al. [24] analyzed electricity spot-prices of the Italian power exchange by comparing traditional methods and computational intelligence techniques NN and SVM models. Also, Nima and Farshid [25] proposed a hybrid method for short-ahead price forecasting composed of NN and evolutionary algorithms.

Several studies discussed the issue of time series prediction using different methodologies including statistical methods, single machine learning models, soft computing on ensemble, and hybrid modeling.

Statistical methods have been investigated for time series prediction in the energy consumption area, such as moving average [26], exponential smoothing [27,28], autoregressive moving average (ARMA) [29], and autoregressive integrated moving average (ARIMA) models [30]. For instance, the ARIMA model has been introduced for natural gas price forecasting [31]. However, these statistical techniques do not yield convincing results for complicated data patterns [32,33]. In this context, the Gray Model (GM) forecast accuracy was enhanced by using a Markov-chain model. The outcome of this study demonstrated that the hybrid GM-Markov-chain model was more accurate and had a higher forecast accuracy than GM (1, 1) [34].

In fact, neural networks offer a promising tool for single machine learning model in time series analysis due to their unique features [35]. To further improve the generalization performance, ANN models were investigated for forecasting future OC [36]. Another study experimented with ANN models to predict the long-term energy consumption [37]. For the same purpose, an ANN model was applied to forecast load demands in future [38].

However, ANNs yield mixed results when dealing with linear patterns [39], it is difficult to obtain high accuracy rates of predictors by using the single method, either statistical or ML techniques individually. In order to avoid the limitations associated with the individual models; researchers suggested a hybrid model which combines linear and nonlinear methods to yield high prediction accuracy rates [32,39]. Several studies investigated hybrid modeling to optimize the parameters of the ANN [40]. Hence the improved performance of artificial bee colony (ABC-LM) over other alternatives has been demonstrated on both benchmark data and OC time series.

Similarly, an NN, combined with three algorithms in a hybrid model, then optimized by using a genetic algorithm was used to estimate OC; the outcome demonstrated the efficiency of the hybrid model overall benchmark models [41]. Moreover, a researcher in [42] proposed a genetic algorithm—gray neural network (GA-GNNM) hybrid model to avoid the problem of over-fitting, by examining hybrid versus a total of 26 combination models. Authors concluded that the hybrid models provided desirable forecasting results, compared to the conventional models. Also, the GA has more flexibility in adapting NN parameters to overcome the performance instability of neural networks [22].

In the same context, hybrid models were investigated to solve prediction intervals and densities problems, and have become more common. As shown in Hansen [43] fuzzy model combined with neural models, this combination increased the computation speed, and the coverage is extended. Thus, the problem of the narrow prediction intervals is resolved. Similarly, in [44] the prediction interval also concerned with blend of neural networks and fuzzy models to determine the optimal order for the fuzzy prediction model and estimate its parameters with greater accuracy. Since prediction intervals and forecast densities have become more popular, many types of research have been done about how to determine the appropriate input lag, for this purpose, the fuzzy time series model suggested increasing accuracy by solving the problems of data size (sampling) and the normality [45]. Regarding the same aspect, Efendi and Deris extended a new adjustment of the interval-length and the partition number of the data set, this study discussed the impact of the proposed interval length in reducing the forecasting error significantly, as well as the main differences between the fuzzy and probabilistic models [46].

Finally, as a conclusion from the above studies, hybrid methods give off an impression of being an astounding way to combine predictions of several learning algorithms. The hybrid regression models give preferred predictive accuracy over any single learner. Nonetheless, there was no distinctive way to merge the outcome forecasts of individual models.

In this paper, the goal is to introduce a novel EL framework that can reduce model uncertainty, enhance model robustness, and enhance forecasting accuracy on oil datasets, improve model accuracy, being defined as having a lower measure of forecasting error. The most important motivation for combining different learning algorithms is based upon the assumption that diverse algorithms using different data representations, dissimilar perceptions, and modelling methods are expected to arrive at outcomes with different prototypes of generalization [47]. In addition, to date, comparatively few researches have addressed ensembles for different regression algorithms [48].

We demonstrate that the OC framework can significantly outperform the current methodologies of utilizing the single and classic ensemble forecasting models in single and multistep performance. Although the idea is straightforward, it is yet a robust approach, as it can outperform the average model, as one does not know a priori which model will perform best. The merits of this proposed methodology are analyzed empirically by first describing the exact study design and after that, assessing the performance of various ensembles of different OC models on the GOC. These outcomes are then compared to the classical approach in the literature, which takes the calibrated model with the lowest measure of forecasting error on the calibration dataset at the horizon (1-ahead) to OC of the same dataset at the horizon $t = n$ (10-ahead).

In summary, the developed ensemble model takes full advantage of each component and eventually achieves final success in energy consumption forecasting. The major contributions of this paper come therefore from three dimensions as follows:

1. In this study, we develop a new ensemble forecasting model that can integrate the merits of single forecasting models to achieve higher forecasting accuracy and stability.
    We have introduced a novel theoretical framework how to predict OC. Although the ensemble concept is more demanding regarding computational requirements, it can significantly outperform the best performing model (SVR) of individual models. While the idea is straightforward, it is yet a robust approach, as it can outperform the linear combination methods, as one does not know a priori which model will perform best.
2. The proposed ensemble forecasting model aims to achieve effective performance in multi-step oil consumption forecasting.
    Multi-step forecasting can effectively capture the dynamic behavior of oil consumption in the future, which is more beneficial to energy systems than one-step forecasting. Thus, this study builds a combined forecasting model to achieve accurate results for multi-step oil consumption forecasting, which will provide better basic for energy planning, production and marketing.
3. The superiority of the proposed ensemble forecasting model is validated well in a real energy consumption data.
    The novel ensemble forecasting displays its superiority compared to the single forecasting model and classic ensemble models, and the prediction validity of the developed combined forecasting model demonstrates its superiority in oil consumption forecasting compared to classical ensemble models (AR, Bagging) and the benchmark single models (SVR, BPNN and LR) as well. Therefore, the new developed forecasting model can be widely used in all temporal data application prediction.
4. A perceptive discussion is provided in this paper to further verify the forecasting efficiency of the proposed model.
    Four discussion aspects are performed, which include the significance of the proposed forecasting model, the comparison with single models, and classical ensemble methods, the superiority of the developed forecasting model's stability, which bridge the knowledge gap for the relevant studies, and provide more valuable analysis and information for oil consumption forecasting.

The structure of the paper is organized into five sections: Section 2 is devoted to describing proposed methods design. Section 3 presents the experimental results. Section 4 offers the consumption prediction analysis and discussion. Section 5 describes the conclusion and further suggestion for future works.

2. Materials and Methods

2.1. Proposed Framework

In Section 1, reviewed the literature in three different areas (i.e., single, hybrid, and soft computing on ensemble). While the hybrid modeling literature advanced significantly over the last 20 years, the research on minimizing forecast error, model uncertainty, and hybrid methods is still relatively limited so far. To the best of our knowledge, no attempts exist yet of combining these different areas, by using EL methods to reduce the issues of OC tasks (see Table 1 for a summary).

In particular, we will outline a very general theoretical framework to calibrate and combine heterogeneous ML models using ensemble methods. Its modularity is displayed in Figure 1 and allows for flexible implementation regarding base models, forecasting techniques, and ensemble architecture. For a practical application of this method, we have split the Stacking Multi-Learning Ensemble (SMLE) framework into four main phases and will describe them including their sub-steps in further detail as follows.

Table 1. Summary of related studies on forecasting OC between 2009 and 2017.

Reference	Method	Type	Duration	Region	Horizon
[36]	ANN	Single	1965–2010	Turkey	Long-term
[37]	MLP	Single	1992–2004	Greek	Long-term
[46]	FTS [1], RTS [2]	Hybrid	1965–2012	Malaysia and Indonesia	Long-term
[45]	FTS, RTS	Hybrid	1965–2012	Malaysia	Long-term
[40]	ABCLM [3]	Hybrid	1981–2006	Jordan, Lebanon, Oman, and Saudi Arabia	Short-term
[41]	ABCNN [4], CSNN [5], GANN [6]	Hybrid	1980–2006	Middle East region	Short-term
[22]	GANN, ABCNN	Hybrid	1980–2006	OPEC [10]	Short-term
[34]	GM [7]	Hybrid	1990–2002	China	Short-term
[44]	ANFIS [8]	Hybrid	1974–2012	U.S.	Short-term
[42]	GA, GNNM [9]	Hybrid	2000–2010	China	Short-term
SMLE *	SVR, BPNN, LR	Ensemble	1965–2016	GOC [11]	Long-term

[1] Fuzzy Time Series; [2] Regression Time Series; [3] Artificial Bee Colony Algorithm; [4] Artificial Bee Colony Neural Network; [5] Cuckoo Search Neural Network; [6] Genetic Algorithm Neural Network; [7] Grey Markov; [8] Adaptive Neuro-Fuzzy Inference Systems; [9] Genetic Algorithm—Gray Neural Network; [10] Organization of the Petroleum Exporting Countries; [11] Global Oil Consumption; * Proposed Method.

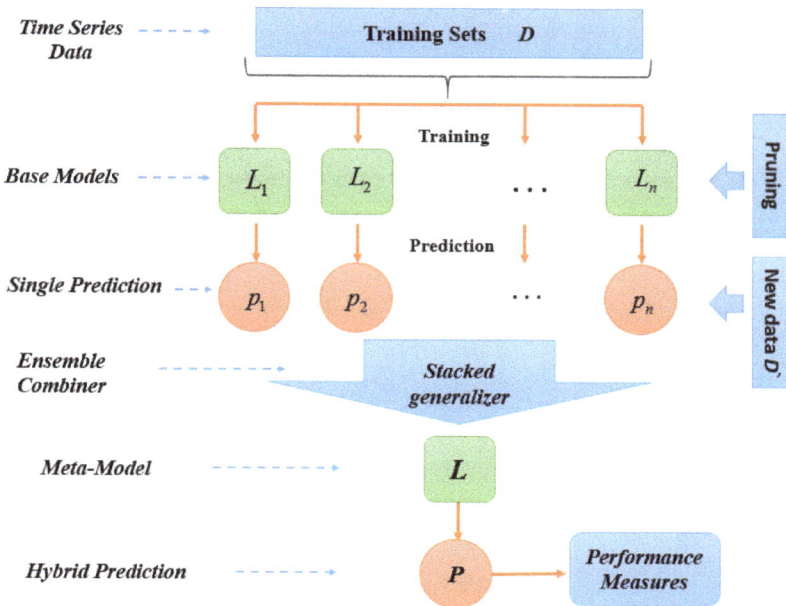

Figure 1. Stacking Multi-Learning Ensemble (SMLE) Framework.

2.2. Ensemble Generation

In the original data set, the initial training data, represented as D, had m observations and n features, so that it is $m \times n$. The modeling procedure can be realized by setting different parameters of the base learners. In this level, some heterogeneous models were trained on D using one method of training (i.e., cross-validation). Moreover, each model offered prediction results $p_i (i \in 1, 2, \ldots, n)$ which were then cast into a second level data; the outcome became the input for the second level as training data.

2.2.1. Ensemble Pruning

Ranking-based subset selection method ranks the candidate models according to criteria, such as the mean absolute percentage error (MAPE), directional accuracy (DA), and Euclidean Distance (ED), and included only the top n models from all candidate models.

2.2.2. Ensemble Integration

This step describes how the selecting models were combined into ensemble forecast. In this context, the stacking method is used to build the second level data, stacking uses a similar idea to K-folds cross-validation to solve two significant issues: Firstly, to create out-of-sample predictions. Secondly, to capture distinct regions, where each model performs the best. The stacking process investigates by inferring the biases of the generalizers concerning the provided base learning set. Then, stacked regression using cross-validation was used to construct the 'good' combination. Consider a linear stacking for the prediction task. The basic idea of stacking is to 'stack' the predictions $f_1, \ldots, f_m$ by linear combination with weights $a_i, \ldots, (i = 1, \ldots, m)$:

$$f_{stacking}(x) = \sum_{i=1}^{m} a_i f_i(x),\qquad(1)$$

where the weight vector a is learned by a meta-learner.

2.2.3. Ensemble Prediction

The second level learner model(s) can be trained on the $D\prime$ data to produce the outcomes which will be used for final predictions. In addition, to select multiple sub-learners, stacking allows the specification of alternative models to learn how to best combine the predictions from the sub-models. Because a meta-model is used to combine the predictions of sub-models best, this method is sometimes termed blending, as in mixing the final predictions.

In brief, Figure 1 demonstrated the general structure of SMLE framework, which consisted of various learning steps, after applying this scheme, three SMLE models were generated, while the difference between the SMLE models were not in structure, but in the type of base model in level #0 and the differences between the three models in the part of base model can be explained as follows:

- 1st SMLE in base layer used SVR learner and in Meta layer LR used as meta learner.
- 2nd SMLE in base layer used BPNN learner and in Meta layer LR used as meta learner.
- 3rd SMLE in base layer used SVR and BPNN learners and in Meta layer LR used as meta learner.

2.3. Experiment Study Design

2.3.1. Data

The GOC data were used as benchmark data; this dataset was downloaded from the website: https://www.bp.com/en/global/corporate/energy-economics/statistical-review-of-worldenergy.html. The data represented total OC in the world; the data was yearly type and had a duration from 1965 to 2016. The data consisted of two factors, thus dependent variable oil consumption (in Million Tonnes), which was a feature over time, and date (in years) was the independent variable in this case study. Therefore, the OC time series for this experiment had 52 data points. For a better explanation, we visualized whole actual time series in Figure 2, with a blue circle in curve.

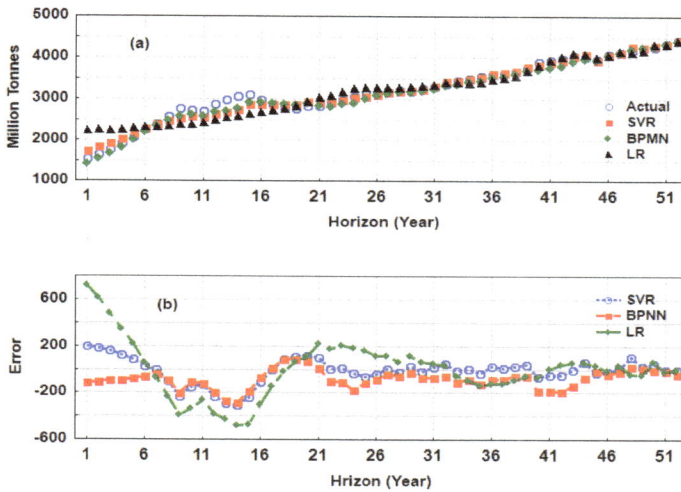

Figure 2. Comparison of (**a**) the actual and predicted consumption with the use of the SVR, BPNN and LR single learners (**b**) errors of all single models.

2.3.2. Models

As above-mentioned, we applied the ensemble SMLE model to predict the GOC data set after combining the heterogeneous models, Table 2 lists the learners' parameters that have been investigated in this paper. These related methods are presented briefly as follows:

1. The BPNN algorithm consists of multiple layers of nodes with nonlinear activation functions and can be considered as the generalization of the singer-layer perceptron. It has been demonstrated to be an effective alternative to traditional statistical techniques in pattern recognition and can be used for approximating any smooth and measurable functions [49]. This method has some superior abilities, such as its nonlinear mapping capability, self-learning and adaptive capabilities, and generalization ability. Besides these features, the ability to learn from experience through training makes MLP an essential type of neural networks and it is widely applied to time series analysis [50].

2. The SVM algorithm is always considered a useful tool for classification and regression problems due to the ability to approximate a function. Furthermore, the kernel function is utilized in the SVR to avoid the calculations in high-dimensional space. As a result, it can perform well when the input features have high dimensionality. It separates the positive and negative examples as much as possible by constructing a hyperplane as the decision surface. The support vector regression (SVR) is the regression extension of SVM, which provides an alternative and promising method to solve time series modeling and forecasting [51,52].

3. LR is a popular statistical method for regression and prediction. It utilizes the ordinary least-squares method or generalized least-squares to minimize the sum of squares of errors (SSE) for obtaining the optimal regression function [53].

Table 2. Summary of parameters setting for all learners.

Model	Parameters
LR	Attribute method selection = Md5, batch Size = 100, and ridge = 1.0×10^{-8}
SVR	Kernel = (Poly), C = 1, exponent = 2 and epsilon = 0.0001.
BPNN	MLP(1-3-1)
Bagging	Base learner = REPTree, bagSizePercent = 100%, No. iteration = 10.
AR	Base learner = linear regression, No. iteration = 10, Shrinkage = 1.0.
1st SMLE	Base learner (SVR(Kernel = (Poly),C = 1, exponent = 2, epsilon = 0.0001)), meta learner (LR), Combination method= Stacked generalization
2nd SMLE	Base learner (MLP(1-3-1)), meta learner (LR), Combination method = Stacked generalization
3rd SMLE	Base learner (SVR (Kernel = (Poly), C = 1, exponent = 2, epsilon = 0.0001) and MLP (1-3-1), meta learner (LR), Combination method= Stacked generalization

2.3.3. Evaluation Measure

This subsection describes several aspects of the evaluation of the different models; the evaluation aspects include the estimation of error rates and pairwise comparisons of classifiers/ensembles.

1. Performance Evaluation

In terms of performance error estimation, the mean absolute percentage error (MAPE) was adopted as an indicator of accuracy for all forecasting methods. The accuracy is expressed as a percentage value, and is defined by the Formula (2) as below:

$$\text{MAPE} = \frac{100}{n} \sum_{i=i}^{n} \left| \frac{\overset{\wedge}{y_i} - y_i}{y_i} \right|, \tag{2}$$

where y_i is the actual value and $\overset{\wedge}{y_i}$ is the forecast value.

2. Time Series Similarity

The distance between time series can be measured by calculating the difference between each point of the series. The Euclidean Distance (ED) between two-time series $Q = \{q_1, q_2, \ldots, q_n\}$ and $S = \{s_1, s_2, \ldots, s_n\}$ is defined as:

$$D(Q, S) = \sqrt{\sum_{i=1}^{n} (q_i - s_i)^2}. \tag{3}$$

This method is moderately easy to calculate, and has complexity of $O(n)$ [54].

3. Continuous Growth Rates (CGR)

Calculating change growth rate in data is useful for average annual growth rates that steadily change. It is famous because it relates the final value in series to the initial value in the same series, rather than just providing the initial and final values separately—it gives the ultimate value in context [30]. The CGR value calculated according to Formula (4) as follows:

$$k = \frac{\ln\left(\frac{y_{t+1}}{y_t}\right)}{t}, \tag{4}$$

where k represents the annual growth rate y_t represents the initial population size, t represents the future time in years and k is CGR.

2.4. The Algorithm for Stacking Multi-Learning Ensemble (SMLE)

In this study, SMLE offers a dynamic EL method. The SMLR method depends on the sequence characteristic of OC data. For accurate OC prediction, we express the algorithm of SMLE when predicting the next mth moment OC at the time t. The general design of the proposed model considered both diversity management and accuracy enhancement for base models. Here the algorithm of SMLE is described below as pseudocode in Algorithm 1:

Algorithm 1: Stacking Multi-Learning Ensemble (SMLE).

Input: Dataset $D = \{(x_1, y_1), (x_2, y_2), \ldots, (x_m, y_m)\}$;
      First-level learning algorithms $L_1, L_2, \ldots, L_n$;
      Second-level learning algorithm L;

Process:

%Train a first-level individual learner h_t by applying the first-level learning algorithm L_t to the original dataset D
    for $t = 1, \ldots, T$:
      $h_t = L_t(D)$
    end;
% generate a new data set
    $D\prime = \phi$;
    for $i = 1, \ldots, m$:
        for $t = 1, \ldots, T$:
           $z_{it} = h_i(x_i)$ *% Use h_t to predict training example x_i*
        end;
        $D\prime = D\prime \cup \{((z_{I1}, z_{i2}, \ldots, z_T), y_i)\}$
    end;
% Train the second-level learner $h\prime$ by applying the second-level learning algorithm L to the new data set $D\prime$
    $h\prime = L(D\prime)$
Output: $H(x) = h\prime(h_1(x_1), \ldots, h_T(x_T))$

---

3. Results

In this section, we evaluated various models on GOC 52-year data sets using BPNN, SVR, and LR as the base models to demonstrate their predictability of both single and EL forecasting. Hence, there were single models used as benchmark model compared to ensemble predictors. In the second experiment, we tested two classic ensemble models include bagging and additive regression (AR). Moreover, the third experiment tested three ensemble models based on SMLE scheme. To establish the validity of the evaluated method, a further procedure was done by comparing the obtained results of single models with the outcome of the ensemble models. Evaluation criteria were used to compare and analyze the prediction, such as T-Time, DA, MAPE, and ED, which are excellent methods for predicting GOC. Meanwhile, we compare the evaluation criteria of multistep (10-ahead) with single step (1-ahead) forecasting to find the better SMLE model for predicting GOC in both short-term and term-long horizon situations. Finally, consumption growth rate evaluated for all prediction outcome.

3.1. Single Models Results

Regarding the experiment design and the overall steps described in Section 2.2, the first test in this experiment was to compare the performance of all base models separately. The output of 10-fold cross-validation tests run on the initial training were used to determine whether each model was sufficient for OC data to make the forecasting results more stable. Figure 2a presents the comparison of the best-obtained results from all base models with the real OC data. It is evident that the results obtained according to the SVR method for the 52 known years (1965–2016) were close to the actual ones and comparable to those produced by the BPNN and LR models.

Similarly, Figure 2b demonstrated the residual errors of the prediction, to make a reliable comparison to quantitatively analyze the performances of the base models; we considered the MAPE measure indices for performance accuracy processes, which are listed in Table 3.

In brief, as seen in Table 3, the MAPE between predicted and actual values for the SVR model is 1.24% given by relative accuracy (DA) 89.9% which indicates clearly that the SVR model is well working and has acceptable accuracy. Regarding the same aspect, we can observe that the SVR had superiority in both run time and similarity (0.01 and 0.034, respectively). However, it is worth mentioning that the LR models scored poor performance compared to other single models. The similarity between actual and predicted data was measured using the Euclidean Distance (ED), as shown in Table 3; the BPNN score 0.034, which was small indicates the best predictive performance, while LR scores 0.074 was the worst similarity across the models.

3.2. Classic Ensemble Models Results

In the second experiment, we empirically tested two classical ensemble models, included bagging and additive regression (AR). To illustrate the behavior of all classical ensemble fitting, they were compared with actual data in Figure 3a,b, for visual comparison of the residual error of each model. The evaluation matrix of single learning, classical ensemble methods, and proposed SMLE models are summarized in Table 3. As observed from Table 3 and Figure 3, the bagging model performed better than the AR model in all evaluation measures, except in DA. Similarly, the bagging model performed better than the best single model (SVR) in performance and similarity while SVR perform best in DA and has least training time. For this dataset, we accordingly developed homogeneous and heterogeneous ensembles of individual models rather than using their hybrid versions.

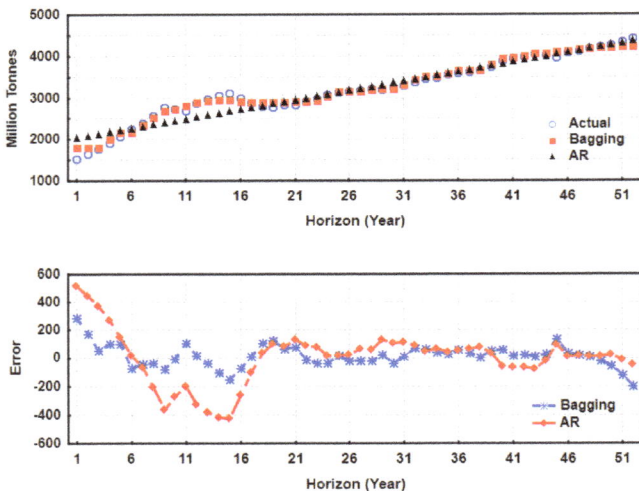

Figure 3. Illustrated (**a**) actual and predicted consumption using classic ensemble learners (**b**) error of classic ensemble models.

3.3. The SMLE Results

In the third experiment, we empirically tested three heterogeneous stacking models, each model was composed of a combination of base and meta-models. The first ensemble model consists of SVR as a base learner and LR as meta-learner. To illustrate the behavior of all SMLE for fitting, they were compared with actual data in Figure 4a,b, for visual comparison of the residual error of each model. The evaluation matrix of single learning methods, and proposed framework is summarized in Table 3.

The outcome of this model, as presented in Table 3, enhanced the forecasting accuracy by 34% when it was compared to the best base learner, SVR. Moreover, the second ensemble model was a mix of BPNN as the base learner and LR as the meta-learner, the combined model increased the forecasting accuracy by decreasing the error by 46%, compared to the best single model as mentioned previously.

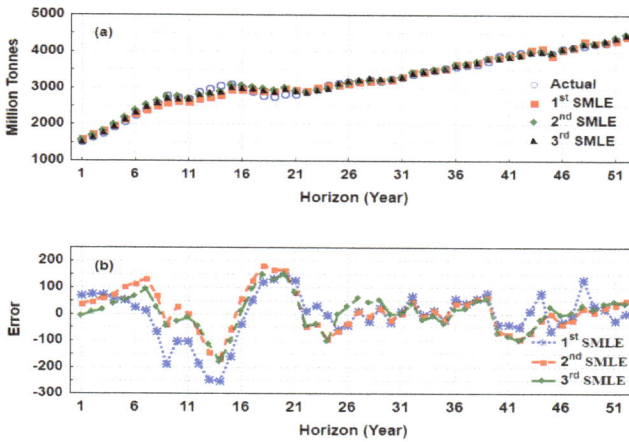

Figure 4. Illustrated (**a**) actual and predicted consumption using SMLE learners (**b**) error of SMLE models.

Table 3. Summary of different evaluating measures among all models on GOC data.

Measures	Single Models			Classic Ensemble Models		SMLE Models		
	LR	SVR	BPNN	Bagging	AR	1st SMLE	2nd SMLE	3rd SMLE
MAPE (%)	6.77	2.82	3.15	2.52	5.19	2.27	2.07	1.65
DA (%)	82.59	89.03	89.90	66.17	82.59	88.50	90.69	91.24
ED	0.074	0.035	0.034	0.026	0.059	0.028	0.024	0.020
T-Time	0.04	0.05	0.03	0.06	0.07	0.09	0.13	0.17

Bold number indicates the best value in all measures.

Finally, the third ensemble model was a combination of SVR and BPNN as base learners and LR as the meta-learner. The forecasting result of this model indicates that the accurate predictive model decreased the error of the best base model by 50%, which led to proof of the superiority of the third model over both the single and combination models. The similarity between actual and predicted data is shown in Table 3, the 3rd SMLE based (SVR-BPNN) model score was 0.020, while 1st SMLE based (SVR) score was 0.028, the worst similarity in across all the models. Also, it can be observed that all ensemble model had less distance compared to single models.

In the same aspect, the 3rd SMLE performed better than the best classic ensemble model (bagging) in all measures, except for training time (T-Time); this was due to the ensemble model learning level which consumed more time and calculation.

4. Discussion

In this subsection, we practically used all single and ensemble forecasters to solve the problem of how to estimate the future OC. For further evaluation of SMLE scheme stability, all models were examined in 1-ahead and 10-ahead horizon predictions.

From Figure 5 and Table 4, it is easy to find that the proposed SMLE method was the best one for OC forecasting in all prediction horizons (i.e., 1, 3, 5, 7, and 10-step-ahead), relative to other models considered in this study. In all the models, the SMLE-based BPNN-SVR model did not only accomplish the highest accuracy at the level estimation, which was measured by the MAPE criteria, it additionally got the highest hit rate in direction prediction, which was estimated by the DA criterion. Then again, among the majority of the models utilized as a part of this investigation, the single LR model performed the poorest in all progression ahead forecasts. LR model not only had the lowest level accuracy, which was measured by MAPE, but also acquired the worst score in direction accuracy, which was measured by the DA criteria. The main reason might be that LR was a class of the typical linear model and it could not capture the nonlinear patterns and occasional characteristics existing in the data series. Apart from the SMLE-based BPNN-SVR and LR models, which performed the best and the poorest, respectively. All models listed in this study produce some interestingly blend results, these outcomes were analyzed by using four estimation criteria (i.e., MAPE, DA, *T*-test, and CGR).

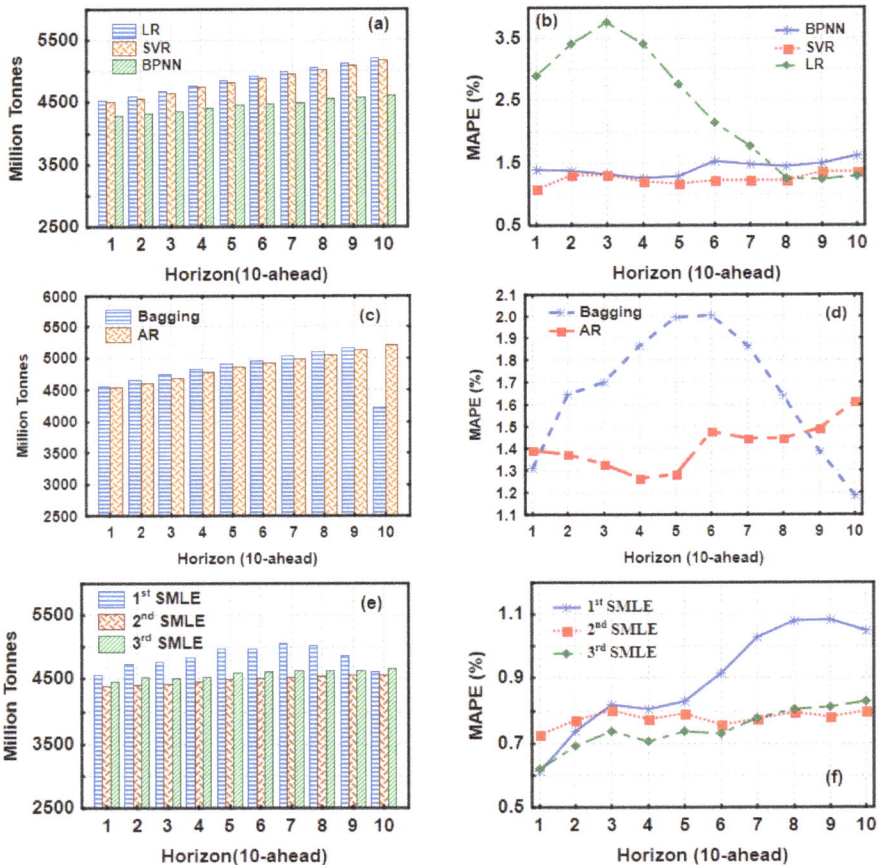

Figure 5. Illustrated 10-ahead consumption prediction and MAPE. (**a**) Single models. (**b**) Error of single models. (**c**) Classic ensemble models. (**d**) Error of classic ensemble models. (**e**) SMLE models. (**f**) error of SMLE models.

Firstly, in the case of level accuracy, the results of the MAPE measure demonstrated that the SMLE-based BPNN-SVR performed the best, followed by SMLE-based BPNN, SMLE-based SVR models, SVR and BPNN, and the weakest model was LR as shown in Figure 5b,f. Moreover, from Table 4, the MAPE values of the SMLE-based BPNN-SVR model were 0.61 in 1-ahead and 0.74 as an average of the 10-step-ahead predictions, which was less than other methods. Also, in the short-term prediction step, better performance was observed when comparing ensemble methods with single models, the results indicate that the ensemble methods outperformed the single and classic ensemble methods in all cases. The principle reason could be that the cross-validation decomposition methodology did efficiently enhance the forecast execution. Interestingly, the 1-step-ahead and multi-step-ahead prediction horizon of single model forecasts were inferior to ensemble models. Focusing on the single methods and classic ensembles, all the ML models outperformed the LR model; the reason may be that LR is a typical linear model, which is not suitable for capturing the nonlinear and seasonal characteristics of OC series. In ML models (i.e., SVR, BPNN), it can be seen that SVR performed slightly better than BPNN in all 10-step-ahead predictions and BPNN perform poorest in all the step prediction. The main reason leading to this may the parameter selection. The MAPE values of LR were from 2.91 to 2.40, which were slightly inferior to SVR and BPNN models. The possible reason was that the prediction results of LR, which was under the linear hypothesis were more volatile than those of the ML models.

Second, the high-level exactness does not necessarily imply that there was a high hit rate in forecasting direction of OC. The correct forecasting direction is essential for the policy manager to make an investment plan in oil-related operations (production, price, and demand).

Table 4. 10-ahead forecasting performance among all models on GOC data.

Model	MAPE (%) over 10-Ahead Horizon					Avg.
	1-ahead	3-ahead	5-ahead	7-ahead	10-ahead	
LR	2.91	3.76	2.76	1.77	1.30	2.40
SVR	1.08	1.30	1.17	1.23	1.36	1.24
BPNN	1.39	1.33	1.28	1.48	1.61	1.42
Bagging	1.31	1.70	1.99	1.86	1.19	1.66
RF	1.39	1.33	1.28	1.45	1.61	1.41
1st SMLE	0.62	0.82	0.83	1.03	1.05	0.90
2nd SMLE	0.73	0.80	0.79	0.77	0.80	0.78
3rd SMLE	0.61	0.74	0.74	0.78	0.83	0.74

Therefore, the DA comparison is necessary. In Figure 6a–c, some similar conclusions can be drawn regarding DA criterion. (i) The proposed 3rd SMLE model performed significantly better than all other models in all cases, followed by the other two ensemble models, then two of the single ML models (i.e., SVR, BPNN), (LR, AR) had equal values, and bagging model had the worst values. Individually, the DA values of all SMLE-based ensembles were similar 92.31% for the 1 step-ahead predictions and showed superiority with 91.24% for average 10-ahead step forecasts for the 3rd SMLE model. (ii) The three ensemble methods mostly outperformed the single prediction models. Furthermore, among the ensemble methods, the SMLE- based BPNN-SVR model performed the best, and SMLE-based BPNN model outperformed SMLE- based SVR model, except for the 2-ahead forecast. (iii) SVR model outperformed other methods, BPNN had the similar performance as SVR in the 2, 3, 5 step-ahead forecasts, except that SVR exceeded BPNN in both 1-ahead and average ahead prediction. The possible reason leading to this phenomenon may be the choice of optimal parameters for the models. We also found that bagging model had the lowest directional accuracy of 66.17%.

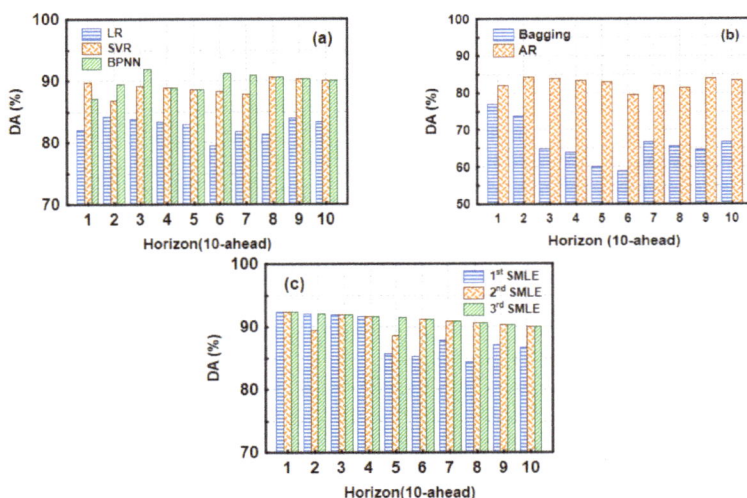

Figure 6. Illustrated 10-ahead consumption directional accuracy (DA) of (**a**) single models, (**b**) classic ensemble models, and (**c**) SMLE ensemble models.

Also, comparing different prediction horizons, the short-term prediction horizon showed better performance for in all the model see Table 4. Taking 1-step-ahead forecasting and an average of the 10-step-ahead predictions for example, for all the SMLE–based ensemble, BPNN, SVR, bagging, AR models, the 1-step-ahead forecasting outperformed the average of the 10-step-ahead forecast, no matter the level accuracy or directional accuracy. Apart from the models mentioned above, SMLE-based ensembles and ML models and classic ensembles performed better in 1-step-ahead prediction given directional accuracy. However, from the point of level accuracy, both these approaches only had slight superiority in 6-step-ahead prediction. Except for the LR, which performed almost poorer in the 1-step-ahead compared to the average of 10-step-ahead prediction as shown in Figure 7a–c.

Third, to further validation of SMLE models forecasting, the *t*-test was used to test the statistical significance of the prediction performance. The *t*-test results presented in Table 5, for all ensemble models under this study were not significant ($df = 51$, p-value > 0.05)). Based on the detailed statistical test, no significant differences were observed between the actual OC and that predicted by the SMLE models. The mean differences in the last column of Table 5, indicate that in the population from where the sample models were drawn, the actual and predicted OC was statistically semi-equal. Therefore, it was possible to prove that the SMLE model was useful in predicting OC based on the heterogeneous models with excellent levels of accuracy (see Table 5). So, we can conclude that the model developed structure is sufficient with more parameters setting (i.e., kernels, neuron) for OC prediction.

Table 5. The *t*-test results of actual and predicted oil consumption using SMLE models.

Model	t	p-Value	Mean Difference
1st SMLE	0.227	0.823	0.8081
2nd SMLE	1.320	0.193	0.6803
3rd SMLE	0.728	0.470	0.6178

The forecasted values for each model and total OC growth rate from 2017 to 2026 is summarized in Table 6. As seen from the table, all models will still be increasing in the period from 2017 to

2026. However, the average annual rates will decrease in all. For the period between 1965 and 2016, the rate of increases was 2.2% for BPNN, 1.3% for LR, 1.8% for SVR, 1.5% for bagging 1.6%, for AR 2.0% for SMLE-based SVR, 2.0% for SMLE-based BPNN, and 2.1% for SMLE-based BPNN-SVR. Additionally, for the forecasted period between 2017 and 2026 the rates were expected to be 0.74%, 1.39%, 1.38%,1.42%, 1.39% 0.13%, 0.38%, and 0.44, respectively. On the other hand, the average annual rate of total oil demand decreased from 1.8% between 1965 and 2016 to 0.91% between 2017 and 2026.

Lastly, the summarized results in Table 6 demonstrate that the annual growth rates of 1-ahead OC were more significant than the total average OC in 10-ahead years. Figure 8, shows the apparent rise in the 1-ahead in both single and classic ensemble models, and for the SMLE models there was a sudden drop from 1- to 2-ahead years, also note the stability in the growth from 2-ahead to 10-ahead, with close values in all models, except for SMLE-based BPNN where there was a few decreasing in the 9-, 10-ahead, sequentially. The decrease in the rate of oil demand may be interpreted as there being other alternative energies that affect oil demand, this will be achieved in the coming decades, as compared with all other energy type consumption. Rates of changes and reserves in the OC of all the models indicate that the SMLE scheme was the best to determine the actual demand of energy globally, which facilitates the planning process, associated with the issue OC prediction. Based on these study findings, we suggested some recommendations.

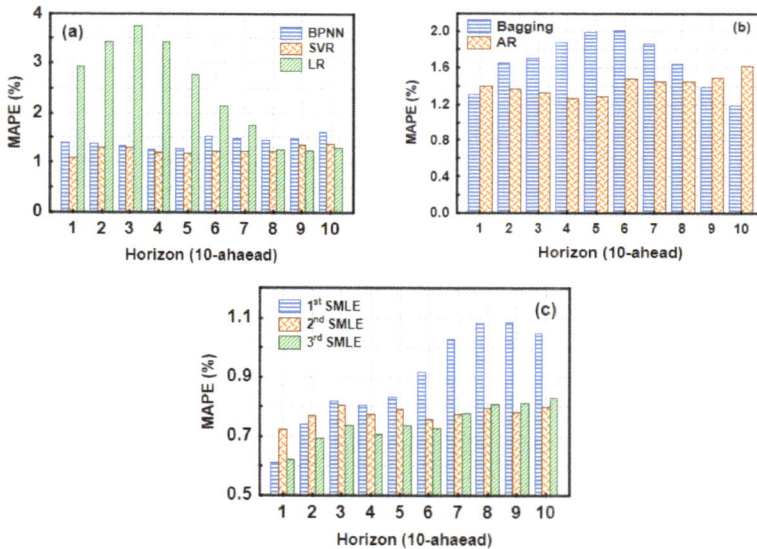

Figure 7. Illustrated 10-ahead consumption prediction errors of (**a**) single models (**b**) classic models (**c**) SMLE ensemble models.

Table 6. Summary of forecasted values and CGR for OC using all models from 2017 to 2026.

Years	Models							
	BPNN	LR	SVR	Bagging	RF	1st SMLE	2nd SMLE	3rd SMLE
2017	4279.16	4531.93	4504.13	4559.73	4531.93	4554.73	4395.00	4459.94
2018	4316.27	4595.00	4567.34	4646.02	4595.00	4737.14	4409.57	4523.12
2019	4349.24	4677.23	4650.48	4748.19	4677.23	4764.54	4426.70	4511.95
2020	4401.42	4771.37	4746.31	4819.88	4771.37	4835.67	4454.33	4530.88
2021	4448.96	4857.41	4821.93	4896.70	4857.41	4967.25	4482.28	4592.90

Table 6. *Cont.*

Years	Models							
	BPNN	**LR**	**SVR**	**Bagging**	**RF**	**1st SMLE**	**2nd SMLE**	**3rd SMLE**
2022	4471.17	4915.79	4889.67	4956.50	4915.79	4967.25	4501.68	4602.70
2023	4496.30	4980.78	4954.70	5025.85	4980.78	5056.31	4519.38	4633.88
2024	4555.86	5052.61	5023.62	5095.53	5052.61	5029.93	4536.88	4635.57
2025	4578.78	5128.13	5097.75	5163.26	5128.13	4861.17	4553.77	4628.92
2026	4606.92	5205.40	5170.20	5215.23	5205.40	4615.31	4566.33	4659.36
2017–2026	0.74%	1.39%	1.38%	1.42%	1.39%	0.13%	0.38%	0.44%
1965–2016	2.2%	1.3%	1.8%	1.5%	1.6%	2.0%	2.0%	2.1%

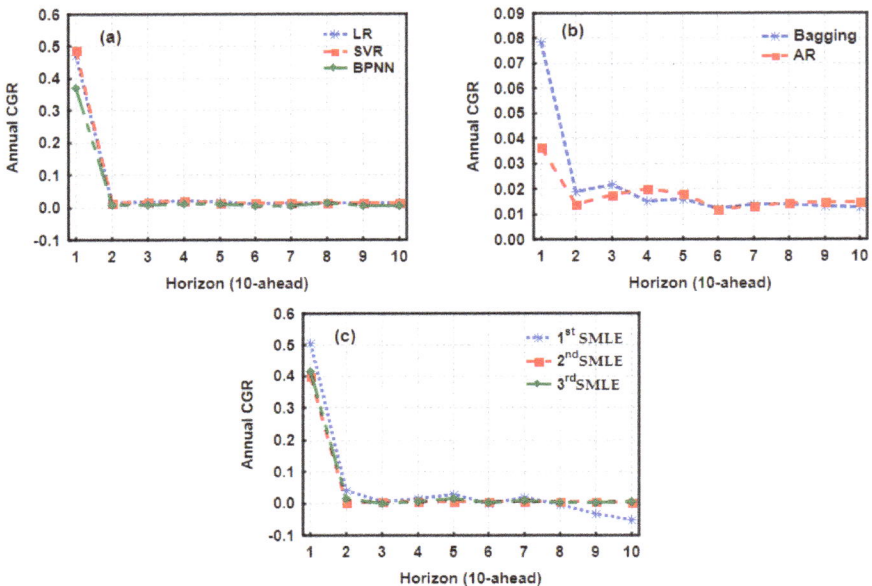

Figure 8. Illustrated annual CGR for 10-ahead consumption prediction using (**a**) single models (**b**) classic ensemble models (**c**) SMLE models.

We summarized all of the above results in Table 7 and Figure 9. In general, combining the forecasters using SMLE will significantly improve the final prediction. Generally, from the analysis of the experiments presented in this study, we can draw several important conclusions as follows: Firstly, the SMLE-based BPNN-SVR model was significantly superior to all models in this study regarding similarity, level accuracy, and direction accuracy. Through performance enhancement, the SMLE-based BPNN-SVR outperformed other models at the 1.17 statistical significance level, compared to the best benchmark models SVR and bagging, respectively. Secondly, the prediction performance of the SMLE-based BPNN-SVR, SMLE-based SVR and SMLE-based BPNN models were better than the single and classic ensemble methods. These results indicate that the hybrid, based on stacking method, can efficiently improve the prediction performance in the case of OC. Thirdly, nonlinear models, with seasonal adjustment, were more suitable as base learners for the ensemble to predict the time series with annual volatility than linear methods, due to properties above of OC (i.e., nonlinear and non-stationary). However, computationally, the new method consumed more time because of its way of segmenting inputs and the use of the ensemble. Fourthly, the average annual rate of total oil demand decreased from 1.8% between 1965 and 2016 to 0.91% between 2017 and 2026. Finally, on one hand, short-term forecasting models, such as BPNN and SVM, provided excellent performance for one-step forecasting task. However, these models performed poorly or suffered severe degradation when applied to the general multistep problems. In general, the performance of ensemble forecasting models (e.g., combining short-term and long-term approaches) was better when compared to single models. Therefore, a forecasting combination can benefit from performance advantages of short-term and long-term models, while avoiding their disadvantages. Furthermore, to overcome the shortcoming of a static combination approach, a dynamic combination of short- and long-term forecasts can be employed by using horizon dependent weights.

Table 7. Summary of evaluation measures among all models on GOC data.

NO.	Model	Evaluation Matrix				Score [1]	Indexed Rank [2]
		T-Time	MAPE (%)	ED	DA (%)		
1	LR	0.04	2.4	0.074	82.59	29	8
2	BPNN	0.05	1.42	0.035	89.03	20	6
3	SVR	0.03	1.24	0.034	89.90	19	5
4	Bagging	0.06	1.66	0.026	66.17	18	4
5	AR	0.07	1.41	0.059	82.59	22	7
6	1st SMLE	0.09	0.9	0.028	88.50	15	3
7	2nd SMLE	0.13	0.78	0.024	90.69	9	2
8	3rd SMLE	0.17	0.74	0.020	91.24	8	1

[1] Score: sum of rank values from (1–8) for each model depends on performance in related measure. [2] Order value for each model depending on total score, for example rank no 1 means the first model.

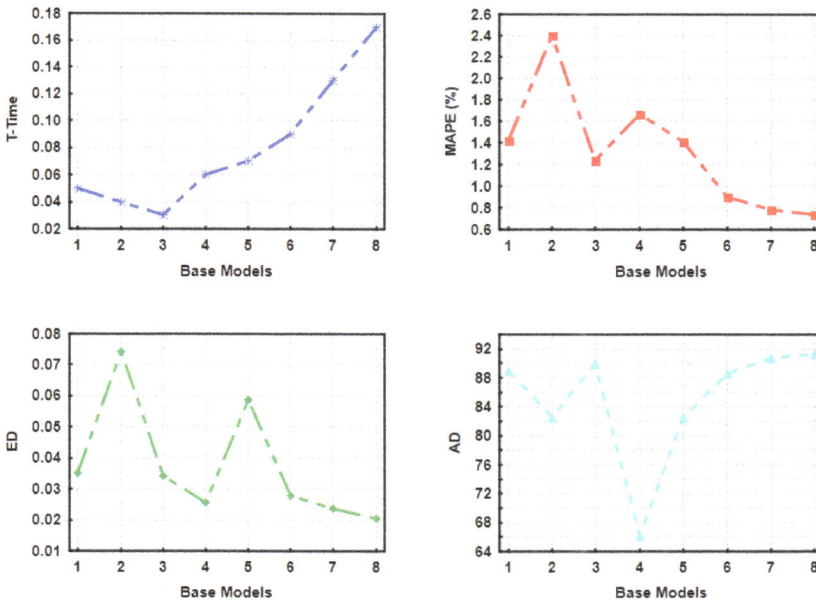

Figure 9. Illustrated T. Time MAPE, ED, and AD evaluation measures of all models on 10-ahead GOC prediction. (The order of models were arranged from 1–8 according to Table 6, as LR, BPNN SVR, bagging, AR, 1st, 2nd, and 3rd SMLE, respectively).

5. Conclusions

Forecasting time series data is considered as one of the most critical applications and has concerned interests of researchers. In this study, we discussed the problem of combining heterogeneous forecasters and showed that ensemble learning methods could be readily adapted for this purpose. We have introduced a novel theoretical ensemble framework integrating BPNN, SVR, and LR, based on the principle of stacking; which was proposed for the GOC forecasting. This framework has been able to reduce uncertainty, improve forecasting performance, and manage the diversity of learning models in empirical analysis.

According to the experimental results and analyses, the proposed ensemble models have been able to outperform the classical ensemble and single models on OC data analyzed results. Furthermore, all ensemble models have been able to exceed the best performing individual models on single-ahead, as well as the multi-ahead horizon.

The advantages of proposed model to the knowledge comes therefore along three aspects as follows:

Firstly, in methodology part, we have introduced a novel theoretical framework based on ensemble learning for OC forecasting. Although the ensemble concept is more demanding regarding computational requirements, it can significantly outperform single models and classical hybrid models. While the idea is straightforward, it is yet a robust approach, as it can outperform linear combination methods, as one does not know a priori which model will perform best.

Secondly, theoretically we have demonstrated that ensemble methods can be successfully used in the context of OC forecasting due to the ambiguity decomposition.

Thirdly, we have conducted a very extensive empirical analysis of advanced machine learning models, as well as ensemble methods. Just the calibration alone of such a wide range of ensemble models is very rare in the literature, considering that the ranking of some evaluation measures per model to run, which was not only due to the limited.

This study has two limitations including: the consideration of the integration of heterogeneous algorithms (SVR, BPNN and LR) without using ensemble pruning for internal hyper-parameters; and the evaluation process investigated on single data set, so that this model can verified in different data sets. All these limitations could be interesting future research.

In future work, homogeneous ensemble model based SVR with different kernels can be developed and evaluated. In addition, investigating ensemble pruning by using evolutionary algorithms that provides an automatic optimization approach to SVR hyper-parameters, could be an interesting future research work in the hybrid-based energy forecasting field. Another direction of future work is to apply ensemble models in other energy prediction problems, such as pricing, production, and load forecasting.

Author Contributions: M.A.K. performed the experiments, analyzed the data, interpreted the results and wrote the paper. X.N. supervised this work, all of the authors were involved in preparing the manuscript.

Funding: This research received no external funding.

Acknowledgments: This research was funded by the Chinese Scholarship Council, China (under Grants CSC No. 2015-736012).

Conflicts of Interest: The authors declare no conflicts of interest.

References

1. Seni, G.; Elder, J.F. Ensemble methods in data mining: Improving accuracy through combining predictions. In *Synthesis Lectures on Data Mining and Knowledge Discovery*; 2010; Volume 2, pp. 1–126.
2. Segal, M.R. *Machine Learning Benchmarks and Random Forest Regression*; Center for Bioinformatics and Molecular Biostatistics: San Francisco, CA, USA, 2004.
3. Grömping, U. Variable importance assessment in regression: Linear regression versus random forest. *Am. Stat.* **2009**, *63*, 308–319. [CrossRef]

4. Youssef, A.M.; Pourghasemi, H.R.; Pourtaghi, Z.S.; Al-Katheeri, M.M. Landslide susceptibility mapping using random forest, boosted regression tree, classification and regression tree, and general linear models and comparison of their performance at Wadi Tayyah Basin, Asir Region, Saudi Arabia. *Landslides* **2016**, *13*, 839–856. [CrossRef]

5. Shrestha, D.L.; Solomatine, D.P. Experiments with AdaBoost. RT, an improved boosting scheme for regression. *Neural Comput.* **2006**, *18*, 1678–1710. [CrossRef] [PubMed]

6. Morra, J.H.; Tu, Z.; Apostolova, L.G.; Green, A.E.; Toga, A.W.; Thompson, P.M. Comparison of AdaBoost and support vector machines for detecting Alzheimer's disease through automated hippocampal segmentation. *IEEE Trans. Med. Imaging* **2010**, *29*, 30–43. [CrossRef] [PubMed]

7. Guo, L.; Ge, P.-S.; Zhang, M.-H.; Li, L.-H.; Zhao, Y.-B. Pedestrian detection for intelligent transportation systems combining adaboost algorithm and support vector machine. *Expert Syst. Appl.* **2012**, *39*, 4274–4286. [CrossRef]

8. Aldave, R.; Dussault, J.-P. Systematic ensemble learning for regression. *arXiv*, 2014.

9. Lemke, C.; Gabrys, B. Meta-learning for time series forecasting and forecast combination. *Neurocomputing* **2010**, *73*, 2006–2016. [CrossRef]

10. Crone, S.F.; Hibon, M.; Nikolopoulos, K. Advances in forecasting with neural networks? Empirical evidence from the NN3 competition on time series prediction. *Int. J. Forecast.* **2011**, *27*, 635–660. [CrossRef]

11. Ardakani, F.J.; Ardehali, M.M. Novel effects of demand side management data on accuracy of electrical energy consumption modeling and long-term forecasting. *Energy Convers. Manag.* **2014**, *78*, 745–752. [CrossRef]

12. Gómez-Gil, P.; Ramírez-Cortes, J.M.; Hernández, S.E.P.; Alarcón-Aquino, V. A neural network scheme for long-term forecasting of chaotic time series. *Neural Process. Lett.* **2011**, *33*, 215–233. [CrossRef]

13. Fonseca-Delgado, R.; Gomez-Gil, P. Selecting and combining models with self-organizing maps for long-term forecasting of chaotic time series. In Proceedings of the 2014 International Joint Conference on Neural Networks, Beijing, China, 6–11 July 2014; pp. 2616–2623.

14. Simmons, L. Time-series decomposition using the sinusoidal model. *Int. J. Forecast.* **1990**, *6*, 485–495. [CrossRef]

15. Abdoos, A.; Hemmati, M.; Abdoos, A.A. Short term load forecasting using a hybrid intelligent method. *Knowl.-Based Syst.* **2015**, *76*, 139–147. [CrossRef]

16. De Gooijer, J.G.; Hyndman, R.J. 25 years of time series forecasting. *Int. J. Forecast.* **2006**, *22*, 443–473. [CrossRef]

17. Stock, J.H.; Watson, M.W. Combination forecasts of output growth in a seven-country data set. *J. Forecast.* **2010**, *23*, 405–430. [CrossRef]

18. Khairalla, M.; Xu, N.; Al-Jallad, N. Modeling and optimization of effective hybridization model for time-series data forecasting. *J. Eng.* **2018**. [CrossRef]

19. Hsiao, C.; Wan, S.K. Is there an optimal forecast combination? *J. Econom.* **2014**, *178*, 294–309. [CrossRef]

20. Barrow, D.; Crone, S. Dynamic model selection and combination in forecasting: An empirical evaluation of bagging and boosting. *Med. Phys.* **2011**, *25*, 435–443.

21. Barrow, D.K.; Crone, S.F.; Kourentzes, N. An evaluation of neural network ensembles and model selection for time series prediction. In Proceedings of the International Joint Conference on Neural Networks, Barcelona, Spain, 18–23 July 2010; pp. 1–8.

22. Chiroma, H.; Abubakar, A.I.; Herawan, T. Soft computing approach for predicting OPEC countries' oil consumption. *Int. J. Oil Gas Coal Technol.* **2017**, *15*, 298–316. [CrossRef]

23. Weron, R. Electricity price forecasting: A review of the state-of-the-art with a look into the future. *Int. J. Forecast.* **2014**, *30*, 1030–1081. [CrossRef]

24. Cincotti, S.; Gallo, G.; Ponta, L.; Raberto, M. Modeling and forecasting of electricity spot-prices: Computational intelligence vs. classical econometrics. *AI Commun.* **2014**, *27*, 301–314.

25. Amjady, N.; Keynia, F. Day ahead price forecasting of electricity markets by a mixed data model and hybrid forecast method. *Int. J. Electr. Power Energy Syst.* **2008**, *30*, 533–546. [CrossRef]

26. Azadeh, A.; Ghaderi, S.; Sohrabkhani, S. Forecasting electrical consumption by integration of neural network, time series and ANOVA. *Appl. Math. Comput.* **2007**, *186*, 1753–1761. [CrossRef]

27. Bianco, V.; Manca, O.; Nardini, S.; Minea, A.A. Analysis and forecasting of nonresidential electricity consumption in Romania. *Appl. Energy* **2010**, *87*, 3584–3590. [CrossRef]

28. Ediger, V.Ş.; Tatlıdil, H. Forecasting the primary energy demand in Turkey and analysis of cyclic patterns. *Energy Convers. Manag.* **2002**, *43*, 473–487. [CrossRef]
29. Manera, M.; Marzullo, A. Modelling the load curve of aggregate electricity consumption using principal components. *Environ. Model. Softw.* **2005**, *20*, 1389–1400. [CrossRef]
30. Ediger, V.Ş.; Akar, S. ARIMA forecasting of primary energy demand by fuel in Turkey. *Energy Policy* **2007**, *35*, 1701–1708. [CrossRef]
31. Mishra, P. Forecasting natural gas price-time series and nonparametric approach. In Proceedings of the World Congress on Engineering, 2012 Vol I WCE 2012, London, UK, 4–6 July 2012; Available online: http://www.iaeng.org/publication/WCE2012/WCE2012_pp490-497.pdf (accessed on 19 June 2018).
32. Zhang, G.P. Time series forecasting using a hybrid ARIMA and neural network model. *Neurocomputing* **2003**, *50*, 159–175. [CrossRef]
33. Lai, K.K.; Yu, L.; Wang, S.; Huang, W. Hybridizing exponential smoothing and neural network for financial time series predication. In *Proceedings of the International Conference on Computational Science, Reading, UK, 28–31 May 2006*; Springer: Berlin/Heidelberg, Germany, 2006; pp. 493–500.
34. Ma, H.; Zhang, Z. Grey prediction with Markov-Chain for Crude oil production and consumption in China. In *Proceedings of the Sixth International Symposium on Neural Networks (ISNN 2009), Wuhan, China, 26–29 May 2009*; Springer: Berlin, Germany, 2009; pp. 551–561.
35. Niska, H.; Hiltunen, T.; Karppinen, A.; Ruuskanen, J.; Kolehmainen, M. Evolving the neural network model for forecasting air pollution time series. *Eng. Appl. Artif. Intell.* **2004**, *17*, 159–167. [CrossRef]
36. Turanoglu, E.; Senvar, O.; Kahraman, C. Oil consumption forecasting in Turkey using artificial neural network. *Int. J. Energy Optim. Eng.* **2012**, *1*, 89–105. [CrossRef]
37. Ekonomou, L. Greek long-term energy consumption prediction using artificial neural networks. *Energy* **2010**, *35*, 512–517. [CrossRef]
38. Buhari, M.; Adamu, S.S. Short-term load forecasting using artificial neural network. In Proceedings of the International Multi-Conference of Engineers and Computer Scientist, Goa, India, 19–22 January 2012.
39. Nochai, R.; Nochai, T. ARIMA model for forecasting oil palm price. In Proceedings of the 2nd IMT-GT Regional Conference on Mathematics, Statistics and Applications, Penang, Malaysia, 13–15 June 2006; pp. 13–15.
40. Chiroma, H.; Abdulkareem, S.; Muaz, S.A.; Abubakar, A.I.; Sutoyo, E.; Mungad, M.; Saadi, Y.; Sari, E.N.; Herawan, T. An intelligent modeling of oil consumption. In *Advances in Intelligent Informatics*; Springer: Berlin, Germany, 2015; pp. 557–568.
41. Chiroma, H.; Khan, A.; Abubakar, A.I.; Muaz, S.A.; Gital, A.Y.U.; Shuib, L.M. Estimation of Middle-East Oil Consumption Using Hybrid Meta-Heuristic Algorithms. Presented at the Second International Conference on Advanced Data and Information Engineering, Bali, Indonesia, 25–26 April 2015.
42. Xia, Y.; Liu, C.; Da, B.; Xie, F. A novel heterogeneous ensemble credit scoring model based on bstacking approach. *Expert Syst. Appl.* **2018**, *93*, 182–199. [CrossRef]
43. Hansen, B.E. Interval forecasts and parameter uncertainty. *J. Econom.* **2006**, *135*, 377–398. [CrossRef]
44. Rubinstein, S.; Goor, A.; Rotshtein, A. Time series forecasting of crude oil consumption using neuro-fuzzy inference. *J. Ind. Intell. Inf.* **2015**, *3*, 84–90. [CrossRef]
45. Efendi, R.; Deris, M.M. Forecasting of malaysian oil production and oil consumption using fuzzy time series. In *Proceedings of the International Conference on Soft Computing and Data Mining, San Diego, CA, USA, 11–14 September 2016*; Springer: Berlin, Germany, 2016; pp. 31–40.
46. Efendi, R.; Deris, M.M. Prediction of Malaysian–Indonesian oil production and consumption using fuzzy time series model. *Adv. Data Sci. Adapt. Anal.* **2017**, *9*, 1750001. [CrossRef]
47. Aho, T.; Enko, B.; Eroski, S.; Elomaa, T. Multi-target regression with rule ensembles. *J. Mach. Learn. Res.* **2012**, *13*, 2367–2407.
48. Xu, M.; Golay, M. Survey of model selection and model combination. *SSRN Electron. J.* **2008**. [CrossRef]
49. Wang, J.-Z.; Wang, J.-J.; Zhang, Z.-G.; Guo, S.-P. Forecasting stock indices with back propagation neural network. *Expert Syst. Appl.* **2011**, *38*, 14346–14355. [CrossRef]
50. Ebrahimpour, R.; Nikoo, H.; Masoudnia, S.; Yousefi, M.R.; Ghaemi, M.S. Mixture of MLP-experts for trend forecasting of time series: A case study of the Tehran stock exchange. *Int. J. Forecast.* **2011**, *27*, 804–816. [CrossRef]

51. Miranian, A.; Abdollahzade, M. Developing a local least-squares support vector machines-based neuro-fuzzy model for nonlinear and chaotic time series prediction. *IEEE Trans. Neural Netw. Learn. Syst.* **2013**, *24*, 207–218. [CrossRef] [PubMed]

52. Kao, L.-J.; Chiu, C.-C.; Lu, C.-J.; Yang, J.-L. Integration of nonlinear independent component analysis and support vector regression for stock price forecasting. *Neurocomputing* **2013**, *99*, 534–542. [CrossRef]

53. Tofallis, C. Least squares percentage regression. *J. Mod. Appl. Stat. Methods* **2009**, *7*, 526–534. [CrossRef]

54. Kianimajd, A.; Ruano, M.G.; Carvalho, P.; Henriques, J.; Rocha, T.; Paredes, S.; Ruano, A.E. Comparison of different methods of measuring similarity in physiologic time series. *IFAC-PapersOnLine* **2017**, *50*, 11005–11010. [CrossRef]

![energies logo] *energies*

MDPI

Article

A Novel Hybrid Interval Prediction Approach Based on Modified Lower Upper Bound Estimation in Combination with Multi-Objective Salp Swarm Algorithm for Short-Term Load Forecasting

Jiyang Wang [1], Yuyang Gao [2,*] and Xuejun Chen [3]

1   Faculty of Information Technology, Macau University of Science and Technology, Macau 999078, China;
    17098531i011001@student.must.edu.mo
2   School of Statistics, Dongbei University of Finance and Economics, Dalian 116025, China
3   Gansu Meteorological Service Centre, Lanzhou 730020, China; xuejunchen1971@163.com
*   Correspondence: gaoyuyang@hotmail.com; Tel.: +86-18340831947

Received: 25 May 2018; Accepted: 10 June 2018; Published: 14 June 2018

Abstract: Effective and reliable load forecasting is an important basis for power system planning and operation decisions. Its forecasting accuracy directly affects the safety and economy of the operation of the power system. However, attaining the desired point forecasting accuracy has been regarded as a challenge because of the intrinsic complexity and instability of the power load. Considering the difficulties of accurate point forecasting, interval prediction is able to tolerate increased uncertainty and provide more information for practical operation decisions. In this study, a novel hybrid system for short-term load forecasting (STLF) is proposed by integrating a data preprocessing module, a multi-objective optimization module, and an interval prediction module. In this system, the training process is performed by maximizing the coverage probability and by minimizing the forecasting interval width at the same time. To verify the performance of the proposed hybrid system, half-hourly load data are set as illustrative cases and two experiments are carried out in four states with four quarters in Australia. The simulation results verified the superiority of the proposed technique and the effects of the submodules were analyzed by comparing the outcomes with those of benchmark models. Furthermore, it is proved that the proposed hybrid system is valuable in improving power grid management.

Keywords: short-term load forecasting; interval prediction; lower upper bound estimation; artificial intelligence; multi-objective optimization algorithm; data preprocessing

---

1. Introduction

Load forecasting is of upmost significance and affects the construction and operation of power systems. In the preparation of the power system planning stage, if the load forecasting result is lower than the real demand, the installed and distribution capacities of the planned power system will be insufficient. The power generated will not be able to meet electricity demand of the community it serves, and the entire system will not be able to operate in a stable manner. Conversely, if the load forecast is too high, it will result in power generation, transmission, and distribution, at a larger scale, that cannot be fully used in the real power system. The investment efficiency and the efficiency of the resource utilization will be reduced in this situation. Therefore, effective and reliable power load forecasting can promote a balanced development of the power system while improving the utilization of energy. There are various power load forecasting methods and, commonly, load forecasting is classified into short-term, medium-term, and long-term, based on the application field and forecasting time. Among these categories, short-term load forecasting (STLF) is an essential tool for the planning and operation [1,2] of energy systems and it has thus been a major area of research during the past few decades.

According to existing research, concern mostly focuses on the point forecasting of STLF. Additionally, the relative algorithms can be mainly classified into three major categories: traditional statistical techniques, computational intelligent methods, and multimodule hybrid models [3].

In the early stages of research, traditional statistical techniques were extensively employed for point forecasting of STLF, such as linear regression methods [4,5], exponential smoothing [6], Kalman filters [7], and other linear time-series models. In general, most of the traditional statistical approaches have been involved in linear analysis and have mainly considered linear factors in time series. However, the short-term load series are a mixture of multiple components which include linear and non-linear factors. Therefore, the traditional statistical approaches encounter difficulties when dealing with the STLF, and the forecasting accuracy is often unsatisfactory. With the development of machine learning and artificial intelligence, an increased number of non-linear computational intelligent methods have been applied to STLF, such as neural network models (NN) [8,9], expert systems [10] and support vector machines (SVM) [11,12]. These approaches have been proved to have advantages in dealing with the non-linear problems of STLF compared to traditional statistical methods, thereby eliciting improved performances in most cases. Most importantly, a key point that influences the performance of computational intelligent methods is the setting of related parameters in algorithms. At this time, efficient hybrid models appeared. In hybrid models, different modules were introduced to improve the performance and accuracy of STLF [13–19]. Among existing reviews in the literature, two popular and efficient modules include the data preprocessing and optimization modules. In the case of the data preprocessing modules, a multiwavelet transform was used in combination with a three-layer feed-forward neural network to extract the training data and predict the load in [13]. Fan et al. [14] used empirical mode decomposition (EMD) to decompose electric load data, generating high-frequency series and residuals for the forecasting of support vector regression (SVR) and autoregression (AR), respectively. The results showed that the hybrid methods can perform well by eliciting good forecasting accuracy and interpretability. In the case of the optimization modules, AlRashidi et al. [15] employed the particle swarm optimizer (PSO) to fine-tune the model parameters, and the forecasting problem was presented in a state space form. Wang et al. [16] proposed a hybrid forecasting model combining differential evolution (DE) and support vector regression (SVR) for load forecasting, where the DE algorithm was used to choose the appropriate parameters for SVR.

However, as mentioned above, the current research on STLF mainly concentrates on point forecasting in which the accuracy is usually measured by the errors between the predicted and the target values. With power system growth and the increase in its complexity, point forecasting might not be able to provide adequate information support for power system decision making. An increasing number of factors, such as load management, energy conversion, spot pricing, independent power producers and non-conventional energy, make point forecasting undependable in practice. In addition to the fact that most of these point forecasting models do not elicit the required precision, they are also not adequately robust. They fail to yield accurate forecasts when quick exogenous changes occur. Other shortcomings are related to noise immunity, portability, and maintenance [20].

In general, point forecasting cannot properly handle uncertainties associated with load datasets in most cases. To avoid such imperfection, interval prediction (IP) of STLF is an efficient way to deal with the forecast uncertainty in electrical power systems. Prediction intervals (PIs) not only provide a range in which targets are highly likely to be covered, but they also provide an indication of their accuracy, known as the coverage probability. Furthermore, the PIs can take into account more uncertain information and the result of (PIs) commonly form a double output (upper bounds and lower bounds) which can reflect more uncertain information and provide a more adequate basis for power system planning.

With the development of artificial intelligence technology, the interval prediction methods based on NN have been proved to be efficient techniques. According to existing research, the popular techniques for constructing PIs are Bayesian [21], delta [22], bootstrap [23], and mean–variance estimation [24]. In the literature, the Bayesian technique [25] is used for the construction of NN-based PIs. Error bars are assigned to the predicted NN values using the Bayesian technique. Even if the theories are effective in the construction of PIs, the calculation of the Hessian matrix will result in the increase of model complexity and computation cost. In [26], the delta technique was applied to construct PIs for STLF, and a simulated annealing (SA) algorithm was introduced to improve the performance of PIs through the minimization of a loss function. In [27], according to bootstrap, error output, resampling, and multilinear regression, were used with STLF for the construction of confidence intervals with NN models. In [24], a mean–variance estimation-based method used NN to estimate the characteristics of the conditional target distribution. Additive Gaussian noise with non-constant variance was the key assumption of the method for PI construction.

Considering most of the existing research studies of PIs by NN mentioned above, the PIs were usually calculated depending on the point forecasting. The NNs were first trained by minimizing an error-based cost function, and the PIs were then constructed depending on the outcomes of trained and tuned NNs. It may be questionable to construct PIs in this way. Furthermore, it is a more reasonable way to output the upper and lower bounds directly [28]. Compared with the Bayesian, delta, and bootstrap techniques, this approach can output the PIs without being dependent on point prediction. However, in traditional research approaches, the cost function mainly aims at guaranteeing coverage probability (CP). However, a satisfactory coverage probability can be achieved easily by assigning sufficiently large and small values to the upper and lower bounds of the PIs. Thus, the prediction interval width (PIW) is another key characteristic which needs to be considered fully. These two goals, that is, achieving a higher CP and a lower PIW, should be considered in a comprehensive manner when the NN parameters are determined.

Therefore, in this study, a hybrid, lower upper bound estimation (LUBE) based on multi-objective optimization is proposed. The requirements for higher CP and lower PIW constitute a typical case of the Pareto optimization problem. In the present study, a significant and valid approach was used to solve the Pareto optimization problem is the multi-objective optimization [29]. There are many algorithms in the literature for solving multi-objective optimizations. For the GA, the most well-regarded multi-objective algorithm is the non-dominated sorting genetic algorithm (NSGA) [30]. Other popular algorithms include the multi-objective particle swarm optimization (MOPSO) [31,32], multi-objective ant colony optimization (MOACO) [33], multi-objective differential evolution (MODE) [34], multi-objective grasshopper optimization (MOGO) [35], multi-objective evolution strategy (MOES) [36], multi-objective sine cosine (MOSC) [37], and multi-objective ant lion [38]. All these algorithms are proved to be effective in identifying non-dominated solutions for multi-objective problems. According to the "no free lunch theorem" for optimization [39,40], there is no algorithm capable of solving optimization algorithms for all types of problems. This theorem logically proves this and proposes new algorithms, or improves the current ones.

In this study, to achieve a better performance in STLF, one of the novel recurrent neural networks, the Elman neural network (ENN) [41], is applied to construct the structure of a modified LUBE. The Elman neural network has already been extensively used in time-series forecasting [42–44]. As a type of recurrent neural network, ENN exhibits superiority on the time delay information because of the existence of the undertaking layer which can connect hidden NN layers and store the historical information in the training process. This structure design of NN commonly leads to a better performance in time-series forecasting.

In traditional STLF, most of the methods construct the training set of the model directly using the original data. However, data in the natural world often receives a lot of noise interference, which will cause more difficulties for desired STLF. Furthermore, improving the signal-to-noise ratio of the training dataset will help the effective training of the model. Amongst the existing denoising methods, empirical mode decomposition (EMD) [45] is extensively used, which is an adaptive method introduced to analyze non-linear and non-stationary signals. In order to alleviate some reconstruction problems, such as "mode mixing" of EMD, some other versions [46–48] are proposed. Particularly, the problem of different number of modes for different realizations of signal and noise need to be considered.

Summing up the above, in this study, a hybrid interval prediction system is proposed to solve the STLF problem based on the modified Lower and Upper bound estimate (LUBE) technique, by incorporating the use of a data preprocessing module, an optimization module, and a prediction module. In order to verify the performance of the proposed model, we choose as the experimental case the power loads of four states in Australia. The elicited results are compared with those from basic benchmark models. In summary, the primary contributions of this study are described below:

(1) *A modified LUBE technique is proposed based on a recurrent neural network, which is able to consider previous information of former observations in STLF.* The contest layer of ENN can store the outputs of a former hidden layer, and then connect the input layer in the current period. Comparison of the traditional interval predictive model with the basic neural network, this mechanism can improve the performance of time series forecasting methods, such as STLF.

(2) *A more convincing optimization technique based on multi-objective optimization is proposed for LUBE.* In LUBE, besides CP, PIW should also be considered in the construction of the cost function. In this study, the novel multi-objective optimization method MOSSA is employed in the optimization module to balance the conflict between higher CP and lower PIW, and to train the parameters in ENN. With this method, the structure of neural networks can provide a better performance in interval prediction.

(3) *A novel and efficient data preprocessing method is introduced to extract the valuable information from raw data.* In order to improve the signal noise ratio (SNR) of the input data, an efficient method is used to decompose the raw data into several empirical modal functions (IMFs). According to the entropy theory, the IMFs with little valuable information are ignored. The performance of the proposed model trained with processed data improves significantly.

(4) *The proposed hybrid system for STLF can provide powerful theoretical and practical support for decision making and management in power grids.* This hybrid system is simulated and tested depending on the abundant samples involving different regions and different times, which indicate its practicability and applicability in the practical operations of power grids compared to some basic models.

The rest of this study is organized as follows: The relevant methodology, including data preprocessing, Elman neural network, LUBE, and multi-objective algorithms, are introduced in Section 2. Section 3 discusses our proposed model in detail. The specific simulation, comparisons and analyses of the model performances are shown in Section 4. In order to further understand the features of the proposed model, several points are discussed in Section 5. According to the results of our research, conclusions are outlined in Section 6.

2. Methodology

In this section, the theory of the hybrid interval prediction model is elaborated, and the methodology of the components in hybrid models, including complete ensemble empirical mode decomposition with adaptive noise (CEEMDAN), Elman neural networks, LUBE, and MOSSA, are explained in detail.

2.1. Data Preprocessing

The EMD technique [45] usually decomposes a signal into several numbers of IMFs. For each IMF, the series have to fulfill two conditions: (i) the number of extrema (maxima and minima) and the number of zero-crossings must be equal or differ at most by one; and (ii) the local mean, defined as the mean of the upper and lower envelopes, must be zero. In order to alleviate mode mixing, the EEMD [46], defines the "true" modes as the average of the corresponding IMFs obtained from an ensemble of the original signal plus different realizations of finite variance white noise. But incompletion of decomposition still exists, and the number of modes will be different due to the noise added. Taking these short comes into account, CEEMDAN is proposed. The details are described as follows: let $E_k(\cdot)$ be the operator which produces the kth mode obtained by EMD and $w^{(i)}$ be a realization of white noise with $N(0,1)$. And then the process of CEEMDAN can be expressed as several stages:

1st step. For every $i = 1, \ldots, I$ decompose each $x^{(i)} = x + \beta_0 w^{(i)}$ by EMD, until the first mode is extracted and compute $\tilde{d}_1$ by:

$$\tilde{d}_1 = \frac{1}{I} \sum_{i=1}^{I} d_1^i = \overline{d}_1 \tag{1}$$

2nd step. At the first stage ($k = 1$) calculate the first residue as $r_1 = x - \tilde{d}_1$.
3rd step. Obtain the first mode of $r_1 + \beta_1 E_1(w^i)$, $i = 1, \ldots I$, by EMD and define the second CEEMDAN mode as:

$$\tilde{d}_2 = \frac{1}{I} \sum_{i=1}^{I} E_1(r_1 + \beta_1 E_1(w^{(i)})) \tag{2}$$

4th step. For $k = 2, \ldots K$ calculate the kth residue:

$$r_k = r_{(k-1)} - \tilde{d}_k \tag{3}$$

5th step. Obtain the first mode of $r_k + \beta_k E_k(w^{(i)})$, $i = 1, \ldots, I$, by EMD until define the $(k+1)$th CEEMDAN mode as:

$$\tilde{d}_{(k+1)} = \frac{1}{I} \sum_{i=1}^{I} E_1(r_k + \beta_k E_k(w^{(i)})) \tag{4}$$

6th step. Go to 4th step for the next k.

Iterate the steps 4 to 6 until the obtained residue cannot be further decomposed by EMD, either because it satisfies IMF conditions or because it has less than three local extremums. Observe that, by construction of CEEMDAN, the final residue satisfies:

$$r_K = x - \sum_{k=1}^{K} \tilde{d}_k \tag{5}$$

with K being the total number of modes. Therefore, the signal of interest x can be expressed as:

$$x = \sum_{k=1}^{K} \tilde{d}_k + r_k \tag{6}$$

which ensures the completeness property of the proposed decomposition and thus providing an exact reconstruction of the original data. The final number of modes is determined only by the data and the stopping criterion. The coefficients $\beta_k = \varepsilon_k std(r_k)$ allow the selection of the SNR at each stage.

The CEEMDAN method can add a limited number of self-use white noises at each stage, which can achieve almost zero reconstruction error with fewer average times. Therefore, CEEMDAN can overcome the "mode-mixing" phenomenon existing in EMD, and can also solve the incompleteness of EEMD decomposition and reduce the computational efficiency by reducing the reconstruction error by increasing the number of integrations.

2.2. Elman Neural Network (ENN)

As an important branch of deep learning, recurrent neural networks have been widely used in academic and industrial fields. The common neural network mainly consists of three layers: input layer, hidden layer and output layer. For the hidden layer, the input information only comes from the input layer. For a recurrent neural network, the input information of the hidden layer will not only come from the input layer, but also from the hidden layer itself and the output layer.

In various structures of the recurrent neural network, Elman neural network (ENN) [49] is typical structure in which the lags of hidden layer are delivered into the current hidden layer by a new layer called the context layer. This structure takes the former information of the hidden layer into account and commonly has a better performance in the time-series forecasting such as STLF, wind speed forecasting, financial time-series forecasting. The structure is showed in Figure 1.

The context layer can feed back the hidden layer outputs in the previous time steps and neurons contained in each layer are used to transmit information from one layer to another. The dynamics of the change in hidden state neuron activations in the context layer is as follows:

$$S_i(t) = g(\sum_{k=1}^{K} V_{ik} S_k(t-1) + \sum_{j=1}^{j} W_{ij} I_j(t-1))$$

(7)

where $S_k(t)$ and $I_j(t)$ denote the output of the context state and input neurons, respectively; V_{ik} and W_{ij} denote their corresponding weights; and $g(\cdot)$ is a sigmoid transfer function. The other related theories such as feed-forward and back propagation are similar with the common back propagation neural network.

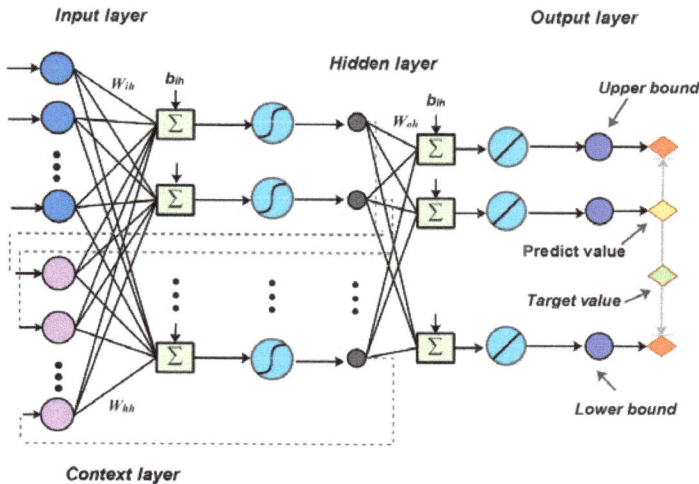

Figure 1. The structure of the lower bound and upper bound estimation (LUBE) based on the Elman neural network.

2.3. Lower Bound and Upper Bound Estimation (LUBE)

In the literature, the traditional interval prediction commonly attempts to construct the PI based on the point prediction. The upper bound and the lower bound are calculated according to the forecasting value and the confidence level. The accuracy of the point forecasting has played a key role in the accuracy of the PI. In this paper, we introduce a novel method of interval prediction called lower bound and upper bound estimation (LUBE). This method directly outputs the lower bound and the upper bound of PI depending on the multi-output neural network. The structure we employed in this paper is shown in Figure 1.

The output of the normal LUBE structure [50] just consist of two neurons which denote the upper bound and the lower bound, while the outputs in our structure of LUBE consist of three neurons. The first output corresponds to the upper bound of the PI, the second output denotes the predicted value, and third output approximates the lower bound of the PI. In the literature, the PI construction techniques attempt to estimate the mean and variance of the targets for construction of PIs. In contrast to existing techniques, the proposed method tries to directly approximate upper and lower bounds of PIs based on the set of inputs. Therefore, in the training process, loss function of this LUBE method based on neural network should be set according to the key criterion of PIs (CP and PIW).

2.4. Multi-Objective Optimization Algorithm

The multi-objective optimization algorithm has been widely used to solve multi-objective optimization problem. In this paper, a novel multi-objective optimization algorithm named Multi-Objective Salp Swarm Algorithm (MOSSA) is introduced.

2.4.1. Multi-Objective Optimization Problem

In multi-objective optimization, all of the objectives are optimized simultaneously. The main concern is formulated as follows:

$$Minimize : F(X) = \{f_1(X), f_2(X), \ldots, f_o(X)\} \tag{8}$$

$$Subject \quad to : g_i(X) \geq 0, i = 1, 2, \ldots, m \tag{9}$$

$$h_i(X) = 0, i = 1, 2, \ldots, p \tag{10}$$

$$lb_i \leq x_i \leq ub_i, i = 1, 2, \ldots, n \tag{11}$$

where o is the number of objectives, m is the number of inequality constraints, p is the number of equality constraints, and lb_i is the lower bound of the ith variable, and ub_i is the upper bound of the ith variable. With one objective we can confidently estimate that a solution is better than another depending on comparing the single criterion, while in a multi-objective problem, there is more than one criterion to compare solutions. The main theory to compare two solutions considering multiple objectives is called Pareto optimal dominance as explained in [51].

There are two main approaches for solving multi-objective problems: a priori and a posteriori [52]. In the priori method, the multi-objective problem is transformed to a single-objective problem by aggregating the objectives with a set of weights determined by experts. The main defect of this method is that the Pareto optimal set and the front need to be constructed by re-running the algorithm and changing the weights [53]. However, the a posteriori method keeps the multi-objective formulation in the solving process, and the Pareto optimal set can be determined in a single run. Without any weight to be defined by experts, this approach can approximate any type of Pareto optimal front. Because of the advantages of a posteriori optimization over the a priori approach, the focus of our research is aimed at a posteriori multi-objective optimization.

2.4.2. Multi-Objective Salp Swarm Algorithm (MOSSA)

As an a posteriori multi-objective optimization, MOSSA [54] is similar to some swarm multi-objective optimization algorithm such as MOPSO [31], MOACO [33] and MOGO [35]. By simulating the biological behavior of ecological communities, the optimal solution is achieved.

Salps belong to the family of Salpidae and have transparent barrel-shaped body. Their tissues are highly similar to jellyfishes. They also move very similar to jellyfish, in which the water is pumped through body as propulsion to move forward. In deep oceans, salps often form a swarm called a salp chain. The main concern about salps in MOSSA is their swarming behavior.

To mathematically model the salp chains, the population is first divided to two groups: leader and followers. The leader is the salp at the front of the chain, whereas the rest of salps are considered as followers. As the name of these salps implies, the leader guides swarm and the followers follow each other.

Similar to other swarm-based techniques, the position of salps is defined in an n-dimensional search space where n is the number of variables of a given problem. Therefore, the positions of all salps are stored in a two-dimensional matrix called x. It is also assumed that there is a food source called F in the search space as the swarm's target.

Definition 1. *To update the position of the leader the following equation is proposed:*

$$x_j^1 = \begin{cases} F_j + c_1\left((ub_j - lb_j)c_2 + lb_j\right) & c_3 \geq 0 \\ F_j - c_1\left((ub_j - lb_j)c_2 + lb_j\right) & c_3 < 0 \end{cases} \tag{12}$$

where x_j^1 shows the position of the first salp (leader) in the jth dimension, F_j is the position of the food source in the jth dimension, ub_j indicates the upper bound of jth dimension, lb_j indicates the lower bound of jth dimension, c_1, c_2, and c_3 are random numbers. Equation (12) shows that the leader only updates its position with respect to the food source.

Definition 2. *The coefficient c_1 is the most important parameter in the Salp swarm algorithm (SSA) because it balances exploration and exploitation is defined as follows:*

$$c_1 = 2e^{-\left(\frac{4l}{L}\right)^2} \tag{13}$$

where l is the current iteration and L is the maximum number of iterations.

The parameter c_2 and c_3 are random numbers uniformly generated in the interval of [0, 1]. In fact, they dictate if the next position in jth dimension should be towards positive infinity or negative infinity as well as the step size.

Definition 3. *To update the position of the followers, the following equations is utilized depending on Newton's law of motion:*

$$x_j^i = \frac{1}{2}a_{ij}t^2 + v_0 t \tag{14}$$

where $i \geq 2$, x_j^i shows the position of ith follower salp in jth dimension, t is time, v_0 is the initial speed, and $a_{ij} = \frac{v_{ij} - v_0}{t}$ where $v_{ij} = \frac{x_{ij} - x_0}{t}, i \geq 2, j \geq 1$.

Because the time in optimization is iteration, the discrepancy between iterations is equal to 1, and considering $v_0 = 0$, this equation can be expressed as follows:

$$x_{j(t)}^i = \frac{1}{2}\left(x_{j(t-1)}^i + x_{j(t-1)}^{i-1}\right) \tag{15}$$

where $i \geq 2$ and $x_{j(t)}^i$ show the position of ith follower salp in jth dimension at t-th iteration.

According to the mathematical emulation explained above, the swarm behavior of salp chains can be simulated vividly.

When dealing with multi-objective problems, there are two issues that need to be adjusted for SSA. First, MOSSA need to store multiple solutions as the best solutions for a multi-objective problem. Second, in each iteration, SSA updates the food source with the best solution, but in the multi-objective problem, single best solutions does not exist.

In MOSSA, the first issue is settled by equipping the SSA algorithm with a repository of food sources. The repository can store a limited number of non-dominated solutions. In the process of optimization, each salp is compared with all the residents in repository using the Pareto dominance operators. If a salp dominates only one solution in the repository, it will be swapped. If a salp dominates a set of solutions in the repository, they all should be removed from the repository and the salp should be added in the repository. If at least one of the repository residents dominates a salp in the new population, it should be discarded straight away. If a salp is non-dominated in comparison with all repository residents, it has to be added to the archive. If the repository becomes full, we need to remove one of the similar non-dominated solutions in the repository. For the second issue, an appropriate way is to select it from a set of non-dominated solutions with the least crowded neighborhood. This can be done using the same ranking process and roulette wheel selection employed in the repository maintenance operator. The pseudo code of MOSSA is showed in Algorithm 1:

Algorithm 1. Pseudo-code of MOSSA.

---

1 *Set* the hyper-parameter:
2 Max_iter: Maximum of iteration
3 ArchiveMaxSize: Max capacity of archive (repository)
4 Dim: The number of parameters on each salp
5 *Ub* and *lb*:   The upper bound and the lower bound of salp population
6 Obj_no: The objective number to be estimated
7 Initialize the salp population $x_i (i = 1, 2, ..., n)$ depending on the *ub* and *lb*
8 Define the objective function (loss function): @ Ob_func
9 *While* (end criterion is not met)
10   Calculate the fitness of each search agent (salp) with Ob_func
11   Determine the non-dominated salps
12   Update the repository considering the obtained non-dominated salps
13   *If* (the repository become full)
14   Call the repository maintenance procedure to remove one repository resident
15   Add the non-dominated salp to the repository
16   *End If*
17   Choose a source of food from repository: F = SelectFood (repository)
18   Update *c1* by $c_1 = 2e^{-\left(\frac{4l}{L}\right)^2}$
19   *For* each salp (x_i):
20   *If* (i==1):
            Update the position of the leading salp by:
21             $x_j^1 = \begin{cases} F_j + c_1\left(\left(ub_j - lb_j\right)c_2 + lb_j\right) & c_3 \geq 0 \\ F_j - c_1\left(\left(ub_j - lb_j\right)c_2 + lb_j\right) & c_3 < 0 \end{cases}$
22   *Else*:
23   Update the position of the leading salp by: $x_{j(t)}^i = \frac{1}{2}\left(x_{j(t-1)}^i + x_{j(t-1)}^{i-1}\right)$
24   *End If*
25   *End For*
26   Amend the salps based on the upper and lower bound of variables ·
27 *End While*
28 Return repository

---

3. Proposed Interval Prediction Model for Short-Term Load Forecasting (STLF)

In this paper, we proposed a hybrid model for interval prediction based on the data preprocessing, multi-objective optimization algorithm and LUBE to solve the problem of STLF. This hybrid model consist of two stages: data de-noising and model prediction.

In the first stage, the main task is to refine the original data. The raw power load data is affected by many internal and external factors in the collection process. Therefore, a lot of unrelated information is integrated in the data. Several pieces of information will further affect the quality of the power load data, and increase the difficulty of accurate forecasting of the power load. In the neural network model, the performance of the model is directly affected by the quality of the data. As a type of machine learning algorithm, the neural network uses its multilayered structure to learn the relevant interdependencies of the data and determine the structural parameters of the prediction model, so as to achieve fitting and forecasting. However, if the input set of the model contains too much noise and "false information", the model will be seriously affected in the training process, and some problems will emerge, such as the overfitting problem. Therefore, we introduced CEEMDAN to eliminate useless information in the raw data. As mentioned above, CEEMDAN can decompose the data series into several IMFs with different frequencies, as shown in Figure 2. Because the IMFs are extracted with envelope curves depending on the extremum, some of the IMFs have higher frequencies, just as the first few IMFs that are shown in Figure 2. In addition, the other IMFs also have lower frequencies and represent the trend factors, thereby formulating the vital basis for time-series prediction. In the actual operations, we can remove the IMFs with higher frequencies, which effectively represent noise to refine the original data. In order to determine which IMFs ought to be abandoned, we calculated the entropy of each IMF and removed the IMFs with lower entropy. After the denoising process, the refined data are transferred to next stage as the input data for training in the predictive model.

In the second stage, the main interval prediction model was proposed. In our hybrid interval prediction model, the PI is output dependent on LUBE, which is based on the multi-output of the Elman neural network (E–LUBE). In the training process, the input set of E–LUBE is constructed as indicated in Formula (16), while the output set is constructed as indicated in Formula (17), where m and s respectively denote the number of features and the numbers of samples, and α denotes the interval width coefficient. In the case of the STLF problem, m indicates the number of previous time-points that we use to forecast the predictive value.

$$\text{Input set}: \begin{bmatrix} x_1 & x_2 & \cdots & x_m \\ x_2 & x_3 & \cdots & x_{m+1} \\ \vdots & \vdots & \ddots & \vdots \\ x_s & x_{s+1} & \cdots & x_{s+m} \end{bmatrix} \tag{16}$$

$$\text{Output set}: \begin{bmatrix} x_{m+1} \times (1-\alpha) & x_{m+1} & x_{m+1} \times (1+\alpha) \\ x_{m+2} \times (1-\alpha) & x_{m+2} & x_{m+2} \times (1+\alpha) \\ \vdots & \vdots & \vdots \\ x_{m+s+1} \times (1-\alpha) & x_{m+s+1} & x_{m+s+1} \times (1+\alpha) \end{bmatrix} \tag{17}$$

According to a trained model, when a new series X_i, $i = 1, \ldots, m$, is input into the model, X_{m+1} with an upper bound X_{m+1}^U and a lower bound X_{m+1}^L will be output. This is the basic mechanism of interval prediction for STLF in this study. However, in traditional multi-output neural networks, the loss function is always the mean-square-error (MSE), which is a key criterion for point forecasting. In this study, we introduced two new criteria (PIW and CP) to construct the loss function, considering the main purpose of our interval prediction. The traditional neural network parameters were determined by using a gradient descent algorithm, but for two of the set criteria, the calculation of the gradient was difficult. Therefore, we employed MOSSA to realize the multi-objective parameter optimization. Furthermore, the optimization problem can be expressed as,

$$argmin \quad \begin{cases} \text{PIW}(\theta) \\ 1/\text{CP}(\theta) \end{cases} \tag{18}$$

where θ is a set of parameters in E–LUBE, including the weight and bias.

When the parameters are determined in the training process, the entire model can be applied to the test set to verify the performance of interval prediction.

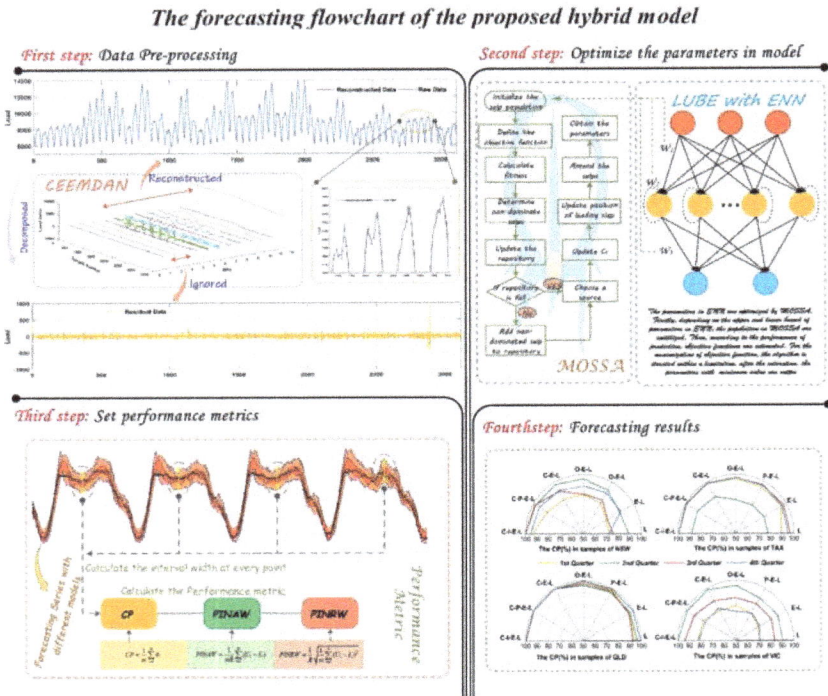

Figure 2. Forecasting flowchart of the proposed hybrid model.

4. Simulations and Analyses

In order to validate the performance of the proposed hybrid model in STLF, four electrical load datasets collected from four states in Australia are used in our research. The four states include New South Wales (NSW), Tasmania (TAX), Queensland (QLD) and Victoria (VIC), and the specific location is showed in Figure 3. The experiments in this study consist of two parts: experiment I and experiment II. For experiment I, the load data of four states are modeled with interval width coefficient $\alpha = 0.05$, and for the experiment II, the interval width coefficient α is set as 0.025 for further analysis. In order to verify the superiority of the proposed hybrid model, several benchmark models which include basic LUBE (LUBE), LUBE with Elman neural network (E–LUBE), E–LUBE with point optimization (PO–E–LUBE), E–LUBE with interval optimization (IO–E–LUBE), and models integrated with CEEMDAN, are exhibited. For persuasive comparability and fairness, the hyper-parameters in each model are consistent, as shown in Table 1. All experiments have been carried out in MATLAB

2016a on a PC with the configuration of Windows 7 64-bit, Inter Core i5-4590 CPU @ 3.30GHz, 8GB RAM.

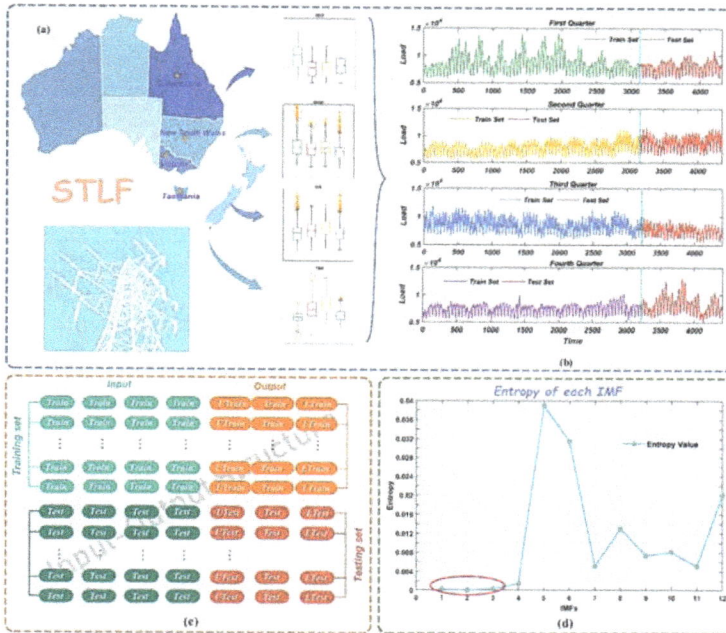

Figure 3. Data description of experiments. (**a**) Location of sample sites; (**b**) Division of train set and test set; (**c**) Structure of input set and output set; and (**d**) Entropy of each IMF).

Table 1. Related parameters in hybrid model.

Submodels and Parameters	Value
Elman Neural Network (ENN)	
Inputnum	6
Hiddennum	13
Outputnum	3
Train.epoch	500
Train.lr	0.1
Train.func	"Adam"
Complete ensemble empirical mode decomposition with adaptive noise (CEEMDAN)	
Nstd	0.2
NR	200
Maxiter	100
Multi-objective salp swarm algorithm (MOSSA)	
Dim	754
Lb	−2
Ub	2
Obj_no	2
Pop_num	50

4.1. Data Descriptions

For each state, we considered the data using half an hour interval in four quarters. The data used in this paper can be obtained on the website of Australian energy market operator (http://www. aemo.com.au/). We chose data from the whole of 2017 from 1 January 2017 0:30 am to 31 December 2017 0:00 am to construct dataset. In each state, the total sample number is 17,520. For each quarter, the number of samples were 4320, 4358, 4416, 4416 respectively. In order to control the comparability, we selected 1200 samples to test the trained model, and used the rest in each quarter to train the models. The proportion of train sets versus the test sets was approximately equal to 3:1. The description of the data characteristics are shown in Figure 4. Considering the structure of the neural network in this study, we set six input neurons, 13 hidden neurons, and three output neurons. Specifically, the output set was formulated in accordance with Formula (17).

During data preprocessing, the input data were divided into several IMFs depending on CEEMDAN, as displayed in Figure 2. According to the energy entropy of each IMF shown in Figure 3, we ignored the IMFs which contained high frequencies, and summed the rest of the IMFs to reconstruct the input set, as shown in Figure 1.

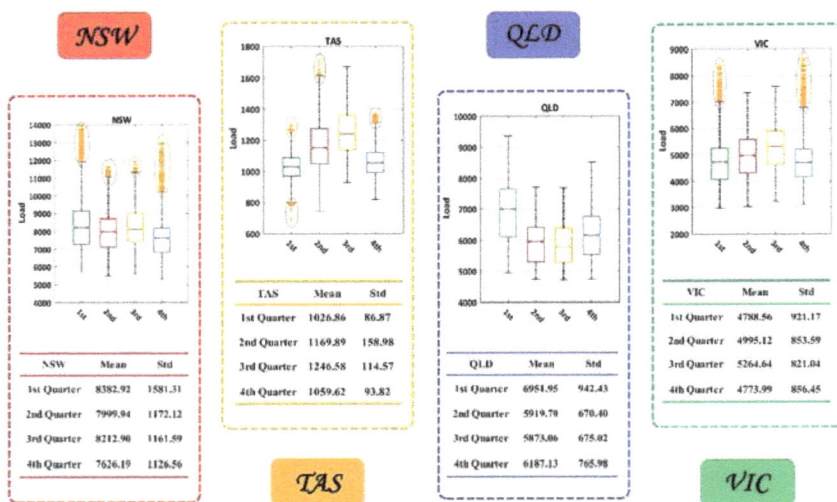

Figure 4. Boxplot of the entire set of data samples.

4.2. Performance Metrics

In order to comprehensively assess the performance of the models, some metrics were employed. These metrics primarily focused on the coverage of the real value in the prediction interval and the width of the interval.

4.2.1. Coverage Probability

Coverage probability [50] is usually considered as a basic feature of PIs and CP is calculated according to the ratio of the number of target values covered by PIs:

$$CP = \frac{1}{m}\sum_{i=1}^{m} \theta_i \tag{19}$$

where m denotes the number of samples, and θ_i is a binary index which measures whether the target value is covered by PIs:

$$\theta_i = \begin{cases} 1 & , \quad y_i^t \in [\hat{L}_i, \hat{U}_i] \\ 0 & , \quad y_i^t \notin [\hat{L}_i, \hat{U}_i] \end{cases} \tag{20}$$

where y_i^t denote the ith target value and $\hat{L}_i$, $\hat{U}_i$ represent the ith lower bound and the upper bound, respectively.

A larger CP means more targets are covered by the constructed PIs and a too small CP indicates the unsatisfied coverage behaviors. To have valid PIs, CP should be larger or at least equal to the nominal confidence level of PIs. Furthermore, in this paper, CP is also an important factor in the process of parameter optimization by the multi-objective optimization algorithm.

4.2.2. Prediction Interval (PI) Normalized Average width and PI Normalized Root-Mean-Square Width

In research studies on interval prediction, more attention is usually paid to CP. However, if the lower and upper bounds of the PIs are expanded from either side, any requirement for a larger CP can be satisfied, even for 100%. However, in some cases, a narrower interval width is necessary for a more precise support for electric power supply. Therefore, the width between the lower and upper bounds should be controlled so that the PIs are more convincing. In this study, the prediction interval width (PIW) is another factor in the process of parameter optimization. With CP and PIW, two objects compose the solution space within which the Pareto solution set is estimated.

In order to eliminate the impact of dimension, some relative indexes should be introduced to improve the comparability of width indicators. Inspired by the mean absolute percentage error (MAPE) in point forecasting, we employed PI normalized average width (PINAW) and PI normalized root-mean-square width (PINRW) [50]:

$$PINAW = \frac{1}{mR} \sum_{i=1}^{m} (U_i - L_i) \tag{21}$$

$$PINRW = \frac{1}{R} \sqrt{\frac{1}{m} \sum_{i=1}^{m} (U_i - L_i)^2} \tag{22}$$

where R equals to the maximum minus minimum of the target values. Normalization by the range R is able to improve comparability of PIs constructed using different methods and for different types of datasets.

4.2.3. Accumulated Width Deviation (AWD)

Accumulated width deviation (AWD) is a criterion that measure the relative deviation degree, and it can be obtained by the cumulative sum of AWD_i [55]. The calculation formula of AWD is expressed as Equations (23) and (24), where α denotes the interval width coefficient and I_i represents the i-th prediction interval.

$$AWD_i = \begin{cases} \frac{L_i^{(\alpha)} - z_i}{U_i^{(\alpha)} - L_i^{(\alpha)}}, z_i < L_i^{(\alpha)} \\ 0, \qquad z_i \in I_i^{(\alpha)} \\ \frac{z_i - U_i^{(\alpha)}}{U_i^{(\alpha)} - L_i^{(\alpha)}}, z_i > U_i^{(\alpha)} \end{cases} \tag{23}$$

$$AWD^{(\alpha)} = \frac{1}{n} \sum_{i=1}^{n} AWD_i^{\alpha} \tag{24}$$

4.3. Experiment I: Cases with Larger Width Coefficients

In this experiment, we set the interval width coefficient $\alpha = 0.05$, which is equivalent to setting the output to $[0.95 \times X, X, 1.05 \times X]$ for a single sample in the training process of the neural network. Based on this structure, the PIs can be output given an input test set. In order to guarantee the diversity of the samples, we studied four different quarterly data for four different states.

The models involved in our research can be divided into three groups for better explanations for the impact of different components. The first group included LUBE and E–LUBE, and the difference between them were the structures of the neural network. The structure of LUBE consisted of three layers which were similar to the traditional BP neural network. Moreover, in the E–LUBE, an extra context layer was added to the structure so that we could validate the impact of the context layer in prediction by comparing the performance of these two models. The second group included the PO–E–LUBE and IO–E–LUBE, and the difference between them included the optimization algorithm in the training process. PO–E–LUBE used the error and variance of point prediction to construct the cost function in MOSSA, whereby the target of minimizing the cost function effectively denotes a requirement for better prediction accuracy. In addition, IO–E–LUBE employed the CP and PIW of the interval prediction to construct the cost function in multi-objective optimization, while the target of minimizing such a cost function denoted the requirements for a better performance in interval coverage, which is more rational for our goal of interval prediction. The comparison between such models can reflect the influence of different cost functions in the parameter optimization process. Furthermore, in the first group, the parameters of the neural network are determined by a conventional gradient descent algorithm, and in the second group, the parameters are determined by a heuristic optimization algorithm. Therefore, the impact of different optimization algorithms can be shown by comparing the models in different groups. In addition, in the third group, the data preprocessing is introduced. Based on the models in the first two groups, CEEMDAN was used to refine the input dataset. The results of the models in this group will display the effect of data preprocessing in the hybrid model.

The simulation results are shown in Tables 2 and 3. Also shown in Figure 5 are the principal indices of interval prediction, namely, CP and PINAW. Based on the conducted comparisons referred to earlier, several conclusions can be inferred:

(1) By comparing the models in the first group, we can conclude that the E–LUBE is superior to LUBE in most cases, such as the fourth quarter in NSW and the first quarter in TAX, as shown in Table 2 and Figure 5. The CP of E–LUBE reached 87.17%, while the CP of LUBE was 72.36% for the fourth quarter in NSW. The rate of improvement was more than 15% with the maintenance of PINAW and PINRW. However, in some cases, the improvement is not remarkable, such as the fourth quarter in QLD, as shown in Table 3 and Figure 5. The performances of these two models are almost the same. In general, the performance of E–LUBE is better than LUBE, which means that E–LUBE with an extra context layer can improve the performance. In theory, the context layers are able to provide more information compared to previous outputs of hidden layers. This superiority has been proved in our experiments. However, owing to the instability of the parameters in the neural network, the improvement is not adequately remarkable in a few cases.

(2) In terms of the optimization methods, and according to the results shown in Figure 5, and Tables 2 and 3, the CPs of the second group (PO–E–LUBE and IO–E–LUBE) perform better than E–LUBE in most cases. E–LUBE uses the gradient descent algorithm, which is sensitive to the initialization, in order to obtain the parameters in NN. Furthermore, the models in the second group use the heuristic swarm optimization algorithm which can synthesize the initialization results using an adequate population size. Thus, the models in the second groups should have elicited better performances in theory unless the random initializations of E–LUBE are perfect. Moreover, within the second group, IO–E–LUBE has a larger CP value than PO–E–LUBE, with low levels of PINAW and PINRW. It is just the influence of the cost function that makes such a difference. The main

object of the interval prediction is a larger CP value along with a narrow width. Therefore, the IO should have an advantage

(3) Incorporation of CEEMDAN in the hybrid models is improved the performances significantly because of the denoising preprocessing. In most cases, the CPs are larger than 80% and 90%, which means more than 80% target load values are covered by the predicted intervals. Furthermore, in some cases, the CPs can reach 100%, such as the second and third quarters in NSW, and the second quarter in QLD. Such accuracy can ensure that the power supply meets the demand. Compared with the original LUBE and E–LUBE, the hybrid model we proposed (CEEMDAN–IO–E–LUBE) elicited a significant improvement in the elicited results of interval prediction.

(4) With a larger width coefficient, the CPs of our models were almost satisfactory. The smallest CP was more than 70%, and the largest CP was able to reach 100%, which is perfect for interval prediction in STLF. However, the PINAW and PINRW were almost all larger than 10, and even reached the value of 20 in second quarter in QLD. But the proposed model still outperforms other models.

(5) Considering the accumulated width deviation (AWD), for a larger width coefficient, the proposed model (CEEMDAN-IO-E-LUBE) has a smaller AWD compared with other benchmark models in most cases. According to the definition of AWD, a smaller AWD means more target values fall into the predicted intervals. For the results in which the target values are over the bounds, the deviations are relatively small. In this experiment, the AWDs of the proposed model are satisfactory in most case. For some cases, the AWDs is even closed to 0, which means almost all target load values fall into the predicted intervals. According to these predicted intervals, load dispatch will be more rational.

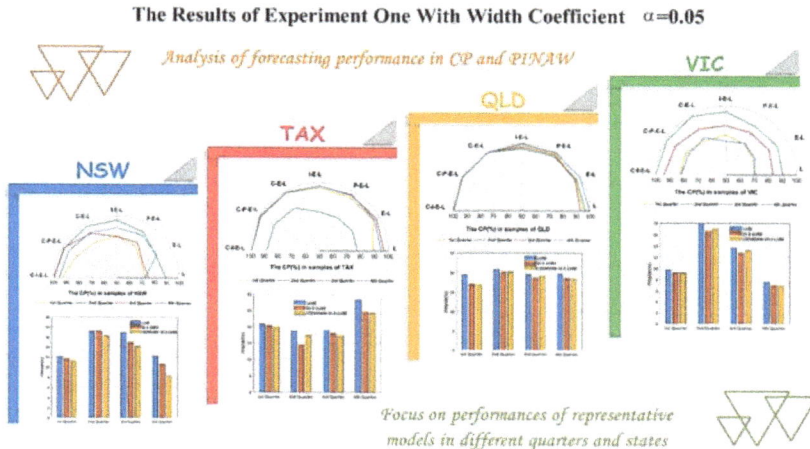

Figure 5. Performance of different samples with the width coefficient 0.05.

Table 2. Results of different models with $\alpha = 0.05$ for sample in New South Wales (NSW) and Tasmania (TAX).

Models		First Quarter				Second Quarter				Third Quarter				Fourth Quarter			
		CP	PINAW	PINRW	AWD	CP	PINAW	PINRW	AWD	CP	PINAW	PINRW	AWD	CP	PINAW	PINRW	AWD
NSW	LUBE	72.66	12.13	12.90	0.030	89.22	17.20	18.01	0.020	75.88	16.97	17.14	0.022	72.36	12.23	13.85	0.085
	E-LUBE	72.75	12.00	12.92	0.037	86.08	17.37	18.20	0.023	74.00	16.14	16.41	0.025	87.17	12.10	13.58	0.021
	PO-E-LUBE	79.67	11.80	12.58	0.010	92.42	17.25	17.83	0.007	82.75	14.98	15.29	0.049	89.75	11.26	12.86	0.033
	IO-E-LUBE	83.00	11.72	12.50	0.032	96.50	17.19	17.68	0.003	83.42	14.96	15.25	0.115	90.25	10.62	11.65	0.015
	CEEMDAN-E-LUBE	83.25	11.67	12.36	0.021	97.83	17.10	17.90	0.002	91.25	14.75	15.28	0.007	92.08	10.48	11.23	0.022
	CEEMDAN-PO-E-LUBE	86.08	11.52	12.24	0.099	97.83	16.66	17.26	0.002	98.33	14.47	14.99	0.005	93.92	10.46	10.97	0.043
	CEEMDAN-IO-E-LUBE	93.58	11.36	12.24	0.016	100.00	16.27	16.66	0.000	100.00	14.18	15.72	0.000	95.75	8.28	10.26	0.007
TAX	LUBE	88.93	20.84	21.25	0.624	77.68	18.75	19.57	0.533	96.88	18.93	19.21	0.021	94.85	28.33	28.71	0.015
	E-LUBE	94.17	20.63	21.10	0.020	78.08	17.45	18.97	0.127	97.58	18.17	18.80	0.008	96.00	25.04	25.52	0.016
	PO-E-LUBE	94.75	20.12	20.63	0.009	78.50	16.55	17.61	0.136	97.67	17.94	18.45	0.011	97.25	25.13	25.48	0.007
	IO-E-LUBE	95.92	20.29	20.65	0.007	78.33	14.38	15.74	0.257	97.42	17.96	18.60	0.006	97.33	24.67	24.98	0.008
	CEEMDAN-E-LUBE	98.50	19.98	20.36	0.005	85.83	16.64	16.93	0.032	99.00	17.35	18.82	0.001	98.92	23.48	23.69	0.002
	CEEMDAN-PO-E-LUBE	99.00	19.92	20.31	0.001	87.92	16.01	16.30	0.034	99.75	17.11	18.56	0.001	99.58	24.79	25.04	0.002
	CEEMDAN-IO-E-LUBE	99.08	19.98	20.30	0.006	88.17	17.25	17.75	0.046	99.25	17.37	19.04	0.001	99.25	24.20	24.49	0.001

The overall span "Criterion (%)" covers all four quarter column groups.

Table 3. Results of different models with $\alpha = 0.05$ for sample in Queensland (QLD) and Victoria (VIC).

Criterion (%)

Models (QLD)	First Quarter				Second Quarter				Third Quarter				Fourth Quarter			
	CP	PINAW	PINRW	AWD	CP	PINAW	PINRW	AWD	CP	PINAW	PINRW	AWD	CP	PINAW	PINRW	AWD
LUBE	94.82	19.47	20.12	0.012	96.58	20.86	21.75	0.004	92.78	19.58	20.11	0.002	99.33	19.69	20.47	0.001
E-LUBE	95.50	17.29	18.38	0.007	96.50	20.96	21.54	0.003	95.17	19.32	19.85	0.001	99.33	19.10	19.80	0.000
PO-E-LUBE	98.83	17.58	18.42	0.001	96.67	20.57	21.07	0.003	97.83	19.07	19.54	0.001	99.75	19.38	20.24	0.001
IO-E-LUBE	99.17	17.03	17.77	0.006	96.78	20.18	21.86	0.003	97.42	18.79	19.27	0.002	99.75	18.53	19.04	0.000
CEEMDAN-E-LUBE	99.75	18.17	18.88	0.000	99.83	20.69	21.60	0.000	99.92	19.32	19.86	0.000	99.83	19.49	20.10	0.000
CEEMDAN-PO-E-LUBE	99.50	18.26	18.97	0.001	99.83	20.19	21.71	0.000	99.92	19.42	19.93	0.000	99.83	19.26	19.80	0.000
CEEMDAN-IO-E-LUBE	99.25	16.96	17.67	0.002	100.00	20.17	21.66	0.000	99.92	19.17	19.53	0.000	99.92	18.42	18.90	0.001

Models (VIC)	First Quarter				Second Quarter				Third Quarter				Fourth Quarter			
	CP	PINAW	PINRW	AWD	CP	PINAW	PINRW	AWD	CP	PINAW	PINRW	AWD	CP	PINAW	PINRW	AWD
LUBE	70.85	9.77	10.62	0.042	90.25	17.85	18.37	0.351	83.42	13.61	14.29	0.013	70.22	7.58	7.90	0.018
E-LUBE	72.08	9.41	10.29	0.020	91.92	16.66	17.41	0.127	85.33	13.25	14.02	0.008	73.67	7.04	7.87	0.016
PO-E-LUBE	76.00	9.41	10.40	0.009	92.75	15.82	16.34	0.136	85.17	13.09	14.12	0.011	76.50	6.94	7.84	0.007
IO-E-LUBE	78.83	9.18	10.18	0.007	95.75	16.61	17.32	0.257	85.58	12.82	13.91	0.006	76.25	7.04	8.22	0.008
CEEMDAN-E-LUBE	78.42	9.29	10.72	0.005	98.83	15.40	16.33	0.032	88.17	12.58	13.30	0.001	80.92	7.23	8.01	0.002
CEEMDAN-PO-E-LUBE	82.67	9.22	10.43	0.001	98.75	17.12	18.12	0.034	91.67	12.70	13.57	0.001	80.50	7.07	7.93	0.002
CEEMDAN-IO-E-LUBE	83.25	9.21	10.13	0.002	99.92	17.00	18.01	0.036	94.08	13.17	13.79	0.001	82.08	6.90	7.78	0.001

4.4. Experiment II: Cases with Smaller Width Coefficients

In this experiment, we set the interval width coefficient $\alpha = 0.025$, which means we set the output to be $[0.925 \times X, X, 1.025 \times X]$ for a single sample in the training process of the neural network. With a narrow width coefficient, the lower and upper bounds were closer to the target value in the training process, which can provide more valuable information in practice. However, a narrow bound might lead to the increase of CP. Thus, a smaller width coefficient requires the models to have better predictive properties. The results of this simulation are shown in Tables 4 and 5, and in Figure 6. Correspondingly, the following conclusions can be drawn:

(1)   As Table 4 and Figure 6 show, the distinction of the models is similar to experiment I. The CPs of the original LUBE and E–LUBE are the smallest among the models in our simulation, and our proposed model CEEMDAN–IO–E–LUBE elicits the best performance

(2)   For some benchmark models in this experiment, with a narrow bound in the training process, the performance was not adequately satisfactory. As the cases of the third quarter in NSW denote and the second quarter in TAX show the CPs of LUBE and E–LUBE are close to 50%, which is not conclusive in practice. However, based on the hybrid mechanism we proposed, the performances were improved significantly. The minimum CP values of CEEMDAN–IO–E–LUBE can reach 70%, and the maximum is close to 100%, such as in the third quarter in QLD. Such results show that the predicted intervals can better cover actual electricity demand data and economize spinning reserve in power grid.

(3)   With a smaller width coefficient, the CPs decreased while the PINAW and PINRW are reduced. For the benchmark models, the results mostly display smaller CPs and larger PINAW or PINRW. However, the proposed model is able to demonstrate larger CPs with smaller PINAW and PINRW values, which is equivalent to a good performance in interval prediction. In some cases, the CP values were larger than 95% with PINAW and PINRW values less than 10. In such cases, the CPs are satisfactory and the widths of the PIs are most appropriate.

(4)   In terms of AWD in this experiment, the proposed model still showed a relatively small AWD compared with other benchmark models, which means the proposed model has a better performance at predicted accuracy. Compared with experiment I, the AWDs in this experiment are bigger. For a smaller width coefficient, the predicted interval will be narrower, which means there will be more target points falling outside the intervals. In some situations, a narrower predicted interval is necessary. The proposed model is able to provide a better performance on the condition of the requirement of a narrower predicted interval of electric load.

4.5. Comparisons and Analyses

According to the comparison of the above two experimental results, the width coefficient has a significant influence on performance, as shown in Figure 7. From one perspective, for most models, a coefficient with a larger width may lead to a larger and more satisfactory CP value, but the index about the width of PI may not be desired. From another perspective, for most models, a narrower width coefficient may elicit the desired PINAW and PINRW values, but the CP is not good enough. Considering such a situation, the proposed models alleviate the contradiction. Even though the CP value of the proposed model will decline when the width coefficient decreases, comprehensive performance is satisfactory. In some exceptional cases, owing to the complexity and instability of the datasets, the performance of the proposed models is not adequate, as the description in Figure 3 shows.

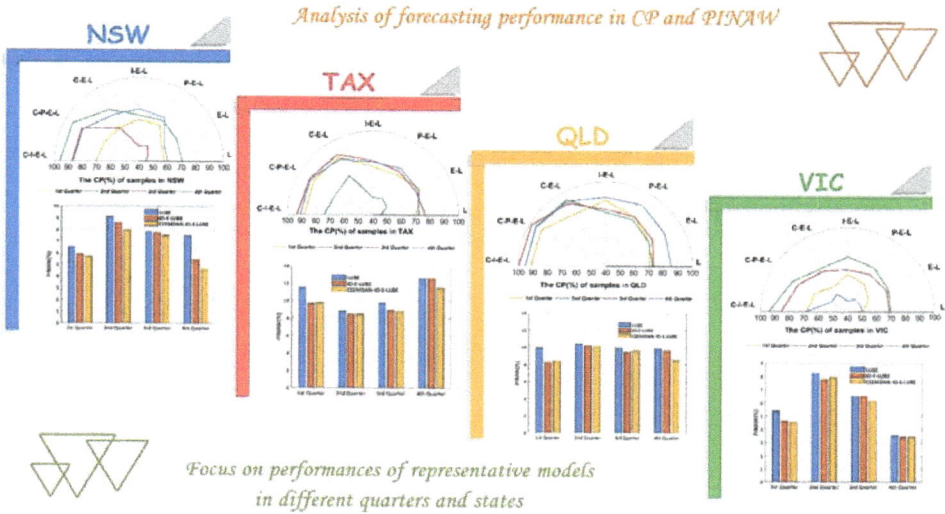

Figure 6. Performance of different samples with the width coefficient 0.025.

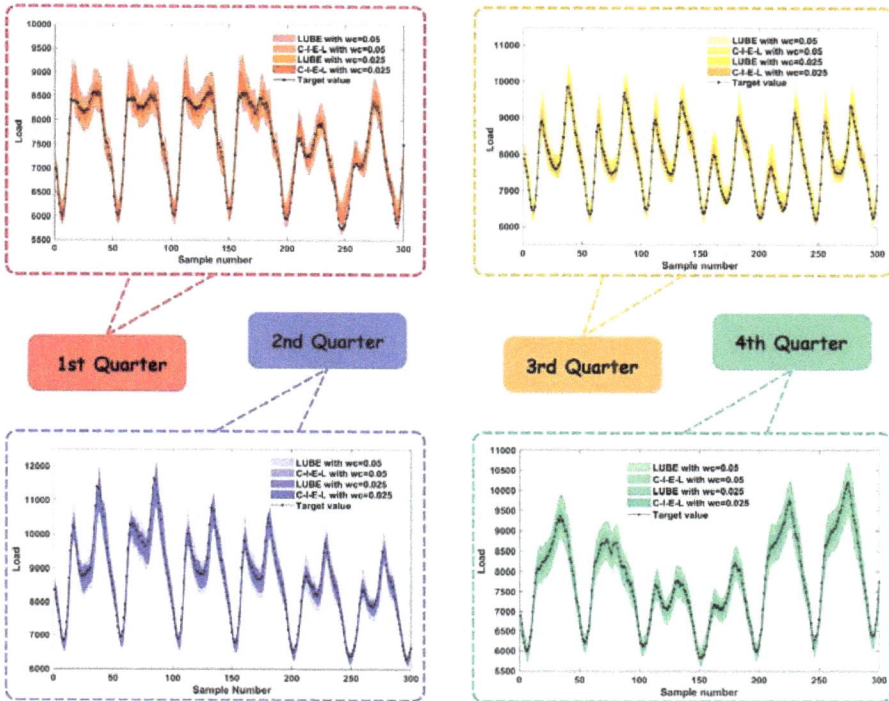

Figure 7. Interval prediction plot of partial samples in NSW.

Table 4. Results of different models with $\alpha = 0.025$ for sample in NSW and TAX.

NSW	First Quarter				Second Quarter				Third Quarter				Fourth Quarter			
	CP	PINAW	PINRW	AWD	CP	PINAW	PINRW	AWD	CP	PINAW	PINRW	AWD	CP	PINAW	PINRW	AWD
LUBE	58.20	6.56	6.92	0.360	70.66	9.08	9.60	0.163	44.85	7.82	8.75	0.531	60.42	7.52	11.83	0.158
E-LUBE	58.58	5.76	6.50	0.286	70.83	9.23	9.82	0.120	46.92	7.60	8.26	0.276	60.75	7.37	12.10	0.146
PO-E-LUBE	67.50	6.00	6.51	0.061	73.67	8.41	8.79	0.027	52.33	7.82	8.64	0.515	76.08	5.66	6.60	0.126
IO-E-LUBE	67.67	5.91	6.51	0.138	73.08	8.58	8.38	0.062	51.00	7.75	8.02	0.231	77.42	5.44	6.07	0.161
CEEMDAN-E-LUBE	69.50	5.80	6.67	0.149	82.50	7.77	8.07	0.057	67.25	7.89	8.65	0.049	77.92	5.37	5.91	0.120
CEEMDAN-PO-E-LUBE	69.75	5.83	6.13	0.118	94.25	7.84	8.86	0.095	86.75	7.28	7.75	0.084	85.67	4.93	5.40	0.069
CEEMDAN-IO-E-LUBE	70.50	5.68	6.24	0.085	96.17	7.96	8.60	0.002	87.25	7.48	7.64	0.014	86.67	4.63	5.13	0.083

TAX	First Quarter				Second Quarter				Third Quarter				Fourth Quarter			
	CP	PINAW	PINRW	AWD	CP	PINAW	PINRW	AWD	CP	PINAW	PINRW	AWD	CP	PINAW	PINRW	AWD
LUBE	70.61	11.52	11.83	0.187	45.78	8.84	10.80	0.534	76.87	9.78	10.85	0.126	72.31	12.61	14.04	0.120
E-LUBE	72.08	10.14	10.39	0.163	51.42	8.67	9.25	0.774	74.67	9.32	9.86	0.073	76.58	12.83	13.65	0.097
PO-E-LUBE	73.08	10.29	11.00	0.099	53.50	7.65	8.87	0.188	76.42	9.04	9.33	0.085	78.75	12.11	12.40	0.050
IO-E-LUBE	74.67	9.65	10.32	0.090	52.67	8.50	9.40	0.278	79.08	8.90	9.75	0.137	78.25	12.60	13.27	0.065
CEEMDAN-E-LUBE	88.08	10.11	10.51	0.036	72.75	9.27	9.71	0.179	90.00	8.53	8.84	0.011	86.25	11.57	11.79	0.018
CEEMDAN-PO-E-LUBE	88.08	10.04	10.24	0.019	72.00	8.57	8.82	0.194	91.50	8.60	8.89	0.009	90.75	11.56	11.84	0.013
CEEMDAN-IO-E-LUBE	88.08	9.78	10.21	0.015	74.25	8.46	8.97	0.171	93.42	8.79	8.30	0.013	91.08	11.49	11.84	0.018

Table 5. Results of different models with $\alpha = 0.025$ for sample in QLD and VIC.

Criterion (%)

Models QLD	First Quarter				Second Quarter				Third Quarter				Fourth Quarter			
	CP	PINAW	PINRW	AWD	CP	PINAW	PINRW	AWD	CP	PINAW	PINRW	AWD	CP	PINAW	PINRW	AWD
LUBE	70.55	9.94	10.82	0.158	72.75	10.44	10.98	0.069	73.36	9.97	10.35	0.075	85.39	9.83	10.76	0.009
E-LUBE	74.75	9.29	10.00	0.024	74.42	10.13	10.43	0.065	77.58	9.57	9.87	0.058	87.17	9.77	10.01	0.008
PO-E-LUBE	80.42	8.71	9.10	0.064	78.08	10.82	11.27	0.049	80.75	9.67	9.89	0.045	90.42	10.11	10.57	0.008
IO-E-LUBE	86.83	8.26	9.71	0.309	78.33	10.16	11.61	0.072	80.50	9.50	9.75	0.046	89.17	9.66	10.11	0.003
CEEMDAN-E-LUBE	82.75	9.54	9.88	0.156	94.50	10.79	11.23	0.012	91.58	9.70	10.18	0.003	93.42	9.25	10.20	0.004
CEEMDAN-PO-E-LUBE	91.00	8.72	9.01	0.085	95.58	10.54	11.01	0.008	98.75	9.77	10.21	0.002	95.17	9.07	9.39	0.003
CEEMDAN-IO-E-LUBE	91.42	8.41	8.77	0.048	95.25	10.13	11.17	0.004	99.75	9.64	9.91	0.002	95.33	8.52	8.84	0.003

Models VIC	First Quarter				Second Quarter				Third Quarter				Fourth Quarter			
	CP	PINAW	PINRW	AWD	CP	PINAW	PINRW	AWD	CP	PINAW	PINRW	AWD	CP	PINAW	PINRW	AWD
LUBE	50.16	5.39	6.83	0.882	70.03	8.25	8.80	0.537	68.91	6.55	7.03	0.622	50.06	3.55	4.10	0.486
E-LUBE	56.75	5.10	6.59	0.923	72.33	8.10	8.63	0.596	70.67	6.63	7.04	0.692	50.00	3.47	4.14	0.575
PO-E-LUBE	62.75	4.83	5.77	0.312	76.25	8.34	8.91	0.081	70.67	6.71	7.31	0.609	50.25	3.53	4.04	0.127
IO-E-LUBE	66.75	4.63	5.47	0.805	79.25	7.82	8.79	0.131	70.00	6.54	7.03	0.013	49.50	3.44	5.22	0.226
CEEMDAN-E-LUBE	60.00	4.75	5.28	0.433	75.83	8.04	8.31	0.059	73.42	6.64	7.04	0.038	54.50	3.68	4.45	0.262
CEEMDAN-PO-E-LUBE	65.83	4.64	5.35	0.123	83.50	8.66	9.27	0.127	78.92	6.98	7.32	0.033	53.25	4.00	4.93	0.473
CEEMDAN-IO-E-LUBE	68.25	4.56	5.64	0.081	94.08	7.93	8.23	0.044	85.75	6.18	6.70	0.043	69.33	3.44	4.45	0.161

5. Discussion

In this section, we discuss some factors which may have an effect on the performances of the proposed models in order to improve the practicability of our hybrid model. The factors involved mainly include the features of the datasets and the setting of the hyperparameters in the algorithm.

5.1. Dataset Features

The feature and quality of the datasets have a significant effect on the performance of the prediction models. In STLF, the data shows periodicity and volatility. The periodicity is attributed to the regularity in the actual use of electricity, and the volatility is attributed to the randomness and occasional use of electricity. Therefore, the linear component and the non-linear components operate simultaneously during the forecasting of the model. Specifically, some outliers may have a negative effect in the process of prediction.

As Figure 4 shows, the dataset features of the different samples are various. According to the boxplot theory, the data points that are larger than Q3 + 1.5IQR or smaller than Q1 − 1.5IQR are regarded as outliers. For the first and fourth quarters in NSW, and the first and fourth quarters in VIC, the distributions of the datasets displayed a number of outliers. Additionally, the results of the models shown in Tables 2–5 demonstrate that the model performance of the sample whose distribution is not desired may be unremarkable. These outliers are important factors that lead to such results, even if the CEEMDAN model has been applied in data preprocessing.

Another set of data features that may cause an unsatisfactory result are the non-linear characteristics of the dataset. It is well known that in traditional research, the prediction of regular and linear time series are easy to reach the desired accuracy. However, unstable and non-linear time series are more difficult to forecast in spite of the applications of novel models, such as the case of machine learning algorithms. A method used to measure the instability of data series is the recurrence plot (RP) [56]. A recurrence plot is an advanced technique of non-linear data analyses. It is the visualization (or a graph) of a square matrix in which the matrix elements correspond to those times at which the state of the dynamical system recurs. Stationary systems will deliver homogeneous recurrence plots, and unstable systems cause changes in the distribution of recurrence points in the plot, which is visible and identifiable by the brightened areas. In this study, we selected VIC as an example to verify the influence of instability. Before drawing the recurrence plot, the time delay and the dimension of the embedded matrix were determined by the C–C method. Depending on the "CRP Toolbox" released by Norbert Marwan [57], the recurrence plot of the four datasets of the different quarters in VIC is shown in Figure 8.

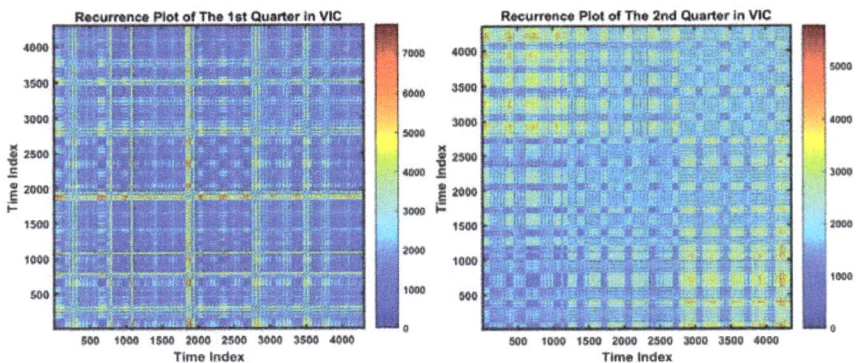

Figure 8. *Cont.*

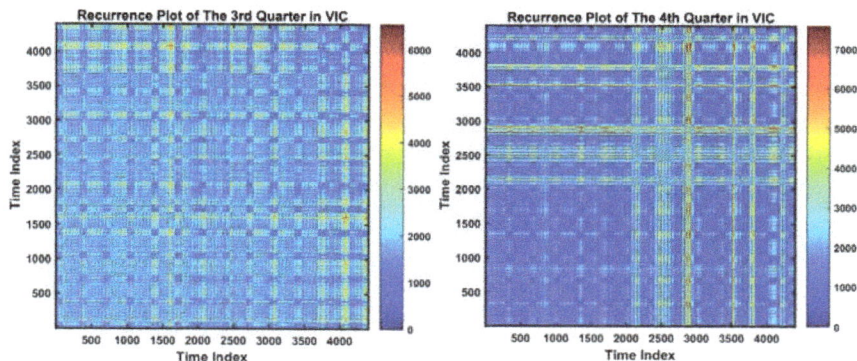

Figure 8. Recurrence plot of the samples obtained from the four quarters in VIC.

As the figure shows, the second and third quarters in VIC display relatively homogeneous distributions, while other quarters display isolated brightened areas. According to the theory of the recurrence plot, the instabilities of the former two samples are weaker, and the other two samples reveal stronger instabilities. Furthermore, we can conclude that the performances of the forecasting models shown in Table 5 are remarkable when the dataset is relatively stable, while the unstable dataset results in an unsatisfactory performance, which cannot be avoided.

5.2. Sensitivity Analysis

The hybrid model proposed in this study is based on the structure of the neural network shown in Figure 1. In the hybrid model, the hyperparameter is a key factor that influences the model's performance. In most studies on machine learning, the setting of the hyperparameters always depends on trials or empirical knowledge. This is the reason why many experimental results cannot be reproduced and why a considerable amount of time and energy is spent on tuning parameters in industrial applications. At present, there is no absolute method to determine the values of all types of hyperparameters. In this study, we also mainly relied on experiences and trials to set the hyperparameters, as shown in Table 1. Among the hyperparameters, several parameters need to be highlighted.

The first one is the number of salp populations in MOSSA. In the swarm heuristic optimization algorithm, the number of swarms is usually a vital factor that needs to be considered. A larger population might provide a larger probability to reach the best individual, but exceeding the desired population may cause an increase in the complexity of the algorithm, which is related to the number of algorithmic iterations. Considering the number of parameters in our proposed model, the population numbers that ranged from 10 to 100 with a step of 10 were evaluated. As a result, we selected the number 50 as the population number (as shown in Table 6) after comprehensively considering the time complexity and model performance.

The second type of hyperparameters that need to be emphasized are the upper and lower bounds of individual parameters in MOSSA. In our simulation, the datasets were normalized within the range of −1 to 1 in order to avoid the influence of dimension and improve the training speed. Therefore, the absolute value of weights and thresholds of neural networks in the training process will not be too large. As Table 6 shows, we set the initial upper and lower bounds to 2 and −2 according to the experiment trials. Excessive range limits may increase the difficulty of searching for the best parameters with a limited number of iterations. Furthermore, the algorithm that operates based on a small range may not elicit the optimal solution.

Table 6. Sensitivity analyses results of different hyper-parameters.

Metrics	The Number of Salp Populations in MOSSA									
	10	20	30	40	50	60	70	80	90	100
CP	95.21	95.32	97.01	96.66	98.69	98.33	98.50	97.85	97.46	98.10
PINAW	17.34	17.63	13.52	13.89	13.10	13.02	13.64	14.05	13.82	13.72
PINRW	18.28	18.05	14.18	14.35	13.84	13.75	14.36	14.92	14.67	14.53
Time(s)	425	452	472	524	548	593	668	734	869	10.45

Metrics	The Initial Threshold of Parameters				
	$[-0.5, 0.5]$	$[-1, 1]$	$[-2, 2]$	$[-3, 3]$	$[-5, 5]$
CP	97.65	98.84	99.00	98.26	96.89
PINAW	14.36	12.82	12.80	13.12	13.68
PINRW	15.32	13.46	13.42	13.94	14.36
Time(s)	433	450	461	453	484

5.3. Consistency Analysis

In this section, in order to verify the consistency of our proposed model, new datasets involving latest dates are introduced. In addition, several basic compared models including long short-term memory (LSTM) networks, function fitting neural networks (FITNET), and least squares support vector machine (LSSVM) which have been proved to provide good results for STLF are employed to verify the advantages of the proposed model.

We chose NSW and VIC randomly as examples. The new datasets are collected from 1 January 2018 0:30 am to 30 May 2018 0:00 am and the total number of samples is 7152. The samples in the second quarter in NSW and the fourth quarter in VIC are chosen as compared datasets. According to the results shown in Table 7, the proposed model also has a good performance on the new datasets. The CP is almost 90%, which means the predicted interval can cover 90% target load value. The consistency of the proposed can be guaranteed, and the change of the dates of dataset will not risk altering the final conclusion.

Considering different basic models for STLF, we chose three widely used artificial intelligence models (LSTM, FITNET, and LSSVM) as comparators to verify the superiority. As shown in Table 7, the proposed models provide a larger CP and smaller PINAW compared with the other three models. In particular, LSTM reveals desired narrower PINAW and PINRW, but the CPs are not satisfactory. Moreover, the proposed model outperformed than other basic models in AWD. Therefore, the proposed approach have a distinct advantage in the performance of short-term power load interval forecasting. It is able to provide a satisfactory CP and restrict the interval width at the same time, which is the most important aspect of superiority of the proposed model.

Table 7. Consistency analysis results of some basic models and new datasets.

Models	NSW-2018-NEW					VIC-2018-NEW				
	CP	PINAW	PINRW	AWD	Time(s)	CP	PINAW	PINRW	AWD	Time(s)
Proposed	89.58	15.51	16.58	0.023	593.29	89.08	11.50	12.66	0.065	564.55
LSSVM	78.67	15.95	17.64	0.677	495.32	86.67	12.16	13.01	0.026	486.85
FITNET	72.08	16.25	17.24	0.043	405.52	74.33	11.66	12.95	0.087	300.72
LSTM	44.00	5.47	5.92	0.382	1199.04	59.83	5.28	5.79	0.250	947.78

Models	NSW-2017-2Q					VIC-2017-4Q				
	CP	PINAW	PINRW	AWD	Time(s)	CP	PINAW	PINRW	AWD	Time(s)
Proposed	100.00	16.27	16.66	0.000	543.20	82.08	6.90	7.78	0.001	526.39
LSSVM	94.42	16.67	17.12	0.038	409.24	71.58	7.60	10.59	0.097	435.50
FITNET	94.33	15.83	17.29	0.012	402.12	74.33	7.88	8.94	0.076	504.31
LSTM	70.67	6.01	6.37	0.101	753.60	65.33	3.10	3.55	0.248	732.21

On the other hand, in order to obtain a better performance and accuracy, the proposed approach is more complex. The algorithm with higher complexity often takes longer in practice. As Table 7 shows, compared with LSSVM and FITNET, the execution times of the proposed model are longer, which is the major disadvantage. However, with the development of hardware, the operational capability of computer can be improved, and the execution time can be reduced. Furthermore, as a kind of artificial intelligence technique, the fine-tuning of hyper-parameters in the proposed model will take time, which is a common situation in academic and industrial fields.

5.4. Further Research Prospect

This paper proposes a hybrid interval prediction model to predict the power load intervals. Compared with other basic models, this model has achieved good results in terms of coverage, interval width, and deviation error of the prediction interval. The model can obtain relatively high coverage under the condition of relatively narrow interval width, and the interval obtained can accurately reflect the changes of future short-term power load and provide more accurate and reliable support for power dispatch. On the other hand, for datasets with more complex changes and non-linear features, although the performance of proposed model is improved compared with the traditional models, it is still not ideal in some cases. For the unfavorable results caused by the characteristics of datasets, we may explore the following two aspects in future:

(a)   Finding and improving prediction methods that can better solve the non-linear characteristics of electrical loads, and improving the performance of predictive models in complex situations;

(b)   Fully analyzing the relevant characteristics in the power load data, selecting different models for different characteristics, and using ensemble learning to integrate and enhance the prediction results.

6. Conclusions

STLF is the basic work of power system planning and operation. However, the power load has regularity and certain randomness at the same time, which increases the difficulty of desired and reliable STLF. Moreover, compared with the prediction of specific points, interval prediction may provide more information for decision making in STLF. In this study, based on LUBE, we developed a novel hybrid model including data preprocessing, a multi-objective salp algorithm, and E–LUBE. In theory, such a hybrid model can reduce the influence of noise in a dataset and the parameter optimization process is more effective and efficient in E–LUBE.

In our proposed approach, we used a multi-objective optimization algorithm to search for the parameters of the neural network and reconstructed the cost function with double interval criterions instead of point criterions (such as MSE) in the traditional method. As Tables 2–5 show, by comparing it with traditional methods, the proposed approach provides a higher CP and a lower interval width at the same time, which makes up for the lower CP and higher interval width of traditional methods.

In order to verify the performance of the proposed model and validate the impact of the constituent components in a hybrid model, we collected 16 samples involving four states using four quarters in Australia, and set several model comparisons in experiments

Furthermore, according to the comparison and analyses results, the conclusions are summarized as follows: (a) an efficient data preprocessing method was applied herein. Depending on the decomposition and reconstruction, this method can significantly improve the model performance in STLF. (b) Compared to the traditional prediction models based on neural networks, the newly developed E–LUBE method has an advantage in terms of comprehensive performance in interval prediction. It can be validated that the context layer with the information of the former hidden layer can improve model performance. (c) The introduction of the novel multi-objective algorithm MOSSA optimized the process of parameter search. The new cost function was based on a double-objective interval index that outperformed the traditional single-objective point error index (such as MSE) in interval prediction. (d) For STLF based on the E–LUBE mechanism, the width coefficient is an important factor. A larger width coefficient may lead to satisfactory CPs, and a smaller width coefficient may result in a satisfactory interval width. Therefore, in practice, the decision maker needs to adjust the width coefficient for specific demands. For example, we chose the width coefficient with a minimum interval width at the same time that the minimum demand of CP was guaranteed. (e) No matter how complex is the dataset, the proposed model always provides the best performance compared to benchmark models. However, because of the complexity of the data itself, some of the performance is not remarkable. In general, the proposed model provided a desired result in most cases.

Furthermore, in a power grid operator the proposed method has a strong practical application significance. A highly accurate forecasting method is one of the most important approaches used in improving power system management, especially in the power market [58]. In actual operation, for secure power grid dispatching, a control center has to make a prediction for the subsequent load. According to historical data, the dataset for the predictive model involved can be constructed. The results of the predictive model are able to provide the upper bound and lower bound of the load at some point in the future. Depending on the upper bound and lower bound, the control center can adjust the quantity of electricity on each charging line. Therefore, such a hybrid approach which can provide more accurate results can ensure the safe operation of the power grid and improve the economic efficiency of power grid operation.

Author Contributions: J.W. carried on the validation and visualization of experiment results; Y.G. carried on programming and writing of the whole manuscript; X.C. provided the overall guide of conceptualization and methodology.

Funding: This research was funded by National Natural Science Foundation of China (Grant number: 71671029) and Gansu science and technology program "Study on the forecasting methods of very short-term wind speeds" (Grant number: 1506RJZA187).

Acknowledgments: This work was supported by the National Natural Science Foundation of China (No. 71671029) and the Gansu science and technology program "Study on the forecasting methods of very short-term wind speeds" (No. 1506RJZA187).

Conflicts of Interest: The authors declare no conflicts of interest.

Abbreviation

STLF	Short-term load forecasting
PI	Prediction intervals
PIW	Prediction intervals width
PINAW	PI normalized average width
ENN	Elman neural network
SNR	Signal to noise ratio
IMF	Intrinsic mode function
Nstd	Noise standard deviation
Pop_num	Total population number
Maxiter	The maximum number of iterations
CEEMDAN	The complete ensemble empirical mode decomposition with adaptive noise
NN	Neural networks
CP	Coverage probability
LUBE	Lower upper bound estimation
PINRW	PI normalized root-mean-square width
Dim	Individual parameter dimension
EMD	Empirical mode decomposition
MSE	Mean square error
NR	Number of realizations
RP	Recurrence plot
MOSSA	Multi-objective salp swarm algorithm
E-LUBE	Lower upper bound estimation with ENN

References

1. Fan, S.; Hyndman, R.J. Short-term load forecasting based on a semi-parametric additive model. *IEEE Trans. Power Syst.* **2012**, *27*, 134–141. [CrossRef]
2. Du, P.; Wang, J.; Yang, W.; Niu, T. Multi-step ahead forecasting in electrical power system using a hybrid forecasting system. *Renew. Energy* **2018**, *122*, 533–550. [CrossRef]
3. Shrivastava, N.A.; Khosravi, A.; Panigrahi, B.K. Prediction Interval Estimation of Electricity Prices using PSO tuned Support Vector Machines. *IEEE Trans. Ind. Inform.* **2015**, *11*. [CrossRef]
4. Hagan, M.T.; Behr, S.M. The Time Series Approach to Short Term Load Forecasting. *IEEE Trans. Power Syst.* **1987**, *2*, 785–791. [CrossRef]
5. Papalexopoulos, A.D.; Hesterberg, T.C. A regression-based approach to short-term system load forecasting. *IEEE Trans. Power Syst.* **1990**, *5*, 1535–1547. [CrossRef]
6. Christiaanse, W.R. Short-Term Load Forecasting Using General Exponential Smoothing. *IEEE Trans. Power Appar. Syst.* **1971**, *PAS-90*, 900–911. [CrossRef]
7. Al-Hamadi, H.M.; Soliman, S.A. Short-term electric load forecasting based on Kalman filtering algorithm with moving window weather and load model. *Electr. Power Syst. Res.* **2004**, *68*, 47–59. [CrossRef]
8. Metaxiotis, K.; Kagiannas, A.; Askounis, D.; Psarras, J. Artificial intelligence in short term electric load forecasting: A state-of-the-art survey for the researcher. *Energy Convers. Manag.* **2003**, *44*, 1525–1534. [CrossRef]
9. Yoo, H.; Pimmel, R.L. Short term load forecasting using a self-supervised adaptive neural network. *IEEE Trans. Power Syst.* **1999**, *14*, 779–784. [CrossRef]
10. Ho, K.L.; Hsu, Y.Y.; Chen, C.F.; Lee, T.E.; Liang, C.C.; Lai, T.S.; Chen, K.K. Short term load forecasting of taiwan power system using a knowledge-based expert system. *IEEE Trans. Power Syst.* **1990**, *5*, 1214–1221. [CrossRef]
11. Mohandes, M. Support vector machines for short-term electrical load forecasting. *Int. J. Energy Res.* **2002**, *26*, 335–345. [CrossRef]
12. Hong, W.C. Electric load forecasting by seasonal recurrent SVR (support vector regression) with chaotic artificial bee colony algorithm. *Energy* **2011**, *36*, 5568–5578. [CrossRef]
13. Liu, Z.; Li, W.; Sun, W. A novel method of short-term load forecasting based on multiwavelet transform and multiple neural networks. *Neural Comput. Appl.* **2013**, *22*, 271–277. [CrossRef]
14. Fan, G.F.; Peng, L.L.; Hong, W.C.; Sun, F. Electric load forecasting by the SVR model with differential empirical mode decomposition and auto regression. *Neurocomputing* **2016**, *173*, 958–970. [CrossRef]
15. AlRashidi, M.R.; EL-Naggar, K.M. Long term electric load forecasting based on particle swarm optimization. *Appl. Energy* **2010**, *87*, 320–326. [CrossRef]
16. Wang, J.; Li, L.; Niu, D.; Tan, Z. An annual load forecasting model based on support vector regression with differential evolution algorithm. *Appl. Energy* **2012**, *94*, 65–70. [CrossRef]
17. Ghayekhloo, M.; Menhaj, M.B.; Ghofrani, M. A hybrid short-term load forecasting with a new data preprocessing framework. *Electr. Power Syst. Res.* **2015**, *119*, 138–148. [CrossRef]
18. Zhang, X.; Wang, J. A novel decomposition-ensemble model for forecasting short-term load-time series with multiple seasonal patterns. *Appl. Soft Comput. J.* **2018**, *65*, 478–494. [CrossRef]
19. Tian, C.; Hao, Y. A Novel Nonlinear Combined Forecasting System for Short-Term Load Forecasting. *Energies* **2018**, *11*, 714.
20. Khotanzad, A.; Hwang, R.C.; Abaye, A.; Maratukulam, D. An Adaptive Modular Artificial Neural Network Hourly Load Forecaster and its Implementation at Electric Utilities. *IEEE Trans. Power Syst.* **1995**, *10*, 1716–1722. [CrossRef]
21. Bishop, C.M. Neural networks for pattern recognition. *J. Am. Stat. Assoc.* **1995**, *92*, 482. [CrossRef]
22. Hwang, J.T.G.; Ding, A.A. Prediction Intervals for Artificial Neural Networks. *J. Am. Stat. Assoc.* **1997**, *92*, 748–757. [CrossRef]
23. Heskes, T. Practical confidence and prediction intervals. *Adv. Neural Inf. Process. Syst.* **1997**, *9*, 176–182.
24. Nix, D.A.; Weigend, A.S. Estimating the mean and variance of the target probability distribution. In Proceedings of the 1994 IEEE International Conference on Neural Networks (ICNN'94), Orlando, FL, USA, 28 June–2 July 1994; Volume 1, pp. 55–60.

25.	Van Hinsbergen, C.P.I.; van Lint, J.W.C.; van Zuylen, H.J. Bayesian committee of neural networks to predict travel times with confidence intervals. *Transp. Res. Part C Emerg. Technol.* **2009**, *17*, 498–509. [CrossRef]
26.	Khosravi, A.; Nahavandi, S.; Creighton, D. Construction of optimal prediction intervals for load forecasting problems. *IEEE Trans. Power Syst.* **2010**, *25*, 1496–1503. [CrossRef]
27.	Da Silva, A.P.A.; Moulin, L.S. Confidence intervals for neural network based short-term load forecasting. *IEEE Trans. Power Syst.* **2000**, *15*, 1191–1196. [CrossRef]
28.	Khosravi, A.; Nahavandi, S.; Creighton, D.; Atiya, A.F. Lower upper bound estimation method for construction of neural network-based prediction intervals. *IEEE Trans. Neural Netw.* **2011**, *22*, 337–346. [CrossRef] [PubMed]
29.	Deb, K.; Agrawal, S.; Pratap, A.; Meyarivan, T. A fast elitist non-dominated sorting genetic algorithm for multi-objective optimization: NSGA-II. *Parallel Probl. Solving Nat. PPSN VI* **2000**, 849–858. [CrossRef]
30.	Deb, K.; Pratap, A.; Agarwal, S.; Meyarivan, T. A fast and elitist multiobjective genetic algorithm: NSGA-II. *IEEE Trans. Evol. Comput.* **2002**, *6*, 182–197. [CrossRef]
31.	Coello Coello, C.A.; Lechuga, M.S. MOPSO: A proposal for multiple objective particle swarm optimization. In Proceedings of the 2002 Congress on Evolutionary Computation, CEC 2002, Honolulu, HI, USA, 12–17 May 2002; Volume 2, pp. 1051–1056.
32.	Padhye, N. Topology Optimization of Compliant Mechanism Using Multi-objective Particle Swarm Optimization. In Proceedings of the 10th Annual Conference Companion on Genetic and Evolutionary Computation, Atlanta, GA, USA, 12–16 July 2008; pp. 1831–1834.
33.	Alaya, I.; Solnon, C.; Ghedira, K. Ant Colony Optimization for Multi-Objective Optimization Problems. In Proceedings of the 19th IEEE International Conference on Tools with Artificial Intelligence (ICTAI 2007), Patras, Greece, 29–31 October 2007; pp. 450–457. [CrossRef]
34.	Xue, F.; Sanderson, A.C.; Graves, R.J. Pareto-based multi-objective differential evolution. In Proceedings of the 2003 Congress on Evolutionary Computation, Canberra, Australia, 8–12 December 2003; Volume 2, pp. 862–869.
35.	Mirjalili, S.Z.; Mirjalili, S.; Saremi, S.; Faris, H.; Aljarah, I. Grasshopper optimization algorithm for multi-objective optimization problems. *Appl. Intell.* **2017**. [CrossRef]
36.	Knowles, J.D.; Corne, D.W. Approximating the Nondominated Front Using the Pareto Archived Evolution Strategy. *Evol. Comput.* **2000**, *8*, 149–172. [CrossRef] [PubMed]
37.	Wang, J.; Yang, W.; Du, P.; Niu, T. A novel hybrid forecasting system of wind speed based on a newly developed multi-objective sine cosine algorithm. *Energy Convers. Manag.* **2018**, *163*, 134–150. [CrossRef]
38.	Du, P.; Wang, J.; Guo, Z.; Yang, W. Research and application of a novel hybrid forecasting system based on multi-objective optimization for wind speed forecasting. *Energy Convers. Manag.* **2017**, *150*, 90–107. [CrossRef]
39.	Wolpert, D.H.; Macready, W.G. No free lunch theorems for optimization. *IEEE Trans. Evol. Comput.* **1997**, *1*, 67–82. [CrossRef]
40.	Service, T.C. A No Free Lunch theorem for multi-objective optimization. *Inf. Process. Lett.* **2010**, *110*, 917–923. [CrossRef]
41.	Rodriguez, P.; Wiles, J.; Elman, J.L. A Recurrent Neural Network that Learns to Count. *Conn. Sci.* **1999**, *11*, 5–40. [CrossRef]
42.	Chandra, R.; Zhang, M. Cooperative coevolution of Elman recurrent neural networks for chaotic time series prediction. *Neurocomputing* **2012**, *86*, 116–123. [CrossRef]
43.	Cacciola, M.; Megali, G.; Pellicanó, D.; Morabito, F.C. Elman neural networks for characterizing voids in welded strips: A study. *Neural Comput. Appl.* **2012**, *21*, 869–875. [CrossRef]
44.	Wang, J.J.; Zhang, W.; Li, Y.; Wang, J.J.; Dang, Z. Forecasting wind speed using empirical mode decomposition and Elman neural network. *Appl. Soft Comput.* **2014**, *23*, 452–459. [CrossRef]
45.	Huang, N.; Shen, Z.; Long, S.; Wu, M.; Shih, H.; Zheng, Q.; Yen, N.; Tung, C.; Liu, H. The empirical mode decomposition and the Hilbert spectrum for nonlinear and non-stationary time series analysis. *Proc. R. Soc. A Math. Phys. Eng. Sci.* **1998**, *454*, 903–995. [CrossRef]
46.	Wu, Z.; Huang, N.E. Ensemble Empirical Mode Decomposition: A Noise-Assisted Data Analysis Method. *Adv. Adapt. Data Anal.* **2009**, *1*, 1–41. [CrossRef]
47.	Yeh, J.R.; Shieh, J.S.; Huang, N.E. Complementary ensemble empirical mode decomposition: A novel noise enhanced data analysis method. *Adv. Adapt. Data Anal.* **2010**, *2*, 135–156. [CrossRef]

48. Torres, M.E.; Colominas, M.A.; Schlotthauer, G.; Flandrin, P. A complete ensemble empirical mode decomposition with adaptive noise. In Proceedings of the 2011 IEEE International Conference on Acoustics, Speech and Signal Processing, Prague, Czech Republic, 22–27 May 2011; Volume 7, pp. 4144–4147. [CrossRef]

49. Elman, J.L. Finding structure in time. *Cogn. Sci.* **1990**, *14*, 179–211. [CrossRef]

50. Quan, H.; Srinivasan, D.; Khosravi, A. Uncertainty handling using neural network-based prediction intervals for electrical load forecasting. *Energy* **2014**, *73*, 916–925. [CrossRef]

51. Coello Coello, C.A. Evolutionary multi-objective optimization: Some current research trends and topics that remain to be explored. *Front. Comput. Sci. China* **2009**, *3*, 18–30. [CrossRef]

52. Branke, J.; Kaußler, T.; Schmeck, H. Guidance in evolutionary multi-objective optimization. *Adv. Eng. Softw.* **2001**, *32*, 499–507. [CrossRef]

53. Deb, K. Advances in Evolutionary Multi-objective Optimization. In *Search Based Software Engineering*; Springer: Berlin/Heidelberg, Germany, 2012; pp. 1–26. ISBN 978-3-642-33119-0.

54. Mirjalili, S.; Gandomi, A.H.; Mirjalili, S.Z.; Saremi, S.; Faris, H.; Mirjalili, S.M. Salp Swarm Algorithm: A bio-inspired optimizer for engineering design problems. *Adv. Eng. Softw.* **2017**, *114*, 163–191. [CrossRef]

55. Wang, J.; Niu, T.; Lu, H.; Guo, Z.; Yang, W.; Du, P. An analysis-forecast system for uncertainty modeling of wind speed: A case study of large-scale wind farms. *Appl. Energy* **2018**, *211*, 492–512. [CrossRef]

56. Eckmann, J.-P.; Kamphorst, S.O.; Ruelle, D. Recurrence Plots of Dynamical Systems. *Europhys. Lett.* **1987**, *4*, 973–977. [CrossRef]

57. Marwan, N.; Wessel, N.; Meyerfeldt, U.; Schirdewan, A.; Kurths, J. Recurrence-plot-based measures of complexity and their application to heart-rate-variability data. *Phys. Rev. E Stat. Phys. Plasmas Fluids Relat. Interdiscip. Top.* **2002**, *66*. [CrossRef] [PubMed]

58. Shu, F.; Luonan, C. Short-term load forecasting based on an adaptive hybrid method. *IEEE Trans. Power Syst.* **2006**, *21*, 392–401.

energies

MDPI

Article

Short-Term Load Forecasting for Electric Bus Charging Stations Based on Fuzzy Clustering and Least Squares Support Vector Machine Optimized by Wolf Pack Algorithm

Xing Zhang

Department of Economic Management, North China Electric Power University, Baoding 071003, China, 51851719@ncepu.edu.cn

Received: 22 May 2018; Accepted: 1 June 2018; Published: 4 June 2018

Abstract: Accurate short-term load forecasting is of momentous significance to ensure safe and economic operation of quick-change electric bus (e-bus) charging stations. In order to improve the accuracy and stability of load prediction, this paper proposes a hybrid model that combines fuzzy clustering (FC), least squares support vector machine (LSSVM), and wolf pack algorithm (WPA). On the basis of load characteristics analysis for e-bus charging stations, FC is adopted to extract samples on similar days, which can not only avoid the blindness of selecting similar days by experience, but can also overcome the adverse effects of unconventional load data caused by a sudden change of factors on training. Then, WPA with good global convergence and computational robustness is employed to optimize the parameters of LSSVM. Thus, a novel hybrid load forecasting model for quick-change e-bus charging stations is built, namely FC-WPA-LSSVM. To verify the developed model, two case studies are used for model construction and testing. The simulation test results prove that the proposed model can obtain high prediction accuracy and ideal stability.

Keywords: short-term load forecasting; electric bus charging station; fuzzy clustering; least squares support vector machine; wolf pack algorithm

---

1. Introduction

In recent years, low-carbon cities have become a common pursuit around the world, which is faced with increasing energy crises and environmental problems [1]. Electric buses (e-buses) have developed quickly with the burgeoning construction of low-carbon cities [2]. As important supporting facilities, charging stations bring new challenges to optimal dispatching and safe operation of the power grid due to great volatility, randomness and intermittence of the load [3]. Therefore, it is of great significance to conduct research on load characteristics analysis and short-term load forecasting. On one hand, this contributes to the optimal combination of generator units in terms of power system, economical dispatch, optimal power flow and electricity market transactions. On the other hand, it provides a decision basis for construction planning, energy management, orderly charging and economical operation for charging stations, which can guarantee and promote the development of low-carbon cities. Therefore, research on short-term load forecasting for quick-change e-bus charging stations has been conducted to provide data support and a theoretical basis for the large-scale construction of charging stations.

Nowadays, scholars have conducted a large amount of research on load forecasting for charging stations. The prediction methods are primarily divided into two categories: traditional forecasting approaches, such as time series [4], regression analysis [5], and fuzzy prediction [6], and artificially intelligent algorithms. Conventional prediction methods aiming at load forecasting for e-bus charging stations are mainly established on the foundation of probability and statistics theory. Ashtari et al. [7] installed GPS equipment on 76 representative plug-in electric vehicles in Winnipeg, Canada, and collected driving data for the whole year. The load forecasting was conducted in consideration of the actual charging sate of battery, stopping time, parking type and vehicle power system. In [8], four variables, including the number of vehicles needing battery change per hour, the starting time of charging, driving distance and charging duration, were taken into account under an uncontrolled swapping and charging scenario. On this basis, the Monte Carlo method integrated non-parameter estimation approach was employed in load forecasting for electric vehicle charging stations. As we can see, traditional prediction methods have the advantages of mature theory, perfect verification approaches and simple calculation, but the weaknesses of a single applicable object and unideal prediction precision are also notable. Accordingly, it is of great importance to apply intelligent forecasting techniques to load forecasting of charging stations with the rapid development of artificial intelligence technology.

Intelligent algorithms for load forecasting chiefly consist of artificial neural networks (ANNs) and support vector machine (SVM) [9]. Back propagation neural network (BPNN), treated as representative of ANNs, is suitable for load prediction of quick-change e-bus charging stations. For example, [10] analyzed the load characteristics and influential factors, as well as executing a BPNN model to predict the short-term load based on the measured data of quick-change e-bus charging stations at the Beijing Olympic Games. The approach in this study proved to be useful. Additionally, some scholars have adopted ANNs for short-term load forecasting of other power systems. Xiao et al [11] combined single spectrum analysis (SSA) with modified wavelet neural network (WNN) to establish a reliable short-term forecasting approach in the field of load, wind speed and electricity price. Reference [12] proposed a novel ensemble prediction method for short-term load forecasting on the foundation of the extreme learning machine (ELM), where four improvements were made to the ELM. The results showed that the prediction accuracy of the proposed technique was superior to the standard ANNs. However, the drawbacks of ANNs include slow convergence and easy trapping into the local minimum, which greatly limit the forecasting precision and stability. SVM model can effectively address these problems [13]; thus, this approach has been widely used in the research of load forecasting. Reference [14] designed an incremental learning model on the basis of SVM to implement load prediction under batch arrival with a large sample. In reference [15], an SVM model based on the selection of similar days for daily load forecasting of electric vehicles was come up with. Correlation analysis was presented to extract the influential factors and grey correlation theory was utilized to obtain a small sample set of similar days. Compared with ANNs, the SVM model achieved better results for load forecasting. Nevertheless, the transformation of the kernel function to convert the problem into quadratic programming reduces the efficiency and accuracy of traditional SVM [16].

Least squares support vector machine (LSSVM) is a modified form for SVM where the least squares linear system serves as the loss function to avoid quadratic programming, and the kernel function is employed to transform prediction problems into equations, as well as to convert inequality constraints into equality ones, which can improve the forecasting accuracy and speed [17]. Reference [18] introduced LSSVM to predict the annual load in China with the rolling mechanism. The good results verified the applicability of LSSVM in load forecasting. Reference [19] built a hybrid model integrated LSSVM with cuckoo search algorithm (CS) for short-term load forecasting. The findings indicated that this proposed technique could obtain good prediction results. Remarkably, LSSVM has not yet been applied to load forecasting for quick-change e-bus charging station. The learning and generalization ability of LSSVM model hinges on the selection of two parameters, namely, regularization parameter γ and kernel parameter σ^2. As a result, it is necessary to utilize an appropriate intelligent algorithm to determine these values. The commonly employed optimization methods include genetic algorithm (GA) [20], particle swarm optimization (PSO) [21], CS [22] and bat algorithm (BA) [23]. However, GA has the disadvantages of premature convergence, complex computation, small processing scale, poor stability and difficulty in coping with nonlinear constraints. The poor accuracy of local search of PSO cannot fully satisfy the need of parameter optimization in LSSVM. The shortcoming of CS and BA is that they easily fall into local optimums, leading to reduction in prediction accuracy. Wolf pack algorithm (WPA), as a new metaheuristic approach, is introduced in this paper to optimize the parameters in LSSVM. This technique possesses good global convergence and computational robustness due to insensitivity of the change of parameters in WPA [24].

As a result of the complexity and diversity of the influential factors for load forecasting in quick-change e-bus charging stations, it is of great necessity to select proper inputs for the prediction, so that redundant data can be reduced and computing efficiency can be improved [25]. Fuzzy clustering (FC) is a mathematical technique that classifies objects according to their characteristics [26]. In view of the fact that the daily load curves with similar influential factors of charging stations are basically consistent, good prediction results can be achieved by the use of samples on similar days. Consequently, a transitive closure algorithm grounded on a fuzzy equivalent matrix in FC is selected in this paper, which can extract samples similar to the predicted day. It can not only avoid the blindness of choosing similar days by experience, but also overcome the adverse effects of unconventional load data caused by sudden change of factors on LSSVM training.

Therefore, the influential factors for the load in quick-change e-bus charging stations are analyzed in this paper, and a load forecasting model combining FC with LSSVM and optimized by WPA (FC-WPA-LSSVM) is established here. The rest of paper is organized as follows: Section 2 conducts an analysis of the daily load characteristics for quick-change e-bus charging stations based on related statistical data and studies various influential factors including day types, meteorological conditions and bus dispatch; Section 3 provides a brief description of FC, LSSVM and WPA, as well as the complete prediction framework; Section 4 introduces an experimental study to validate the proposed method; and Section 5 makes further validation. In Section 6, conclusions are obtained.

2. Analysis of Load Characteristics of E-Bus Charging Stations

The load of a large quick-change e-bus charging station in Baoding, China, is provided in this paper. When the bus comes into the station, the battery with electricity depletion is changed by the quick-change robot, which is further connected to the charging platform. Then, a battery filled with electricity is installed in the bus. After that, the e-bus goes into a specific area to wait for dispatch instructions. According to the dispatch, the e-bus appears at the charging station after 8:00 a.m. each day, which leads to a rise in load. The chargers will not stop working until the battery charging of the last e-bus is completed. At that time, the load decreases to the lowest point.

A typical daily load curve of the e-bus charging station is shown in Figure 1, which displays the active power per hour in a day. In common with the traditional load curve, there exist obvious crests and troughs. However, the curve of the e-bus charging station fluctuates greatly, and apparent distinctions appear among different curves, whereby the load in winter and summer is high, while the load in spring and autumn is low. All of these characteristics create difficulties for the daily load forecasting of the charging station.

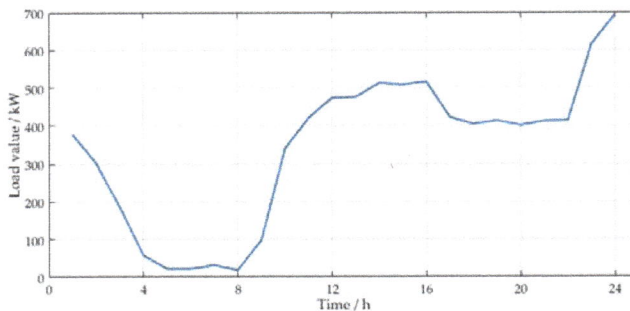

Figure 1. Typical daily load curve of an e-bus charging station in Baoding.

The load is influenced by various factors. Here, three variables, including day types, meteorological conditions and e-bus dispatching, are selected. Unlike traditional motor vehicles, the source of power for electric buses is all electric power. When there is a traffic jam, there is no energy loss for electric buses. Therefore, traffic congestion factors do not affect the load of charging stations.

2.1. Day Types

E-bus charging stations serve the electricity supply of urban e-buses. In accordance with the habits and demands of citizens, the scheduling of e-buses between weekdays and weekends is different across the week, which also results in obvious differences in the load curve.

Table 1 displays the annual mean of daily maximum load and daily average load for the e-bus charging station in Baoding in 2016 on the basis of day types. It can be seen that the loads on workdays are relatively higher than those on weekends. Thus, a week can be divided into two categories, namely workdays, including Monday to Friday, and weekends, which contain Saturday and Sunday. Special holidays, such as Dragon Boat Day, Labor Day or National Day, can be separated as a new type alone.

Table 1. Load characteristics of different day types.

Day Type	Annual Average Daily Maximum Load/kW	Annual Average Daily Average Load/kW
Monday	669.16	386.70
Tuesday	663.28	377.07
Wednesday	649.63	376.95
Thursday	647.03	366.23
Friday	636.54	370.55
Saturday	573.46	338.97
Sunday	590.45	349.94

2.2. Meteorological Conditions

Data related to meteorological conditions and the power load of Baoding from August 16 to September 15, 2017 (31 days in total) are collected and shown in Figure 2. The meteorological conditions include the daily maximum temperature, daily weather, daily average wind speed and daily average humidity. In the daily weather condition, "1" is used to represent a sunny day, "2" is used to represent cloudy day, and "3" is used to represent a rainy or snowy day. As can be seen in Figure 2, there is a significant positive correlation between daily maximum temperature and power load, and weather and power load show a negative correlation. However, there is no obvious relationship between the average wind speed factor and load, and the average humidity factor is similar. Thus, it can be found that the load of e-bus charging stations is remarkably affected by temperature, as well as by rainy and snowy days, while the influence of other meteorological conditions such as humidity and wind speed is so weak that they can be omitted. Therefore, temperature and rainy and snowy days are selected as influential indicators in this paper.

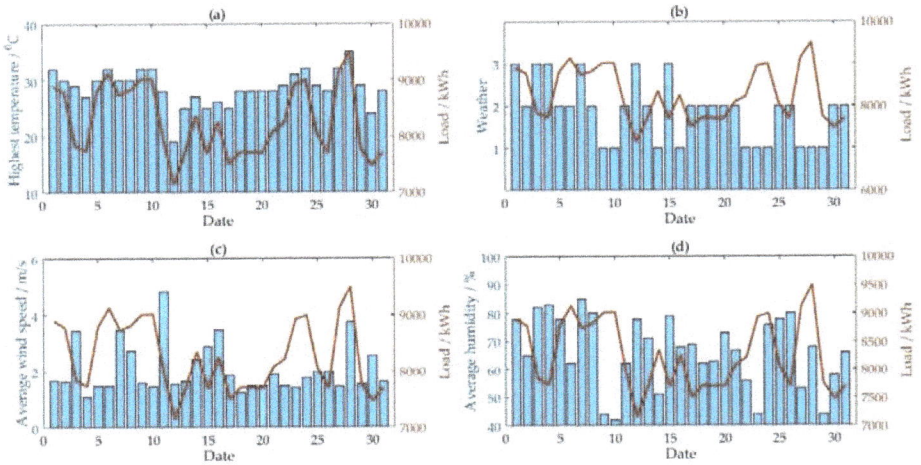

Figure 2. The relationship between meteorological conditions and power load. (**a**) the highest temperature and load; (**b**) the weather and load; (**c**) the average wind speed and load; (**d**) the average humidity and load.

Similar to traditional power loads, the daily load of the charging station increases owing to the use of air conditioners on e-buses when the temperature change of coldness and warmth is aggravated. Since temperature has an important influence on battery capacity, as well as on the charging and discharging process, the charging time is diverse at different temperatures, which also leads to distinct trends of load. The daily load curves from September 12 to 14, 2017 are taken as an example, in which the total number of charged e-buses in these three days was about 60 and the maximum temperature dropped from 35 to 24. As seen in Figure 3, the violent fluctuation of air temperature in adjacent days causes great changes in daily load curves. Thus, it is necessary to take temperature as an influential factor in the selection of subsequent similar day samples.

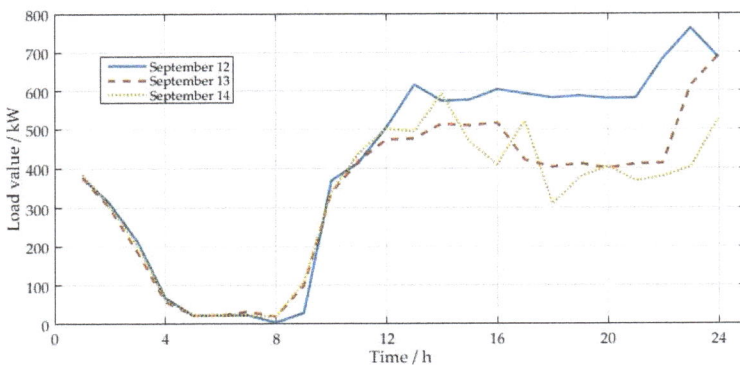

Figure 3. Relationship between temperature and daily load.

Taking the daily load curves on August 29 and August 30 in 2017 as an example, weather conditions can be divided into sunny days and rainy days. Figure 4 illustrates the relationship between weather conditions and the daily load of the charging station. It proves that daily maximum load decreases on rainy and snowy days on account of the deceleration of e-buses, which leads to a decrease in the daily driving mileage and charging times as well as the reduction of total load in the charging station. To this end, rainy and snowy days are another vital factor that affects the load characteristics of e-bus charging stations.

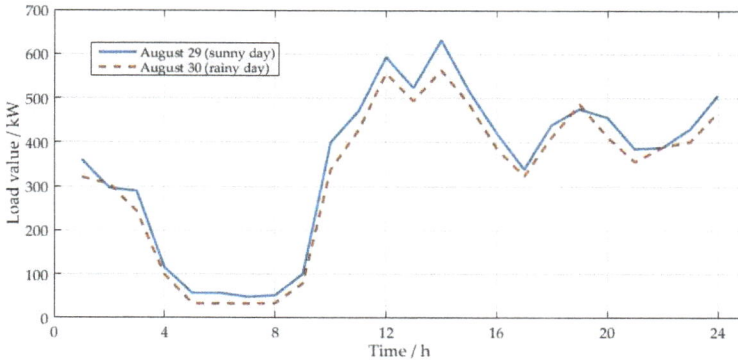

Figure 4. Relationship between weather conditions and daily load.

2.3. Bus Dispatching

The scheduling of departure time and off-running time is a momentous task for bus operation companies. In light of the daily plan of bus dispatching, different charging intensities of e-buses in the station cause changes in the daily load curve in the charging station at different periods. Moreover, diverse demands of the public, traffic jams, and sudden situations require the addition of temporary e-buses to enhance transport capacity, which brings about changes in bus scheduling on different days. Bus dispatching is one of the direct reasons for the fluctuation of daily load curve and the distinction of load curves among days. According to the dispatch plan made in advance, the total number of e-buses that need to be charged on a predicted day can be estimated; namely, the accumulated number of e-buses charged daily, which is used as an indicator to reflect the effect of bus dispatching on the load of the quick-change e-bus charging station.

3. Methodology

3.1. Fuzzy Clustering

FC analysis is a mathematical technique that achieves classification of objects through the establishment of fuzzy similarity relations based on their characteristics, familiarity and comparability. The fuzzy equivalent matrix dynamic clustering method is implemented in this paper.

Suppose n samples on the predicted day, that is $\mathbf{X} = [\mathbf{x}_1, \mathbf{x}_2, ..., \mathbf{x}_n]$. Each sample $\mathbf{x}_j$ comprises m indicators, expressed as $x_j = [x_{j1}, x_{j2}, ..., x_{jm}]^T, j = 1, 2, ..., n$.

The specific steps of FC can be explained as follows:

(1) Data standardization. Considering different dimensions and orders of magnitude, the data must be standardized as Equation (1) [27].

$$x'_{jk} = (x_{jk} - x_{kmin})/(x_{kmax} - x_{kmin}), \quad (j = 1, 2, ..., n; k = 1, 2, ..., m) \tag{1}$$

where x_{jk} is the raw data, x_{kmin} and x_{kmax} are the minimum and maximum of $x_{1k}, x_{2k}, \cdots, x_{nk}$, respectively, x'_{jk} is the standardized data.

(2) Establishment of fuzzy similarity relation matrix. In order to measure the comparability of the classified samples, a fuzzy similarity relation matrix $\mathbf{R} = \{r_{ij}\}$ needs to be constructed by similarity of

coefficient approach, distance or closeness. An absolute value index method is introduced here [28], as expressed in Equation (2).

$$r_{ij} = \exp\left(-\sum_{k=1}^{m} \left| x'_{ik} - x'_{jk} \right|\right), \ (i = 1, 2, ..., n; \ j = 1, 2, ..., n; \ k = 1, 2, ..., m) \tag{2}$$

Then the transitive closure $\mathbf{R}^*$ of $\mathbf{R}$ can be obtained by square synthesis.

(3) Dynamic clustering. Select an appropriate threshold L to separate $\mathbf{R}^*$. The clustering results are up to the level of L. When L drops from 1 to 0, a dynamic clustering graph is obtained by changing the rough classification to a fine one. The best value of L can be acquired based on its change rate [29].

$$C_i = \frac{L_{i-1} - L_i}{n_i - n_{i-1}} \tag{3}$$

where i is the clustering order of L in a descending form; n_i and n_{i-1} are the number of elements in i and $i-1$ clusters, respectively; L_i and L_{i-1} are the confidence levels in i and $i-1$ clusters, respectively. If $C_i = \max(C_j)$, L_i is treated as the best threshold. Thus, n samples can be separated into several categories and each type contains a different number of samples.

(4) Classification recognition. The category consistent with the forecasted day needs to be identified after sample classification. The Euclidean distance is calculated between the predicted day and the above categories one by one [26]:

$$d_{ij} = \frac{1}{\sqrt{m}} \sqrt{\sum_{k=1}^{m} (x'_{ik} - x'_{jk})^2} \tag{4}$$

where x'_{ik} is the characteristic vector on the predicted day, x'_{jk} represents the characteristic vector of each category. This paper takes the type with the shortest Euclidean distance as the classification of the forecasted day to make the prediction.

3.2. Least Squares Support Vector Machine

As an extension of SVM, LSSVM transforms the inequality constraints into equality ones and converts quadratic programming problems into linear equation ones, which is conducive to the improvement of convergence speed [30].

Set the training samples as $T = \{(x_i, y_i)\}_{i=1}^{N}$, where N is the total number of samples. The regression model can be expressed as follows [31]:

$$y(x) = \boldsymbol{w}^T \times \varphi(x) + b \tag{5}$$

where $\varphi()$ is a function that maps the training samples into a highly dimensional space, $\boldsymbol{w}$ and b represent the weight and bias, respectively.

For LSSVM, the optimization problem can be defined as Equation (6) [32]:

$$min \frac{1}{2} \boldsymbol{w}^T \boldsymbol{w} + \frac{1}{2} \gamma \sum_{i=1}^{N} \xi_i^2 \tag{6}$$

$$s.t. \quad y_i = \boldsymbol{w}^T \phi(x_i) + b + \xi_i, \ i = 1, 2, ..., N \tag{7}$$

where γ is the regularization parameter that balances the complexity and precision of the model. ξ_i equals the error.

To obtain the solution, the Lagrange function can be established as Equation (8).

$$L(w, b, \xi_i, \alpha_i) = \frac{1}{2} w^T w + \frac{1}{2} \gamma \sum_{i=1}^{N} \xi_i^2 - \sum_{i=1}^{N} \alpha_i \left[w^T \varphi(x_i) + b + \xi_i - y_i \right] \tag{8}$$

where α_i is the Lagrange multipliers. Take the derivatives of each variable in the function and make them equal zero:

$$\begin{cases} \frac{\partial L}{\partial w} = 0 \to w = \sum_{i=1}^{N} \alpha_i \varphi(x_i) \\ \frac{\partial L}{\partial b} = 0 \to \sum_{i=1}^{N} \alpha_i = 0 \\ \frac{\partial L}{\partial \xi} = 0 \to \alpha_i = \gamma \xi_i \\ \frac{\partial L}{\partial \alpha} = 0 \to w^T \varphi(x_i) + b + \xi_i - y_i = 0 \end{cases} \tag{9}$$

Eliminate w as well as ξ_i and transform it into the following problem:

$$\begin{bmatrix} 0 & e_n^T \\ e_n & \Omega + \gamma^{-1} \cdot I \end{bmatrix} \times \begin{bmatrix} b \\ a \end{bmatrix} = \begin{bmatrix} 0 \\ y \end{bmatrix} \tag{10}$$

where

$$\Omega = \varphi^T(x_i)\varphi(x_i) \tag{11}$$

$$e_n = [1, 1, ..., 1]^T \tag{12}$$

$$\alpha = [\alpha_1, \alpha_2, ..., \alpha_n] \tag{13}$$

$$y = [y_1, y_2, ..., y_n]^T \tag{14}$$

The solution can be obtained based on the linear equations above:

$$y(x) = \sum_{i=1}^{N} \alpha_i K(x_i, x) + b \tag{15}$$

where $K(x_i, x)$ is the kernel function that meets Mercer's condition. The radial basis function (RBF) is employed as the kernel function here on the basis of its wide convergence region and extensive application scope, as shown in Equation (16).

$$K(x_i, x) = \exp\left\{ -\|x - x_i\|^2 / 2\sigma^2 \right\} \tag{16}$$

where σ^2 represents the kernel parameter that reflects the characteristic of training samples and has influence on generalization ability of the technique.

As we can see, the performance improvement of LSSVM model is greatly dependent on the appropriate setting of the following parameters: regularization parameter γ and kernel parameter σ^2 [33].

3.3. Wolf Pack Algorithm

In consideration of the blindness of manual selection in LSSVM model parameters, the optimal value of regularization parameter γ and kernel parameter σ^2 of LSSVM is obtained through the wolf pack algorithm. The WPA technique is inspired by research on the hunting behaviors of wolves [34]. According to their roles in hunting, wolves can be divided into three types: head wolves, safari wolves and feral wolves, who work together to complete the task. Random walk, call to action and siege are three main behaviors of wolves, which are simulated in the WPA model. The determination of the head wolf and the replacement of the wolf pack follow the common rules that the "winner is the king" and "the survival of the fittest", respectively [35]. WPA is illustrated in Figure 5.

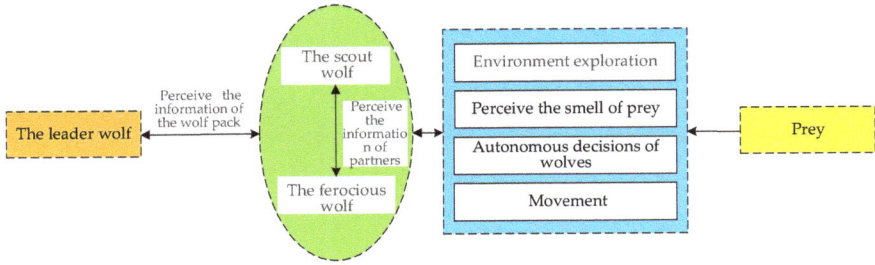

Figure 5. Bionic graph of WPA.

The principle and steps of WPA are summarized as follows [36]:

(1) Initialize wolf pack. Suppose in D dimensional space, there are N wolves, wherein the location of the i-th wolf is set as:

$$X_i = (x_{i1}, x_{i2}, ..., x_{id}), 1 \le i \le N, 1 \le d \le D \qquad (17)$$

The initial position is generated as Equation (18):

$$x_{id} = x_{\min} + rand \times (x_{\max} - x_{\min}) \qquad (18)$$

where *rand* represents random numbers within the range [0,1], and $x_{\max}$ and $x_{\min}$ are the upper limit and lower limit of the search space, respectively.

(2) Generate the head wolf. The wolf at Y_{lead} with the best target function is selected as the head one. The head wolf does not update its position in the hunting process or participate in hunting; instead, it is directly iterated. If $Y_{lead} < Y_i$, $Y_{lead} = Y_i$, where Y_i represents the location of the safari wolf i. Otherwise, the safari wolf i randomly walks in h directions until the maximum value H is achieved or the location cannot be further optimized; then the search is stopped. y_{ijd} is the location at j-th point in d-th dimension of the i-th wolf.

$$y_{ijd} = y_{id} + rand \times step_a \qquad (19)$$

(3) Keep close to the prey. The head wolf pushes the wolf pack to update their positions through call to action. The new position of the i-th wolf in d-dimension is described as Equation (20):

$$z_{id} = x_{id} + rand \times step_b \times (x_{id} - x_{lid}) \qquad (20)$$

where $step_a$ is the step length of wolves in search, $step_b$ represents the step length of wolves towards the target, x_{id} and x_{lid} are the location of the i-th wolf and the corresponding head wolf in d-dimension, respectively.

(4) Encircle the prey. The head wolf sends signals to the surrounding wolf pack after finding the prey so that the encirclement and suppression of the target prey can be completed, as shown in Equations (21) and (22):

$$X_i^{t+1} = \begin{cases} X_i^t, & r_m < \theta \\ X_i + rand \times ra, & r_m > \theta \end{cases} \qquad (21)$$

$$ra(t) = ra_{\min} \times (x_{\max} - x_{\min}) \times e^{\frac{\ln(ra_{\min}/ra_{\max})}{maxt}} \qquad (22)$$

where t equals the number of iterations, ra is the step length at the time of encirclement and suppression, X_i is the location of the head wolf that sends the signal, and X_i^t is the location of the i-th wolf in the t-th iteration.

(5) The mechanism of competition and regeneration of the wolf pack. In encirclement and suppression, the wolves that fail to get food will be eliminated and the rest of wolves will be retained. Simultaneously, new wolves are randomly generated in the same number as the eliminated ones.

(6) Judge whether the maximum number of iterations has been reached. If the maximum number of iterations has been reached, the position of the wolf is output; that is, the optimal value of the LSSVM's parameters. If the maximum number of iterations has not been reached, then return to step 2.

3.4. Establishment of the Hybrid Forecasting Model

This paper firstly analyzes the influential load factors for quick-change e-bus charging stations, and FC is implemented to extract similar days to the predicted one as the training samples. Then, WPA is integrated with the LSSVM model to obtain the optimal values of γ and σ^2. Finally, an analysis is performed on the forecasting results. The framework of the proposed hybrid approach is displayed in Figure 6.

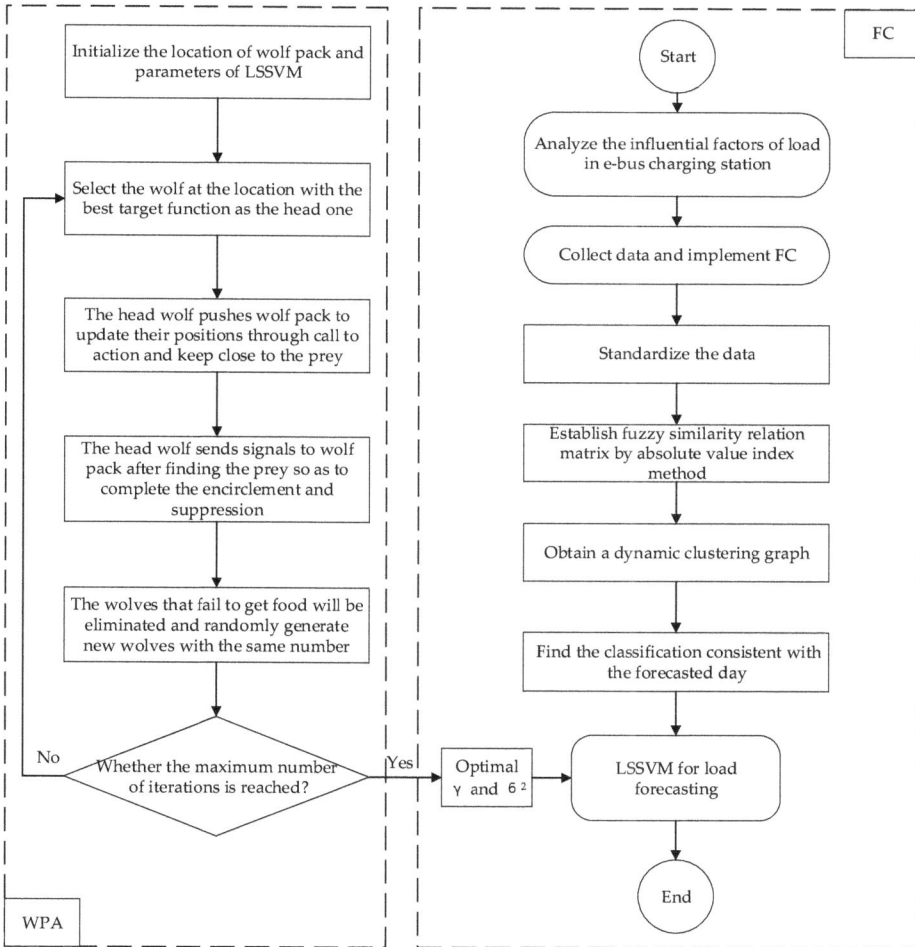

Figure 6. The flow chart of the proposed forecasting model.

4. Case Study

Base on the daily load, meteorological data and operation information of an e-bus charging station in Baoding, China, in 2017, a case study was carried out for the purpose of demonstrating the efficiency of the proposed model in load forecasting for e-bus charging station. The load data was provided by State Grid Hebei Electric Power Company in China, and the input data was provided by the local meteorological department. This paper adopts Matlab R2014b (Gamax Laboratory Solutions Kft., Budapest, Hungary) to program, and as for the test platform environment, an Intel Core i5-6300U (Intel Corporation, Santa Clara, CA, USA), 4G memory and Windows 10 Professional (Microsoft corporation, Redmond, WA, USA) Edition system was used. In order to eliminate the particularity of the target days and examine the generalization performance of the established technique, the data for one day from each of the four seasons was selected as test samples; that is, April 15, July 15, October 15 and January 15 were chosen as test samples for spring, summer, autumn and winter, respectively.

4.1. Input Selection and Pre-Processing

Based on the analysis of load characteristics in the e-bus charging station in Section 2, a set of eight variables was used as the input, including day type, maximum temperature, minimum temperature, weather condition, the accumulated daily number of charged e-buses and the loads at the same moment in the previous three days. Days can be divided into three categories: workdays (Monday to Friday), weekends (Saturday and Sunday) and legal holidays were valued at 1, 0.5 and 0, respectively. Weather conditions were separated into two types, where sunny and cloudy days were valued at 1, and rainy and snowy days were valued at 0.5. The loads at the same moment in the previous three days refer to those nearest the predicted day in similar samples after clustering according to the rule that "Everything looks small in the distance and is big on the contrary." The temperature, load data, and daily accumulated charged e-buses should be normalized as presented in Equation (1).

4.2. Model Performance Evaluation

It's important to effectively evaluate the load forecasting results for e-bus charging stations, and the performance of the prediction models is usually assessed by statistical criteria: the relative error (RE), root mean square error (RMSE), mean absolute percentage error (MAPE) and average absolute error (AAE). The smaller the values of these four indicators are, the better the forecasting performance is. In addition, the indicators named RMSE, MAPE and AAE can reflect the overall error of the prediction model and the degree of error dispersion. The smaller the values of these three indicators are, the more concentrated the distribution of errors is. The four generally adopted error criteria are displayed as follows:

(1) Relative error (RE)

$$RE = \frac{\hat{x}_i - x_i}{x_i} \times 100\% \tag{23}$$

(2) Root mean square error (RMSE)

$$RMSE = \sqrt{\frac{1}{n}\sum_{i=1}^{n}\left(\frac{\hat{x}_i - x_i}{x_i}\right)^2} \tag{24}$$

(3) Mean absolute percentage error (MAPE)

$$MAPE = \frac{1}{n}\sum_{i=1}^{n}|(\hat{x}_i - x_i)/x_i| \cdot 100\% \tag{25}$$

(4) Average absolute error (AAE)

$$AAE = \frac{1}{n}(\sum_{i=1}^{n}|\hat{x}_i - x_i|) / (\frac{1}{n}\sum_{i=1}^{n} x_i) \tag{26}$$

where x and $\hat{x}$ are the actual load and the forecasted one of charging station, respectively; n equals the number of groups in the dataset. The smaller these evaluation indicators are, the higher the prediction accuracy is.

4.3. Results Analysis

The parameters of the proposed model are set as: the total wolf pack $N = 50$, iteration number $t = 100$, $step_a = 1.5$, $step_b = 0.8$, $q = 6$, $h = 5$. The forecasting results are shown in Figure 7.

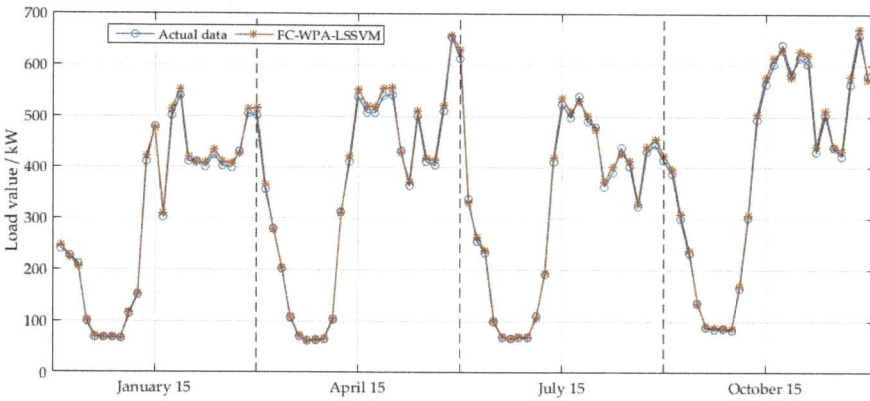

Figure 7. Forecasting results of the proposed model.

As can be seen from Figure 7, the proposed model is very close to the actual load curve in each season and has a good degree of fit. Figure 8 shows the relative error of the prediction results. It can be seen that the relative error of the prediction results of the FC-WPA-LSSVM model is controlled within the range $[-3\%, 3\%]$, and the degree of deviation is acceptable.

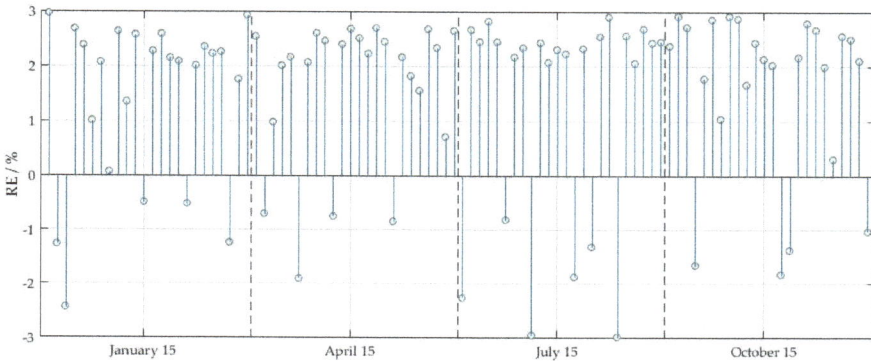

Figure 8. The RE of the proposed model.

4.4. Discussion

In order to verify the performance of the forecasting approach, three basic techniques, including WPA-LSSVM [37], LSSVM [38], and BPNN [39], were introduced to make a comparison. The parameter settings in WPA-LSSVM were consistent with those in the established model. In LSSVM, the regularization parameter γ and the kernel parameter σ^2 were valued at 12.6915 and 12.0136, respectively. In BPNN, tansig was utilized as the transfer function in the hidden layer, and purelin was employed as the transfer function in the output layer. The maximum number of convergence was 200, the error was equal to 0.0001, and the learning rate was set as 0.1. The determination of the initial weights and thresholds depend on their own training. Figure 9 illustrates the load forecasting results of FC-WPA-LSSVM, WPA-LSSVM, LSSVM and BPNN. Figure 10 presents the values of RE for each prediction method.

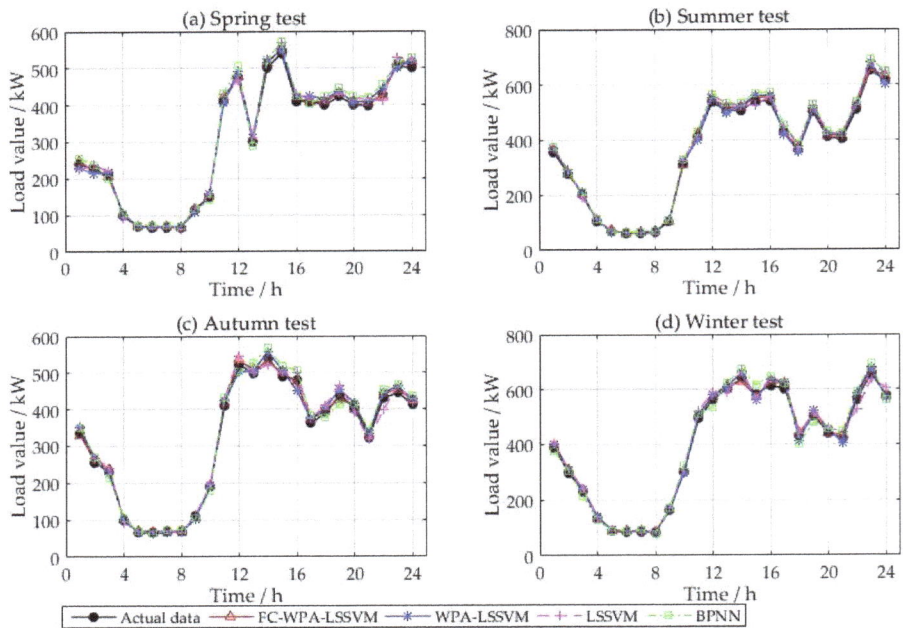

Figure 9. Forecasting results: (**a**) forecasting results of Spring test; (**b**) forecasting results of Summer test; (**c**) forecasting results of Autumn test; (**d**) forecasting results of Winter test.

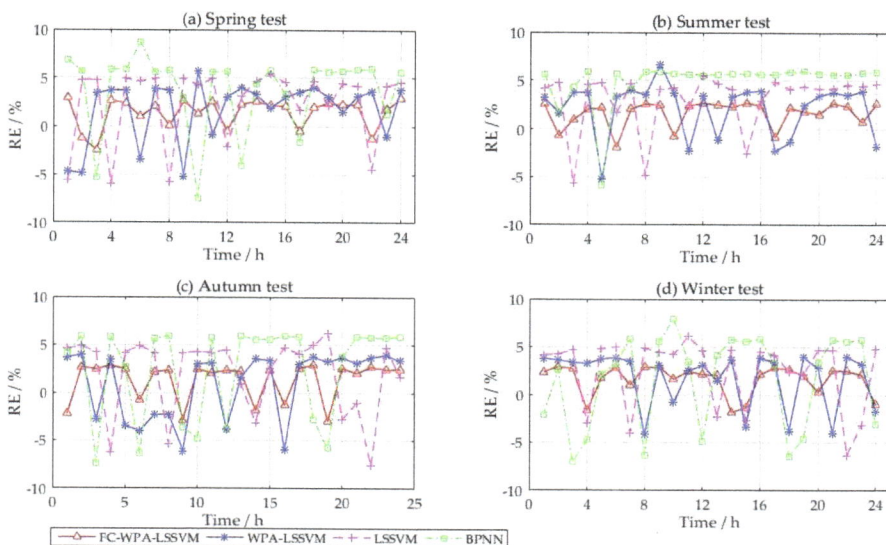

Figure 10. RE of forecasting approaches: (**a**) RE of Spring test; (**b**) RE of Summer test; (**c**) RE of Autumn test; (**d**) RE of Winter test.

From Figures 9 and 10, it can be seen that the prediction error range of FC-WPA-LSSVM was controlled to within [−3% + 3%], where the minimum error (7:00 in the spring test) and the maximum error (18:00 in the autumn test) were 0.08% and −2.98%, respectively. Among them, 10 error points of the results were within [−1%, 1%], namely 7:00, 11:00 and 16:00 in the spring test, 1:00, 2:00, 9:00, 16:00, 23:00 in the summer test, 6:00 in the autumn test, 19:00 in the winter test; the corresponding values of RE were 0.08%, −0.49%, −0.52%, −0.71%, −0.98%, −0.74%, −0.85%, 0.71%, −0.81% and 0.31%, respectively. In addition, 19 error points of WPA-LSSVM were controlled to within [−3%, 3%], while the corresponding number for LSSVM was 17, of which 2 points of WPA-LSSVM were within the range [−1%, 1%], namely at 10:00 in the spring test (RE = −0.86%) and 9:00 in the winter test (RE = − 0.79%), but all error points of LSSVM were outside the range [−1%, 1%]. The minimum errors of WPA-LSSVM and LSSVM were −0.79% and −1.07% respectively, while their maximum errors were 6.6% and −7.59%, respectively. The errors of the BPNN model were mostly within the ranges [−6%, −4%] or [4%, 6%], where the maximum and minimum of RE were individually equal to 1.36% and 8.73%, respectively. In this regard, the forecasting accuracy ranked from the highest to the lowest was: FC-WPA-LSSVM, WPA-LSSVM, LSSVM, and BPNN. Hence, FC can effectively avoid the blindness in the selection of similar days through experience. In contrast with LSSVM, administering WPA improves the prediction precision by virtue of the parameter optimization of LSSVM. It is without doubt that the forecasting accuracy of some points in FC-WPA-LSSVM is worse than the other three approaches; for instance, the error of FC-WPA-LSSVM was 1.76% at 22:00 in the spring test, which was greater than WPA-LSSVM and BPNN.

The performance comparison results of the forecasting models were measured by RMSE, MAPE and AAE, as presented in Figure 11. This demonstrates that the proposed approach outperforms the other models in terms of all the evaluation criteria, of which RMSE, MAPE and AAE of FC-WPA-LSSVM were equal to 2.20%, 2.09% and 2.09%, respectively. This is mainly due to the fact that FC can overcome the adverse effects of unconventional load data caused by factor mutation on LSSVM training, and WPA improves the generalization ability and prediction accuracy by parameter optimization in LSSVM model. In comparison with BPNN, LSSVM can avoid the drawbacks of premature convergence and easily falling into local optimum.

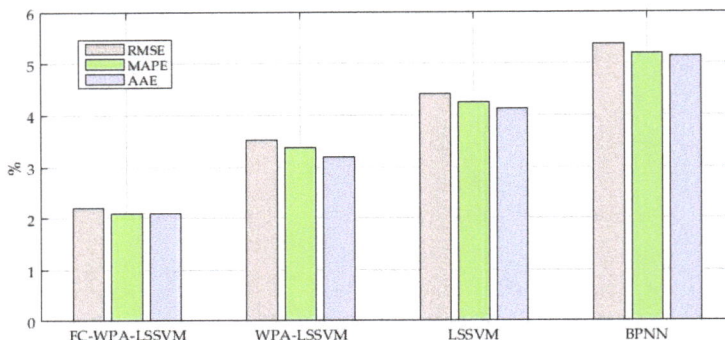

Figure 11. RMSE, MAPE and AAE of the forecasting results (I).

5. Further Study

In order to further verify the validity of the proposed method, another e-bus charging station in Baoding, China, was selected for an experimental study. The load data of the station from January, 2016 to December, 2016 are provided in this paper, where seven successive days in each season were taken as test samples and the remaining data were used as training samples. The setting of parameters in WPA-LSSVM was consistent with the proposed method. In LSSVM, γ and σ^2 were equal to 10.2801 and 11.2675, respectively. The values of the parameters in the BPNN model were same as those in the previous case study. Figure 12 displays the values of RMSE, MAPE and AAE.

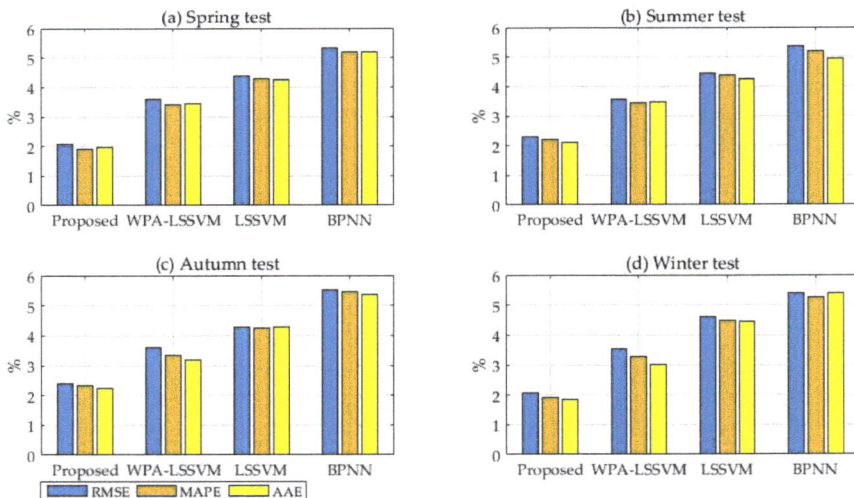

Figure 12. RMSE, MAPE and AAE of the forecasting results (II): (**a**) Spring test; (**b**) Summer test; (**c**) Autumn test; (**d**) Winter test.

From Figure 12, it can be seen that FC-WPA-LSSVM presents the lowest RMSE, MAPE and AAE, with corresponding values of 2.07%, 1.92% and 1.97 in the spring test, 2.29%, 2.20% and 2.11% in the summer test, 2.39%, 2.35% and 2.25% in the autumn test, and 2.08%, 1.90% and 1.84% in the winter test. It can be seen that the overall prediction performance of the forecasting approach was optimal due to the advantages of FC, WPA and LSSVM. In conclusion, the load forecasting model for e-bus charging stations based on FC-WPA-LSSVM can provide accurate data support for the economical operation of the station. In addition, the proposed model can also be applied to the load forecasting of other charging stations, and its prediction accuracy will not be affected by changes in the number of electric vehicles and other factors.

Since this forecasting model is based on MATLAB development, if the transportation company wants to use this model to predict the load in the future, they can also use it easily and obtain the forecast results without additional costs.

6. Conclusions

In view of the load characteristics for e-bus charging stations, this paper selected eight variables, including day type, maximum temperature, minimum temperature, weather condition, the number of accumulated daily number of charged e-buses and the loads at the same moment in the previous three days, as the input. A novel short-term load forecasting technique for e-bus charging stations based on FC-WPA-LSSVM was proposed, in which FC was used to extract similar dates as training samples, and WPA was introduced to optimize the parameters in LSSVM to improve the prediction accuracy. Two case studies were carried out to verify the developed approach in comparison with WPA-LSSVM, LSSVM and BPNN. The experimental results showed that the forecasting precision of the proposed model was better than the contrasting models. Hence, FC-WPA-LSSVM provides a new idea and reference for short-term load forecasting of e-bus charging stations.

The load of e-bus charging stations is a kind of power load with complex change rules and diverse influential factors. With the large-scale application of electric vehicles, more and more e-bus charging stations will start to be put into use. At that time, research on actual operation of charging stations will be more abundant. It is necessary to make further efforts to seek more suitable load forecasting approaches for e-bus charging stations based on the study of load variation rules and the internal relationships between the load and influential factors.

Funding: This work is supported by the Fundamental Research Funds for the Central Universities (Project No. 2018MS144).

Acknowledgments: Thanks for State Grid Hebei Electric Power Company providing the relevant data supporting.

Conflicts of Interest: The authors declare no conflict of interest.

References

1. Tan, S.; Yang, J.; Yan, J.; Lee, C.; Hashim, H.; Chen, B. A Holistic low Carbon city Indicator Framework for Sustainable Development. *Appl. Energy* **2017**, *185*, 1919–1930. [CrossRef]
2. Yan, J.; Chou, S.K.; Chen, B.; Sun, F.; Jia, H.; Yang, J. Clean, Affordable and Reliable Energy Systems for low Carbon city Transition. *Appl. Energy* **2017**, *194*, 305–309. [CrossRef]
3. Majidpour, M.; Qiu, C.; Chu, P.; Pota, H.; Gadh, R. Forecasting the EV Charging load Based on Customer Profile or Station Measurement? *Appl. Energy* **2016**, *163*, 134–141. [CrossRef]
4. Paparoditis, E.; Sapatinas, T. Short-Term Load Forecasting: The Similar Shape Functional Time-Series Predictor. *IEEE Trans. Power Syst.* **2013**, *28*, 3818–3825. [CrossRef]
5. Yildiz, B.; Bilbao, J.I.; Sproul, A.B. A Review and Analysis of Regression and Machine Learning Models on Commercial Building Electricity load Forecasting. *Renew. Sust. Energy Rev.* **2017**, *73*, 1104–1122. [CrossRef]
6. Cerne, G.; Dovzan, D.; Skrjanc, I. Short-term load forecasting by separating daily profile and using a single fuzzy model across the entire domain. *IEEE Trans. Ind. Electron.* **2018**, *65*, 7406–7415. [CrossRef]
7. Ashtari, A.; Bibeau, E.; Shahidinejad, S.; Molinski, T. PEV Charging Profile Prediction and Analysis Based on Vehicle Usage Data. *IEEE Trans. Smart Grid* **2012**, *3*, 341–350. [CrossRef]

8. Dai, Q.; Cai, T.; Duan, S.; Zhao, F. Stochastic Modeling and Forecasting of Load Demand for Electric Bus Battery-Swap Station. *IEEE Trans. Power Deliv.* **2014**, *29*, 1909–1917. [CrossRef]
9. Tarsitano, A.; Amerise, I.L. Short-term load Forecasting Using a Two-Stage Sarimax Model. *Energy* **2017**, *133*, 108–114. [CrossRef]
10. Zhang, W.G.; Xie, F.X.; Huang, M.; Juan, L.; Li, Y. Research on Short-Term Load Forecasting Methods of Electric Buses Charging Station. *Power Syst. Prot. Control* **2013**, *41*, 61–66.
11. Xiao, L.; Shao, W.; Yu, M.; Ma, J.; Jin, C. Research and Application of a Hybrid Wavelet Neural Network Model with the Improved Cuckoo Search Algorithm for Electrical Power SYSTEM forecasting. *Appl. Energy* **2017**, *198*, 203–222. [CrossRef]
12. Li, S.; Wang, P.; Goel, L. A Novel Wavelet-Based Ensemble Method for Short-Term Load Forecasting with Hybrid Neural Networks and Feature Selection. *IEEE Trans. Power Syst.* **2016**, *31*, 1788–1798. [CrossRef]
13. Xiong, X.; Chen, L.; Liang, J. A New Framework of Vehicle Collision Prediction by Combining SVM and HMM. *IEEE Trans. Intell. Transp. Syst.* **2017**, *19*, 1–12. [CrossRef]
14. Yang, Y.L.; Che, J.X.; Li, Y.Y.; Zhao, Y.J.; Zhu, S.L. An incremental electric load forecasting model based on support vector regression. *Energy* **2016**, *113*, 796–808. [CrossRef]
15. Liu, W.; Xiaobo, X.U.; Xi, Z. Daily load forecasting based on SVM for electric bus charging station. *Electr. Power Autom. Equip.* **2014**, *34*, 41–47.
16. Deo, R.C.; Kisi, O.; Singh, V.P. Drought forecasting in eastern Australia using multivariate adaptive regression spline, least square support vector machine and M5Tree model. *Atmos. Res.* **2017**, *184*, 149–175. [CrossRef]
17. Lin, W.M.; Tu, C.S.; Yang, R.F.; Tsai, M.T. Particle swarm optimisation aided least-square support vector machine for load forecast with spikes. *IET Gener. Trans. Distrib.* **2016**, *10*, 1145–1153. [CrossRef]
18. Li, C.; Li, S.; Liu, Y. A least squares support vector machine model optimized by moth-flame optimization algorithm for annual power load forecasting. *Appl. Intell.* **2016**, *45*, 1–13. [CrossRef]
19. Liang, Y.; Niu, D.; Ye, M.; Hong, W.C. Short-Term Load Forecasting Based on Wavelet Transform and Least Squares Support Vector Machine Optimized by Improved Cuckoo Search. *Energies* **2016**, *9*, 827. [CrossRef]
20. Padilha, C.A.D.A.; Barone, D.A.C.; Neto, A.D.D. A multi-level approach using genetic algorithms in an ensemble of Least Squares Support Vector Machines. *Knowl.-Based Syst.* **2016**, *106*, 85–95. [CrossRef]
21. Dong, R.; Xu, J.; Lin, B. ROI-based study on impact factors of distributed PV projects by LSSVM-PSO. *Energy* **2017**, *124*, 336–349. [CrossRef]
22. Sun, W.; Sun, J. Daily PM2.5 Concentration Prediction Based on Principal Component Analysis and LSSVM Optimized by cuckoo Search Algorithm. *J. Environ. Manag.* **2016**, *188*, 144. [CrossRef] [PubMed]
23. Niu, D.; Liang, Y.; Wang, H.; Wang, M.; Hong, W.C. Icing Forecasting of Transmission Lines with a Modified Back Propagation Neural Network-Support Vector Machine-Extreme Learning Machine with Kernel (BPNN-SVM-KELM) Based on the Variance-Covariance Weight Determination Method. *Energies* **2017**, *10*, 1196. [CrossRef]
24. Chen, X.; Tang, C.; Wang, J.; Zhang, L.; Liu, Y. A Novel Hybrid Based on Wolf Pack Algorithm and Differential Evolution Algorithm. *Int. Symp. Comput. Intell. Des.* **2017**, 69–74. [CrossRef]
25. Xue, B.; Zhang, M.; Browne, W.N.; Yao, X. A Survey on Evolutionary Computation Approaches to Feature Selection. *IEEE Trans. Evolut. Comput.* **2016**, *20*, 606–626. [CrossRef]
26. Hassanpour, H.; Zehtabian, A.; Nazari, A.; Dehghan, H. Gender classification based on fuzzy clustering and principal component analysis. *IET Comput. Vis.* **2016**, *10*, 228–233. [CrossRef]
27. Kumar, M.R.; Ghosh, S.; Das, S. Frequency dependent piecewise fractional-order modelling of ultracapacitors using hybrid optimization and fuzzy clustering. *J. Power Sources* **2016**, *335*, 98–104. [CrossRef]
28. Alban, N.; Laurent, B.; Mitherand, N.; Ousman, B.; Martin, N.; Etienne, M. Robust and Fast Segmentation Based on Fuzzy Clustering Combined with Unsupervised Histogram Analysis. *IEEE Intell. Syst.* **2017**, *32*, 6–13. [CrossRef]
29. Bai, X.; Wang, Y.; Liu, H.; Guo, S. Symmetry Information Based Fuzzy Clustering for Infrared Pedestrian Segmentation. *IEEE Trans. Fuzzy Syst.* **2017**. [CrossRef]
30. Zhu, B.; Han, D.; Wang, P. Forecasting carbon price using empirical mode decomposition and evolutionary least squares support vector regression. *Appl. Energy* **2017**, *191*, 521–530. [CrossRef]
31. Yuan, X.; Tan, Q.; Lei, X.; Yuan, Y.; Wu, X. Wind Power Prediction Using Hybrid Autoregressive Fractionally Integrated Moving Average and Least Square Support Vector Machine. *Energy* **2017**, *129*, 122–137. [CrossRef]

32. Niu, D.; Li, Y.; Dai, S.; Kang, H.; Xue, Z.; Jin, X.; Song, Y. Sustainability Evaluation of Power Grid Construction Projects Using Improved TOPSIS and Least Square Support Vector Machine with Modified Fly Optimization Algorithm. *Sustainability* **2018**, *10*, 231. [CrossRef]
33. Niu, D.; Dai, S. A Short-Term Load Forecasting Model with a Modified Particle Swarm Optimization Algorithm and Least Squares Support Vector Machine Based on the Denoising Method of Empirical Mode Decomposition and Grey Relational Analysis. *Energies* **2017**, *10*. [CrossRef]
34. Hu-Sheng, W.U.; Zhang, F.M.; Lu-Shan, W.U. New swarm intelligence algorithm—Wolf pack algorithm. *Syst. Eng. Electr.* **2013**, *35*. [CrossRef]
35. Wu, H.S.; Zhang, F.M. Wolf Pack Algorithm for Unconstrained Global Optimization. *Math. Probl. Eng.* **2014**, *1*, 1–17. [CrossRef]
36. Li, C.M.; Du, Y.C.; Wu, J.X.; Lin, C.H.; Ho, Y.R.; Lin, Y.J.; Chen, T. Synchronizing chaotification with support vector machine and wolf pack search algorithm for estimation of peripheral vascular occlusion in diabetes mellitus. *Biomed. Signal Process. Control* **2014**, *9*, 45–55. [CrossRef]
37. Mustaffa, Z.; Sulaiman, M.H. Price predictive analysis mechanism utilizing grey wolf optimizer-Least Squares Support Vector Machines. *J. Eng. Appl. Sci.* **2015**, *10*, 17486–17491.
38. Li, X.; Lu, J.H.; Ding, L.; Xu, G.; Li, J. In Building Cooling Load Forecasting Model Based on LS-SVM. In Proceedings of the IEEE Internatonal Asia-Pacific Conference on Information Processing, Shenzhen, China, 18–19 July 2009.
39. Bin, H.; Zu, Y.X.; Zhang, C. A Forecasting Method of Short-Term Electric Power Load Based on BP Neural Network. *Appl. Mech. Mater.* **2014**, *538*, 247–250. [CrossRef]

![energies logo] *energies*

MDPI

Article

A Hybrid BA-ELM Model Based on Factor Analysis and Similar-Day Approach for Short-Term Load Forecasting

Wei Sun and Chongchong Zhang *

Department of Business Administration, North China Electric Power University, Baoding 071000, China; bdsunwei@126.com
* Correspondence: mr_zhangcc@126.com

Received: 9 April 2018; Accepted: 2 May 2018; Published: 17 May 2018

Abstract: Accurate power-load forecasting for the safe and stable operation of a power system is of great significance. However, the random non-stationary electric-load time series which is affected by many factors hinders the improvement of prediction accuracy. In light of this, this paper innovatively combines factor analysis and similar-day thinking into a prediction model for short-term load forecasting. After factor analysis, the latent factors that affect load essentially are extracted from an original 22 influence factors. Then, considering the contribution rate of history load data, partial auto correlation function (PACF) is employed to further analyse the impact effect. In addition, ant colony clustering (ACC) is adopted to excavate the similar days that have common factors with the forecast day. Finally, an extreme learning machine (ELM), whose input weights and bias threshold are optimized by a bat algorithm (BA), hereafter referred as BA-ELM, is established to predict the electric load. A simulation experience using data deriving from Yangquan City shows its effectiveness and applicability, and the result demonstrates that the hybrid model can meet the needs of short-term electric load prediction.

Keywords: short-term load forecasting; factor analysis; ant colony clustering; extreme learning machine; bat algorithm

---

1. Introduction

Short-term load forecasting is an important component of smart grids, which not only can achieve the goal of saving cost but also ensure a continuous flow of electricity supply [1]. Moreover, against the background of energy-saving and emission-reduction, accurate short-term load prediction plays an important role in avoiding a waste of resources in the process of power dispatch. Nevertheless, it should be noted that the inherent irregularity and linear independence of the loading data present a negative effect on the exact power load prediction.

Since the 1950s, short-term load forecasting has been attracting considerable attention from scholars. Generally speaking, the methods for load forecasting can be classified into two categories: traditional mathematical statistical methods and approaches which are based on artificial intelligence. The conventional methods like regression analysis [2,3] and time series [4] are mainly based on mathematical statistic models such as the vector auto-regression model (VAR) and auto-regressive moving average model (ARMA). With the development of science and technology, the shortcomings of statistical models, such as the effect of regression analysis based on historical data that will be weakened with the extension of time or the results of time-series prediction that are not ideal when the stochastic factors are large, are beginning to appear and are criticized by researchers for their low non-linear fitting capability.

Owing to the characteristic of strong self-learning, self-adapting ability and non-linearity, artificial intelligence methods such as back propagation neural networks (BPNN), support vector machine (SVM) as well as the least squares support vector machine (LSSVM) etc. have obtained greater attention and have had a wide application in the field of power load forecasting during the last decades [5,6]. Park [7] and his partners first used the artificial neural network in electricity forecasting. The experimental results demonstrated the higher fitting accuracy of the artificial neural network (ANNs) compared with the fundamental methods. Hernandez et al. [8] successfully presented a short-term electric load forecast architectural model based on ANNs and the results highlighted the simplicity of the proposed model. Yu and Xu [9] proposed a combinational approach for short-term gas-load forecasting including the improved BPNN and the real-coded genetic algorithm which is employed for the parameter optimization of the prediction model, and the simulation illustrated its superiority through the comparisons of several different combinational algorithms. Hu et al. [10] put forward a generalized regression neural network (GRNN) optimized by the decreasing step size fruit fly optimization algorithm to predict the short-term power load, and the proposed model showed a better performance with a stronger fitting ability and higher accuracy in comparison with traditional BPNN.

Yet, the inherent feature of BPNN may cause low efficiency and local optimal. Furthermore, the selection of the number of BPNN hidden nodes depends on trial and error. As a consequence, it is difficult to obtain the optimal network. On the basis of structural risk, empirical risk and vapnik–chervonenkis (VC) dimension bound minimization principle, the support vector machine (SVM) showed a smaller practical risk and presented a better performance in general [11]. Zhao and Wang [12] successfully conducted a SVM for short-term load forecasting, and the results demonstrated the excellence of the forecasting accuracy as well as computing speed. Considering the difficulty of the parameter determination that appeared in SVM, the least squares support vector machine (LSSVM) was put forward as an extension, which can transform the second optimal inequality constraints problem in original space into an equality constraints' linear system in feature space through non-linear mapping and further improve the speed and accuracy of the prediction [13]. Nevertheless, how to set the kernel parameter and penalty factor of LSSVM scientifically is still a problem to be solved.

Huang et al. [14] proposed a new single-hidden layer feed forward neural network and named it as the extreme learning machine (ELM) in 2009, in which one can randomly choose hidden nodes and then analytically determine the output weights of single-hidden layer feed-forward neural network (SLFNs). The extreme learning machine tends to have better scalability and achieve similar (for regression and binary class cases) or much better (for multi-class cases) generalization performance at much faster learning speed (up to thousands of times) than the traditional SVM and LSSVM [15]. However, it is worth noting that the input weights matrix and hidden layer bias assigned randomly may affect the generalization ability of the ELM. Consequently, employing an optimization algorithm so as to obtain the best parameters of both the weight of input layer and the bias of the hidden layer is vital and necessary. The bat algorithm (BA), acknowledged as a new meta-heuristic method, can control the mutual conversion between local search and global search dynamically and performs better convergence [16]. Because of the excellent performance of local search and global search in comparison with existing algorithms like the genetic algorithm (GA) and particle swarm optimization algorithm (PSO), researchers and scholars have applied BA in diverse optimization problems extensively [17–19]. Thus, this paper adopted the bat algorithm to obtain the input weight matrix and the hidden layer bias matrix of ELM corresponding to the minimum training error, which can not only maximize the merit of BA's global and local search capability and ELM's fast learning speed, but also overcome the inherent instability of ELM.

The importance of forecasting methods is self-evident, yet the analysis and processing of the original load data also cannot be ignored. Some predecessors have supposed historical load and weather as the most influential factors in their research [20–22]. However, selecting the historical load data scientifically or not can cause a strong impact on the accuracy of prediction. In addition, there are still many other external weather factors that may also potentially influence the power load. Only considering the temperature as the input variable may be not enough [23–25], and other meteorological factors such as humidity, visibility and air pressure etc. also should be taken into consideration. Besides, it is necessary to analyze and pretreat the influence factors on the premise of considering the influence factors synthetically so as to achieve the goal of improving the generalization ability and the precision of the prediction model. Therefore, this paper applied factor analysis (FA) and the similar-day approach (SDA) for input data pre-processing, where the former is utilized to extract the latent factors that essentially affect the load and the SDA is adopted to excavate the similar days that have common factors with the forecast day.

To sum up, the load forecasting process of the ELM optimized by the bat algorithm can be elaborated in four steps. Firstly, based on 22 original influence factors, factor analysis is adopted to extract the latent factors which essentially affect load. To further explore the relationship between historical load and current load, a partial auto correlation function (PCAF) is applied to demonstrate the significance of previous data. Then, in accordance with the latent factors and the loads of each day, ant colony clustering is used to divide the load to different clusters.

The rest of the paper is organized as follows: Section 2 gives a brief description about the material and methods, including bat algorithm (BA), extreme learning machine (ELM), ant colony clustering algorithm (ACC) as well as the framework of the whole model. Data analysis and processing are considered in Sections 3 and 4 which present an empirical analysis of the power load forecasting. Finally, conclusions are drawn in Section 5.

2. Methodology

2.1. Bat Algorithm

Based on the echolocation of micro-bats, Yang [26] proposed a new meta-heuristic method and called it the bat algorithm, one that combines the advantages both the genetic algorithm and particle swarm optimization with the superiority of parallelism, quick convergence, distribution and less parameter adjustment. In the d dimensions of search space during the global search, the bat i has the position of x_i^t, and velocity v_i^t at the time of t, whose position and velocity will be updated as Equations (1) and (2), respectively:

$$x_i^{t+1} = x_i^t + v_i^{t+1}; \tag{1}$$

$$v_i^{t+1} = v_i^t + \left(x_i^t - \hat{x}\right) \cdot F_i \tag{2}$$

where $\hat{x}$ is the current global optimal solution; and F_i is the sonic wave frequency which can be seen in Equation (3):

$$F_i = F_{min} + (F_{max} - F_{min})\beta \tag{3}$$

where β is a random number within [0, 1]; F_{max} and F_{min} are the max and min sonic wave frequency of the bat I. In the process of flying, each initial bat is assigned one random frequency in line with $[F_{min}, F_{max}]$.

In local search, once a solution is selected in the current global optimal solution, each bat would produce a new alternative solution in the mode of random walk according to Equation (4):

$$x_n(i) = x_0 + \mu A^t \tag{4}$$

where x_0 is a solution that is chosen in current optimal disaggregation randomly; A^t is the average volume of the current bat population; and μ is a D dimensional vector within in [−1, 1].

The balance of bats is controlled by the impulse volume A(i) and impulse emission rate R(i). Once the bat locks the prey, the volume A(i) will be reduced and the emission rate R(i) will be increased at the same time. The update of A(i) and R(i) are expressed as Equations (5) and (6), respectively:

$$A^{t+1}(i) = \gamma A^t(i) \tag{5}$$

$$R^{t+1} = R^0(i) \cdot (1 - e^{-\theta t}) \tag{6}$$

where γ and θ are both constants that γ is within [0, 1] and $\theta > 0$. This paper set the two parameters as $\gamma = \theta = 0.9$. The basic steps of the standard bat algorithm can be summarized as the pseudo code seen in the following:

Bat algorithm.

1: Initialize the location of bat populations x_i ($i = 1, 2, 3, \ldots, n$) and velocity v_i
2: Initialize frequency F_i pulse emission rate R_i and loudness A_i
3: While (t < the maximum number of iterations)
4:      Generate new solutions by adjusting the frequency
5:      Generate new velocity and location
6:      If (rand >R_i)
7:      Select a solution among best solutions
8:      Generate new local solution around the selected best solution
9: End if
10:      Get a new solution through flying randomly
11: If (rand < A_i & $f(x_i) < f(x^*)$)
12:      Accept the new solution
13:      Increase ri and decrease A_i
14: End if
15:      Rank the bats and find the current best x^*.
16: End

2.2. Extreme Learning Machine

After setting the input weights and hidden layer biases randomly, the output weights of the ELM can be analytically determined by solving a linear system in accordance with the thinking of the Moore–Penrose (MP) generalized inverse. The only two parameters needed to be assigned allow the extreme learning machine to generate the input weights matrix and hidden layer biases automatically at fast running speed. Consequently, the extreme learning machine expresses the advantages of a fast learning speed, small training error and strong generalization ability compared with the traditional neural networks in solving non-linearity problems [27]. The concrete framework of ELM is shown in Figure 1 and the computational steps of the standard ELM can be illustrated as follows:

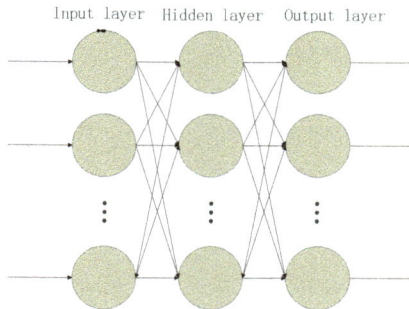

Figure 1. The framework of the extreme learning machine.

The connection weights both between input layer and hidden layer and between hidden layer and output layer as well as the hidden layer neuron threshold are shown in the following:

$$\omega = [\omega_{i1}, \omega_{i2}, \cdots, \omega_{in}]_{L \times n} \tag{7}$$

where ω is the connection weights between input layer and hidden layer; n is the input layer neuron number, and L is the hidden layer neuron number, and,

$$\beta = [\beta_{i1}, \beta_{i2}, \cdots, \beta_{im}]_{L \times m} \tag{8}$$

where β is the connection weights between hidden layer and output layer and m is the output layer neuron number, and,

$$X = [x_{i1}, x_{i2}, \cdots, x_{iQ}]_{n \times Q} \tag{9}$$

$$Y = \left[y_{i1}, y_{i2}, \cdots, y_{iQ} \right]_{m \times Q} \tag{10}$$

where X is the input vector and Y is the corresponding output vector, and,

$$H = \begin{bmatrix} g(\omega_1 x_1 + b_1) & g(\omega_2 x_1 + b_2) & \cdots & g(\omega_1 x_1 + b_1) \\ g(\omega_1 x_2 + b_1) & g(\omega_2 x_2 + b_2) & \cdots & g(\omega_1 x_2 + b_1) \\ \vdots & \vdots & & \vdots \\ g(\omega_1 x_Q + b_1) & g(\omega_2 x_Q + b_2) & \cdots & g(\omega_1 x_Q + b_1) \end{bmatrix} \tag{11}$$

where H is the hidden layer output matrix, b is the bias which is generated randomly in the process of network initialization, and g(x) is the activation function of the ELM.

2.3. Ant Colony Clustering Algorithm

When processing the large number of samples, the traditional clustering learning algorithm often has the disadvantages of slow clustering speed, falling easily into local optimal, and it is difficult to obtain the optimal clustering result. At the same time, the clustering algorithm involves the selection of the number of clustering K, which directly affects the clustering result. Using ant colony clustering to pre-process the load samples can reduce the number of input samples on the premise of including all sample features, and also can effectively simplify the network structure and reduce the calculation effort. The flowchart of the ant colony clustering algorithm is shown in Figure 2.

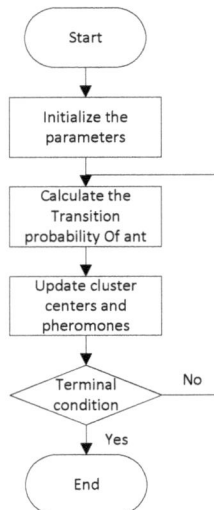

Figure 2. The flowchart of the ant colony clustering algorithm.

2.4. Introduction of Factor Analysis-Ant Colony Clustering-Bat Algorithm-Extreme Learning Machine (FA-ACC-BA-ELM) Model

Since the ELM has less ability to respond to samples of the training set, its generalization ability is insufficient. So we propose BA-ELM. In this paper, the flowchart of the factor analysis-similar day-bat algorithm-extreme learning machine (FA-SD-BA-ELM) model is shown in Figure 3. As discussed in part 1, auto correlation and the partial correlation function (PACF) are executed to analyze the inner relationships between the history loads. Based on the influencing factors of load, factor analysis (FA) is used for extracting input variables. According to the result of factors analysis and previous load, the ant colony clustering algorithm (ACC) is used to find historical days that have common factors similar to the forecast day. Part 2 is the bat optimization algorithm (BA) and part 3 is the forecasting of the extreme learning machine (ELM).

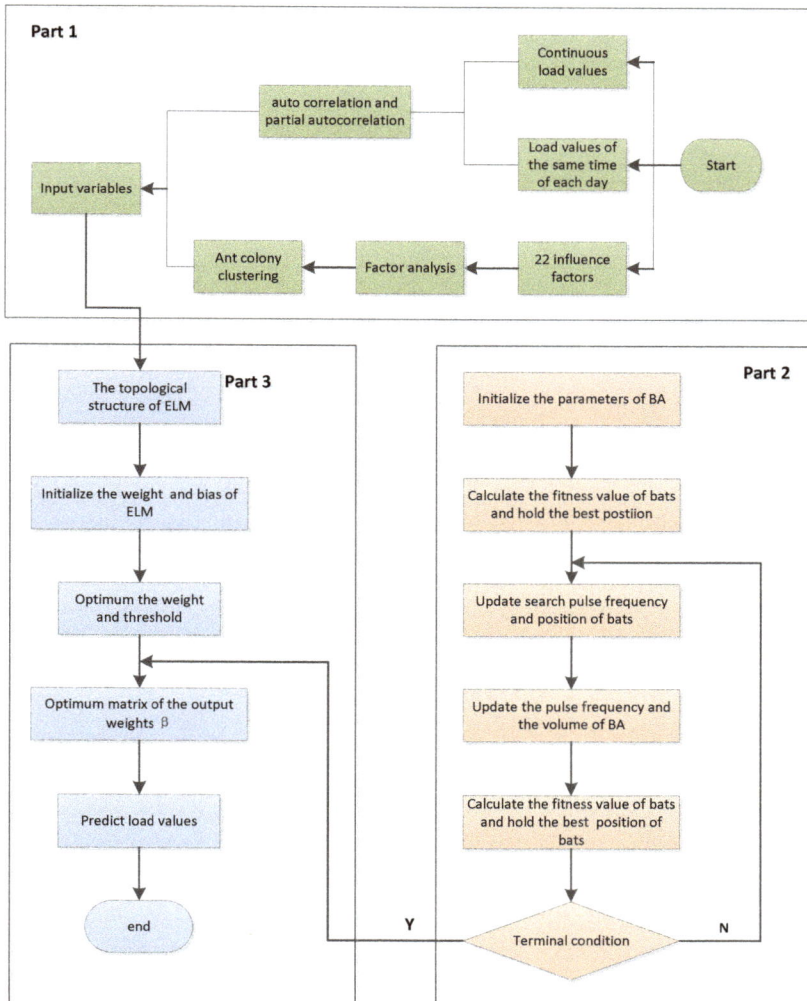

Figure 3. The flowchart of the factor analysis-ant colony clustering-bat algorithm-extreme learning machine (FA-ACC-BA-ELM) model.

3. Data Analysis and Preprocessing

3.1. Selection of Influenced Indexes

Considering that the human activities are always disturbed by many external factors and then the power load is affected, some effective features are selected as factors. In this paper, the selection of factors is mainly based on four aspects:

(1) The historical load. Generally speaking, the historical load impacts on the current load in short-term load forecasting. In this paper, the daily maximum load, daily minimum load, average daily load, peak average load of previous day, valley average load of previous day, average load of the day before, average load of 2 days before, average load of 3 days before, average load of 4 days before, average load of 5 days before and average load of 6 days before are taken into consideration.

(2) The temperature. As people use temperature-adjusting devices to adapt to the temperature, in a previous study [23–25], temperature was considered as an essential input feature and the forecasting results were accurate enough. In this paper, the maximum temperature, the minimum temperature and the average temperature are selected as factors.

(3) The weather condition. We mainly take into account the seasonal patterns, humidity, visibility, weather patterns, air pressure and wind speed. The four seasons are represented as 1, 2, 3 and 4 respectively. For different weather patterns, we set different weights: {sunny, cloudy, overcast, rainy} = {0, 1, 2, 3}.

(4) The day type. In this aspect, the type of day and date are taken into consideration. The type of date means the days are divided into workdays (Monday–Friday), weekend (Saturday–Sunday), and holidays. The weights of three types of date are 0, 1 and 2 respectively. For the date, we set different weight: {Monday, Tuesday, Wednesday, Thursday, Friday, Saturday, Sunday} = {1, 2, 3, 4, 5, 6, 7}.

3.2. Factor Analysis

Originally proposed by British psychologist C.E. Spearman, factor analysis is the study of statistical techniques for extracting highly interrelated variables into one group, and each type of group becomes a factor that reflects most of the original information with fewer factors. Not only does factor analysis reduce indicators' dimensions and improve the generalization of the model but also the common factors it elicited to portray and replace primitive variables can commendably mirror and explain the complicated relationship between variables, keeping data messages with essentially no less information. In this paper, factor analysis is used to extract factors that can reflect the most information of the original 22 influencing variables, whose result is shown in Table Table 2.

First of all, Table 1 gives the result of Kaiser-Meyer-Olkin (KMO) and the Barlett test of sphericity that can serve as a criteria to judge whether the data is suitable for the factor analysis. The statistic value more than 0.7 can illustrate the compatibility and the 0.74 obtained from the power load data confirms the correctness of factor analysis.

Table 2 shows six factors that are extracted from 22 original variables. The accumulative contribution rate at 84.434%, more than 80%, reflects that the new six factors can deliver the most information of the original indicators. It can be seen from Table 2 that factor 1 that mainly represents the history load accounts for the largest proportion at 35.128%. In addition, considering that the variables in factor 1 may not be sufficient on behalf of the historical load, the paper carried out a further analysis of the previous data by means of the correlation analysis which can be seen in part 3.2. Factor 2 which mainly represents meteorology element accounts for 19.646%, and the remaining four factors are 10.514%, 7.746%, 6.087%, and 5.313%, respectively.

Table 1. KMO and Barlett test of sphericity.

KMO		Value	0.740
Barlett test of sphericity	Approximate chi-square value		1525.304
	Degrees of freedom (Df.)		231
	Significance (Sig.)		0.000

Table 2. Results of factor analysis.

Indicator	Variable	Load	Contribution Rate (%)
Factor 1	Minimum temperature	0.732	35.128
	Daily maximum load	0.714	
	Daily minimum load	0.726	
	Average daily load	0.870	
	Season patterns	0.736	
	Peak average load of previous day	0.922	
	Valley average load of previous day	0.801	
	Average load of the day before	0.917	
	Average load of 2 days before	0.830	
	Average load of 3 days before	0.695	
Factor 2	Maximum temperature	−0.732	19.646
	Average temperature	−0.697	
	Humidity	0.810	
	Visibility	−0.724	
	Weather patterns	0.724	
	Average load of 4 days before	0.547	
Factor 3	Type of date	0.622	10.514
	Average load of 5 days before	0.612	
	Average load of 6 days before	0.609	
Factor 4	Air pressure	0.563	7.746
Factor 5	Date	0.883	6.087
Factor 6	Wind speed	−0.533	5.313

3.3. The Analysis of Correlation

Additionally, this paper conducted a further analysis of the correlation between the amount of historical load and the target load from two different viewpoints so as to eliminate the internal correlation. On the one hand, the partial auto correlation function (PACF) was carried out throughout the overall power load to dig out the correlation between the target load and the previous load. On the other hand, the whole load data with the same time interval were also implemented by PACF individually to seek the relationship among the load with the same time. The results of partial auto correlation can be seen in Figures 4 and 5, respectively.

For instance, under the confidence level of 90%, it can be seen from Figure 4 that the lags of the first 2 h are significant to the current data. That is to say, the loads of the first two hours are influential to the current load. As for Figure 5, it is known that only the first lag 1 is prominent to the current load data except the load of 00:00 (Lag 2). Consequently, it can be concluded that the four factors including the first two hours before 00:00 and the same time power load that occurred yesterday and the day before yesterday were selected as the input factors at the time of 00:00.

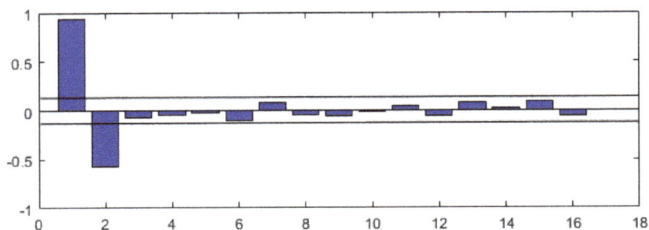

Figure 4. The partial auto correlation result of the overall power load.

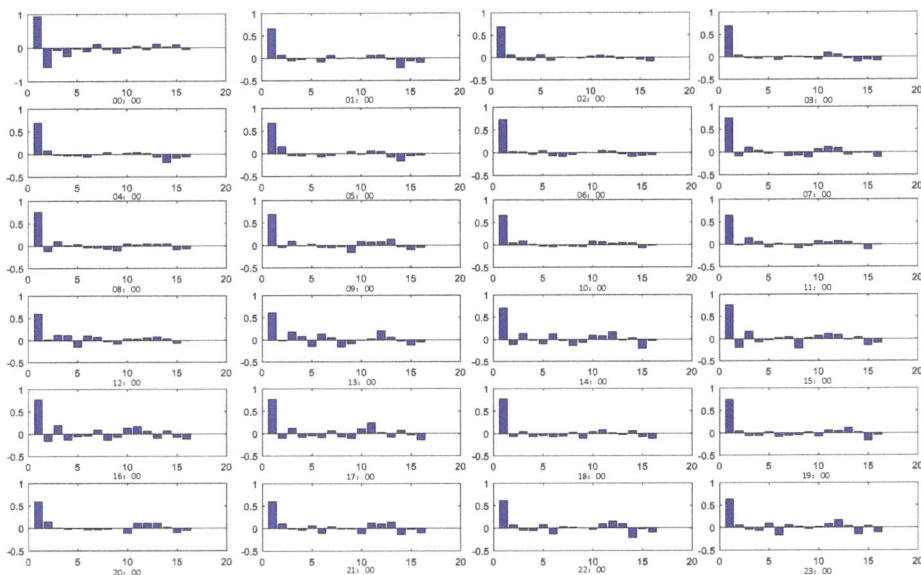

Figure 5. The partial auto correlation result of the load with the same interval.

3.4. Clustering with Ant Colony Algorithm

Selecting the exogenous features as input directly may lead the prediction model to a slow convergence and to poor prediction accuracy. Thus, the paper employs the similar day load which is clustered by the ant colony clustering algorithm for the prediction so as to improve the forecasting accuracy. According to the load every day and the six factors extracted from 22 variables, the 60 days from 1 May 2013 to 30 June 2013 are named with numbers from 1 to 60 and are divided into four clusters by the ant colony algorithm. The parameters of the ACC algorithm can be seen in Table 3, and the clustering result is expressed in Table 4. As a consequence, it can be known that the three test days whose numbers are 58, 59, and 60 belong to class 4, class 1, and class 3, respectively.

Table 3. Parameters of the ant colony clustering algorithm.

Parameter	m	Alpha	Beta	Rho	N	NC_max
Value	30	0.5	0.5	0.1	4	100

Table 4. Results of ant colony clustering algorithm.

Classification	Date Number
Class 1	3→21→25→28→45→51→54→56→59
Class 2	1→7→8→9→10→15→16→26→39→43→44→49→53→57
Class 3	5→12→13→17→19→20→29→31→34→35→37→40→41→42→46→47→48→55→60
Class 4	2→4→6→11→14→18→22→23→24→27→30→32→33→36→38→50→52→58

3.5. Application of BA-ELM

To verify the rationality of data processing, the BA-ELM model was conducted on Yangquan City load forecasting. In this paper, the relative error (RE), mean absolute percentage error (MAPE), mean absolute error (MAE) and root-mean-square error (RMSE) are employed to validate the performance of the model. The formulas definition are expressed as follows, respectively:

$$RE(i) = \frac{\widehat{y}_i - y_i}{y_i} \times 100\% \tag{12}$$

$$AE(i) = \left| \frac{\widehat{y}_i - y_i}{y_i} \right| \times 100\% \tag{13}$$

$$MAPE = \frac{1}{n} \sum_{i=1}^{n} \left| \frac{\widehat{y}_i - y_i}{y_i} \right| \tag{14}$$

$$RMSE = \sqrt{\frac{1}{n} \sum_{i=1}^{n} (\widehat{y}_i - y_i)^2} \tag{15}$$

$$MAE = \frac{1}{n} \sum_{i=1}^{n} |\widehat{y}_i - y_i| \tag{16}$$

where n stands for the quantity of the test sample, $\widehat{y}_i$ is the real load, while y_i is the corresponding predicted output.

Moreover, the paper compared the ELM with the benchmark model's LSSVM and the BPNN to demonstrate the superiority of the proposed model. The parameters of the models are shown in Table 5. Figure 6 shows the iterations process of BA. From the figure we can see that BA achieves convergence at 350 times. The optimal values of the parameters are shown in Table 6.

Table 5. Parameters of models.

Model	Parameters
BA-ELM	n = 10, N_iter = 500, A = 1.6, r = 0.0001, f = [0, 2]
ELM	N = 10, g(x) = 'sig'
LSSVM	$\gamma = 50$; $\sigma^2 = 2$
BPNN	G = 100; hidden layer node = 5; learning rate = 0.0004

Table 6. The optimal parameters.

Parameter		Value								
The input weight matrix	$\omega_{ij} =$	−5.12	−5.12	−5.12	−2.62	−5.11	5.12	5.12	−5.05	−5.12
		−3.61	−0.52	−1.50	5.12	5.12	−5.11	−0.13	−5.12	−5.12
		1.14	−5.12	4.77	−5.12	5.12	−0.06	−0.61	2.08	−3.05
		−2.03	5.12	4.26	4.92	0.03	5.12	2.74	3.37	2.28
		−0.44	2.33	5.12	−1.72	5.12	0.54	1.38	3.48	4.83
		5.12	−4.59	−5.12	−5.12	2.56	0.49	1.32	4.03	1.46
		3.18	4.87	5.12	5.10	2.65	2.19	−5.12	1.06	4.63
		2.66	−5.12	−3.91	−5.12	5.12	2.16	5.12	−5.12	−2.09
		3.86	−5.12	1.85	5.12	−1.44	−5.12	5.12	1.97	5.00
		0.30	5.12	−4.42	−5.12	4.08	−4.79	5.12	−5.12	−5.12

Table 6. *Cont.*

Parameter		Value		
The bias matrix	$\beta_{ik} =$	$\begin{pmatrix} -5.12 \\ 5.12 \\ 3.19 \\ 5.12 \\ -1.84 \\ -1.37 \\ 2.81 \\ -2.42 \\ -5.12 \\ 3.61 \end{pmatrix}$	$\begin{matrix} -5.12 \\ 5.12 \\ 3.19 \\ 5.12 \\ -1.84 \\ -1.37 \\ 2.81 \\ -2.42 \\ -5.12 \\ 3.61 \end{matrix}$	$\begin{pmatrix} -5.12 \\ 5.12 \\ 3.19 \\ 5.12 \\ -1.84 \\ -1.37 \\ 2.81 \\ -2.42 \\ -5.12 \\ 3.61 \end{pmatrix}$
The output weight matrix	$\rho = ($	0.34 −0.45 −0.48 0.38 0.41 −0.28 0.40 −0.23 −0.21 0.24 $)$		

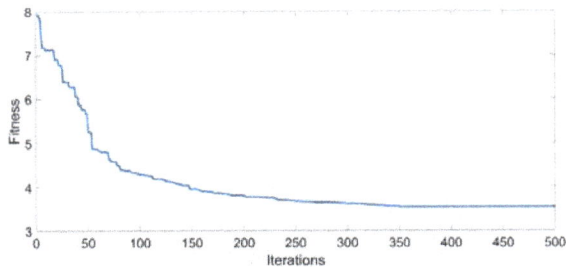

Figure 6. The iterations process of the bat algorithm (BA).

4. Case Study

In order to testify the feasibility of the proposed model, the 24-h power load data of Yangquan City are selected for two months. It can be seen that there is nearly no apparent regularity to be obtained from the actual load curves showed in Figure 7 which represents the four classes of load curve. As mentioned above, the three testing days belong to classes 4, 1, 3 respectively and the prediction model is built for the power load forecasting at the same time.

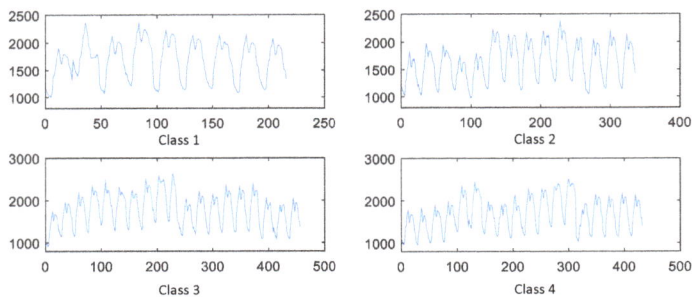

Figure 7. The four types of power load curve.

The program runs in MATLAB R2015b under the WIN7 system. The short-term electric load forecasting results of three days of the BA-ELM, ELM, BP and LSSVM models are shown in Tables 7–9, respectively. For the purpose of explaining the results more clearly, the forecasting values curve of the proposed model and comparisons are shown in Figures 8–10. In addition, Figures 11–13 reflect the comparisons of relative errors between the proposed model and the others. According to Figures 8–10, the deviation can be captured between the actual value and the forecasting results. It can be seen that the forecasting results' curve of the BA-ELM method are close to the actual data in all testing days, which indicates its higher fitting accuracy.

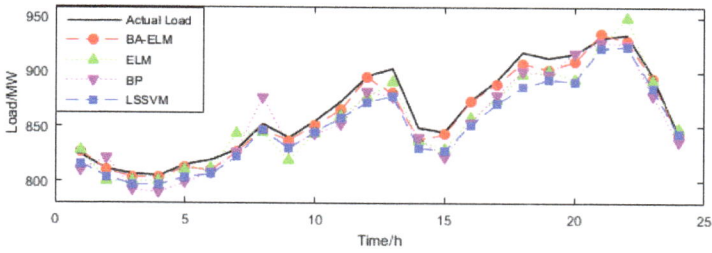

Figure 8. Compared load forecasting results on 28 June.

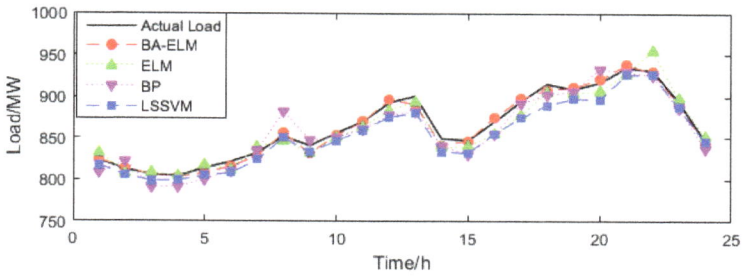

Figure 9. Compared load forecasting results on 29 June.

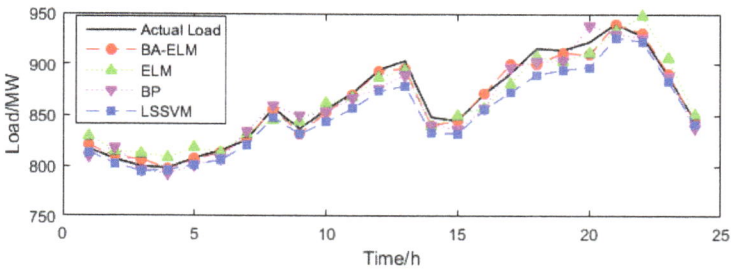

Figure 10. Compared load forecasting results on 30 June.

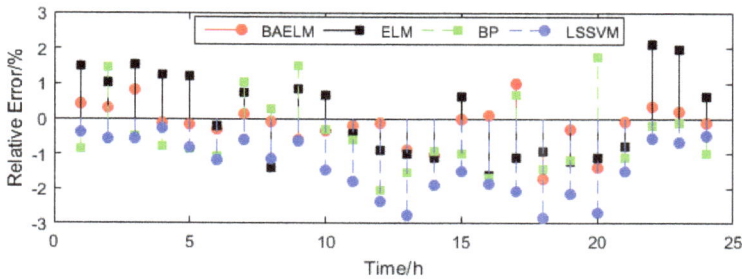

Figure 11. Compared relative errors of four models on 28 June.

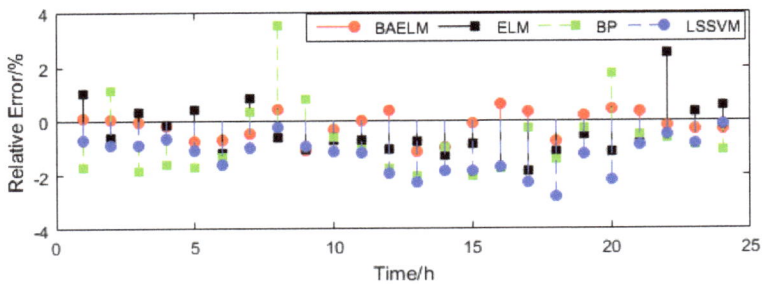

Figure 12. Compared relative errors of four models on 29 June.

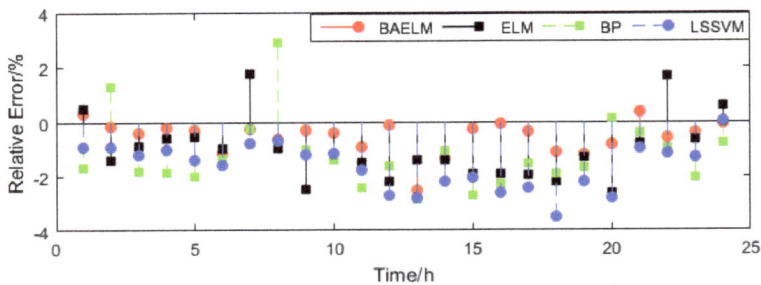

Figure 13. Compared relative errors of four models on 30 June.

Table 7. Actual load and forecasting results on Day 1 (Unit: MV).

	Time/h	Actual Data	BA-ELM	ELM	BP	LSSVM
D1	0:00	816.47	819.89	828.77	809.44	813.29
D1	1:00	810.47	808.58	814.55	817.98	801.67
D1	2:00	795.42	805.98	811.75	795.65	794.95
D1	3:00	793.99	797.15	807.93	792.02	795.95
D1	4:00	809.73	806.25	817.20	800.71	801.11
D1	5:00	813.95	812.47	813.37	806.33	805.36
D1	6:00	832.92	826.51	831.65	833.89	820.42
D1	7:00	839.06	855.99	845.01	859.13	847.20
D1	8:00	829.00	831.28	843.41	848.80	830.78
D1	9:00	848.10	852.50	861.05	852.81	842.98
D1	10:00	865.43	870.18	868.09	866.61	856.15
D1	11:00	882.36	893.75	886.89	876.41	873.40
D1	12:00	881.99	895.85	894.77	889.86	878.92
D1	13:00	828.12	839.33	838.77	840.03	831.76
D1	14:00	824.73	844.35	849.89	835.96	831.65
D1	15:00	856.02	871.74	857.20	856.50	854.95
D1	16:00	868.95	900.32	881.47	897.30	872.83
D1	17:00	904.87	900.41	907.67	902.96	889.97
D1	18:00	905.26	911.81	903.64	903.92	894.79
D1	19:00	902.23	909.76	912.14	938.68	897.51
D1	20:00	920.87	939.86	933.37	930.14	926.57
D1	21:00	925.12	931.54	948.12	926.59	923.08
D1	22:00	893.86	891.45	907.02	888.47	883.78
D1	23:00	843.04	844.05	850.39	836.67	841.05

Table 8. Actual load and forecasting results on Day 2 (Unit: MV).

Time/h		Actual Data	BA-ELM	ELM	BP	LSSVM
D2	0:00	813.56	823.65	831.48	808.98	817.28
D2	1:00	809.75	813.14	807.71	821.76	805.37
D2	2:00	814.06	805.71	808.58	791.03	798.56
D2	3:00	794.74	802.96	803.47	791.70	799.16
D2	4:00	809.89	807.84	817.35	800.06	805.06
D2	5:00	816.16	815.76	811.90	810.62	808.21
D2	6:00	828.37	827.97	839.11	834.82	823.81
D2	7:00	844.26	855.64	846.80	881.84	849.91
D2	8:00	824.92	831.49	831.84	847.35	832.55
D2	9:00	852.17	853.25	850.02	850.91	846.54
D2	10:00	863.06	870.05	864.05	860.95	859.72
D2	11:00	880.26	896.07	883.27	877.19	875.25
D2	12:00	883.78	891.19	894.20	882.91	880.90
D2	13:00	828.22	840.99	838.46	840.79	833.57
D2	14:00	821.18	846.60	839.96	830.01	831.78
D2	15:00	851.78	875.29	854.88	854.81	855.43
D2	16:00	871.49	897.56	878.00	892.13	874.37
D2	17:00	899.60	908.66	905.04	902.64	890.10
D2	18:00	901.80	910.90	904.57	906.42	897.73
D2	19:00	898.35	920.69	906.55	933.13	896.98
D2	20:00	908.94	938.02	927.70	929.86	926.43
D2	21:00	931.82	929.26	954.66	925.09	926.55
D2	22:00	891.29	892.19	898.24	887.12	887.74
D2	23:00	839.30	843.91	851.50	837.50	845.48

Table 9. Actual load and forecasting results on Day 3 (Unit: MV).

Time/h		Actual Data	BA-ELM	ELM	BP	LSSVM
D3	0:00	812.83	826.59	828.03	810.38	816.37
D3	1:00	801.64	810.06	799.93	821.78	804.09
D3	2:00	801.97	803.68	799.95	792.22	797.19
D3	3:00	796.35	803.46	800.56	790.13	797.01
D3	4:00	808.94	812.67	810.88	798.79	803.98
D3	5:00	816.21	810.10	811.44	808.49	806.53
D3	6:00	828.45	826.87	843.63	827.00	822.53
D3	7:00	847.85	846.77	844.31	877.13	846.64
D3	8:00	831.33	837.25	819.12	831.35	829.91
D3	9:00	853.77	851.47	843.37	843.03	845.06
D3	10:00	851.61	865.18	860.53	852.02	857.88
D3	11:00	878.35	895.21	876.79	881.66	872.19
D3	12:00	884.54	880.56	891.03	877.67	877.97
D3	13:00	832.52	837.68	837.29	839.94	830.52
D3	14:00	826.76	842.95	829.22	822.08	828.02
D3	15:00	857.72	873.55	857.38	853.94	851.38
D3	16:00	870.69	889.24	874.85	878.75	870.82
D3	17:00	897.52	907.94	898.13	900.71	886.03
D3	18:00	891.26	902.31	901.23	897.71	893.15
D3	19:00	891.92	909.41	892.96	917.94	891.46
D3	20:00	911.87	934.60	923.71	927.50	921.99
D3	21:00	929.45	928.95	949.86	925.33	923.44
D3	22:00	890.98	893.84	891.63	879.08	885.75
D3	23:00	842.39	842.59	848.01	836.70	843.36

We commonly consider the RE in the range of [−3%, 3%] and [−1%, 1%] as a standard to testify the performance of the proposed model. Based on these tables and figures, we can determine that: (1) on 28 June, the relative errors of the proposed model and others were all in the range of [−3%, 3%]; only one point (3.52%) of BPNN on 29 June and one point (−3.50%) of LSSVM on 30 June are beyond the range of [−3%, 3%], which indicates that the accuracy is increased after the process of reducing dimensions and clustering. (2) Most relative error points of the BA-ELM locate in the range of [−1%, 1%] on all three days. By contrast, most points of the ELM are beyond the range of [−1%, 1%], which can demonstrate that the BA applied in ELM increases the accuracy and stability of ELM. (3) On 28 June, called Day 1 in this paper, the ELM has 14 predicted points exceed the range of [−1%, 1%], and there is only one point (2.12%) beyond the range of [−2%, 2%] at 21:00; the BP has a dozen predicted points outside the range of [−1%, 1%], and there is one predicted point (−2.05%) beyond the range of [−2%, 2%] at 11:00; the LSSVM has 14 predicted points beyond the range of [−1%, 1%], and there are six predicted points beyond the range of [−2%, 2%], which are −2.38% at 11:00, −2.76% at 12:00, −2.07% at 16:00, −2.85% at 17:00, −2.17% at 18:00 and −2.7% at 19:00. (4) On 29 June, called Day 2 in this paper, the ELM has 10 predicted points exceed the range of [−1%, 1%], and there is only one points beyond the range of [−2%, 2%], which is 2.52% at 21:00; the BP has 16 predicted points exceeding the range of [−1%, 1%], and there are three predicted points beyond the range of [−2%, 2%], which are 3.52% at 7:00, −2.03% at 12:00 and −2.03% at 14:00; the LSSVM has 13 predicted points beyond the range of [−1%, 1%], and there are four predicted points outside the range of [−2%, 2%], which are −2.25% at 12:00, −2.27% at 16:00, −2.77% at 15:00 and −2.17% at 19:00. (5) On 30 June, called Day 3 in this paper, the ELM has 15 predicted points exceed the range of [−1%, 1%], and there are three points beyond the range of [−2%, 2%], which are −2.48% at 8:00, −2.19% at 17:00 and −2.61% at 19:00; the BP has 19 predicted points exceed the range of [−1%, 1%], and there are six predicted points beyond the range of [−2%, 2%], which are 2.91% at 7:00, −2.43% at 10:00, −2.85% at 12:00, −2.73% at 14:00, −2.3% at 15:00 and −2.05% at 22:00; the LSSVM has 18 predicted points beyond the range of [−1%, 1%], and there are nine predicted points outside the range of [−2%, 2%], which are −2.17% at 12:00, −2.03% at 13:00, −2.59% at 14:00, −2.41% at 15:00, −3.5% at 16:00, −2.19% at 17:00 and −2.78% at 18:00. From the global view of relative errors, the forecasting accuracy of BA-ELM is better than the other models, since it has the most predicted points in the ranges [−1%, 1%], [−2%, 2%] and [−3%, 3%]. Compared with BPNN and LSSVM, the relative errors of ELM are low. The reason is that the BPNN can have advantages when dealing with the big sample, but its forecasting results are not very good when dealing with a small sample problem like short-term load forecasting. The kernel parameter and penalty factor setting manually of LSSVM are difficult to confirm, which has a significant influence on the forecasting accuracy.

The number of points that are less than 1%, 2%, 3% and more than 3% and the corresponding percentage of them in the predicted points are accounted for, respectively. The statistical results are shown in Table 10. It can be seen that there are 61 predicted points whose the AE of the BA-ELM model is less than 1%, which accounts for 84.72% of the total amount; and 10 predicted points in the range of [1%, 2%], accounting for 13.89% of the total amount; and only 1 predicted point in the range of [2%, 3%], accounting for 1.39% of the total amount. Moreover, there are no predicted points whose AE is more than 3%, accounting for 0% of the total amount. It can be concluded that the forecasting performance of the proposed model is superior, and its accuracy is higher, which means the BA-ELM model is suitable for short-term load forecasting.

The average RMSE and MAPE of the BA-ELM, ELM, BPNN and LSSVM models are listed in Table 11. In order to show the comparisons clearly, the RMSE, MAE and MAPE of four forecasting models in three testing days are show in Figures 14–16. It can be concluded that both of the RMSE, MAE and MAPE of BA-ELM are lower on three testing days. On 28 June, the RMSE, MAE and MAPE of ELM are slightly bigger than BP, but smaller than that of LSSVM. On 29 June, the RMSE, MAE and MAPE of ELM are smaller than that of BP and LSSVM. The RMSE, MAE and MAPE of BP are close to that of LSSVM. On 30 June, the RMSE, MAE and MAPE of ELM are smaller than BP and LSSVM's, and that of BP are smaller than LSSVM's. To sum up, combining this with the Table 11, the average behavior of four models are BA-ELM, ELM, BPNN and LSSVM from low to high successively.

Table 10. Accuracy estimation of the prediction point for the test set.

Prediction Model	<1%		>1% and <2%		>2% and <3%		>3%	
	Number	Percentage	Number	Percentage	Number	Percentage	Number	Percentage
BA-ELM	61	84.72%	10	13.89%	1	1.39%	0	0
ELM	33	45.83%	33	45.83%	6	8.34%	0	0
BPNN	24	33.33%	37	51.39%	10	14.29%	1	1.39%
LSSVM	27	37.50%	26	36.11%	18	25%	1	1.39%

Table 11. Average forecasting results of four models.

Model ╲ Index	**BA-ELM**	**ELM**	**BPNN**	**LSSVM**
RMSE (MW)	5.89	11.08	12.74	14.47
MAPE (%)	0.49	1.13	1.29	1.43
MAE (MW)	4.27	9.81	11.14	12.51

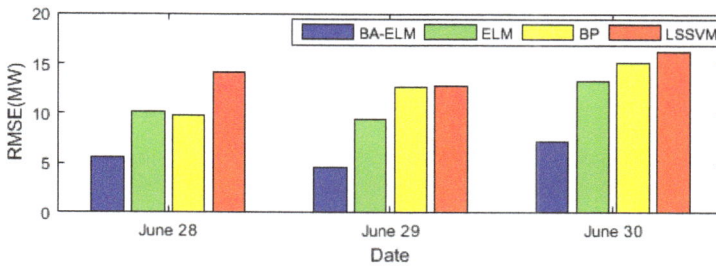

Figure 14. Root-mean-square error (RMSE) of different models in testing period.

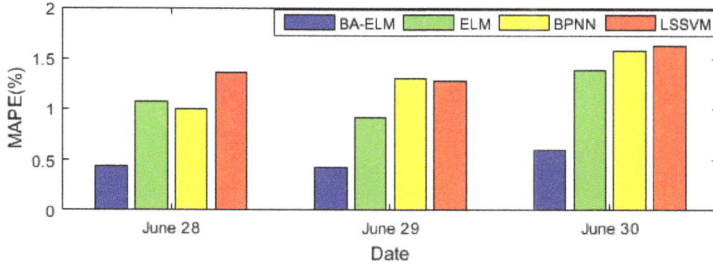

Figure 15. Mean absolute percentage error (MAPE) of different models in testing period.

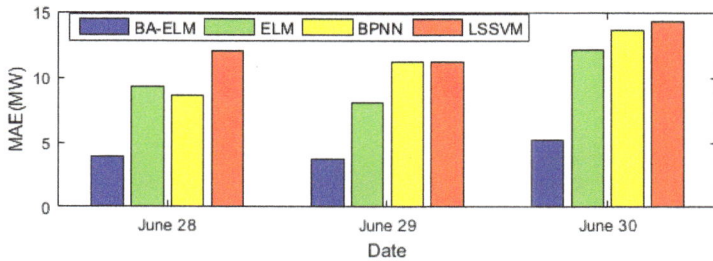

Figure 16. Mean absolute error (MAE) of different models in testing period.

5. Conclusions

With the development of society and technology, research to improve the precision of load forecasting has become necessary because short-term power load forecasting can be regarded as a vital component of smart grids that can not only reduce electric power costs but also ensure the continuous flow of electricity supply. This paper selected 22 original indexes as the influential factors of power load and factor analysis was employed to discuss their correlation and economic connotations, from which it can be seen that the historical data occupied the largest contribution rate and the meteorological factor followed thereafter. Consequently, the paper introduced the auto correlation and partial auto correlation function to further explore the relationship between historical load and current load. Considering the influence of similar day, ant colony clustering was adopted to cluster the sample for the sake of searching the days with analogous features. Finally, the extreme learning machine optimized by a bat algorithm was conducted to predict the days that are chosen to test. The simulation experiment carried out in Yangquan City in China verified the effectiveness and applicability of the proposed model, and a comparison with benchmark models illustrated the superiority of the novel hybrid model successfully.

Author Contributions: W.S. conceived and designed this paper. C.Z. wrote this paper.

Conflicts of Interest: The authors declare no conflict of interest.

References

1. Hernandez, L.; Baladron, C.; Aguiar, J.M.; Carro, B.; Sanchez-Esguevillas, A.J.; Lloret, J.; Massana, J. A Survey on Electric Power Demand Forecasting: Future Trends in Smart Grids, Microgrids and Smart Buildings. *IEEE Commun. Surv. Tutor.* **2014**, *16*, 1460–1495. [CrossRef]
2. Lv, Z.J. Application of regression analysis in power load forecasting. *Hebei Electr. Power* **1987**, *1*, 17–23.
3. Li, P.O.; Li, M.; Liu, D.C. Power load forecasting based on improved regression. *Power Syst. Technol.* **2006**, *30*, 99–104.
4. Li, X.; Zhang, L.; Yao, S.; Huang, R.; Liu, S.; Lv, Q.; Zhang, L. A New Algorithm for Power Load Forecasting Based on Time Series. *Power Syst. Technol.* **2006**, *31*, 595–599.
5. Metaxiotis, K.; Kagiannas, A. Artificial intelligence in short term electric load forecasting: A state-of-the-art survey for the researcher. *Energy Convers. Manag.* **2003**, *44*, 1525–1534. [CrossRef]
6. Hippert, H.S.; Pedreira, C.E.; Souza, R.C. Neural Networks for Short-Term Load Forecasting: A Review and Evaluation. *IEEE Trans. Power Syst.* **2001**, *16*, 44–55. [CrossRef]
7. Park, D.C.; El-Sharkawi, M.A.; Marks, R.J.; Atlas, L.E.; Damborg, M.J. Electric Load Forecasting Using an Artificial Network. *IEEE Trans. Power Syst.* **1991**, *6*, 422–449. [CrossRef]
8. Hernandez, L.; Baladrón, C.; Aguiar, J.M.; Carro, B.; Sanchez-Esguevillas, A.J.; Lloret, J. Short-Term Load Forecasting for Microgrids Based on Artificial Neural Networks. *Energies* **2013**, *6*, 1385–1408. [CrossRef]
9. Yu, F.; Xu, X. A short-term load forecasting model of natural gas based on optimized genetic algorithm and improved BP neural network. *Appl. Energy* **2014**, *134*, 102–113. [CrossRef]
10. Hu, R.; Wen, S.; Zeng, Z.; Huang, T. A short-term power load forecasting model based on the generalized regression neural network with decreasing step fruit fly optimization algorithm. *Neurocomputing* **2017**, *221*, 24–31. [CrossRef]
11. Li, Y.C.; Fang, T.J.; Yu, E.K. Study on short—Term load forecasting using support vector machine. *Proc. CSEE* **2003**, *23*, 55–59.
12. Zhao, D.F.; Wang, M. Short—Term load forecasting based on support vector machine. *Proc. CSEE* **2002**, *22*, 26–30.
13. Mesbah, M.; Soroush, E.; Azari, V.; Lee, M.; Bahadori, A.; Habibnia, S. Vapor liquid equilibrium prediction of carbon dioxide and hydrocarbon systems using LSSVM algorithm. *J. Supercrit. Fluids* **2015**, *97*, 256–267. [CrossRef]
14. Huang, G.B.; Zhu, Q.Y.; Siew, C.K. Extreme learning machine: Theory and applications. *Neurocomputing* **2006**, *70*, 489–501. [CrossRef]

15. Huang, G.B.; Zhou, H.; Ding, X.; Zhang, R. Extreme learning machine for regression and multiclass classification. *IEEE Trans. Syst. Man Cybern. Part B* **2012**, *42*, 513–529. [CrossRef] [PubMed]

16. Yang, X.S.; Hossein Gandomi, A. Bat algorithm: A novel approach for global engineering optimization. *Eng. Comput.* **2012**, *29*, 267–289. [CrossRef]

17. Zhang, J.W.; Wang, G.G. Image Matching Using a Bat Algorithm with Mutation. *Appl. Mech. Mater.* **2012**, *203*, 88–93. [CrossRef]

18. Mishra, S.; Shaw, K.; Mishra, D. A New Meta-heuristic Bat Inspired Classification Approach for Microarray Data. *Procedia Technol.* **2012**, *4*, 802–806. [CrossRef]

19. Nakamura, R.Y.; Pereira, L.A.; Costa, K.A.; Rodrigues, D.; Papa, J.P.; Yang, X.S. BBA: A Binary Bat Algorithm for Feature Selection. In Proceedings of the 25th SIBGRAPI Conference on Graphics, Patterns and Images, Ouro Preto, Brazil, 22–25 August 2012; pp. 291–297.

20. Niu, D.; Dai, S. A Short-Term Load Forecasting Model with a Modified Particle Swarm Optimization Algorithm and Least Squares Support Vector Machine Based on the Denoising Method of Empirical Mode Decomposition and Grey Relational Analysis. *Energies* **2017**, *10*, 408. [CrossRef]

21. Liang, Y.; Niu, D.; Ye, M.; Hong, W.C. Short-Term Load Forecasting Based on Wavelet Transform and Least Squares Support Vector Machine Optimized by Improved Cuckoo Search. *Energies* **2016**, *9*, 827. [CrossRef]

22. Sun, W.; Liang, Y. Least-Squares Support Vector Machine Based on Improved Imperialist Competitive Algorithm in a Short-Term Load Forecasting Model. *J. Energy Eng.* **2014**, *141*, 04014037. [CrossRef]

23. Hooshmand, R.A.; Amooshahi, H.; Parastegari, M. A hybrid intelligent algorithm based short-term load forecasting approach. *Int. J. Electr. Power Energy Syst.* **2013**, *45*, 313–324. [CrossRef]

24. Bahrami, S.; Hooshmand, R.A.; Parastegari, M. Short term electric load forecasting by wavelet transform and grey model improved by PSO (particle swarm optimization) algorithm. *Energy* **2014**, *72*, 434–442. [CrossRef]

25. Yeom, C.U.; Kwak, K.C. Short-Term Electricity-Load Forecasting Using a TSK-Based Extreme Learning Machine with Knowledge Representation. *Energies* **2017**, *10*, 1613. [CrossRef]

26. Yang, X.S. A New Metaheuristic Bat-Inspired Algorithm. *Comput. Knowl. Technol.* **2010**, *284*, 65–74.

27. Deng, W.Y.; Zheng, Q.H.; Chen, L.; Xu, X.B. Study on fast learning method of neural network. *Chin. J. Comput.* **2010**, *33*, 279–287. [CrossRef]

MDPI

Article

Short-Term Load Forecasting for Electric Vehicle Charging Station Based on Niche Immunity Lion Algorithm and Convolutional Neural Network

Yunyan Li *, Yuansheng Huang and Meimei Zhang

Department of Economic Management, North China Electric Power University, Baoding 071000, China; 51850962@ncepu.edu.cn (Y.H.); 51851539@ncepu.edu.cn (M.Z.)
* Correspondence: liyunyanbd@126.com

Received: 20 April 2018; Accepted: 10 May 2018; Published: 14 May 2018

Abstract: Accurate and stable prediction of short-term load for electric vehicle charging stations is of great significance in ensuring economical and safe operation of electric vehicle charging stations and power grids. In order to improve the accuracy and stability of short-term load forecasting for electric vehicle charging stations, an innovative prediction model based on a convolutional neural network and lion algorithm, improved by niche immunity, is proposed. Firstly, niche immunity is utilized to restrict over duplication of similar individuals, so as to ensure population diversity of lion algorithm, which improves the optimization performance of the lion algorithm significantly. The lion algorithm is then employed to search the optimal weights and thresholds of the convolutional neural network. Finally, a proposed short-term load forecasting method is established. After analyzing the load characteristics of the electric vehicle charging station, two cases in different locations and different months are selected to validate the proposed model. The results indicate that the new hybrid proposed model offers better accuracy, robustness, and generality in short-term load forecasting for electric vehicle charging stations.

Keywords: electric vehicle (EV) charging station; short-term load forecasting; niche immunity (NI); lion algorithm (LA); convolutional neural network (CNN)

---

1. Introduction

The development of the electric vehicle (EV) industry has attracted broad attention from governments, auto manufacturers, and energy enterprises. Electric vehicles are regarded as an effective way to cope with the depletion of fossil energy and increasingly serious environmental pollution [1]. Charging stations, serving as the infrastructure, have been extensively built along with the advance of EVs. However, the volatility, randomness, and intermittence of the load bring new challenges to optimal dispatching and safe operation of power grids [2]. The establishment of a scientific and reasonable short-term load forecasting model for EV charging stations will not only improve the prediction precision for optimal dispatching, but will also promote the rational construction of charging stations, and boost the popularity rate of EVs. Accordingly, focus on the research of load forecasting for EV charging stations is of great significance.

The current methods of load forecasting for EV charging stations can be divided into two parts, namely: statistical approaches and artificial intelligent algorithms. Statistical forecasting models are based on the theory of probability and statistics, such as the Monte Carlo method [3]. Concretely, on the foundation of a residents' traffic behavior database, the Monte Carlo approach exploits a definite probability distribution function to fit the users' driving behaviors, and establishes a mathematical model with random probability to forecast the charging time, location, and load demand of EVs in the future [4]. Simple though it is, this kind of method is not suited to address load forecasting for inaccurate estimation, considering the randomly selected distribution parameters [5]. Additionally, Ref. [6] carried out charging load prediction of EVs based on the statistical analysis of vehicle data from the perspective of time and space. In order to simulate the driving patterns of EVs, Ref. [7] outlined an improved charging load calculation model, where charging probability was proposed to illustrate the uncertainty of charging behaviors and kernel density functions. Multidimensional probability distribution functions were utilized to replace deterministic ones, and a random number was generated to present the coupling characteristics of driving discipline. The view of big data was indicated in the literature [8], which calculated the load of every EVs at the charging station, and summed them up; thus, load forecasting results were obtained. Nevertheless, these statistical approaches are criticized by researchers for their weakness of universality, due to the difficulty of parameter determination.

With the rapid development of artificial intelligence (AI) technology, intelligent algorithms, which mainly include artificial neural networks (ANNs) and support vector machines (SVM), are gradually applied to load forecasting of EV charging stations by scholars [9]. Ref. [10] employed back propagation neural network (BPNN) models to predict the daily load curve of EV charging stations, with consideration of various factors. Here, fuzzy clustering analysis based on transfer closure methods was adopted to select the historical load similar to the predicted one as the training samples, so as to improve forecasting accuracy. The drawbacks of BPNN are the existence of many parameters to set, and trapping into the local minimum or over-fitting easily. To address these problems, Ref. [11] studied a short term load forecasting model for EV charging stations on the basis of radial basis function neural networks (RBFNN), and modified it by the use of fuzzy control theory. The results showed that prediction accuracy was further improved. In [12], particle swarm optimization and spiking neural networks were combined to forecast the short term load of EV charging stations. The findings revealed that the prediction accuracy of the proposed model was superior to BPNN. An SVM integrated with genetic algorithms was exploited in short term load forecasting for EV charging stations in [13], which illustrated that it was difficult for SVMs to deal with large-scale training samples and achieve ideal prediction accuracy. The aforementioned algorithms belong to shallow learning with weak ability in processing complex functions, and cannot completely reflect the characteristics of information based on prior knowledge. To this end, deep learning algorithms provide better ways to present data feature by abstracting the bottom feature combination into high-level [14].

At present, deep learning algorithms have been widely applied in various fields, especially in the field of prediction. Ref. [15] executed an advertising click rate prediction method based on a deep convolutional neural network (CNN). This model accomplished feature learning through the simulation of human thinking, and analyzed the role of different features in forecasting. Ref. [16] successfully introduced deep structure networks into ultra short term photovoltaic power predictions. A deep belief network with restricted Boltzmann machine was presented to extract deep features to finish the unsupervised learning, and the supervised BPNN was taken as the conventional fitting layer to obtain the forecasting results. Ref. [17] built deep CNN for bioactivity prediction of small molecules in drug discovery applications. These studies have demonstrated that deep learning algorithms have better prediction accuracy in comparison to shallow learning. CNN allows the existence of deformed data and reduces parameters through local connection and weight sharing; thus, forecasting precision and efficiency can be greatly improved [18]. As a result, CNN is selected as the prediction model in this paper. Notably, the fitting accuracy of CNN is influenced by its two parameters' selection, namely: weight and threshold. Consequently, it's vital to apply an appropriate intelligent algorithm to determine theses values. Several traditional optimization algorithms have been used to select parameters for CNN, such as genetic algorithms, particle swarm optimizations and ant colony algorithms. Although the above algorithms have their own advantages, they also have corresponding shortcomings. For example, genetic algorithm cannot guarantee the convergence to the best, and is easy to fall into the local optimum, which leads to a decrease in prediction accuracy [19]. Particle swarm optimization will appear in premature convergence in different situations [20]. Ant colony algorithms have low searching efficiency and long calculation times, and local search accuracy is not high. Also, it cannot fully meet the needs of the CNN parameter optimization problem [21]. The Lion algorithm (LA), based on the social behavior of lions, was introduced by B.R. Rajakumar in 2012 [22]. Compared with preceding models, this approach shows strong robustness and good abilities in global optimization, and fast convergence. Nevertheless, inbreeding appears among the lions with large fitness during the iterative process, which leads to premature convergence and diversity reduction. To settle this problem, niche immune algorithms are employed in this paper to optimize LA, namely NILA. Here, niche immune algorithms can restrict over-duplication of similar individuals, so as to ensure the diversity of the population, and improve the optimization effect of the lion algorithm for selecting the parameters of CNN. This hybrid optimization method is used to automatically determine the appropriate values in CNN model.

This paper combines NILA with the CNN model for load forecasting of EV charging stations, with scientific analysis of influential factors. The rest of the paper is organized as follows: Section 2 shows a brief description of LA, NILA, and CNN, as well as the framework of the proposed technique; Section 3 presents an analysis of the influential factors and determines the input; Section 4 introduces an experiment study to test the accuracy and robustness of the established model; Section 5 makes further validation on this method, and Section 6 concludes this paper.

The innovations of this paper are as follows:

(1) The construction of the forecasting model

Firstly, it is the first time to combine CNN and lion algorithm improved by niche immunity and employ this model for the load forecasting of electric vehicle charging stations. Furthermore, the CNN model used for load forecasting cannot only allow the existence of deformed data, but also improve the load forecasting efficiency and accuracy by parameter reduction through local connection and shared weight. Finally, niche immunity is used in this paper to restrict over duplication of similar individuals, so as to ensure the diversity of population, and it effectively improves the optimization effect of the lion algorithm, as we can conclude from the case study.

(2) The input selection of the forecasting

In order to produce a scientific and reasonable input index system for the forecasting model, this paper fully analyzes the load characteristics in an EV charging station. And it can be concluded that the load in the EV charging station is heavily influenced by meteorological conditions, seasonal variation, and day types, which are more comprehensive and effective for forecasting.

In summary, this paper not only creatively combines various prediction theories to construct a comprehensive forecasting model, but also conducts the study of influential factors affecting the load of EV charging stations so that a scientific and reasonable input index system is produced.

2. Methodology

2.1. Lion Algorithm Improved by Niche Immune (NILA)

2.1.1. Lion Algorithm (LA)

Lion algorithm is a social behavior-based bionic algorithm developed by B. R. Rajakumar in 2012. The iteration and generation of optimal solutions can be realized through territorial lion's breeding, and its defense to other nomadic lions. In this approach, every single solution corresponds to "Lion".

LA proceeds through four main steps: population initialization, mating and mutation, territorial defense, and territorial takeover. The objective function is set as Equation (1):

$$\min f(x_1, x_2, \cdots, x_n), \quad (n \geq 1) \tag{1}$$

Step 1: Population initialization

In the first stage of this algorithm, $2n$ lions are averagely assigned to two groups as the candidate population, namely male lions $A^m = [\psi_1^m, \psi_2^m, \psi_3^m, \cdots, \psi_l^m]$ and female lions $A^f = [\psi_1^f, \psi_2^f, \psi_3^f, \cdots, \psi_l^f]$. l represents the length of the solution vector.

Step 2: Mating

Mating is an essential process to update and maintain the stability of the lion group via crossover, mutation, cluster or killing the sick and weak, thus new solutions can be continually delivered through iteration.

Dual probabilities based crossover is introduced in this paper, that is, crossover is implemented with two different probabilities. The lion A^m and lioness A^f generate a new cub $A^{cub} = [\psi_1^{cub}, \psi_2^{cub}, \psi_3^{cub}, \cdots, \psi_l^{cub}]$ through mating. Then, four cubs $A_{1\sim4}^{cub}$ are generated according to two randomly selected crossover points by ψ_i^m and ψ_j^f.

Random mutation with p is enabled to generate $A_{5\sim8}^{cub}$, resulting in 8 cubs after crossover and mutation.

The cubs are separated into male cubs (A^{m_cub}) and female cubs (A^{f_cub}) by K-means clustering.

Then, in light of health status, the weak cubs in larger group are killed off to ensure an equal number in the two cubs. After population regeneration, the age of the cub is initialized as 0.

Step 3: Territorial defense

During breeding, it will be attacked by the nomadic lion. At this time, the male lion will defend and protect the cubs, and occupy the territory, as illustrated in Figure 1.

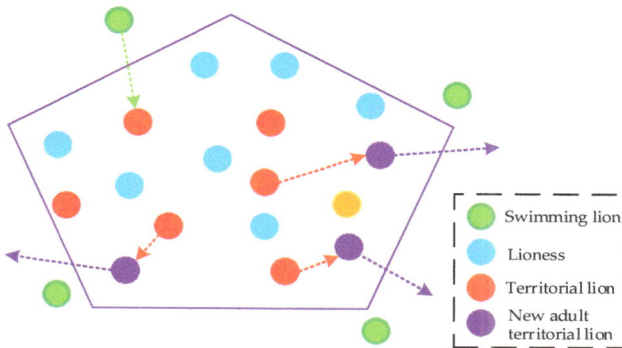

Figure 1. Lion defense process.

The nomadic lion ψ^{nomad} is generated in the way that is the same as territorial lion. Then the new solution ψ^{nomad} is used to attack the male lion ψ_i^m. If ψ^{nomad} is superior to other solutions in the pride, replace ψ_i^m with ψ^{nomad}. The new lion will continue to mate, and the old, as well as the cubs, will be killed off. Otherwise, the old lion will continue with territory defense, and the cubs will be one year older than before.

$f(\cdot)$ is the objective function and $f\left(\psi^{pride}\right)$ is the value of the whole population which can be calculated as Equation (2).

$$f\left(\psi^{pride}\right) = \frac{1}{2\left(1 + \|\psi^{m\_cub}\|\right)} \left(f(\psi^m) + f\left(\psi^f\right) + \frac{age_{mat}}{age_{cub} + 1} \cdot \sum_{k=1}^{\|\psi^{m\_cub}\|} \frac{f\left(\psi_k^{m\_cub}\right) + f\left(\psi_k^{f\_cub}\right)}{\|\psi^{m\_cub}\|} \right) \quad (2)$$

where $f(\psi^m)$ and $f(\psi^f)$ represent the values of lion and lioness, respectively; $f(\psi_k^{m\_cub})$ and $f(\psi_k^{f\_cub})$ equals the values of male cub and female cub, respectively; $\|\psi^{m\_cub}\|$ means the number of male cubs; age_{mat} is employed to designate the time required for mating.

Step 4: Territorial takeover

In this step, the optimal solutions among the lion and lioness are found to replace the inferior one. Mating will not end until the terminating conditions are reached. The best lion ψ_{best}^m and lioness ψ_{best}^f are determined according to the following criteria:

$$f(\psi_{best}^m) < f\left(\psi_{pride}^m\right), \ \psi_{best}^m \neq \psi_{pride}^m, \ \psi_{pride}^m = \left\{\psi^m, \psi^{m\_cub}\right\} \tag{3}$$

$$f\left(\psi_{best}^f\right) < f\left(\psi_{pride}^f\right), \ \psi_{best}^f \neq \psi_{pride}^f, \ \psi_{pride}^f = \left\{\psi^f, \psi^{f\_cub}\right\} \tag{4}$$

In the pseudo code, κ represents the number of breeding and $\kappa_{strenth}$ describes the female's optimal breeding ability, generally set to 5. $\kappa_{strenth}$ is set as 0 at the time of initial pride generation, and should be incremented. If the female lion is replaced, κ has to be started from 0. On the other hand, if the old lioness continually existed, κ should be accumulated. When the previous steps are completed, go back to Step 2 until the termination condition is satisfied. The best lion responds to the optimal solution.

2.1.2. LA Improved by Niche Immune

LA is a parallel combination of self-adaption, group search and a heuristic random search, while inbreeding appears among the lions with large fitness during the iterative process, resulting in premature convergence and diversity reduction. Niche immunity is exploited in this paper to restrict over duplication of similar individuals, so as to ensure the diversity of population. The detailed steps of NI algorithm are displayed in [23]. LA optimized by NI can be performed as follow:

Step 1: According to the value of objective function, M cloned lions can be obtained in the center of the location at a specified iteration interval.

$$M_j = M_{max} \times \left(1 - \frac{\rho_j}{\sum\limits_{j=1}^{N} \rho_j}\right) \tag{5}$$

where M_j is the clone number of the j-th lion, M_{max} represents the maximum clone number that is set to 40 here. ρ_j is the objective function value of the j-th lion.

Step 2: M lions are mutated by single parent after clone. For the lion with low objective function value, mutation is carried out by the parthenogenetic lions, as given in Equations (6) and (7).

$$x_{i+1} = x_i + r \times randn(1) \tag{6}$$

$$r = \frac{2 \times P_{max}}{N} \tag{7}$$

where x_i represents the lion, x_{i+1} is the offspring generation after parthenogenesis, P_{max} is the maximum value of lion location, N equals the number of lions.

Step 3: Make comparison among the M mutated lions and select the one with the maximum objective function value as the new lion.

2.2. Convolutional Neural Network (CNN)

As a kind of ANN with deep learning ability, the CNN achieves local connections and shares the weights of neurons in the same layer [24]. The network consists of 1~3 feature extraction layers and fully connected layers. Each feature extraction layer includes a convolutional one and a subsampling one. The structure of CNN containing a feature extraction layer is shown in Figure 2.

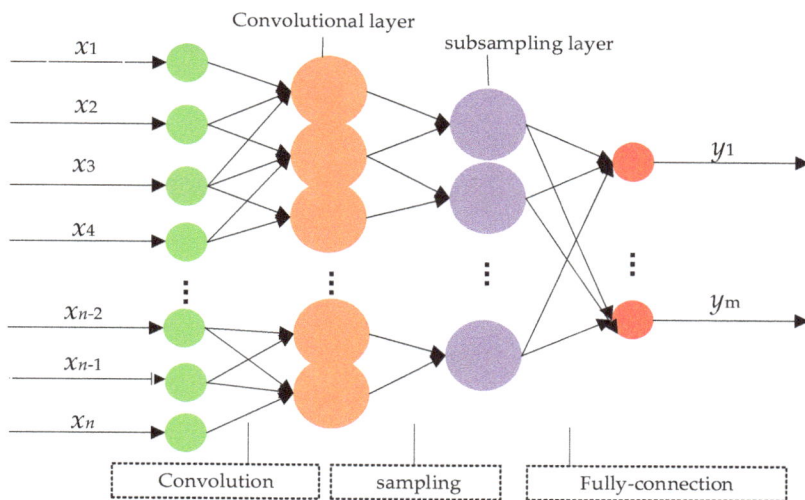

Figure 2. Convolutional neural network model.

In convolutional layer, the original data is processed by the convolutional kernel to obtain the output, as described in Equation (8).

$$x_j^l = f\left(\sum_{j=m}^{k} x_j^{l-1} w_j^l + \theta_j^l\right) \quad (j = 1, 2, \cdots, n; \, 0 < m \le k \le n) \tag{8}$$

where $f(I) = \frac{1}{1+e^{-I}}$, $I = \sum_{j=m}^{k} x_j^{l-1} w_j^l + b_j^l$ $(1, 2, \cdots, n; 0 < m \le k \le n)$. x_j^l and x_j^{l-1} represent the output in Layer l and the input in Layer $l-1$, respectively. j is the local connection ranging from m to k. w_j^l equals the weight and θ_j^l is the bias.

The subsampling process can be expressed as follows:

$$x_j^l = g(x_j^{l-1}) + \theta_j^l \tag{9}$$

where $g(\sim)$ represents the function that selects the average or maximum value.

Then, the obtained data is linked to the fully connected layer as presented in Equation (10).

$$x^l = f(I^l), I^l = W^l x^{l-1} + \theta^l \tag{10}$$

where W_l is the weight from Layer $l-1$ to Layer l and x^l represents the output data.

In the above calculation, each convolutional kernel plays a role in all the input via the slide. Different convolutional kernels corresponding to multiple sets of output where the weight of the same convolutional kernel is identical. The output of different groups are combined and then transferred to the subsampling layer. Here, the output in the previous convolutional layer is treated as the input data. At this time, set the range of values and use the average or maximum as the specific values in the range. The data needs to be combined to satisfy a dimensionality reduction. Finally, the results can be derived from the fully connected layer [25].

The application of the CNN model has two main advantages: (a) the existence of deformed data is allowed; (b) the load forecasting efficiency and accuracy can be improved by parameter reduction through local connection and shared weight. However, the stability of the prediction results can not be guaranteed, due to the subjective determination of the weights and thresholds [26]; thus, NILA is proposed to complete the optimal parameter selection in this paper to overcome this shortcoming.

2.3. The Forecasting Model of NILA-CNN

The short-term load forecasting approach for EV charging stations incorporating NILA and CNN is constructed as Figure 3 shows.

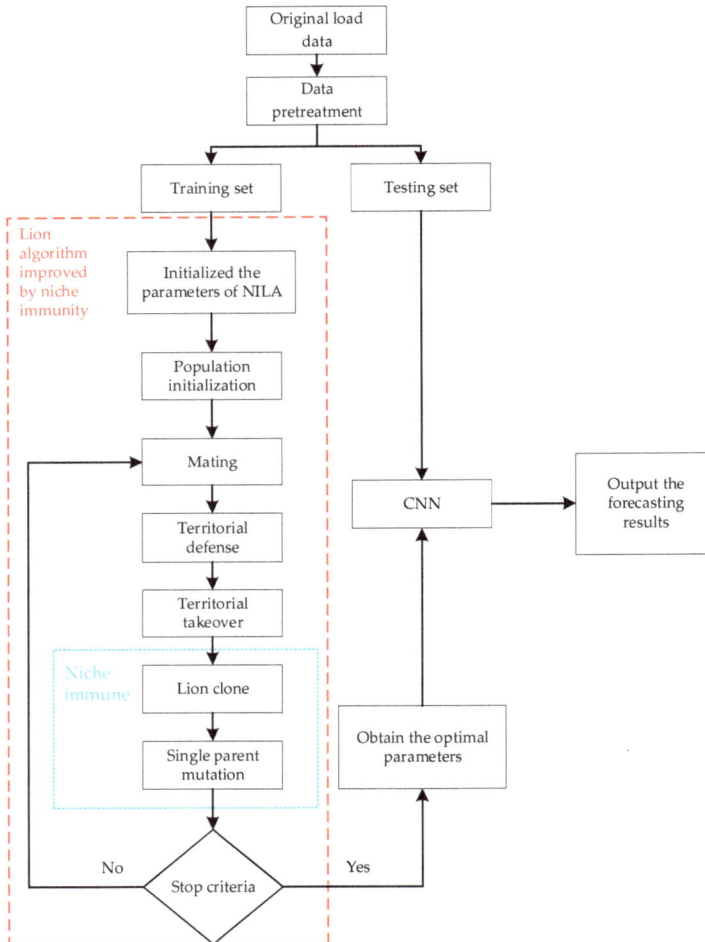

Figure 3. Flowchart of Lion Algorithm Improved by Niche Immune (NILA) - Convolutional Neural Network (CNN) algorithm.

On the basis of NILA-CNN model, the optimal parameters of CNN can be derived as follows:

(1) Input selection (xi) and data pre-processing. The initial input set is formed based on the load analysis of EV charging stations and needs to be quantified and normalized. The specific data preprocessing method is shown in Section 4.1.

(2) Parameters initialization. Randomly determine the weights and thresholds of all layers in CNN model from the smaller numerical set.

(3) NILA optimization. Search the optimal weights and thresholds of CNN on the basis of NILA. If the maximum iteration number is reached, the optimal parameters are obtained; if not, repeat the optimization steps until the condition is satisfied.

(4) CNN training. After initialization including the neuron numbers in the input layer, convolutional layer, and subsampling layer, respectively, train the CNN optimized by NILA, and derive the optimal forecasting model.

(5) Simulation and prediction. Forecast the short-term load of EV charging stations based on the trained approach and analyze the results.

3. Analysis of Load Characteristics in Electric Vehicle (EV) Charging Station

The study of influential factors that affect the load in charging station contribute to load forecasting accuracy improvement. This paper selects an EV charging station in Beijing as a case study. It can be seen that the load is heavily influenced by meteorological conditions, seasonal variation, and day types.

3.1. Seasonal Variation

Seasonal variation has an obvious effect on the load characteristics in EV charging station [27]. Therefore, the typical daily load curves in spring, summer, autumn and winter are compared in Figure 4. It should be noted that these four days are all Tuesday, and are all sunny days.

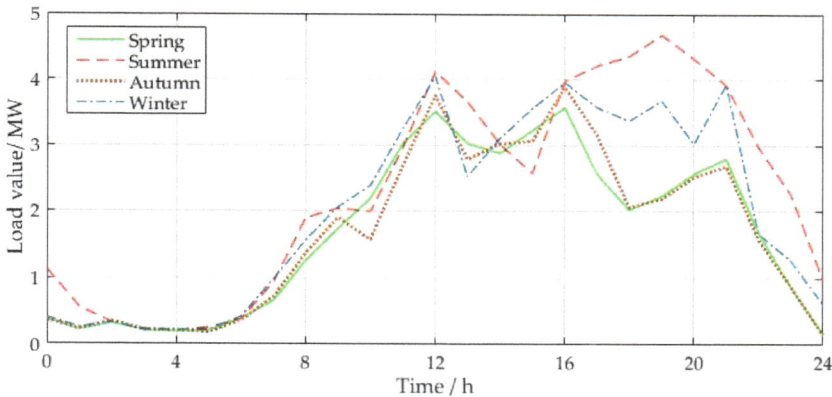

Figure 4. Typical daily load curves in four seasons.

As presented in Figure 4, the load of the EV charging station is relatively high in winter and summer, mainly due to increasing use of air conditioning in these two seasons, which leads to more energy consumption. As a result, air conditioning load can be considered as a vital influencing factor.

3.2. Meteorological Conditions

The load in EV charging station is greatly affected by temperature and weather type, while wind and humidity play insignificant roles [28,29]. Here, take the daily load curves on 1 June, 8 June and 15 June in 2017 as examples. The average daily temperatures are 23.5 °C, 27 °C and 31 °C, respectively. It can be seen that there is a positive relationship between temperature and daily load, as shown in Figure 5. Therefore, temperature is selected as the influential factor in this paper.

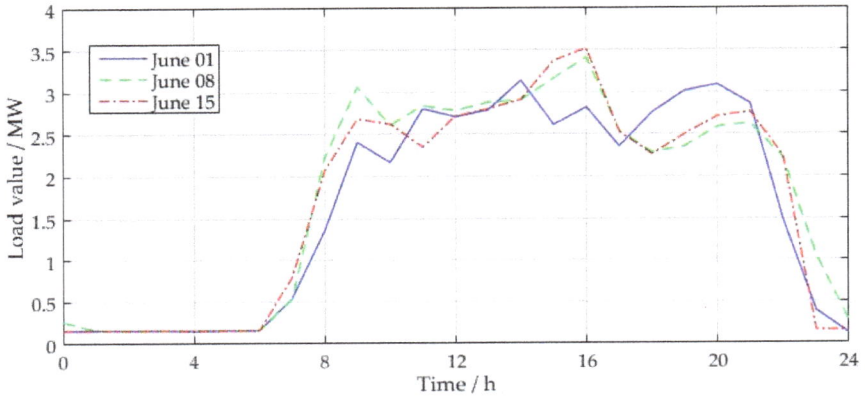

Figure 5. Relationship between temperature and daily load of electric vehicle (EV) charging station.

Divide the weather conditions into two categories: sunny days and rainy days. Figure 6 illustrates the relationship between weather conditions and the daily load of the EV charging station on 21 February and 22 February in 2017. It is sunny on 21 February and it is rainy on 22 February. It proves that snow days can reduce the daily maximum load as a result of vehicle's deceleration, which leads to the decrease of daily driving mileage and charging. Hence, snow is another important influential factor.

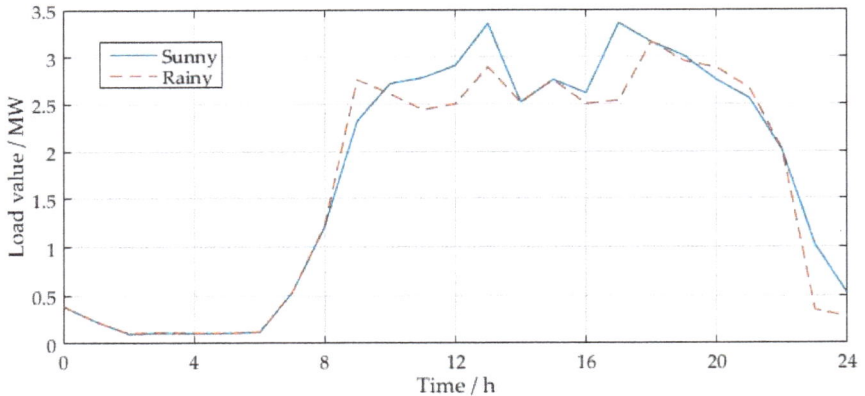

Figure 6. Relationship between weather condition and daily load of EV charging station.

3.3. Day Types

Divide the days into workdays, Saturday and Sunday. Figure 7 describes the relationship between day types and daily load of the EV charging station based on the data from 14 August to 20 August in 2017. It is Monday to Friday from 14 August to 18 August. 19 August and 20 August are Saturday and Sunday respectively. The loads on workdays are slightly lower than those of the weekends. From Monday to Friday, the use of EVs focuses on the period that people go to and from work, while the abundant outdoor activities on Saturday and Sunday increase the use of EVs. To this end, the day type is chosen as an influential indicator in this paper.

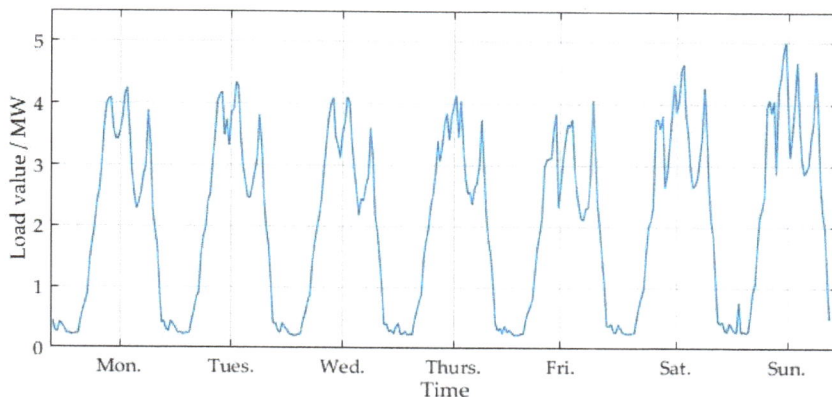

Figure 7. Relationship between day type and daily load of EV charging station.

4. Case Study

The selected EV charging station in this paper is composed of 5 large power chargers which can be used by batteries with high capacity in a single box, or series batteries with low capacity in three boxes, and 10 small power chargers that can be only employed by a battery with low capacity in a single box. The load data every 30 min from 1 June 2016 to 30 November 2017 are collected from the charging station. The data from 1 June 2016 to 29 November 2017 are selected as training set, and the remaining data on 30 November 2017 are utilized as test set.

4.1. Input Selection and Processing

According to the analysis of the load characteristics for EV charging station, ten influential factors including seasonal category, maximum temperature, minimum temperature, weather condition, day type, and the loads at the same moment in the previous five days are selected as input in this paper. The input features are discussed as follows: (a) the season can be divided into four categories: spring (March, April and May), summer (June, July and August), autumn (September, October, November) and winter (December, January and February), which are set as $\{1, 2, 3, 4\}$. (b) Weather conditions are decomposed into two types: sunny and cloudy days, valued at 1, and rainy and snowy days, valued at 0.5. (c) Days can be divided into workdays (Monday to Friday) and weekends (Saturday and Sunday). When quantifying the day type, workdays are valued at 1, and weekends at 0.5. Because the collecting data is not publically available, statistically significant parameters are presented in Table 1.

Table 1. Statistically significant parameters of the collecting data.

Statistics	Total Days	Maximum Load (MW)	Minimum Load (MW)	Maximum Temperature (°C)	Minimum Temperature (°C)
Value	547	5.212	0.006	36	−13
Statistics	Number of days in spring (day)	Number of days in summer (day)	Number of days in autumn (day)	Number of days in winter (day)	Number of precipitation days (day)
Value	92	184	182	89	76

The temperature and load data should be normalized in accordance with Equation (11).

$$Y = \{y_i\} = \frac{x_i - x_{min}}{x_{max} x_{min}} \quad i = 1, 2, 3, \ldots, n \tag{11}$$

where x_i is the actual value, x_{min} and x_{max} equals the minimum and maximum values in the samples, respectively, y_i represents the normalized load.

4.2. Model Performance Evaluation

This paper assesses the forecasting model by using the following appropriate indicators.

(1)   Relative error (*RE*):

$$RE = \frac{x_i - \hat{x}_i}{x_i} \times 100\% \tag{12}$$

(2)   Root mean square error (*RMSE*):

$$RMSE = \sqrt{\frac{1}{n}\sum_{i=1}^{n}(\frac{x_i - \hat{x}_i}{x_i})^2} \tag{13}$$

(3)   Mean absolute percentage error (*MAPE*):

$$MAPE = \frac{1}{n}\sum_{i=1}^{n}|(x_i - \hat{x}_i)/x_i| \cdot 100\% \tag{14}$$

(4)   Average absolute error (*AAE*):

$$AAE = \frac{1}{n}(\sum_{i=1}^{n}|x_i - \hat{x}_i|)/(\frac{1}{n}\sum_{i=1}^{n}x_i) \tag{15}$$

where x is the actual load of charging station and $\hat{x}$ is the corresponding forecasted load, n represents the groups in the dataset. The smaller these evaluation indicators are, the higher the prediction accuracy.

4.3. Results Analysis

In NILA, set $age_{mat} = 3$, $\kappa_{strenth} = 5$, the maximum iteration number is 100, $p = 0.5$, and the specific iteration process is shown in Figure 8. As can be seen in Figure 8, the optimal parameter of CNN is obtained at the thirty-fifth iteration. In order to validate the performance of the proposed technique NILA-CNN, comparisons are made with the final forecasting results from different algorithms involving LA-CNN, single CNN, SVM, and time series (TS). The parameter settings in LA-CNN model are consistent with those in NILA-CNN. The CNN model consists of one feature extraction layer which includes a convolutional layer with 12 neurons, and a subsampling layer with 5 neurons. The maximum number of training times, and the training error, are 200 and 0.0001, respectively. In SVM, the regularization parameter is 9.063, the kernel parameter equals 0.256, and the loss parameter is equal to 3.185. In Table 2, load forecasting results are derived from five different techniques.

Table 2. Actual load and forecasting results in 30 November 2017 (Unit: MW).

Time/h	Actual Data	NILA-CNN	LA-CNN	CNN	SVM	TS
0:00	0.374	0.384	0.387	0.361	0.364	0.354
0:30	0.408	0.398	0.422	0.399	0.427	0.432
1:00	0.282	0.282	0.277	0.272	0.292	0.302
1:30	0.262	0.255	0.254	0.271	0.247	0.245
2:00	0.402	0.411	0.414	0.418	0.381	0.431
2:30	0.330	0.321	0.341	0.342	0.315	0.353
3:00	0.269	0.267	0.260	0.258	0.280	0.284
3:30	0.247	0.242	0.244	0.241	0.257	0.261
4:00	0.251	0.254	0.243	0.242	0.257	0.240
4:30	0.253	0.245	0.245	0.262	0.265	0.267
5:00	0.246	0.252	0.255	0.256	0.233	0.226
5:30	0.269	0.276	0.277	0.259	0.254	0.285
6:00	0.503	0.510	0.519	0.510	0.515	0.537
6:30	0.696	0.715	0.719	0.668	0.721	0.743
7:00	0.850	0.832	0.824	0.882	0.889	0.910
7:30	1.003	1.013	0.987	1.038	0.957	1.059
8:00	1.560	1.518	1.507	1.615	1.521	1.653
8:30	1.999	2.055	2.066	2.071	1.901	2.109
9:00	2.100	2.159	2.170	2.025	2.185	1.980
9:30	2.316	2.374	2.387	2.283	2.396	2.450
10:00	3.757	3.687	3.628	3.618	3.932	3.995
10:30	3.761	3.671	3.784	3.806	3.598	4.000
11:00	3.612	3.519	3.486	3.752	3.780	3.928
11:30	3.821	3.923	3.706	3.971	3.883	4.120
12:00	2.635	2.679	2.595	2.736	2.760	2.503
12:30	2.882	2.955	2.783	2.985	3.004	3.043
13:00	3.354	3.403	3.470	3.220	3.153	3.582
13:30	3.832	3.930	3.707	3.686	4.008	4.094
14:00	4.335	4.225	4.189	4.487	4.531	4.643
14:30	3.867	3.876	3.897	4.013	4.028	4.136
15:00	4.063	3.942	3.931	4.121	3.889	4.330
15:30	4.559	4.688	4.707	4.741	4.363	4.879
16:00	4.654	4.708	4.799	4.830	4.438	4.988
16:30	3.819	3.710	3.936	3.906	3.593	4.079
17:00	3.498	3.472	3.379	3.623	3.566	3.303
17:30	2.959	2.886	2.858	2.856	3.081	3.170
18:00	2.647	2.710	2.686	2.595	2.762	2.829
18:30	2.695	2.753	2.783	2.591	2.551	2.846
19:00	2.795	2.773	2.890	2.898	2.651	2.950
19:30	3.158	3.068	3.253	3.044	3.020	3.003
20:00	3.479	3.407	3.594	3.396	3.565	3.684
20:30	4.271	4.381	4.130	4.114	4.449	4.511
21:00	3.577	3.673	3.454	3.437	3.752	3.829
21:30	2.605	2.583	2.625	2.697	2.489	2.787
22:00	2.059	2.006	1.988	2.136	1.980	2.200
22:30	1.831	1.876	1.891	1.904	1.754	1.958
23:00	1.135	1.165	1.170	1.091	1.101	1.071
23:30	0.447	0.438	0.462	0.463	0.428	0.478

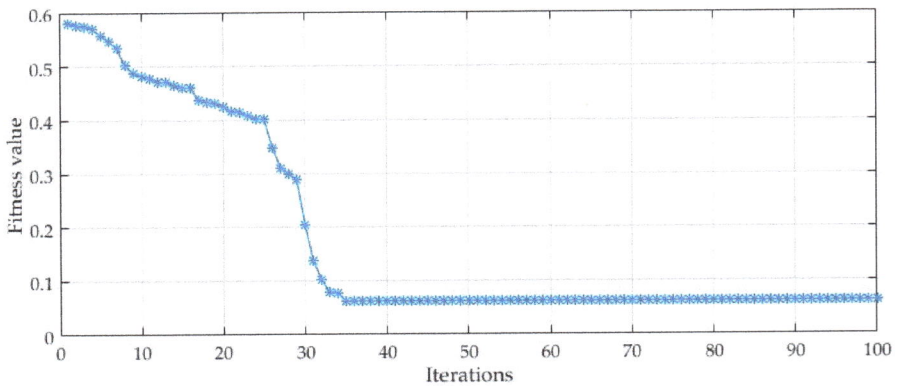

Figure 8. The iterative process of NILA.

Figure 9 displays the prediction results of Table 2, shown for more intuitive analysis. The values of *RE* obtained from the forecasting models are illustrated in Figure 10. Under the circumstance of electricity market, the error range between short-term load forecasting and the actual value should be [−3%, +3%]. It can be seen that the prediction error range of NILA-CNN is controlled within [0.23%, 2.86%] while the prediction error ranges of LA-CNN and CNN are [0.62%, 3.47%] and [−4%, 2.28%], respectively. Among them, 6 error points of NILA-CNN are controlled in [−1%, 1%], while the corresponding number of LA-CNN and CNN are 3 and 0. The errors of SVM model mostly range from [−6%, −4%] or [4%, 6%], and additionally, the errors of TS present a large fluctuation ranging, from [−8%, −5%] and [5%, 8%]. Thus, the prediction precision from the superior to the inferior can be ranked as follows: NILA-CNN, LA-CNN, CNN, SVM, TS. This demonstrates that NI can effectively improve the performance of LA. Further, NILA is conducive to high forecasting accuracy, due to the optimal parameter setting in the CNN model. Although the prediction results of NILA-CNN model are greater than other four methods in some points, such as at 10: 30, the overall errors perform the best.

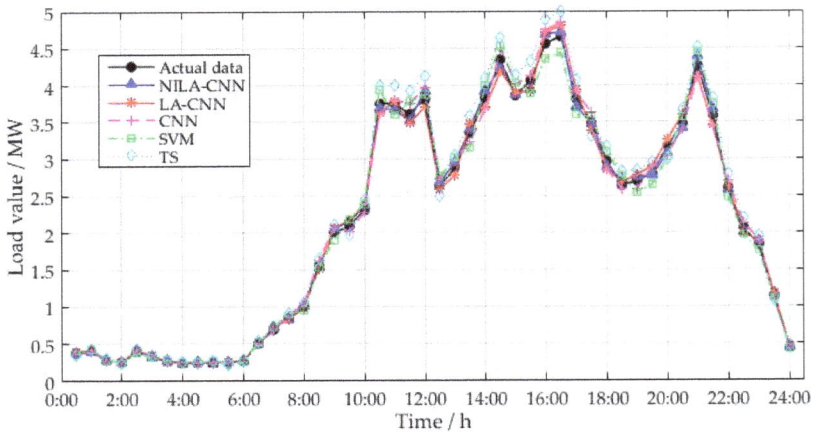

Figure 9. Prediction results.

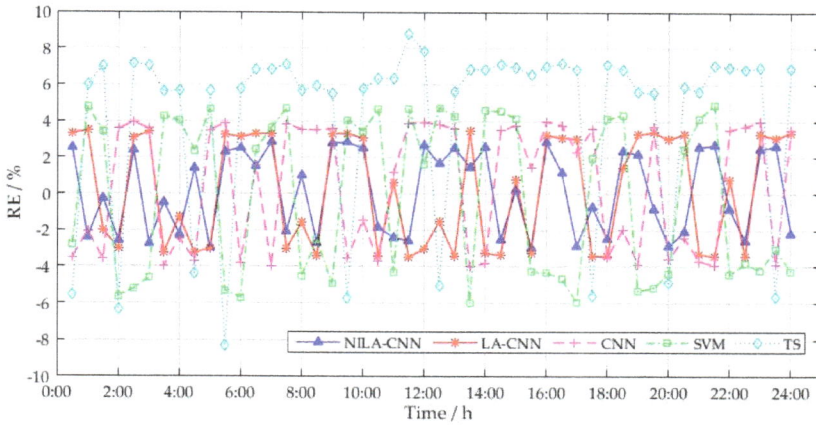

Figure 10. *RE* of prediction methods.

The statistical errors of the five prediction models are displayed in Figure 11. The analysis shows that: (a) NILA-CNN model outperforms other four techniques in terms of *RMSE* (2.27%), *MAPE* (2.14%) and *AAE* (2.096%). (b) Compared with LA-CNN, NI avoids premature convergence based on increasing the diversity of lion population. (c) The generalization ability and prediction accuracy of the CNN model can be improved by parameter optimization. (d) the CNN model can make a deep excavation of the internal relationship between the influential factors and the load of EV charging station in comparison with SVM. (e) ANN can reflect the non-linear relationship more accurately than TS methods.

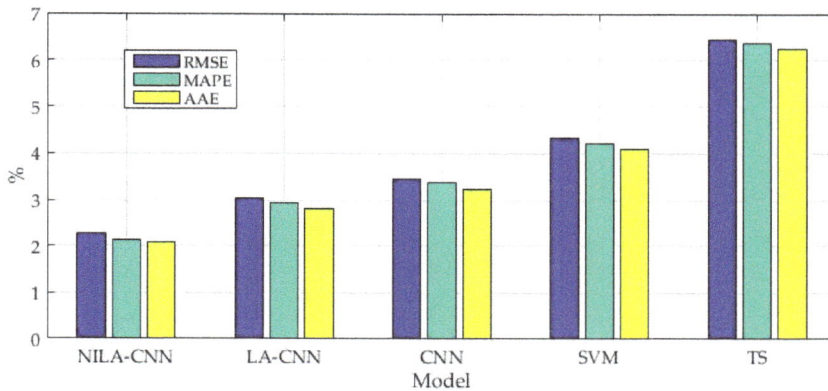

Figure 11. *RMSE*, *MAPE* and *AAE* of prediction methods (I).

5. Further Study

In order to further verify the effectiveness of the proposed model, one more case which selects the data from another EV charging station is provided in this paper. The study is carried out with data from 1 June 2016 to 31 May 2017. To reflect the influence of seasonal factors on load, data from 7 days of each season are selected as a test set, with the rest as a training set. The specific data division is shown in Table 3.

Table 3. The data division of case two.

Data Type	Data Range	Season Type
Training set	1 June 2016–24 August 2016	Autumn
	1 September 2016–23 November 2016	Winter
	1 December 2016–21 February 2017	Spring
	1 March 2017–24 May 2017	Summer
Test set	25 August 2016–31 August 2016	Autumn
	24 November 2016–30 November 2016	Winter
	22 February 2017–28 February 2017	Spring
	25 May 2017–31 May 2017	Summer

The five models shown above are still used in this experiment, where the parameter settings of NILA-CNN, LA-CNN and CNN are consistent. In SVM, the regularization parameter is 2.0153, the kernel parameter is 0.015, and the loss parameter is 0.013. The statistical errors including *RMSE*, *MAPE* and *AAE* are displayed in Figure 12.

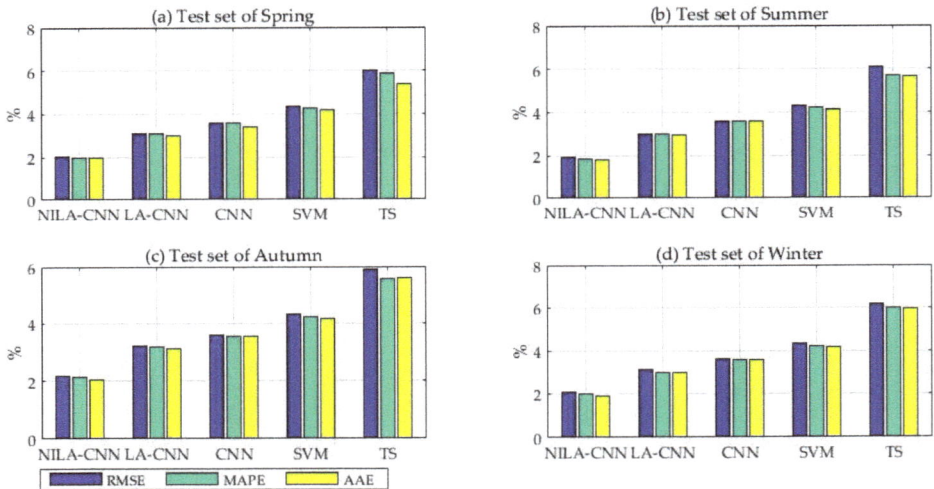

Figure 12. *RMSE*, *MAPE* and *AAE* of prediction methods (II). ((**a**) is the error results of test set in Spring; (**b**) is the error results of test set in Summer; (**c**) is the error results of test set in Autumn; (**d**) is the error results of test set in Winter).

As demonstrated in Figure 12, the values of *RMSE*, *MAPE* and *AAE* of NILA-CNN in four seasons are all the lowest among the forecasting techniques, namely 2.010, 2.00% and 1.97% in Spring, 1.93%, 1.86% and 1.80% in Summer, 2.16%, 2.14% and 2.04% in Autumn, 2.07%, 2.00% and 1.90% in Winter. Meanwhile, it can be noted that the overall prediction accuracy of LA-CNN is better than that of the CNN model, and CNN-based approaches are superior to SVM and TS, which proves the advantages of NI, LA and CNN. Therefore, the short-term load forecasting for EV charging stations based on the NILA-CNN model is efficient enough to compete with existing approaches in prediction precision. As a hybrid algorithm, the proposed model is able to provide accurate data support for economic operation of the charging station.

6. Conclusions

In recent years, with the gradually worsening energy crisis and the intensification of global warming, EVs have become one of the main development directions for new energy vehicles due, to their energy savings and emission reductions. EV charging stations are an important part of the power load; thus, research on their short-term load forecasting is not only of great significance for economic dispatch in the grid, but also contributes to stable operation of the charging station. In this paper, a short-term load forecasting method for EV charging stations combining NILA with CNN is established, where NI is used to improve the optimization performance of LA, and the hybrid technique NILA is introduced to determine the optimal parameters of CNN model, so as to obtain better prediction accuracy. Through analysis of load characteristics in the charging station, ten influential factors are selected as input, including seasonal category, maximum temperature, minimum temperature, weather condition, day type, and the loads at the same moment in previous five days. According to the case studies, CNN integrated with NILA outperforms other models in terms of prediction precision, indicating that NILA-CNN model is a promising technique for short-term load forecasting of EV charging station.

Author Contributions: Y.L. designed this research and wrote this paper; Y.H. provided professional guidance; M.Z. processed the data and revised this paper.

Funding: This research was funded by [the Fundamental Research Funds for the Central Universities] grant number [2014MS146].

Acknowledgments: This work is supported by the Fundamental Research Funds for the Central Universities (Project No. 2014MS146).

Conflicts of Interest: The authors declare no conflict of interest.

Abbreviations

EV	Electric vehicle
CNN	Convolutional neural network
LA	Lion algorithm
NI	Niche immunity
NILA	Lion algorithm improved by niche immunity
ANN	Artificial neural network
SVM	Support vector machine
RBFNN	Radial basis function neural network
TS	time series
RE	Relative error
RMSE	Root mean square error
MAPE	Mean absolute percentage error
AAE	Average absolute error
LA-CNN	Convolutional neural network optimized by lion algorithm
NILA-CNN	Convolutional neural network optimized by niche immunity lion algorithm

References

1. Tribioli, L. Energy-Based Design of Powertrain for a Re-Engineered Post-Transmission Hybrid Electric Vehicle. *Energies* **2017**, *10*, 918. [CrossRef]
2. Tan, K.M.; Ramachandaramurthy, V.K.; Yong, J.Y.; Padmanaban, S.; Mihet-Popa, L.; Blaabjerg, F. Minimization of Load Variance in Power Grids—Investigation on Optimal Vehicle-to-Grid Scheduling. *Energies* **2017**, *10*, 1880. [CrossRef]
3. Zhang, Y.; Su, X.; Yan, X.; Li, M.; Li, D.D. A method of charging load forecast based on electric vehicle time-space characteristics. *Electr. Power Constr.* **2015**, *7*, 75–82.

4. Soares, F.J.; Lopes, J.A.P.; Almeida, P.M.R. A Monte Carlo method to evaluate electric vehicles impacts in distribution networks. In Proceedings of the 2010 IEEE Conference on Innovative Technologies for an Efficient and Reliable Electricity Supply (CITRES), Waltham, MA, USA, 27–29 September 2010; pp. 365–372.

5. Huang, H.; Chung, C.Y.; Chan, K.W.; Chen, H. Quasi-Monte Carlo Based Probabilistic Small Signal Stability Analysis for Power Systems with Plug-In Electric Vehicle and Wind Power Integration. *IEEE Trans. Power Syst.* **2013**, *28*, 3335–3343. [CrossRef]

6. Ashtari, A.; Bibeau, E.; Shahidinejad, S.; Molinski, T. PEV charging profile prediction and analysis based on vehicle usage data. *IEEE Trans. Smart Grid* **2012**, *3*, 341–350. [CrossRef]

7. Bing, Y.; Wang, L.; Liao, C. Charging Load Calculation Method of Large-scale Electric Vehicles with Coupling Characteristics. *Autom. Electr. Power Syst.* **2015**, *39*, 76–82.

8. Huang, X.Q.; Chen, J.; Chen, Y.; Yang, H.; Cao, Y.; Jiang, L. Load forecasting method for electric vehicle charging station based on big data. *Autom. Electr. Power Syst.* **2016**, *12*, 68–74.

9. Kachoosangi, F.T. How Reliable Are ANN, ANFIS, and SVM Techniques for Predicting Longitudinal Dispersion Coefficient in Natural Rivers? *J. Hydraul. Eng.* **2016**, *142*, 04015039.

10. Zhang, W.; Xie, F.; Huang, M.; Li, J.; Li, Y. Research on short-term load forecasting methods of electric buses charging station. *Power Syst. Prot. Control* **2013**, *41*, 61–66.

11. Chang, D.; Ren, J.; Zhao, J.; Duan, X.; Gong, W.; Zhang, Z. Study on short term load forecasting of electric vehicle charging station based on RBF-NN. *J. Qingdao Univ. (Eng. Technol. Ed.)* **2014**, *4*, 44–48.

12. Wang, Z.; Dai, B.; Li, X. Research of Short-Term Load Forecasting Model for Electrical Vehicle Charging Stations based on PSO-SNN. *Electr. Eng.* **2016**, *17*, 46–50.

13. Luo, Z.; Hu, Z.; Song, Y.; Yang, X.; Zhan, K.; Wu, J. Study on plug-in electric vehicles charging load calculating. *Autom. Electr. Power Syst.* **2011**, *35*, 36–42.

14. Schmidhuber, J. Deep learning in neural networks: An overview. *Neural Netw.* **2015**, *61*, 85–117. [CrossRef] [PubMed]

15. Shi, J.; Zhang, J. Ultra Short-Term Photovoltaic Refined Forecasting Model Based on Deep Learning. *Electr. Power Constr.* **2017**, *38*, 28–35.

16. Li, S.; Lin, L.; Sun, C. Click-Through Rate Prediction for Search Advertising based on Convolution Neural Network. *Intell. Comput. Appl.* **2015**, *5*, 22–25.

17. Wallach, I.; Dzamba, M.; Heifets, A. AtomNet: A Deep Convolutional Neural Network for Bioactivity Prediction in Structure-based Drug Discovery. *Math. Z.* **2015**, *47*, 34–46.

18. Sss, K.; Ayush, K.; Babu, R.V. DeepFix: A Fully Convolutional Neural Network for Predicting Human Eye Fixations. *IEEE Trans. Image Process.* **2017**, *26*, 4446–4456.

19. Ruiz, G.R.; Bandera, C.F.; Temes, T.G.; Gutierrez, A.S. Genetic algorithm for building envelope calibration. *Appl. Energy* **2016**, *168*, 691–705. [CrossRef]

20. Chen, Z.; Xiong, R.; Cao, J. Particle swarm optimization-based optimal power management of plug-in hybrid electric vehicles considering uncertain driving conditions. *Energy* **2016**, *96*, 197–208. [CrossRef]

21. Ye, K.; Zhang, C.; Ning, J.; Liu, X. Ant-colony algorithm with a strengthened negative-feedback mechanism for constraint-satisfaction problems. *Inf. Sci.* **2017**, *406*, 29–41. [CrossRef]

22. Rajakumar, B.R. The Lion's Algorithm: A New Nature-Inspired Search Algorithm. *Procedia Technol.* **2012**, *6*, 126–135. [CrossRef]

23. Liu, J.; Wang, H.; Sun, Y.; Li, L. Adaptive niche quantum-inspired immune clonal algorithm. *Nat. Comput.* **2016**, *15*, 297–305. [CrossRef]

24. Lawrence, S.; Giles, C.L.; Tsoi, A.C. Face recognition: a convolutional neural-network approach. *IEEE Trans. Neural Netw.* **1997**, *8*, 98–113. [CrossRef] [PubMed]

25. Dieleman, S.; Willett, K.W.; Dambre, J. Rotation-invariant convolutional neural networks for galaxy morphology prediction. *Mon. Not. R. Astron. Soc.* **2015**, *450*, 1441–1459. [CrossRef]

26. Li, M.; Ling, C.; Xu, Q.; Gao, J. Classification of G-protein coupled receptors based on a rich generation of convolutional neural network, N-gram transformation and multiple sequence alignments. *Amino Acids* **2017**, *50*, 255–266. [CrossRef] [PubMed]

27. Ovalle, A.; Fernandez, J.; Hably, A.; Bacha, S. An Electric Vehicle Load Management Application of the Mixed Strategist Dynamics and the Maximum Entropy Principle. *IEEE Trans. Ind. Electron.* **2016**, *63*, 3060–3071. [CrossRef]

28. Hu, Z.; Zhan, K.; Zhang, H.; Song, Y. Pricing mechanisms design for guiding electric vehicle charging to fill load valley. *Appl. Energy* **2016**, *178*, 155–163. [CrossRef]

29. Xiang, Y.; Liu, J.; Li, R.; Li, F.; Gu, C.; Tang, S. Economic planning of electric vehicle charging stations considering traffic constraints and load profile templates. *Appl. Energy* **2016**, *178*, 647–659. [CrossRef]

![energies logo] *energies*

MDPI

Article

Short-Term Load Forecasting with Multi-Source Data Using Gated Recurrent Unit Neural Networks

Yixing Wang [ID], Meiqin Liu * [ID], Zhejing Bao and Senlin Zhang

College of Electrical Engineering, Zhejiang University, Hangzhou 310027, China; wangyixing@zju.edu.cn (Y.W.);
zjbao@zju.edu.cn (Z.B.); slzhang@zju.edu.cn (S.Z.)
* Correspondence: liumeiqin@zju.edu.cn; Tel.: +86-139-5800-7313

Received: 29 March 2018; Accepted: 1 May 2018; Published: 3 May 2018

Abstract: Short-term load forecasting is an important task for the planning and reliable operation of power grids. High-accuracy forecasting for individual customers helps to make arrangements for generation and reduce electricity costs. Artificial intelligent methods have been applied to short-term load forecasting in past research, but most did not consider electricity use characteristics, efficiency, and more influential factors. In this paper, a method for short-term load forecasting with multi-source data using gated recurrent unit neural networks is proposed. The load data of customers are preprocessed by clustering to reduce the interference of electricity use characteristics. The environmental factors including date, weather and temperature are quantified to extend the input of the whole network so that multi-source information is considered. Gated recurrent unit neural networks are used for extracting temporal features with simpler architecture and less convergence time in the hidden layers. The detailed results of the real-world experiments are shown by the forecasting curve and mean absolute percentage error to prove the availability and superiority of the proposed method compared to the current forecasting methods.

Keywords: short-term load forecasting; artificial intelligence; gated recurrent unit; recurrent neural network; power grid

---

1. Introduction

Load forecasting is an essential part for energy management and distribution management in power grids. With the continuous development of the power grids and the increasing complexity of grid management, accurate load forecasting is a challenge [1,2]. High-accuracy power load forecasting for customers can make the reasonable arrangements of power generation to maintain the safety and stability of power supply and reduce electricity costs so that the economic and social benefit is improved. Moreover, forecasting at individual customer level can optimize power usage and help to balance the load and make detailed grid plans. Load forecasting is the process of estimating the future load value at a certain time with historical related data, which can be divided into long-term load forecasting, medium-term load forecasting and short-term load forecasting according to the forecasting time interval. Short-term load forecasting, which this paper focuses on, is the daily or weekly forecasting [3,4]. It is used for the daily or weekly schedule including generator unit control, load allocation and hydropower dispatching. With the increasing penetration of renewable energies, short-term load forecasting is fundamental for the reliability and economy of power systems.

Models of short-term load forecasting can be classified into two categories consisting of tradition statistic models and artificial intelligent models. Statistic models, such as regression analysis models and time sequence models, are researched and used frequently were previously limited by computing capability. Taylor et al. [5] proposed an autoregressive integrated moving average (ARIMA) model with an extension of Holt–Winters exponential smoothing for short-term load forecasting. Then, an power autoregressive conditional heteroskedasticity (PARCH) method was presented for better performance [6]. These statistic models need fewer historical data and have a small amount of calculation. However, they require a higher stability of the original time sequences and do not consider the uncertain factors such as weather and holidays. Therefore, artificial intelligent models about forecasting, such as neural networks [7,8], fuzzy logic method [9] and support vector regression [10], were proposed with the development of computer science and smart grids. Recently, neural networks are becoming an active research topic in the area of artificial intelligence for its self-learning and fault tolerant ability. Some effective methodologies for load forecasting based on neural networks have been proposed in recent years. A neural network based method for the construction of prediction intervals was proposed by Quan et al. [7]. Lower upper bound estimation was applied and extended to develop prediction intervals using neural network models. The method resulted in higher quality for different types of prediction tasks. Ding [8] used separate predictive models based on neural networks for the daily average power and the day power variation forecasting in distribution systems. The prediction accuracy was improved with respect to naive models and time sequence models. The improvement of forecasting accuracy cannot be ignored, but, with the increasing complexity and scale of power grids, high-accuracy load forecasting with advanced network model and multi-source information is required.

Deep learning, proposed by Hinton [11,12], made a great impact on many research areas including fault diagnosis [13,14] and load forecasting [15–17] by its strong learning ability. Recurrent neural network (RNNs), a deep learning framework, are good at dealing with temporal data because of its interconnected hidden units. It has proven successful in applications for speech recognition [18,19], image captioning [20,21], and natural language processing [22,23]. Similarly, during the process of load forecasting, we need to mine and analyse large quantities of temporal data to make a prediction of time sequences. Therefore, RNNs are an effective method for load forecasting in power grids [24,25]. However, the vanishing gradient problem limits the performance of original RNNs. The later time nodes' perception of the previous ones decreases when RNNs become deep. To solve this problem, an improved network architecture called long short-term memory (LSTM) networks [26] were proposed, and have proven successful in dealing with time sequences for power grids faults [27,28]. Research on short-term load forecasting based on LSTM networks was put forward. Gensler et al. [29] showed the compared results for solar power forecasting about physical photovoltaic forecasting model, multi-layer perception, deep belief networks and auto-LSTM networks. It proved the LSTM networks with autoencoder had the lowest error. Zheng et al. [30] tackled the challenge of short-term load forecasting with proposing a novel scheme based on LSTM networks. The results showed that LSTM-based forecasting method can outperform traditional forecasting methods. Aiming at short-term load forecasting for both individual and aggregated residential loads, Kong et al. [31] proposed an LSTM recurrent neural network based framework with the input of day indices and holiday marks. Multiple benchmarks were tested in the real-world dataset and the proposed LSTM framework achieved the best performance. The research works mentioned above indicate the successful application of LSTM for load forecasting in power grids. However, load forecasting needs to be fast and accurate. The principle and structure of LSTM are complex with input gate, output gate, forget gate and cell, so the calculation is heavy for forecasting in a large scale grid. Gated recurrent unit (GRU) neural networks was proposed in 2014 [32], which combined the input gate and forget gate to a single gate called update gate. The model of a GRU is simpler compared with an LSTM block. It was proved on music datasets and ubisolf datasets that GRU's performance is better with less parameters about convergence time and required training epoches [33]. Lu et al. [34] proposed a multi-layer self-normalizing GRU model for short-term electricity load forecasting to overcome the exploding and vanishing gradient problem. However, short-term load forecasting for customers is influenced by factors including date, weather and temperature, which previous research did not consider seriously. People may need more energy when the day is cold or hot. Enterprises or factories may reduce their power consumption on holidays.

In this paper, a method based on GRU neural networks with multi-source input data is proposed for short-term load forecasting in power grids. Moreover, this paper focuses on the load forecasting for individual customers, which is an important and tough problem because of the high volatility and uncertainty [30]. Therefore, before training the networks, we preprocess the customers' load data with clustering analysis to reduce the interference of the electricity use characteristics. Then, the customers are classified into three categories to form the training and test samples by K-means clustering algorithm. To obtain not only the load measurement data but also the important factors including date, weather and temperature, the input of the network are set as two parts. The temporal features of load measurement data are extracted by GRU neural networks. The merge layer is built to fuse the multi-source features. Then, we can get the forecasting results by training the whole network. The methodologies are described in detail in Section 2. The main contributions of this paper are as follows.

1. Trained samples are formed by clustering to reduce the interference of different characteristics of customers.
2. Multi-source data including date, weather and temperature are quantified for input so that the networks obtain more information for load forecasting.
3. The GRU units are introduced for more accurate and faster load forecasting of individual customers.

In general, the proposed method uses the clustering algorithm, quantified multi-source information and GRU neural network for short-term load forecasting, which past research did not consider comprehensively. The independent experiments in the paper verify the advantages of the proposed method. The rest of the paper is organized as follows. The methodology based on GRU Neural Networks for short-term load forecasting is proposed in Section 2. Then, the results and discussion of the simulation experiments are described to prove the availability and superiority of the proposed method in Section 3. Finally, the conclusion is made in Section 4.

2. Methodology Based on GRU Neural Networks

In this section, the methodology is proposed for short-term load forecasting with multi-source data using GRU Neural Networks. First, the basic model of GRU neural networks are introduced [32]. Then, data description and processing are elaborated. The load data are clustered by K-means clustering algorithm so that the load samples with similar characteristics in a few categories are obtained. This helps improve the performance of load forecasting for individual customers. In the last subsection, the whole proposed model based on GRU neural networks is shown in detail.

2.1. Model of GRU Neural Networks

Gated recurrent unit neural networks are the improvement framework based on RNNs. RNNs are improved artificial neural networks with the temporal input and output. Original neural networks only have connections between the units in different layers. However, in RNNs, there are connections between hidden units forming a directed cycle in the same layer. The network transmits the temporal information through these connections. Therefore, the RNNs outperform conventional neural networks in extracting the temporal features by these connections. A simple structure for an RNN is shown in Figure 1. The input and output are time sequences, which is different from original neural networks. The process of forward propagation is shown in Figure 1 and given by Equations (1)–(3).

$$a_h^t = \sum_{i=1}^{I} w_{ih} x_i^t + \sum_{h'=1}^{H} w_{h'h} s_{h'}^{t-1} \tag{1}$$

$$s_h^t = f_h(a_h^t) \tag{2}$$

$$a_o^t = \sum_{h=1}^{H} w_{ho} s_h^t \tag{3}$$

where w is the weight; a is the sum calculated through weights; f is the activation function; s is the value after calculation by the activation function; t represents the current time of the network; i is the number of input vectors; h is the number of hidden vectors in t is time; h' is the number of hidden vectors in $t-1$ time; and o is the number of output vectors.

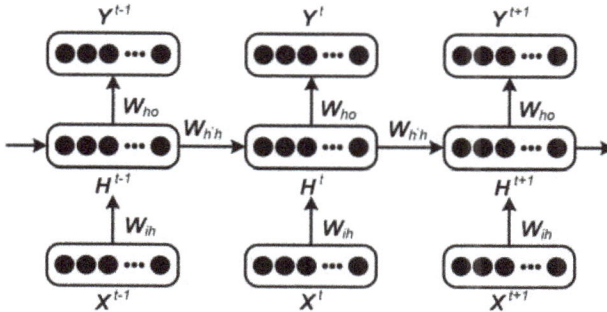

Figure 1. A simple RNN structure, where X is the input unit, H is the hidden unit, Y is the output unit, and W is the weight matrix.

Similar to conventional neural networks, RNNs can be trained by back-propagation through time [35] with the gradient descent method. As shown in Figure 1, each hidden layer unit receives not only the data input but also the output of the hidden layer in the last time step. The temporal information can be recorded and put into the calculation of the current output so that the dynamic changing process can be learned with this architecture. Therefore, RNNs are reasonable to predict the customer load curves in power grids. However, when the time sequence is longer, the information will reduce and disappear gradually through transferring in hidden units. The original RNNs have the vanishing gradient problem and the performance declines when dealing with long time sequences.

The vanishing gradient problem can be solved by adding control gates for remembering information in the process of data transfer. In LSTM networks, the hidden units of RNNs are replaced with LSTM blocks consisting of cell, input gate, output gate and forget gate. Moreover, the forget gate and input gate are combined into a single update gate in GRU neural network. The structure of GRU is shown in Figure 2.

The feedforward deduction process for GRU units is shown in Figure 2 and given by Equations (4)–(10).

$$a_u^t = \sum_{i=1}^{I} w_{iu} x_i^t + \sum_{h=1}^{H} w_{hu} s_h^{t-1} \tag{4}$$

$$s_u^t = f(a_u^t) \tag{5}$$

$$a_r^t = \sum_{i=1}^{I} w_{ir} x_i^t + \sum_{h=1}^{H} w_{hr} s_h^{t-1} \tag{6}$$

$$s_r^t = f(a_r^t) \tag{7}$$

$$\tilde{a}_{h'}^t = s_u^t \sum_{h=1}^{H} w_{hh'} s_h^{t-1} + \sum_{i=1}^{I} w_{ih'} x_i^t \tag{8}$$

$$\tilde{s}_{h'}^t = \phi(\tilde{a}_{h'}^t) \tag{9}$$

$$s_h^t = (1 - s_u^t) \tilde{s}_{h'}^t + s_u^t s_h^{t-1} \tag{10}$$

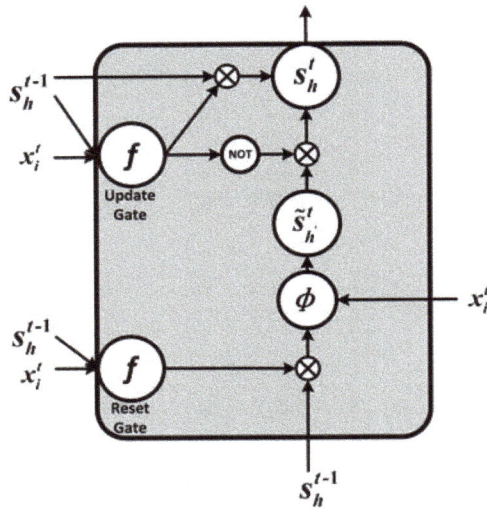

Figure 2. Inner structure of GRU, where all arrows represent the weights between gates and units and the units of f and ϕ are the activation functions. The parameters are explained in detail after the Equations (4)–(10).

where u is the number of update gate vector; r is the number of reset gate vector; h is the number of hidden vectors at t time step; h' is the number of hidden vectors at $t-1$ time step; f and ϕ are the activation functions; f is the sigmoid function and ϕ is the tanh function generally; and $\tilde{s}_{h'}^t$ means the new memory of hidden units at t time step.

According to Figure 2, the new memory $\tilde{s}_{h'}^t$ is generated by the input x_i^t at the current time step and the hidden unit state s_h^{t-1} at the last time step, which means the new memory can combine the new information and the historical information. The reset gate determines the importance of s_h^{t-1} to $\tilde{s}_{h'}^t$. If the historical information s_h^{t-1} is not related to new memory, the reset gate can completely eliminate the information in the past. The update gate determines the degree of transfer from s_h^{t-1} to s_h^t. If $s_u^t \approx 1$, s_h^{t-1} is almost completely passed to s_h^t. If $s_u^t \approx 0$, $\tilde{s}_{h'}^t$ is passed to s_h^t. The structure shown in Figure 2 results in a long memory in GRU neural networks. The memory mechanism solves the vanishing gradient problem of original RNNs. Moreover, compared to LSTM networks, GRU neural networks merge the input gate and forget gate, and fuse the cell units and hidden units in LSTM block. It maintains the performance with simpler architecture, less parameters and less convergence time [33]. Correspondingly, GRU neural networks are trained by back-propagation through time as RNNs [35].

2.2. Data Description

The real-world load data of individual customers in Wanjiang area is recorded from Dongguan Power Supply Bureau of China Southern Power Grid in Guangdong Province, China during 2012–2014. The topology structure of Wanjiang area is shown in Figure 3. There are 36 feeders connecting to the load sides in the Wanjiang area, i.e., Feeders 1—36. The active power is extracted for load forecasting from these feeders. The sampling period is 15 min as the meter record data. The load curve of a customer, No. 53990001, from Feeder 2 during a month is shown in Figure 4, where the different load characteristics of the customer on each day can be concluded.

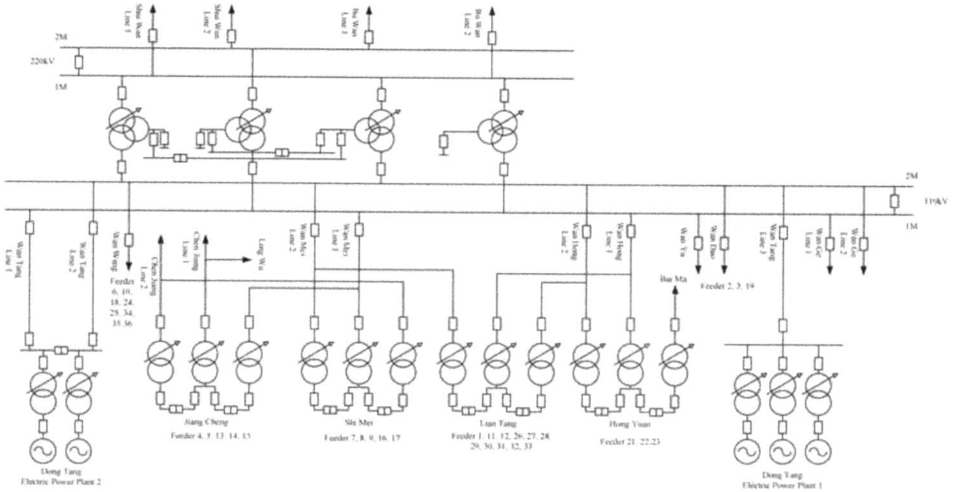

Figure 3. Primary electrical system in Wanjiang area above 110 kv, including electric power plants, transmission buses, converting stations, and user loads. The feeders are marked under their corresponding load sides.

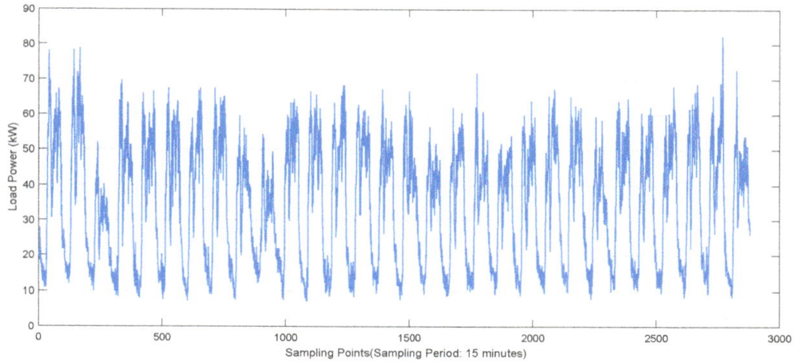

Figure 4. Load curve of Customer 53990001 from Feeder 2 during a month, where the sampling period is 15 min.

Besides the historical load curves, short-term load forecasting is influenced by the factors of date, weather and temperature. The real historical data of weather and temperature in the corresponding area in Dongguan City were obtained online from the weather forecast websites. The categories of weather include sunny, cloud, overcast, light rain, shower, heavy rain, typhoon and snow. The date features can be found in calendars.

2.3. Clustering and Quantization

The custom of electricity use and characteristics of load curve are different among the different categories of customers such as industrial customers, residential customers and institution customers. The different characteristics would affect the performance of forecasting. Training forecasting networks with each customer separately would be a huge computation and storage problem. Therefore, in the proposed method, the load curve samples are divided into certain categories using K-means clustering algorithm. Samples with similar characteristics form a certain category, which form the input of GRU neural networks for the corresponding customers. K-means clustering algorithm is a simple and available method for clustering through unsupervised learning with fast convergence and less parameters. The only parameter, K, number of clustering category, can be determined by Elbow method with the turning point of loss function curve.

Suppose the input sample is $S = x_1, x_2, ..., x_m$. The algorithm is shown as follows.

1. Randomly initialize K clustering centroids $c_1, c_2, ..., c_K$.
2. For $i = 1, 2, ..., m$, label each sample x_i with the clustering centroid closest to x_i, getting K categories noted by G_k.

$$label_i = \underset{1 \le k \le K}{\arg\min} \ \|x_i - c_k\|, i = 1, 2, ..., m \tag{11}$$

3. For $k = 1, 2, ..., K$, average the samples assigned to G_k to update c_k.

$$c_k = \frac{1}{|G_k|} \sum_{i \in G_k} x_i, k = 1, 2, ..., K \tag{12}$$

4. Repeat Steps 2 and 3 until the change of clustering centroid or the loss function of clustering less than a set threshold. The loss function is given by Equation (13), where x_j is the samples in categories $G_k, j = 1, 2, ..., n_k$ and n_k is the number of samples in categories G_k.

$$J(c_1, c_2, ..., c_K) = \frac{1}{2} \sum_{k=1}^{K} \sum_{j=1}^{n_k} \|x_j - c_k\| \tag{13}$$

Moreover, the factors of date, weather and temperature should be added into input with quantization. First, the power consumption should be different between weekdays and weekends. The official holidays are also an important factor, so we quantify the date index as shown in Table 1, where the index of official holidays is 1 no matter what day it is. Similarly, the weather and temperature are quantified according to their inner relations, as shown in Tables 2 and 3.

Table 1. Quantization for the factors of date.

Date (D)	Mon.	Tues.	Wed.	Thur.	Fri.	Sat.	Sun.	Official Holidays
Index	0	0.02	0.04	0.06	0.08	0.6	0.8	1

Table 2. Quantization for the factors of weather.

Weather (W)	Sunny	Cloud	Overcast	Light Rain	Shower	Heavy Rain	Typhoon	Snow
Index	0	0.1	0.2	0.4	0.5	0.6	0.8	1

Table 3. Quantization for the factors of temperature.

Temperature (T/°C)	T ≤ 0	0 < T ≤ 10	10 < T ≤ 20	20 < T ≤ 30	30 < T ≤ 40	T ≥ 40
Index	0	0.2	0.4	0.6	0.8	1

2.4. The Proposed Framework Based on GRU Neural Networks

The schematic diagram of proposed framework based on GRU neural networks for short-term load forecasting is shown in Figure 5. The individual customers are clustered into a few categories for more accurate forecasting. The samples are recorded from the categories where the customer to be predicted locates in. The load measurement data of individual customers in one day is extracted as a sample for short-term load forecasting, noted by P. The dimension of P is 96 with the 15 min sampling period. Then, the samples are reshaped into two-dimension for the input of GRU neural networks. Considering the influencing factors date D, weather W and temperature T, date D_p, weather W_p and temperature T_p on the forecasting day are added to the another input of the GRU neural networks. Considering the general factor of date, the prediction interval is set to seven days. Therefore, the load measurement data P_l on the day in the last week from the forecasting day, D_p, W_p and T_p, are recorded as the overall input. The load measurement data P_p on the forecasting day are recorded as the output, whose dimension is 96. Therefore, the input X and output Y of samples are given by Equations (14) and (15).

$$X = \{P_l;\ D_p,\ W_p,\ T_p\} \tag{14}$$

$$Y = \{P_p\} \tag{15}$$

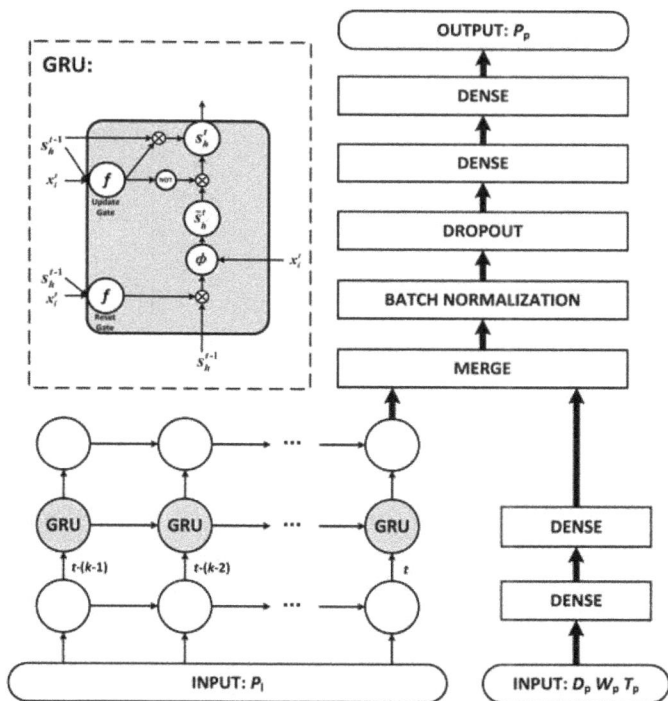

Figure 5. Schematic diagram of proposed framework based on GRU Neural Networks for short-term load forecasting, where k is the number of hidden units and t is the time step. The parameters of GRU units are clarified in Section 2.1. The input and output parameters are explained in the next subsection.

The features from GRU neural networks and fully connected neural network are merged with the concatenating mode and passes through batch normalization and dropout layer to avoid overfitting and increase the learning efficiency. The principle is that batch normalization can avoid the gradient vanishing of falling into the saturated zone, and that the better performance in fixed combination is avoided when random neurons do not work in a dropout layer. Then, two-layer fully connected neural network are added before the output for learning and generalization ability. With training by back-propagation through time, the whole network implements the short-term load forecasting for individual customers. The structure can be extended if there is more information in the practical situation. The basic theory is also acceptable for medium-term load forecasting and long-term load forecasting, but different influence factors should be considered and the model should be changed with different input, output, and inner structure for good performance.

3. Experiments and Results

In this section, the experiments are described in detail and the results are shown in figures and tables. The specific discussion for results is elaborated after the results and prove the improved performance compared to other methods. The data for experiments are recorded in Section 2.2.

3.1. Clustering Analysis for Load Curve of Individual Customers

Before the short-term load forecasting using GRU neural networks, the load curves of individual customers are clustered to different categories for samples with K-means clustering algorithm. The parameter K is selected as 3 by Elbow method. There are 746 customers in the Wanjiang area in Dongguan city. The load measurement data should be processed with 0–1 standardization to the same scale for clustering to reduce the impact of different magnitudes and dimensions. The clustering is done for 10 times with load curves in 10 days for the individual customers. The clustering results are obtained with the average results in 10 days and the number of each clustering category is shown in Table 4. The standardized curves for 30 selected customers in three categories on a weekday are shown in Figures 6–8.

Table 4. Number of each clustering category.

Categories	Category 1	Category 2	Category 3
Number of Customers	221	308	217

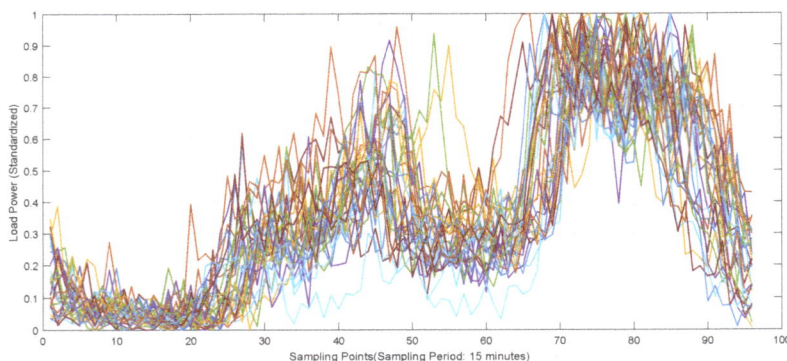

Figure 6. Load curves of 30 customers in Category 1.

As can be seen in Figures 6–8, different customers have different characteristics of electricity use. According to Figure 6, there are two electric peaks in a day. The evening peak is higher than the noon peak. The classic representation of this characteristic in Figure 6 is residential customers. Different from Figure 6, Figure 7 maintains the peak from 9 a.m. to late at night except noon. They are the general load curves of industry and business customers. In Figure 8, there are two electric peaks in the morning and afternoon. It should belong to the government and institutional customers. Even though a few customers have differences with the overall curve, this is the best clustering for them and it does not influence the overall performance greatly. With the clustering of individual customers, the networks can be trained with samples in the same category according to the customer to be predicted, so that the interference of electricity use characteristics can be reduced.

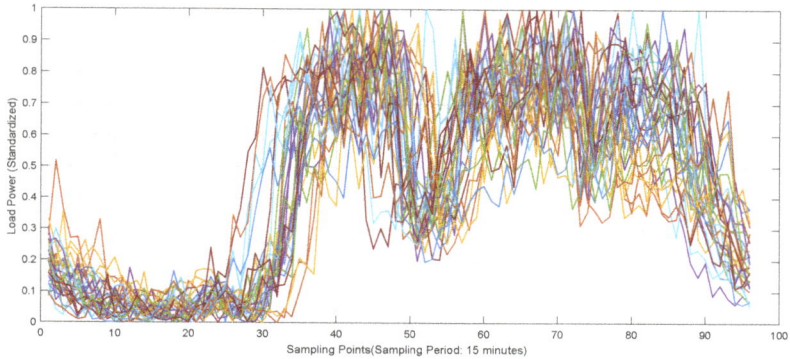

Figure 7. Load curves of 30 customers in Category 2.

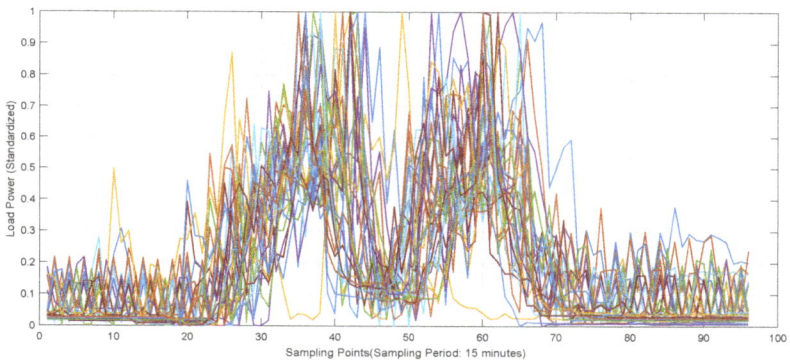

Figure 8. Load curves of 30 customers in Category 3.

3.2. The Detailed Network Structure and Parameters

The detailed structure of whole network are shown in Table 5. The parameters of the network are set as shown in Table 6. The structure and parameters are set for better performance according to the multiple experiments for customers in Wanjiang area. The "RMSprop" optimizer is chosen for its better performance in recurrent neural networks. The parameters can be adjusted for the different practical situations. In this paper, the number of epoch is set to 200 for the proposed method and can be adjusted for the compared methods. The training is stopped when the error decreases to a steady state.

Table 5. Number of units in the proposed network.

Layer	Number of Units	Layer	Number of Units
INPUT: P_1	(6, 16)	INPUT: $D_p\ W_p\ T_p$	3
BATCH NORMALIZATION	(6, 16)	DENSE1	100
GRU	300	DENSE2	100
MERGE	400		
DROPOUT	400		
DENSE3	100		
DENSE4	100		
OUTPUT:P_p	96		

Table 6. Parameter setting in the proposed network.

Parameters	Value
Input Time Steps	6
Input Dimension	16
Batch Size	30
Epoch	200
Optimizer	RMSprop
Learning Rate	0.001
Decay	0.9
Dropout Rate	0.6

3.3. Comparison of Results of Proposed Method

The results of the proposed method are shown as follows. In the experiments, the training samples are recorded from the load data in the period from October 2012 to September 2013 while the test samples are recorded from load data in the period from October to December of 2013. The number of recorded training samples and test samples of each categories is 36,000 and 9000, respectively, with 100 customers in a category. The ratio of sample number is 4:1. Mean absolute percentage error (MAPE) is the classic evaluation index for load forecasting. The computational formula is given by Equations (16) and (17), where $n = 96$ represents the dimension of samples and m represents the number of test samples.

$$MAPE = \frac{\sum_{j=1}^{m} |E_j|}{m} \times 100\% \tag{16}$$

$$E_j = \frac{\sum_{i=1}^{n} |P_{p,ij} - P_{l,ij}|}{n} \tag{17}$$

Customer 53990001 is selected from Category 2 for the forecasting customer. The MAPEs during a training period for Category 2 are shown in Figure 9 when the parameters are set as shown in Table 6. The compared curves of actual load and forecasting load using the proposed method on 18 November for Customer 53990001 are shown in Figure 10. The MAPE for Customer 53990001 on 18 November is 10.23%. The compared curves of actual load and forecasting load from 18 to 24 November for Customer 53990001 are shown in Figure 11. The MAPE for Customer 53990001 in this week is 10.97%. In Figures 10 and 11, the error in sample points of one day is basically average and becomes larger when the curve comes to a peak. It is reasonable because the high or low peak is not reachable in most cases. The network should balance the prediction results for most situations during the training process. According to Figure 9, the MAPE decreases to a steady state as the epoch increases to 200. According to Figures 10 and 11, the forecasting curve is close to the actual curve, which proves the availability of the proposed method.

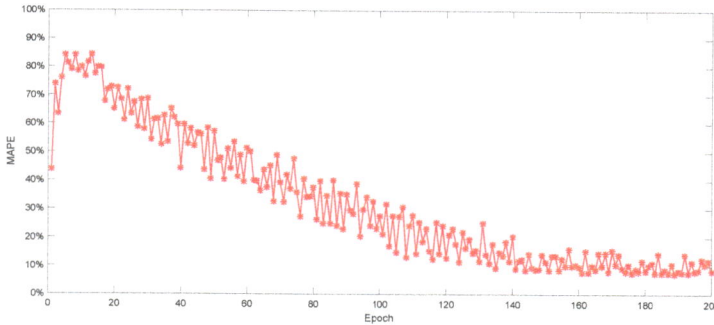

Figure 9. MAPEs during a training period for the Category 2.

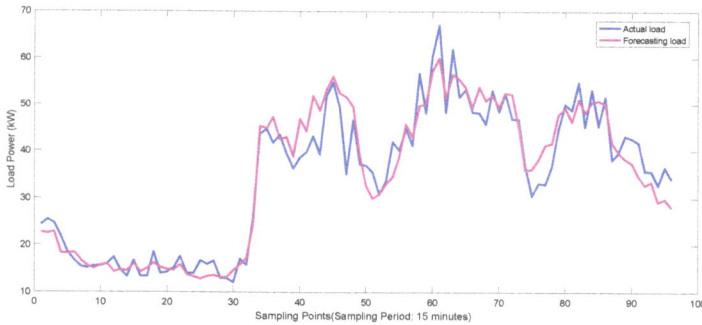

Figure 10. Compared curves of actual load and forecasting load in a day for Customer 53990001.

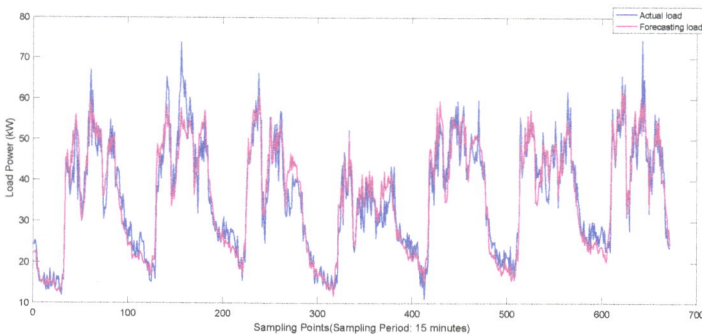

Figure 11. Compared curves of actual load and forecasting load in a week for Customer 53990001.

The samples are preprocessed by K-means clustering algorithm to form three categories for training. We performed a comparative experiment with variable-controlling approach about clustering. The compared results of Customer 53990001 on four different days of November 2013 are shown in Figure 12. The compared MAPEs of prediction on 18 November for nine customers in three categories from different feeders are shown in Table 7. It can be concluded that the forecasting curve without clustering deviates from the actual curve and that its MAPE is larger. The reason is that different characteristics of electricity use create a bad effect for short-term load forecasting. The effect reduces when we use corresponding trained networks for different customers. Therefore, the performance is generally improved by clustering.

Table 7. Compared MAPEs for nine customers in three categories with or without clustering.

Customer	Category	Feeder	MAPE with Clustering	MAPE without Clustering
37148000	1	2	10.80%	22.98%
51690000	1	7	9.25%	21.17%
37165000	1	7	11.12%	20.12%
53990001	2	2	10.07%	27.31%
54265001	2	3	11.91%	22.72%
54265002	2	3	10.76%	28.45%
31624001	3	35	13.56%	21.12%
41661001	3	34	12.23%	24.85%
76242001	3	33	9.98%	28.38%

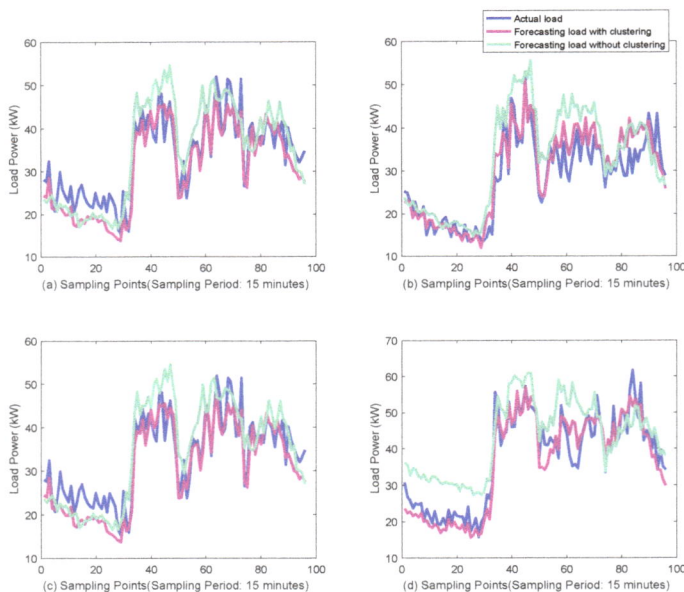

Figure 12. Compared curves of actual load and forecasting load of Customer 53990001 with or without clustering: (**a–d**) the results for four different days in November 2013.

The input of proposed network includes D_p, W_p, T_p and P_l, which means that the network obtains and fuses the previous load changing process and other environmental information. In this case, we removed the input layer and the following fully connected layers in the network. The comparison results of Customer 53990001 with multi-source or only load data input are shown in Figure 13. The compared MAPEs for nine customers in three categories from different feeders are shown in Table 8. The experimental condition is the same as the one above. It can be concluded that the performance of only using load data is obviously poorer. Although the change shape is similar to actual, the curves deviate from the actual curves. Correspondingly, the MAPEs are larger. The reason is that date, weather and temperature are necessary factors to consider during short-term load forecasting processing. People would raise their load on a hot or cold day, even a rainy or snowy day.

Resident customers may increase electricity consumption on weekends but business customers may not. These are some obvious reasons why we should consider the environment factors.

Table 8. Compared MAPEs for nine customers in three categories with multi-source data or only load data.

Customer	Category	Feeder	MAPE with Multi-Source Data	MAPE with only Load Data
37148000	1	2	10.80%	15.09%
51690000	1	7	9.25%	15.06%
37165000	1	7	11.12%	16.50%
53990001	2	2	10.07%	18.56%
54265001	2	3	11.91%	17.36%
54265002	2	3	10.76%	15.99%
31624001	3	35	13.56%	15.89%
41661001	3	34	12.23%	16.46%
76242001	3	33	9.98%	17.33%

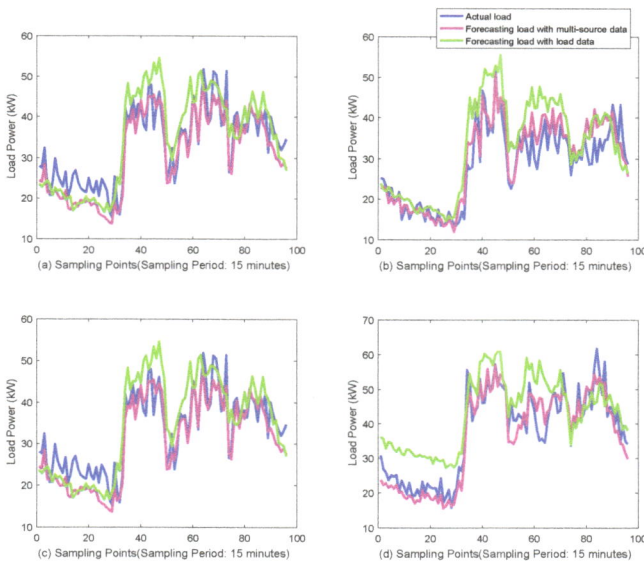

Figure 13. Comparison curve of actual load and forecasting load of Customer 53990001 with or without multi-source data: (**a–d**) the results for the same four days in November 2013 as the experiment in Figure 12.

It can be concluded from the two experiments that the MAPEs are floating in a certain degree. The maximal MAPEs of all samples in the conditions of the two experiments are shown in Table 9. The maximal MAPE without clustering and with only load data is significantly larger than the proposed method with clustering and multi-source data. The maximal MAPE of proposed method is 15.12%, which is acceptable for load forecasting of individual customers.

Table 9. Maximal MAPEs in different conditions.

Conditions	Forecast without Clustering	Forecast with only Load Data	Proposed Method
Maximal MAPE	30.25%	21.87%	15.12%

The performances are good with LSTM networks in dealing with time sequence but there are more parameters to train compared with GRU neural networks. In the proposed network, the GRU layers have 285,300 parameters to train while the LSTM layers have 380,400 parameters with the same architecture. The cost time for training with LSTM network is about 20% longer than training with GRU neural networks in the experiments in this paper. The MAPEs of network with LSTM and GRU layer in the same architecture with the same samples in Category 2 during the training process are shown in Figure 14. We can conclude that GRU neural networks do better in both convergence speed and training time, which depends on the improved single structure of GRU units.

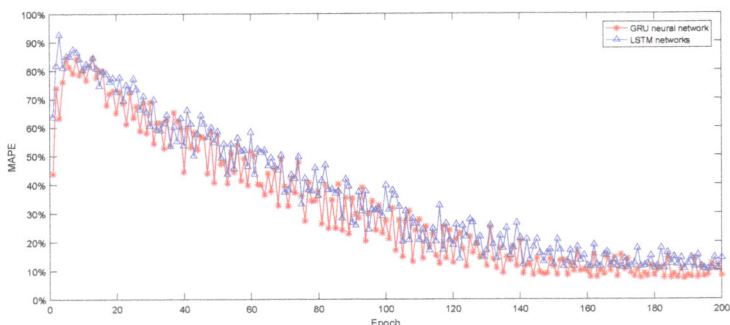

Figure 14. The MAPEs of network with LSTM and GRU layers in the same architecture with same samples during the training process.

We also performed the experiments to compare with current methods such as back-propagation neural networks (BPNNs) [7,8], stacked autoencoders (SAEs) [17], RNNs [24,25], and LSTM [29–31]. Their parameters and structures are set as described in Section 3.2. The compared average MAPEs of these methods, trained and tested with all samples described at the beginning of this subsection, are shown in Figure 15. The specific values of average and maximal MAPEs are shown in Table 10. Moreover, the results of nine customers are shown to validate the better performance of the proposed methods. The MAPEs for 30 November 2013 are shown in Table 11. It can be concluded that the proposed method results in smaller error in both average and maximal MAPEs. The proposed method performs better compared to the other current methods in most cases for short-term load forecasting in Wanjiang area.

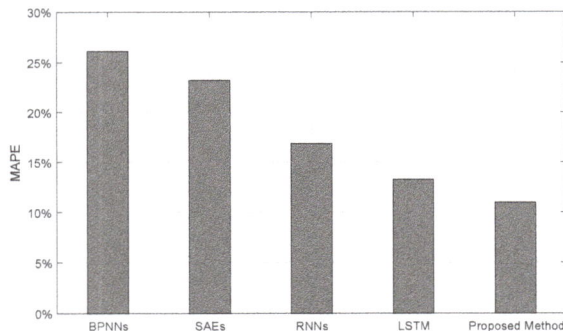

Figure 15. Compared average MAPEs trained and tested with all samples in three categories.

Table 10. Average and maximal MAPEs of the proposed and current methods for short-term load forecasting.

Method	BPNNs	SAEs	RNNs	LSTM	Proposed Method
Average MAPE	26.12%	23.23%	16.85%	13.26%	**10.98%**
Maximal MAPE	32.42%	28.51%	20.97%	17.45%	**15.12%**

Table 11. MAPEs of compared methods for nine customers' short-term load forecasting on 30 November 2013.

Customer	Category	Feeder	BPNNs	SAEs	RNNs	LSTM	Proposed Method
37148000	1	2	26.96%	23.28%	16.56%	13.09%	**10.80%**
51690000	1	7	27.56%	25.61%	15.56%	10.77%	**9.25%**
37165000	1	7	28.55%	24.81%	13.35%	14.55%	**11.12%**
53990001	2	2	24.23%	22.23%	17.87%	11.27%	**10.07%**
54265001	2	3	26.33%	27.56%	17.23%	12.22%	**11.91%**
54265002	2	3	29.23%	24.89%	15.63%	14.63%	**10.76%**
31624001	3	35	30.45%	22.56%	19.93%	**12.29%**	13.56%
41661001	3	34	32.23%	25.65%	16.72%	14.65%	**12.23%**
76242001	3	33	25.36%	23.22%	14.66%	13.27%	**9.98%**

In detail, the forecasting load curves of Customers 37148000 and 53990001 on 30 November 2013 based on these methods are shown in Figures 16 and 17. We can observe the closest curve to the actual curve is the proposed method in the results of these experiments. Time information is important in short-term load forecasting which the BPNNs and SAEs cannot extract. Therefore, they get poorer performance in the experiments. The vanishing gradient problem limits the performance of RNNs because of the decreasing perception of nodes. The architecture is simpler and the parameters are fewer in GRU neural network compared to LSTM networks (Section 2.1). Therefore, the performances of GRU neural networks are better than the other current methods. In general, the availability and improvement of the proposed method are proven by the real-world experiments.

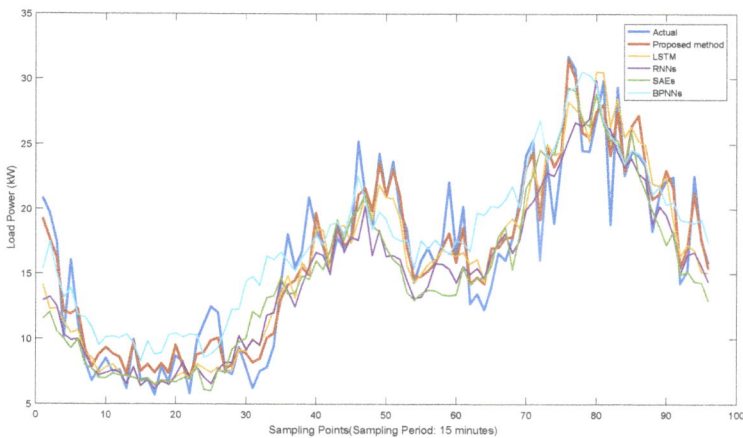

Figure 16. The load curves of Customer 37148000 based on the proposed method and the other current methods.

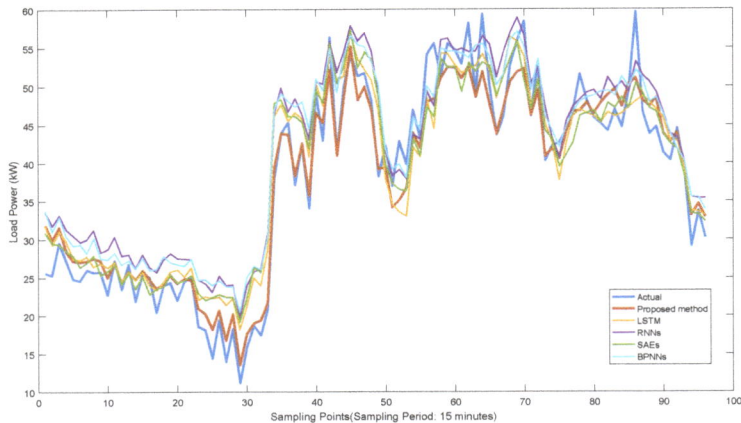

Figure 17. The load curves of Customer 53990001 based on the proposed method and the other current methods.

4. Conclusions

To increase the stability and economy of power grids, a method for short-term load forecasting with multi-source data using GRU neural networks is proposed in this paper, which focuses on individual customers. The proposed structure of the whole network is shown in Figure 5. The real-world load data of individual customers is recorded from Dongguan Power Supply Bureau of China Southern Power Grid in Guangdong Province, China. Before training, the customers with load data are clustered into three categories by K-means clustering algorithm to reduce the interference of different electricity use characteristics. Then, the environment factors are quantified and put into the input of the proposed networks for more information. The GRU units are introduced into the network for its simpler structure and faster convergence compared to LSTM blocks. The results in Figures 12 and 13 show that clustering and multi-source input can help to improve the performance of load forecasting. The average MAPE can be low as 10.98% for the proposed method, which outperforms the other current methods such as BPNNs, SAEs, RNNs and LSTM. The improvement is notable (Figures 15–17). In general, the availability and superiority of the proposed method are verified in this paper. In the future, combining with the technique of peak prediction could be a subject worth studying for load forecasting. Moreover, since the load forecasting for the customers in all power grid areas is a large-scale task, transfer learning and continuous learning will be considered based on the proposed framework for high-efficiency load forecasting.

Author Contributions: Y.W. and M.L. conceived the main idea; Z.B. and S.Z. collected the data and designed the experiments; Y.W., M.L. and Z.B. performed the experiments; and Y.W. wrote the paper.

Acknowledgments: This work is supported by the Zhejiang Provincial Natural Science Foundation of China under Grant LZ15F030001.

Conflicts of Interest: The authors declare no conflict of interest.

References

1. Espinoza, M.; Suykens, J.A.; Belmans, R.; De Moor, B. Electric load forecasting. *IEEE Contr. Syst. Mag.* **2007**, *27*, 43–57. [CrossRef]
2. Hong, T. Energy forecasting: Past, present, and future. *Foresight Int. J. Appl. Forecast.* **2014**, *32*, 43–48.
3. Gross, G.; Galiana, F.D. Short-term load forecasting. *Proc. IEEE* **1987**, *75*, 1558–1573. [CrossRef]

4. Hong, T.; Fan, S. Probabilistic electric load forecasting: A tutorial review. *Int. J. Forecast.* **2016**, *32*, 914–938. [CrossRef]
5. Taylor, J.W.; McSharry, P.E. Short-term load forecasting methods: An evaluation based on european data. *IEEE Trans. Power Syst.* **2008**, *22*, 2213–2219. [CrossRef]
6. Chen, H.; Wan, Q.; Zhang, B.; Li, F.; Wang, Y. Short-term load forecasting based on asymmetric ARCH models. In Proceedings of the Power and Energy Society General Meeting, Providence, RI, USA, 25–29 July 2010; pp. 1–6.
7. Quan, H.; Srinivasan, D.; Khosravi, A. Short-term load and wind power forecasting using neural network-based prediction intervals. *IEEE Trans. Neural Netw. Learn. Syst.* **2014**, *22*, 303–315. [CrossRef] [PubMed]
8. Ding, N.; Benoit, C.; Foggia, G.; Besanger, Y.; Wurtz, F. Neural network-based model design for short-term load forecast in distribution systems. *IEEE Trans. Power Syst.* **2016**, *31*, 72–81. [CrossRef]
9. Cevik, H.H.; CunkaS, M. Short-term load forecasting using fuzzy logic and ANFIS. *Neural Comput. Appl.* **2015**, *26*, 1355–1367. [CrossRef]
10. Lee, C.W.; Lin, B.Y. Applications of the chaotic quantum genetic algorithm with support vector regression in load forecasting. *Energies* **2017**, *10*, 1832. [CrossRef]
11. Hinton, G.E.; Salakhutdinov, R.R. Reducing the dimensionality of data with neural networks. *Science* **2006**, *313*, 504–507. [CrossRef] [PubMed]
12. Hinton, G.E.; Osindero, S.; Teh, Y.W. A fast learning algorithm for deep belief nets. *Neural Comput.* **2006**, *18*, 1527–1554. [CrossRef] [PubMed]
13. Chen, K.; Huang, C.; He, J. Fault detection, classification and location for transmission lines and distribution systems: A review on the methods. *High Volt.* **2016**, *1*, 25–33. [CrossRef]
14. Wang, Y.; Liu, M.; Bao, Z. Deep learning neural network for power system fault diagnosis. In Proceedings of the 35th Chinese Control Conference, Chengdu, China, 27–29 July 2016; pp. 6678–6683.
15. Ryu, S.; Noh, J.; Kim, H. Deep neural network based demand side short term load forecasting. *Energies* **2016**, *10*, 3. [CrossRef]
16. Qiu, X.; Ren, Y.; Suganthan, P.N.; Amaratunga, G.A. Empirical mode decomposition based ensemble deep learning for load demand time series forecasting. *Appl. Soft Comput.* **2017**, *54*, 246–255. [CrossRef]
17. Tong, C.; Li, J.; Lang, C.; Kong, F.; Niu, J.; Rodrigues, J.J. An efficient deep model for day-ahead electricity load forecasting with stacked denoising auto-encoders. *J. Parallel Distrib. Comput.* **2017**, 1–7. [CrossRef]
18. Liao, Y.; Chen, S. A modular RNN-based method for continuous Mandarin speech recognition. *IEEE Trans. Speech Aud. Proc.* **2016**, *9*, 252–263. [CrossRef]
19. Nakashika, T.; Takiguchi, T.; Ariki, Y. Voice conversion using RNN pre-trained by recurrent temporal restricted Boltzmann machines. *IEEE/ACM Trans. Aud. Speech Lang. Proc.* **2015**, *23*, 580–587. [CrossRef]
20. Wu, H.; Prasad, S. Convolutional recurrent neural networks forhyperspectral data classification. *Remote Sens.* **2017**, *9*, 298. [CrossRef]
21. Zuo, H.; Fan, H.; Blasch, E.; Ling, H. Combining convolutional and recurrent neural networks for human skin detection. *IEEE Signal Proc. Lett.* **2017**, *24*, 289–293. [CrossRef]
22. Chien, J.T.; Ku, Y.C. Bayesian recurrent neural network for language modeling. *IEEE Trans. Neural Netw. Learn. Syst.* **2016**, *27*, 361–374. [CrossRef] [PubMed]
23. Ororbia, A.G., II; Mikolov, T.; Reitter, D. Learning simpler language models with the differential state framework. *Neural Comput.* **2017**, *29*, 3327–3352. [CrossRef] [PubMed]
24. Shi, H.; Xu, M.; Li, R. Deep learning for household load forecasting—A novel pooling deep RNN. *IEEE Trans. Smart Grid* **2017**, 1–10. [CrossRef]
25. Wei, L.Y.; Tsai, C.H.; Chung, Y.C.; Liao, K.H.; Chueh, H.E.; Lin, J.S. A study of the hybrid recurrent neural network model for electricity loads forecasting. *Int. J. Acad. Res. Account. Fin. Manag. Sci.* **2017**, *7*, 21–29.
26. Hochreiter, S.; Schmidhuber, J. Long short-term memory. *Neural Comput.* **1997**, *9*, 1735–1780. [CrossRef] [PubMed]
27. Wei, D.; Wang, B.; Lin, G.; Liu, D.; Dong, Z.; Liu, H.; Liu, Y. Research on unstructured text data mining and fault classification based on RNN-LSTM with malfunction inspection report. *Energies* **2017**, *10*, 406. [CrossRef]
28. Zhang, S.; Wang, Y.; Liu, M.; Bao, Z. Data-based line trip fault prediction in power systems using LSTM networks and SVM. *IEEE Access* **2017**, *6*, 7675–7686. [CrossRef]

29. Gensler, A.; Henze, J.; Sick, B.; Raabe, N. Deep Learning for solar power forecasting—An approach using AutoEncoder and LSTM Neural Networks. In Proceedings of the International Conference on Systems, Man, and Cybernetics, Budapest, Hungary, 9–12 October 2016; pp. 2858–2865.

30. Zheng, J.; Xu, C.; Zhang, Z.; Li, X. Electric load forecasting in smart grids using long-short-term-memory based recurrent neural network. In Proceedings of the 51st Annual Conference on Information Sciences and Systems, Baltimore, MD, USA, 22–24 March 2017; pp. 1–6.

31. Kong, W.; Dong, Z.Y.; Jia, Y.; Hill, D.J.; Xu, Y.; Zhang, Y. Short-term residential load forecasting based on LSTM recurrent neural network. *IEEE Trans. Smart Grid* **2017**, 1–11. [CrossRef]

32. Cho, K.; Van Merrienboer, B.; Gulcehre, C.; Bahdanau, D.; Bougares, F.; Schwenk, H.; Bengio, Y. Learning phrase representations using RNN encoder-decoder for statistical machine translation. *arXiv* **2014**, arXiv:1406.1078.

33. Chung, J.; Gulcehre, C.; Cho, K.; Bengio, Y. Empirical evaluation of gated recurrent neural networks on sequence modeling. *arXiv* **2014**, arXiv:1412.3555.

34. Lu, K.; Zhao, Y.; Wang, X.; Cheng, Y.; Pang, X.; Sun, W.; Jiang, Z.; Zhang, Y.; Xu, N.; Zhao, X. Short-term electricity load forecasting method based on multilayered self-normalizing GRU network. In Proceedings of the Conference on Energy Internet and Energy System Integration, Beijing, China, 26–28 November 2017; pp. 1–5.

35. Werbos, P.J. Backpropagation through time: What it does and how to do it. *Proc. IEEE* **1990**, *78*, 1550–1560. [CrossRef]

energies

MDPI

Article

Deep Belief Network Based Hybrid Model for Building Energy Consumption Prediction

Chengdong Li [1,*], Zixiang Ding [1], Jianqiang Yi [2] ⬤, Yisheng Lv [2]⬤ and Guiqing Zhang [1]

[1] School of Information and Electrical Engineering, Shandong Jianzhu University, Jinan 250101, China; zixiang.ding@foxmail.com (Z.D.); qqzhang@sdjzu.edu.cn (G.Z.)
[2] Institute of Automation, Chinese Academy of Sciences, Beijing 100190, China; jianqiang.yi@ia.ac.cn (J.Y.); yisheng.lv@ia.ac.cn (Y.L.)
[*] Correspondence: lichengdong@sdjzu.edu.cn; Tel.: +86-0531-8636-1056

Received: 15 December 2017; Accepted: 16 January 2018; Published: 19 January 2018

Abstract: To enhance the prediction performance for building energy consumption, this paper presents a modified deep belief network (DBN) based hybrid model. The proposed hybrid model combines the outputs from the DBN model with the energy-consuming pattern to yield the final prediction results. The energy-consuming pattern in this study represents the periodicity property of building energy consumption and can be extracted from the observed historical energy consumption data. The residual data generated by removing the energy-consuming pattern from the original data are utilized to train the modified DBN model. The training of the modified DBN includes two steps, the first one of which adopts the contrastive divergence (CD) algorithm to optimize the hidden parameters in a pre-train way, while the second one determines the output weighting vector by the least squares method. The proposed hybrid model is applied to two kinds of building energy consumption data sets that have different energy-consuming patterns (daily-periodicity and weekly-periodicity). In order to examine the advantages of the proposed model, four popular artificial intelligence methods—the backward propagation neural network (BPNN), the generalized radial basis function neural network (GRBFNN), the extreme learning machine (ELM), and the support vector regressor (SVR) are chosen as the comparative approaches. Experimental results demonstrate that the proposed DBN based hybrid model has the best performance compared with the comparative techniques. Another thing to be mentioned is that all the predictors constructed by utilizing the energy-consuming patterns perform better than those designed only by the original data. This verifies the usefulness of the incorporation of the energy-consuming patterns. The proposed approach can also be extended and applied to some other similar prediction problems that have periodicity patterns, e.g., the traffic flow forecasting and the electricity consumption prediction.

Keywords: building energy consumption prediction; deep belief network; contrastive divergence algorithm; least squares learning; energy-consuming pattern

1. Introduction

With the growth of population and the development of economy, more and more energy is consumed in the residential and office buildings. Building energy conservation plays an important role in the sustainable development of economy. However, some ubiquitous issues, e.g., the poor building management and the unreasonable task scheduling, are impeding the efficiency of the energy conservation policies. To improve the building management and the task scheduling of building equipment, one way is to provide accurate prediction of the building energy consumption.

Nowadays, numerous data-driven artificial intelligence approaches have been proposed for building energy consumption prediction. In [1], the random forest and the artificial neural network

(ANN) were applied to the high-resolution prediction of building energy consumption, and their experimental results demonstrated that both models have comparable predictive power. In [2], a hybrid model combining different machine learning algorithms was presented for optimizing energy consumption of residential buildings under the consideration of both continuous and discrete parameters of energy. In [3], the extreme learning machine (ELM) was used to estimate the building energy consumption, and simulation results indicated that the ELM performed better than the genetic programming (GP) and the ANN. In [4], the clusterwise regression method, also known as the latent class regression, which integrates clustering and regression, was utilized to the accurate and stable prediction of building energy consumption data. In [5], the feasibility and applicability of support vector machine (SVM) for building energy consumption prediction were examined in a tropical region. Moreover, in [6–9], a variation of SVM, the support vector regressor (SVR) was proposed for forecasting the building energy consumption and the electric load. Furthermore, in [10], a novel machine learning model was constructed for estimating the commercial building energy consumption.

The historical building energy consumption data have high levels of uncertainties and randomness due to the influence of the human distribution, the thermal environment, the weather conditions and the working hours in buildings. Thus, there still exists the need to improve the prediction precision for this application. To realize this objective, we can take two strategies into account. The first strategy is to adopt the more powerful modeling methods to learn the information hidden in the historical data, while the other one is to incorporate the knowledge or patterns from our experience or data into the prediction models.

On the one hand, the deep learning technique provides us one very powerful tool for constructing the prediction model. In the deep learning models, more representative features can be extracted from the lowest layer to the highest layer [11,12]. Until today, this miraculous technique has been widely used in various fields. In [13], a novel predictor, the stacked autoencoder Levenberg–Marquardt model was constructed for the prediction of traffic flow. In [14], an extreme deep learning approach that integrates the stacked autoencoder (SAE) with the ELM was proposed for building energy consumption prediction. In [15], the deep learning was employed as an ensemble technique for cancer detection. In [16], the deep convolutional neural network (CNN) was utilized for face photo-sketch recognition. In [17], a deep learning approach, the Gaussian–Bernoulli restricted Boltzmann machine (RBM) was applied to 3D shape classification through using spectral graph wavelets and the bag-of-features paradigm. In [18], the deep belief network (DBN) was applied to solve the natural language understanding problem. Furthermore, in [19], the DBN was utilized to fuse the virtues of multiple acoustic features for improving the robustness of voice activity detection. As one popular deep learning method, the DBN has shown its superiority in machine learning and artificial intelligence. This study will adopt and modify the DBN to make it be suitable for the prediction of building energy consumption.

On the other hand, knowledge or patterns from our experience can provide additional information for the design of the prediction models. In [20–22], different kinds of prior knowledge were incorporated into the SVM models. In [23], the knowledge of symmetry was encoded into the type-2 fuzzy logic model to enhance its performance. In [24,25], the knowledge of monotonicity was incorporated into the fuzzy inference systems to assure the models' monotonic input–output mappings. In [26–29], how to encode the knowledge into neural networks was discussed. As shown in these studies, through incorporating the knowledge or pattern, the constructed machine learning models will yield better performance and have significantly improved generalization ability.

From the above discussion, both the deep learning method and the domain knowledge are helpful for the prediction models' performance improvement. Following this idea, this study tries to present a hybrid model that combines the DBN model with the periodicity knowledge of the building energy consumption to further improve the prediction accuracy. The final prediction results of the proposed hybrid model are obtained by combining the outputs from the modified DBN model and the energy-consuming pattern model. Here, the energy-consuming pattern represents the periodicity property of building energy consumption and can be extracted from the observed historical energy consumption data. In this study, firstly, the structure of the proposed hybrid model will be presented, and how to extract the energy-consuming pattern will be demonstrated. Then, the training algorithm for the modified DBN model will be provided. The learning of the DBN model mainly includes two steps, which firstly optimizes the hidden parameters by the contrastive divergence (CD) algorithm in a pre-train way, and then determines the output weighting vector by the least squares method. Furthermore, the proposed hybrid model will be applied to the prediction of the energy consumption in two kinds of buildings that have different energy-consuming patterns (daily-periodicity and weekly-periodicity). Additionally, to show the superiority of the proposed hybrid model, comparisons with four popular artificial intelligence methods—the backward propagation neural network (BPNN), the generalized radial basis function neural network (GRBFNN), the extreme learning machine (ELM), and the support vector regressor (SVR) will be made. From the comparison results, we can observe that all the predictors (DBN, BPNN, GRBFNN, ELM and SVR) designed using both the periodicity knowledge and residual data perform much better than those designed only by the original data. Hence, we can judge that the periodicity knowledge is quite useful for improving the prediction performance in this application. The experiments also show that, among all the prediction models, the proposed DBN based hybrid model has the best performance.

The rest of this paper is as follows. In Section 2, the deep belief network will be reviewed. In Section 3, the proposed hybrid model will be presented firstly, and then the modified DBN will be provided. In Section 4, two energy consumption prediction experiments for buildings that have different energy-consuming patterns will be done. In addition, the experimental and comparison results will be given. Finally, in Section 5, the conclusions of this paper will be drawn.

2. Introduction of DBN

The DBN is a stack of restricted Boltzmann machine (RBM) [11,30]. Therefore, for better understanding, we will introduce the RBM before the introduction of the DBN in this section.

2.1. Restricted Boltzmann Machine

The structure of a typical RBM model is shown in Figure 1. The RBM is an undirected, bipartite graphical model, which consists of the visible (input) layer and the hidden (output) layer. The visible layer and the hidden layer are respectively made up of n visible units and m hidden units, and there is a bias in each unit. Moreover, there are no interconnection within the visible layer or the hidden layer [31].

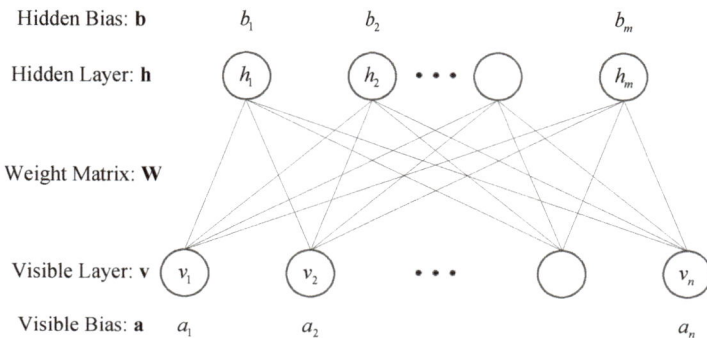

Figure 1. The structure of a typical RBM model.

The activation probability of the jth hidden unit can be computed as follows when a visible vector $\boldsymbol{v}(v_1, \ldots, v_i, \ldots, v_n)$ is given [32]

$$p(h_j = 1|\boldsymbol{v}) = \sigma(b_j + \sum_{i=1}^{n} v_i w_{ij}), \tag{1}$$

where $\sigma(\cdot)$ is the sigmoid function, w_{ij} is the connection weight between the ith visible unit and jth hidden unit, and b_j is the bias of the jth hidden unit.

Similarly, when a hidden vector $\boldsymbol{h}(h_1, \ldots, h_j, \ldots, h_m)$ is known, the activation probability of the ith visible unit can be computed as follows:

$$p(v_i = 1|\boldsymbol{h}) = \sigma(a_i + \sum_{j=1}^{m} h_j w_{ij}), \tag{2}$$

where $i = 1, 2, \ldots, n$, and a_i is the bias of the ith visible unit.

Hinton et al. [33] have proposed the contrastive divergence (CD) algorithm to optimize the RBM. The CD algorithm based RBM's iterative learning procedures for binomial units are listed as follows [32].

Step 1: Initialize the number of visible units n, the number of hidden units m, the number of training data N, the weighting matrix $\boldsymbol{W}$, the visible bias vector $\boldsymbol{a}$, the hidden bias vector $\boldsymbol{b}$ and the learning rate ϵ.

Step 2: Assign a sample $\boldsymbol{x}$ from the training data to be the initial state $\boldsymbol{v_0}$ of the visible layer.

Step 3: Calculate $p(h_{0j} = 1|\boldsymbol{v_0})$ according to Equation (1), and extract $h_{0j} \in \{0, 1\}$ from the conditional distribution $p(h_{0j} = 1|\boldsymbol{v_0})$, where $j = 1, 2, \ldots, m$.

Step 4: Calculate $p(v_{1i} = 1|\boldsymbol{h_0})$ according to Equation (2), and extract $v_{1i} \in \{0, 1\}$ from the conditional distribution $p(v_{1i} = 1|\boldsymbol{h_0})$, where $i = 1, 2, \ldots, n$.

Step 5: Calculate $p(h_{1j} = 1|\boldsymbol{v_1})$ according to Equation (1).

Step 6: Update the parameters according to the following equations:

$$\boldsymbol{W} = \boldsymbol{W} + \epsilon(p(\boldsymbol{h_0} = 1|\boldsymbol{v_0})\boldsymbol{v_0}^{\mathrm{T}} - p(\boldsymbol{h_1} = 1|\boldsymbol{v_1})\boldsymbol{v_1}^{\mathrm{T}}),$$
$$\boldsymbol{a} = \boldsymbol{a} + \epsilon(\boldsymbol{v_0} - \boldsymbol{v_1}),$$
$$\boldsymbol{b} = \boldsymbol{b} + \epsilon(p(\boldsymbol{h_0} = 1|\boldsymbol{v_0}) - p(\boldsymbol{h_1} = 1|\boldsymbol{v_1})).$$

Step 7: Assign another sample from the training data to be the initial state $\boldsymbol{v_0}$ of the visible layer, and iterate Steps 3 to 7 until all the N training data have been used.

2.2. Deep Belief Network

As aforementioned, the DBN as a miraculous deep model is a stack of RBMs [11,30,34,35]. Figure 2 illustrates the architecture of the DBN with k hidden layers and its layer-wise pre-training process.

The activation of the kth hidden layer with respect to input sample $\boldsymbol{x}$ can be computed as

$$A_k(\boldsymbol{x}) = \sigma\left(\boldsymbol{b}_k + \boldsymbol{W}_k \sigma\left(\cdots + \boldsymbol{W}_2 \sigma\left(\boldsymbol{b}_1 + \boldsymbol{W}_1 \boldsymbol{x}\right)\right)\right), \tag{3}$$

where $\boldsymbol{W}_u$ and $\boldsymbol{b}_u$ ($u = 1, 2, \ldots, k$) are, respectively, the weighting matrices and hidden bias vectors of the uth RBM. Furthermore, σ is the logistic sigmoid function $\sigma(x) = 1/(1 + e^{-x})$.

In order to obtain better feature representation, the DBN utilizes deep architecture and adopts the layer-wise pre-training to optimize the inter-layer weighting matrix [11]. The training algorithm of the DBN will be given in the next section in detail.

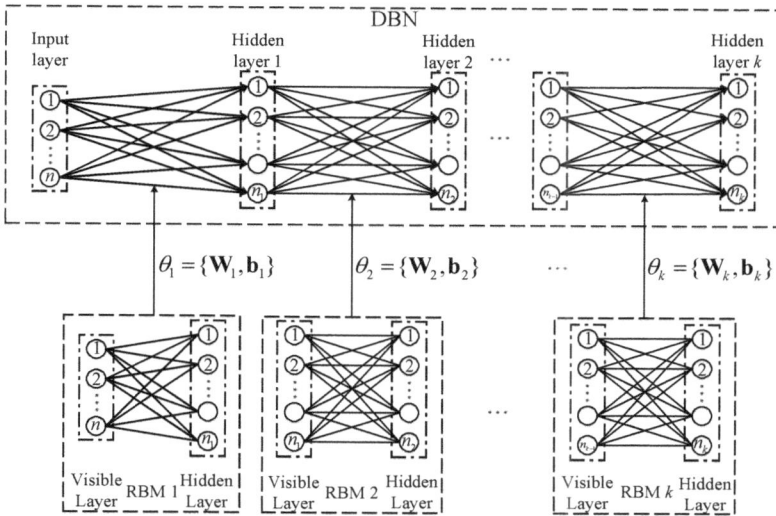

Figure 2. The architecture of the DBN with *k* hidden layers.

3. The Proposed Hybrid Model

In this section, the structure of the hybrid model will be proposed first. Then, the extraction of the energy-consuming pattern and the generation of the residual data will be given. Finally, the modified DBN (MDBN) and its training algorithm will be presented.

To begin, we assume that we have collected the sampling data for *M* consecutive days, and, in each day, we collected *T* data points. Then, sampled time series of energy consumption data can be written as a series of 1D vectors as

$$Y = \{Y_1, Y_2, \ldots, Y_M\}, \tag{4}$$

where

$$Y_1 = [y_1(1), y_1(2), \ldots, y_1(T)],$$
$$\vdots \tag{5}$$
$$Y_M = [y_M(1), y_M(2), \ldots, y_M(T)],$$

and *T* is the sampling number per day.

3.1. Structure of the Hybrid Model

The hybrid model combines the modified DBN (MDBN) model with the periodicity knowledge of the building energy consumption to obtain better prediction accuracy. The design procedure of the proposed model is depicted in Figure 3 and is also given as follows:

Step 1: Extract the energy-consuming pattern as the periodicity knowledge from the training data.
Step 2: Remove the energy-consuming pattern from the training data to generate the residual data.
Step 3: Utilize the residual data to train the MDBN model.
Step 4: Combine the outputs from the MDBN model with the periodicity knowledge to obtain the final prediction results of the hybrid model.

It is obvious that the extraction of the energy-consuming pattern, the generation of the residual data and the construction of the MDBN model are crucial in order to build the proposed hybrid model. Consequently, we will introduce them in detail in the following subsections.

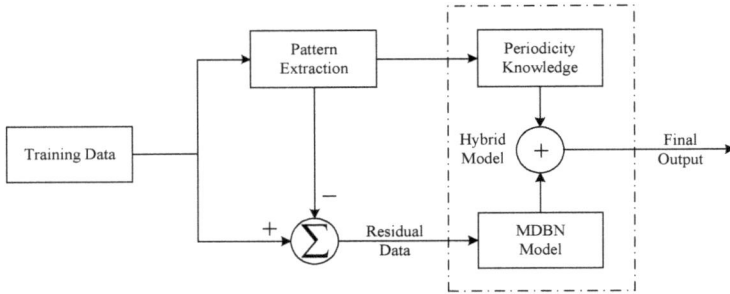

Figure 3. The structure of the hybrid model.

3.2. Extraction of the Energy-Consuming Patterns and Generation of the Residual Data

Obviously, various regular patterns of energy consumption (e.g., daily-periodicity, weekly-periodicity, monthly-periodicity and even yearly-periodicity) exist in different kinds of buildings. In this study, we will take the daily-periodic and the weekly-periodic energy-consuming patterns as examples to introduce the method for extracting them from the original data.

3.2.1. The Daily-Periodic Pattern

For daily-periodic energy-consuming pattern, it can be extracted from the original time series by the following equation:

$$\bar{Y}_{Ave} = \left[\frac{1}{M} \sum_{z=1}^{M} y_z(1), \frac{1}{M} \sum_{z=1}^{M} y_z(2), \ldots, \frac{1}{M} \sum_{z=1}^{M} y_z(T) \right]. \tag{6}$$

Then, the residual time series Y_{Res} of the data set after removing the daily-periodic pattern can be generated as

$$Y_{Res} = \left\{ Y_1 - \bar{Y}_{Ave}, Y_2 - \bar{Y}_{Ave}, \cdots, Y_M - \bar{Y}_{Ave} \right\}. \tag{7}$$

3.2.2. The Weekly-Periodic Pattern

Being different from the daily-periodic energy-consuming pattern, the weekly-periodic energy-consuming pattern includes two parts, which are the patterns of weekdays and weekends. The weekday pattern and the weekend pattern can be respectively computed as

$$\bar{Y}_{Weekday} = \left[\frac{1}{M_1} \sum_{z=1}^{M_1} p_z(1), \frac{1}{M_1} \sum_{z=1}^{M_1} p_z(2), \ldots, \frac{1}{M_1} \sum_{z=1}^{M_1} p_z(T) \right], \tag{8}$$

$$\bar{Y}_{Weekend} = \left[\frac{1}{M_2} \sum_{z=1}^{M_2} q_z(1), \frac{1}{M_2} \sum_{z=1}^{M_2} q_z(2), \ldots, \frac{1}{M_2} \sum_{z=1}^{M_2} q_z(T) \right], \tag{9}$$

where

$$P = \left\{ P_1 = [p_1(1), \ldots, p_1(T)], \ldots, P_{M_1} = [p_{M_1}(1), \ldots, p_{M_1}(T)] \right\}, \tag{10}$$
$$Q = \left\{ Q_1 = [q_1(1), \ldots, q_1(T)], \ldots, Q_{M_2} = [q_{M_2}(1), \ldots, q_{M_2}(T)] \right\}, \tag{11}$$

are, respectively, the data sets of weekdays and weekends, and $M_1 + M_2 = M$.

Then, to generate the residual time series Y_{Res} for the building energy consumption data set, we use the following rules:

$$If \quad Y_z \in P, \qquad\qquad then \quad Y_{z,Res} = Y_z - \bar{Y}_{Weekday}, \tag{12}$$

$$If \quad Y_z \in Q, \qquad\qquad then \quad Y_{z,Res} = Y_z - \bar{Y}_{Weekend}, \tag{13}$$

where $z = 1, 2, \ldots, M$.

Subsequently, Y_{Res} can be written as

$$Y_{Res} = \{Y_{1,Res}, Y_{2,Res}, \ldots, Y_{M,Res}\}. \tag{14}$$

3.3. Modified DBN and Its Training Algorithm

In this subsection, the structure of the MDBN will be shown firstly. Then, the pre-training process of the DBN part will be described in detail. At last, the least squares method will be employed to determine the weighting vector of the regression part.

3.3.1. Structure of the MDBN

In the parameter optimization of the traditional DBNs, the CD algorithm is adopted to pre-train the parameters of multiple RBMs, and the BP algorithm is used to finely tune the parameters of the whole network. In this paper, we add an extra layer as the regression part to the DBN to realize the prediction function. Thus, we call it the modified DBN (MDBN). The structure of the MDBN is demonstrated in Figure 4. In addition, we propose a training algorithm that combines the CD algorithm with the least squares method for the learning of the MDBN model.

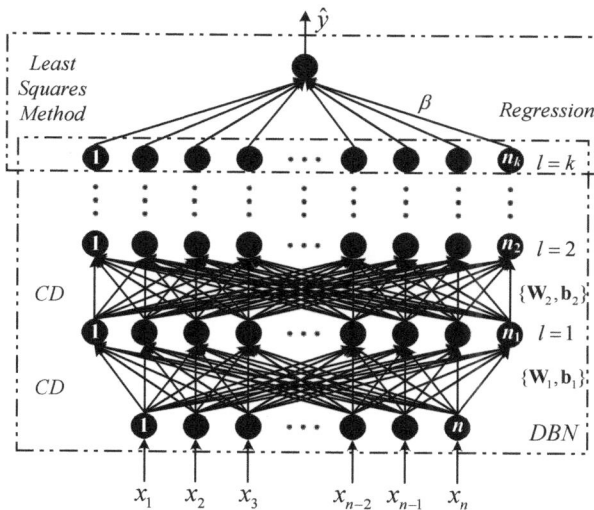

Figure 4. The structure of the modified DBN.

We divide the training process of the MDBN into two steps. The first step adopts the contrastive divergence algorithm to optimize the hidden parameters in a pre-train way, while the second one determines the output weighting vector by the least squares method. The detailed description will be given as below.

3.3.2. Pre-Training of the DBN Part

Generally speaking, with the number of hidden layers increasing, the effectiveness of the BP algorithm for optimizing the parameters of the deep neural network is getting lower and lower because of the gradient divergence. Fortunately, Hinton et al. [11] proposed a fast learning algorithm for the DBN. This novel approach realizes layer-wise pre-train of the multiple RBMs in the DBN in a bottom-up way as described below:

Step 1: Initialize the number of hidden layers k, the number of the training data N and the initial sequence number of hidden layer $u = 2$.

Step 2: Assign a sample x from the training data to be the input data of the DBN.

Step 3: Regard the input layer and the first hidden layer of the DBN as an RBM, and compute the activation $A_1(x)$ by Equation (3) when the training process of this RBM is finished.

Step 4: Regard the uth and the $(u + 1)$th hidden layer as an RBM with the input $A_{u-1}(x)$, and compute the activation $A_u(x)$ by Equation (3) when the training process of this RBM is completed.

Step 5: Let $u = u + 1$, and iterate Step 4 until $u > k$.

Step 6: Use the $A_k(x)$ as the input of the regression part.

Step 7: Assign another sample from the training data as the input data of the DBN, and iterate Step 3 to 7 until all the N training data have been assigned.

3.3.3. Least Squares Learning of the Regression Part

Suppose that the training set is $\aleph = \{(x^{(l)}, y^{(l)}) | x^{(l)} \in \mathbf{R}^n, y^{(l)} \in \mathbf{R}, l = 1, \cdots, N\}$. As aforementioned, once the pre-training of the DBN part is completed, the activation of the final hidden layer of the MDBN with respect to the input $x^{(l)}$ can be obtained to be $A_k(x^{(l)})$, where $l = 1, 2, \ldots, N$. Furthermore, the activation of the final hidden layer of the MDBN with respect to all the N training data can be written in the matrix form as

$$A_k(X) = [A_k(x^{(1)}), A_k(x^{(2)}), \cdots, A_k(x^{(N)})]^T$$

$$= \begin{bmatrix} \sigma\left(b_k + w_k\sigma\left(\cdots + w_2\sigma\left(b_1 + w_1x^{(1)}\right)\right)\right) \\ \sigma\left(b_k + w_k\sigma\left(\cdots + w_2\sigma\left(b_1 + w_1x^{(2)}\right)\right)\right) \\ \vdots \\ \sigma\left(b_k + w_k\sigma\left(\cdots + w_2\sigma\left(b_1 + w_1x^{(N)}\right)\right)\right) \end{bmatrix}_{N \times n_k}, \tag{15}$$

where n_k is the number of neurons of the kth hidden layer.

We always expect that each actual value $y^{(l)}$ with respect to $x^{(l)}$ can be approximated by the output $\hat{y}^{(l)}$ of the predictor with no error. This expectation can be mathematically expressed as

$$\sum_{l=1}^{N} \|\hat{y}^{(l)} - y^{(l)}\| = 0, \tag{16}$$

where $\hat{y}^{(l)}$ is the output of the MDBN and can be computed as

$$\hat{y}^{(l)} = A_k(x^{(l)})\beta \tag{17}$$

in which β is the output weighting vector and can be expressed as

$$\beta = [\beta_1, \beta_2, \cdots, \beta_{n_k}]^T_{n_k \times 1}. \tag{18}$$

Then, Equation (16) can be rewritten in the matrix form as

$$A_k(\boldsymbol{X})\boldsymbol{\beta} = \boldsymbol{Y}, \tag{19}$$

where

$$\boldsymbol{Y} = [y^{(1)}, y^{(2)}, \ldots, y^{(N)}]_{N \times 1}^{\mathrm{T}}. \tag{20}$$

From Equation (19), the output weighting vector $\boldsymbol{\beta}$ can be derived by the least squares method as [36–39]

$$\boldsymbol{\beta} = A_k(\boldsymbol{X})^\dagger \boldsymbol{Y}, \tag{21}$$

where $A_k(\boldsymbol{X})^\dagger$ is the Moore–Penrose generalized inverse of $A_k(\boldsymbol{X})$.

4. Experiments

In this section, first of all, four comparative artificial intelligence approaches will be introduced briefly. Next, the applied data sets and experimental setting will be discussed. Then, the proposed hybrid model will be applied to the prediction of the energy consumption in a retail store and an office building that respectively have daily-periodic and weekly-periodic energy-consuming patterns. Finally, we will give the comparisons and discussions of the experiments.

4.1. Introduction of the Comparative Approaches

To make a quantitative assessment of the proposed MDBN based hybrid model, four popular artificial intelligence approaches, the BPNN, GRBFNN, ELM, and SVR, are chosen as the comparative approaches and introduced briefly below.

4.1.1. Backward Propagation Neural Network

The structure of BPNN with L hidden layers is demonstrated in Figure 5. The BPNN as one popular kind of ANN adopts back propagation algorithm to obtain the optimal weighting parameters of the whole network [40–42].

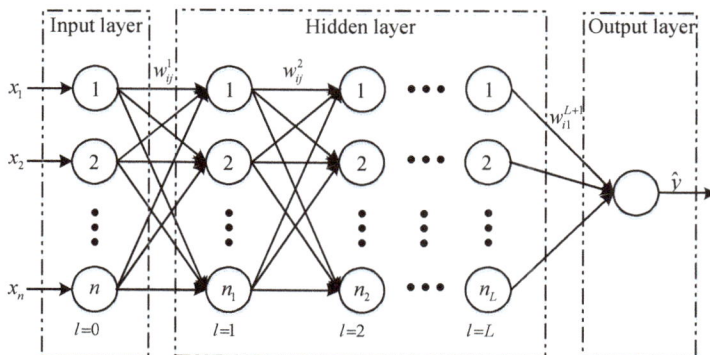

Figure 5. The structure of BPNN with L hidden layers.

As shown in Figure 5, the final output of the network can be expressed as [40–42]

$$\hat{y} = f(\sum_{s=1}^{n_L} w_{s1}^{L+1} \cdots f(\sum_{j=1}^{n_1} w_{jk}^2 f(\sum_{i=1}^{n} w_{ij}^1 x_i))), \tag{22}$$

where w_{ij}^k is the connection weight between the ith unit of kth layer and the jth unit of $(k+1)$th layer, and $f(\cdot)$ is the logistic sigmoid function.

In order to obtain the optimal parameters of the BPNN, the Backward Propagation (BP) algorithm is adopted to minimize the following cost function for each training data point

$$E(t, \boldsymbol{w}) = (\hat{y}^{(t)} - y^{(t)})^2, \tag{23}$$

where $\hat{y}^{(t)}$ and $y^{(t)}$ are the predicted and actual values with respect to the input $\boldsymbol{x}^{(t)}$.

The update rule for the weight w_{ij}^k can be expressed as

$$w_{ij}^k(t+1) = w_{ij}^k(t) - \eta \frac{\partial E(t, \boldsymbol{w})}{\partial w_{ij}^k}, \tag{24}$$

where η is the learning rate, and $\frac{\partial E(t, \boldsymbol{w})}{\partial w_{ij}^k}$ is the gradient of the parameter w_{ij}^k, and can be calculated by the backward propagation of the errors.

The BP algorithm has two phases—forward propagation and weight update. In the forward propagation stage, when an input vector is input to the NN, it is propagated forward through the whole network until it reaches the output layer. Then, the error between the output of the network and the desired output is computed. In the weight update phase, the error is propagated from the output layer back through the whole network, until each neuron has an associated error value that can reflect its contribution to the original output. These error values are then used to calculate the gradients of the loss function that are fed to the update rules to renew the weights [40–42].

4.1.2. Generalized Radial Basis Function Neural Network

The radial basis function (RBF) NN is a feed-forward NN with only one hidden layer whose structure is demonstrated in Figure 6. The RBFNN has Gaussian functions as its hidden neurons. The GRBFNN is a modified RBFNN and adopts the generalized Gaussian functions as its hidden neurons [43,44].

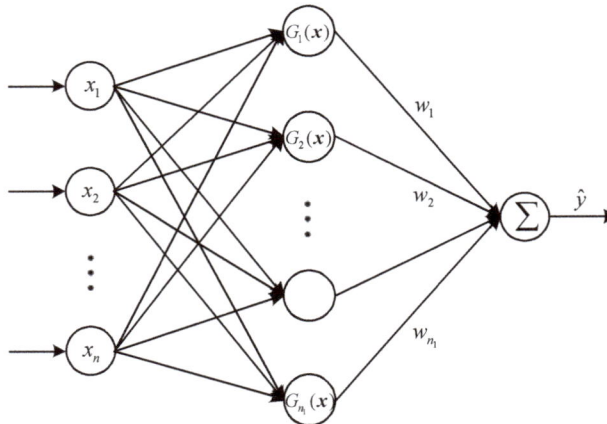

Figure 6. The topological structure of the feed-forward single-hidden-layer NN.

The output of the GRBFNN can be expressed as [43,44]

$$\hat{y} = \sum_{j=1}^{n_1} w_j G_j(x) = \sum_{j=1}^{n_1} w_j exp\left(-\frac{\|x - c_j\|^{\tau_j}}{d_j^{\tau_j}}\right), \tag{25}$$

where n_1 is the number of hidden neurons, τ_j is the shape parameter of the jth radial basis function in the hidden layer, and c_j and d_j are, respectively, the center and width of the jth radial basis function.

In order to determine the parameters τ, c and d in the hidden layer and the connection weight w_j, the aforementioned BP algorithm can also be employed.

4.1.3. Extreme Learning Machine

The ELM is also a feed-forward neural network with only one hidden layer as demonstrated in Figure 6. However, the ELM and GRBFNN have different parameter learning algorithms and different activation functions in the hidden neurons.

In the ELM, the activation functions in the hidden neurons can be the hard-limiting activation function, the Gaussian activation function, the Sigmoidal function, the Sine function, etc. [36,37].

In addition, the learning algorithm for the ELM is listed below:

- Randomly assign input weights or the parameters in the hidden neurons.
- Calculate the hidden layer output matrix H, where

$$H = \begin{pmatrix} G_1(x^{(1)}) & \cdots & G_{n_1}(x^{(1)}) \\ \vdots & \ddots & \vdots \\ G_1(x^{(N)}) & \cdots & G_{n_1}(x^{(N)}) \end{pmatrix}_{N \times n_1}. \tag{26}$$

- Calculate the output weights $w = [w_1, w_2, \cdots, w_{n_1}]^T = H^+ Y$, where $Y = [y^{(1)}, y^{(2)}, \cdots, y^{(N)}]^T$ and H^+ is the Moore–Penrose generalized inverse of the matrix H.

This learning process is very fast and can lead to excellent modeling performance. Hence, the ELM has found lots of applications in different research fields.

4.1.4. Support Vector Regression

The SVR is a variant of SVM. It can yield improved generalization performance through minimizing the generalization error bound [45]. In addition, the kernel trick is adopted to realize the nonlinear transformation of input features.

The model of the SVR can be defined by the following function

$$\hat{y} = f(x, w) = w^T \varphi(x) + b, \tag{27}$$

where $w = [w_1, \cdots, w_n]$, $\varphi(x)$ is the nonlinear mapping function.

Using the training set $\aleph = \{(x^{(l)}, y^{(l)})\}_{l=1}^{N}$, we can determine the parameters w and b, and then obtain the SVR model as

$$\hat{y} = f(x) = \sum_{l=1}^{N} w^{*T} \varphi(x) + b^*, \tag{28}$$

where

$$\begin{cases} w^* = \sum_{l=1}^{N} (\alpha_l^* - \alpha_l) \varphi(x^{(l)}), \\ b^* = \frac{1}{y_l} - w^{*T} \varphi(x^{(l)}), \end{cases} \tag{29}$$

in which α_l and α_l^* are the Langrange multipliers and can be determined by solving the following dual optimization problem [46]:

$$
\begin{cases}
\max\limits_{\alpha,\alpha^*} -\varepsilon \sum\limits_{l=1}^{N} (\alpha_l^* + \alpha_l) + \sum\limits_{l=1}^{N} (\alpha_l^* - \alpha_l) y^{(l)} - \dfrac{1}{2} \sum\limits_{l,t=1}^{N} (\alpha_l^* - \alpha_l)(\alpha_t^* - \alpha_t) \varphi^{\mathrm{T}}(\boldsymbol{x}^{(l)}) \varphi(\boldsymbol{x}^{(t)}), \\[2mm]
\sum\limits_{l=1}^{N} \alpha_l^* = \sum\limits_{l=1}^{N} \alpha_l, \quad 0 < \alpha_l, \alpha_l^* < C,
\end{cases}
\tag{30}
$$

where C is the regularization parameter and ε is the error tolerance parameter.

4.2. Applied Data Sets and Experimental Setting

In this subsection, first of all, the building energy consumption data sets will be described. Next, three design factors that are utilized to determine the optimal structure of the MDBN will be shown. Finally, five indices will be given to evaluate the performances of the predictive models.

4.2.1. Applied Data Sets

Two kinds of building energy consumption data sets were downloaded from [47]. The first data set includes 34,848 samples from 2 January 2010 to 30 December 2010. The data in this data set were collected every 15 min in one retail store in Fremont, CA, USA. We then aggregated them to generate the hourly energy consumption data. The second data set contains 22,344 samples from 4 April 2009 to 21 October 2011. The data in this set were collected every 60 min in one office building in Fremont, CA, USA. Parts of the samples of the two data sets are depicted in Figure 7.

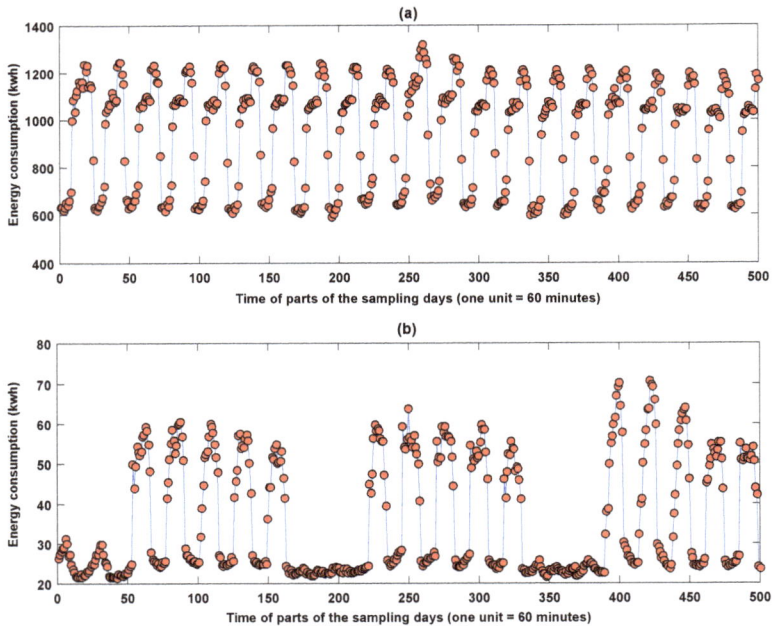

Figure 7. Parts of the samples of two data sets: (**a**) the first 500 data points of the retail store; (**b**) the first 500 data points of the office building.

4.2.2. Design Factors for MDBN

To determine the optimal structure of the MDBN for building energy consumption prediction, we will take three design factors, the number of hidden layers, hidden neurons and input variables, with their corresponding levels into account. The three design factors and their corresponding levels are presented in Table 1 and discussed in detail below.

Table 1. Design factors and their corresponding levels.

Design Factors	Level		
	1	2	3
i	2 hidden layers	3 hidden layers	4 hidden layers
ii	50 hidden units	100 hidden units	150 hidden units
iii	4 input variables	5 input variables	6 input variables

- *Design Factor i: the number of hidden layers k*
  The number of hidden layers determines how many RBMs are stacked. In this study, we consider the number of hidden layers 2, 3 and 4 as Levels 1, 2 and 3, respectively.
- *Design Factor ii: the number of uth hidden units n_u*
  The number of hidden units is an important factor that greatly influences the performance of the MDBN model. Here, we assume that the numbers of neurons in all hidden layers are equal, i.e., $n_1 = n_2 = \cdots = n_k$. In this paper, we set the number of neurons 50, 100 and 150 as Levels 1, 2 and 3, respectively.
- *Design Factor iii: the number of input variables r*
  In this paper, we utilize r energy consumption data in the building energy consumption time series before time t to predict the value at time t. In other words, we utilize $x = [y(t-1), y(t-2), \ldots, y(t-r)]$ to predict the value of $y = y(t)$. Here, we consider the number of input variables 4, 5 and 6 as Levels 1, 2 and 3, respectively.

4.2.3. Comparison Setting

In this study, the performances of all the predictors constructed by utilizing the energy-consuming patterns are compared with those designed by the original data. To evaluate the performances of the models, we utilize the following two kinds of indices.

We first consider the mean absolute error (MAE), the root mean square error (RMSE), and the mean relative error (MRE), and calculate them as

$$MAE = \frac{1}{K} \sum_{l=1}^{K} \left| \hat{y}^{(l)} - y^{(l)} \right|, \tag{31}$$

$$RMSE = \sqrt{\frac{\sum_{l=1}^{K} (\hat{y}^{(l)} - y^{(l)})^2}{K}}, \tag{32}$$

$$MRE = \frac{1}{K} \sum_{l=1}^{K} \frac{\left| \hat{y}^{(l)} - y^{(l)} \right|}{y^{(l)}} \times 100\%, \tag{33}$$

where K is the number of training or testing data pairs, and $\hat{y}^{(l)}, y^{(l)}$ are, respectively, the predicted value and actual value with respect to the input $x^{(l)}$.

The MAE, RMSE and MRE are common measures of forecasting errors in time series analysis. They serve to aggregate the magnitudes of the prediction errors into a single measure. The MAE is an average of the absolute errors between the predicted values and actual observed values. In addition, the RMSE represents the sample standard deviation of the differences between the predicted values and the actual observed values. As larger errors have a disproportionately large effect on MAE and RMSE, they are sensitive to outliers. The MRE, also known as the mean absolute percentage deviation, can remedy this drawback, and it expresses the prediction accuracy as a percentage through dividing the absolute errors by their corresponding actual values. For prediction applications, the smaller the values of MAE, RMSE and MRE are, the better the forecasting performance will be.

To better show the validity of the models, we also consider another two statistical indices, which are, respectively, the Pearson correlation coefficient, denoted as r, and the coefficient of determination, denoted as R^2. These two indices can be calculated as

$$r = \frac{K(\sum_{l=1}^{K} \hat{y}^{(l)} \cdot y^{(l)}) - (\sum_{l=1}^{K} \hat{y}^{(l)}) \cdot (\sum_{l=1}^{K} y^{(l)})}{\sqrt{(K \sum_{l=1}^{K} (\hat{y}^{(l)})^2 - (\sum_{l=1}^{K} \hat{y}^{(l)})^2) \cdot (K \sum_{l=1}^{K} (y^{(l)})^2 - (\sum_{l=1}^{K} y^{(l)})^2)}}, \tag{34}$$

$$R^2 = \frac{\left[\sum_{l=1}^{K} (\hat{y}^{(l)} - \hat{y}_{Ave}) \cdot (y^{(l)} - y_{Ave}) \right]^2}{\sum_{l=1}^{K} (\hat{y}^{(l)} - \hat{y}_{Ave}) \cdot \sum_{l=1}^{K} (y^{(l)} - y_{Ave})}, \tag{35}$$

where K is also the number of training or testing data pairs, and $\hat{y}_{Ave}, y_{Ave}$ are, respectively, the averages of the predicted and actual values.

The statistic r is a measure of the linear correlation between the actual values and the predicted values. It ranges from -1 to 1, where -1 means the total negative linear correlation, while 1 is total positive linear correlation. The statistic R^2 provides a measure of how well actual observed values are replicated by the predicted values. In other words, it is a measure of how good a predictor might be constructed from the observed training data [48]. The value of R^2 ranges from 0 to 1. In regression applications, the larger the values of r and R^2 are, the better the prediction performances will be.

4.3. Energy Consumption Prediction for the Retail Store

In this subsection, the energy-consuming pattern of the retail store will be extracted from the retail store data set firstly. Then, the configurations of the five prediction models for predicting the retail store energy consumption will be shown in detail. At last, the experimental results will be given.

4.3.1. Energy-Consuming Pattern of the Retail Store

We utilize Equations (6) and (7) to obtain the daily-periodic energy-consuming pattern and the residual time series of the retail store.

Figure 8a shows the daily-periodic energy-consuming pattern. In addition, the residual time series of the retail store, which is used to optimize the MDBN is demonstrated in Figure 8b.

4.3.2. Configurations of the Prediction Models

As aforementioned, we will take three design factors, the number of hidden layers, hidden neurons and input variables, with their corresponding levels into account to determine the optimal structure of the MDBN model for building energy consumption prediction. Consequently, $3^3 = 27$ trials are ran. In addition, the experimental results are shown in Table 2. It is obvious that trail 19 can obtain the best performance. In other words, the optimal structure of the MDBN for retail store energy consumption prediction has four hidden layers, 150 hidden units and four input variables.

Furthermore, the parameter configurations of the other four comparative predictors for retail store energy consumption prediction are listed in detail as follows.

- For the BPNN, there were 110 neurons in the hidden layer that can realize the nonlinear transformation of features by the sigmoid function. Additionally, the algorithm was ran for 7000 iterations to achieve the learning objective.
- For the GRBFNN, the 6-fold cross-validation was adopted to determine the optimized spread of the radial basis function. Furthermore, the spread was chosen from 0.01 to 2 with the 0.1 step length.

- For the ELM, there were 100 neurons in the hidden layer, and the hardlim function was chosen as the activation function for converting the original features into another space.
- For the SVR, the penalty coefficient was set to be 80, and the radial basis function was chosen as the kernel function to realize the nonlinear transformation of input features.

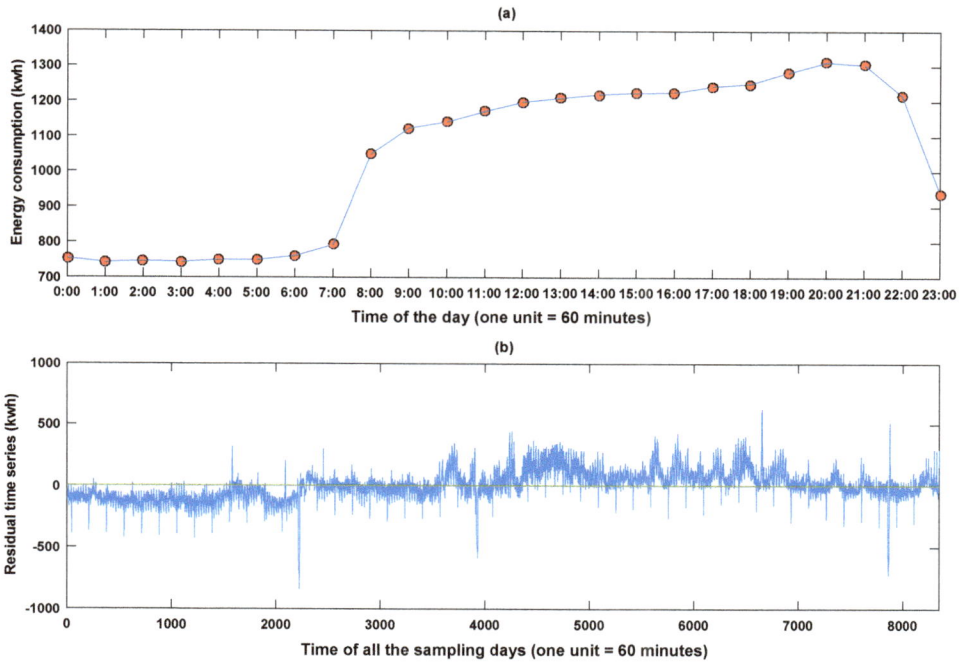

Figure 8. Periodicity knowledge and the residual time series of the retail store data set: (**a**) the daily-periodic energy-consuming pattern; (**b**) the residual time series.

Table 2. Experimental results of the MDBN in 27 trails under the consideration of three design factors and their corresponding levels.

Trials	Factor			Residual Data			Trials	Factor			Residual Data		
	i	*ii*	*iii*	MAE (kwh)	MRE (%)	RMSE (kwh)		*i*	*ii*	*iii*	MAE (kwh)	MRE (%)	RMSE (kwh)
1	1	1	1	49.21	5.26	80.79	15	2	2	3	49.65	5.31	80.03
2	1	1	2	48.74	5.18	80.03	16	2	3	1	50.18	5.36	82.13
3	1	1	3	48.73	5.19	78.06	17	2	3	2	48.43	5.11	78.24
4	1	2	1	49.12	5.24	81.20	18	2	3	3	48.33	5.12	77.96
5	1	2	2	48.25	5.16	79.39	19	3	1	1	**47.71**	**5.03**	**76.83**
6	1	2	3	49.16	5.24	79.36	20	3	1	2	48.37	5.11	77.63
7	1	3	1	49.42	5.28	81.85	21	3	1	3	48.13	5.11	77.60
8	1	3	2	49.33	5.25	81.40	22	3	2	1	48.72	5.18	79.16
9	1	3	3	48.65	5.18	78.69	23	3	2	2	49.66	5.28	79.84
10	2	1	1	48.73	5.20	79.65	24	3	2	3	49.08	5.22	78.03
11	2	1	2	49.61	5.29	81.24	25	3	3	1	51.07	5.50	83.35
12	2	1	3	47.95	5.08	77.96	26	3	3	2	48.81	5.18	79.22
13	2	2	1	48.83	5.17	79.93	27	3	3	3	48.33	5.09	77.50
14	2	2	2	49.97	5.33	81.33							

4.3.3. Experimental Results

For the testing data of the retail store, parts of the prediction results of the five predictors constructed by utilizing the energy-consuming pattern are illustrated in Figure 9. Furthermore, for better visualization, the prediction error histograms of the five predictors are shown in Figure 10. It is obvious that the more the prediction errors float around zero, the better the forecasting performance of the predictor will be.

Then, to examine the superiority of the hybrid model for the retail store energy consumption prediction, the five prediction models are compared considering different data types (the original and residual data). The original data means that the predictors are learned using the original data series, while the residual data means that the predictors are constructed by both the energy-consuming pattern and the residual data series. Experimental results are demonstrated in detail in Table 3.

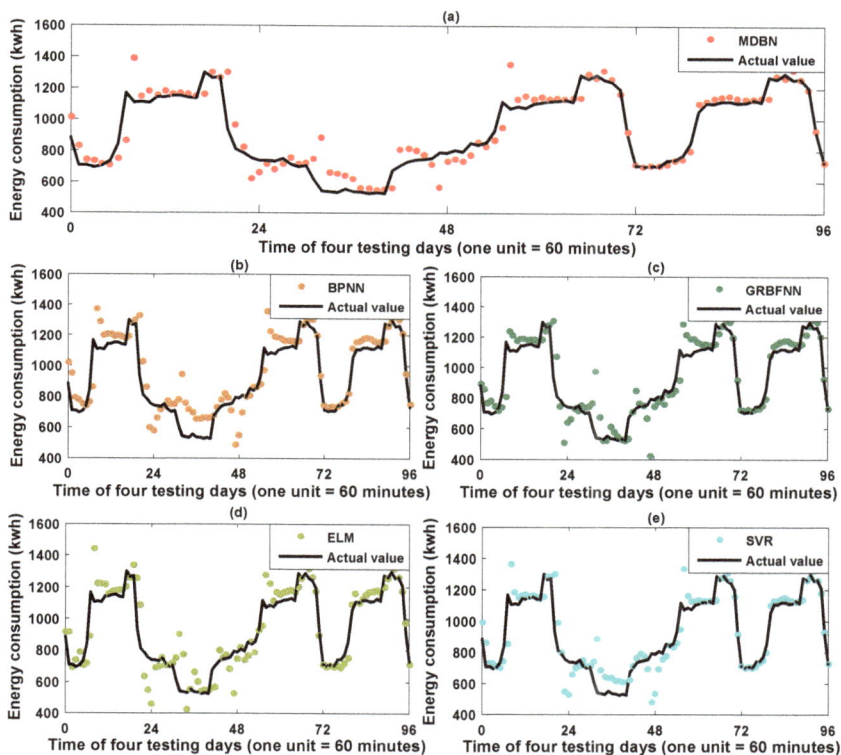

Figure 9. Parts of prediction results of the five predictors constructed by utilizing the energy-consuming pattern: (**a**) hybrid DBN model; (**b**) BPNN; (**c**) GRBFNN; (**d**) ELM; and (**e**) SVR.

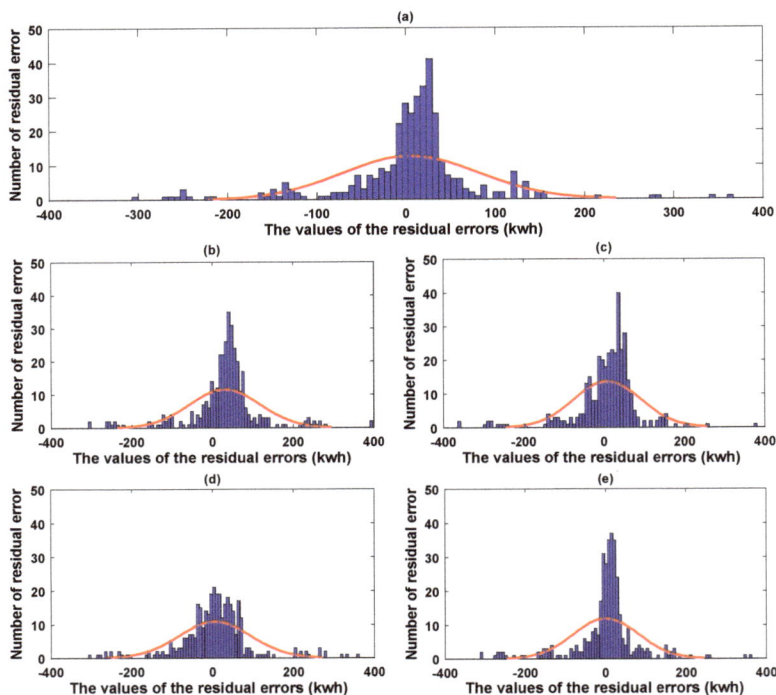

Figure 10. Prediction error histograms of the five predictors constructed by utilizing the energy-consuming pattern: (**a**) hybrid DBN model; (**b**) BPNN; (**c**) GRBFNN; (**d**) ELM; and (**e**) SVR.

Table 3. The performances of the five models for the retail store energy consumption prediction.

Methods	Data Type	MAE (kwh)	MRE (%)	RMSE (kwh)	r	R^2
MDBN	Residual data	**47.71**	**5.03**	**76.83**	**0.94**	**0.89**
	Original data	54.38	5.59	86.43	0.93	0.86
BPNN	Residual data	65.69	7.24	93.38	0.92	0.85
	Original data	75.45	8.20	100.40	0.94	0.87
GRBFNN	Residual data	54.60	5.75	83.87	0.93	0.87
	Original data	52.51	5.62	87.54	0.93	0.86
ELM	Residual data	58.54	6.29	88.62	0.93	0.86
	Original data	78.86	8.34	113.02	0.89	0.79
SVR	Residual data	48.28	5.19	81.31	0.93	0.87
	Original data	52.19	5.42	89.93	0.92	0.85

4.4. Energy Consumption Prediction for the Office Building

In this subsection, first of all, the energy-consuming pattern of the office building will be extracted from the office building data set. Then, the configurations of the five prediction models for predicting the office building energy consumption will be shown in detail. Finally, the experimental results will be given.

4.4.1. Energy-Consuming Pattern of the Office Building

Being similar to the retail store experiment, we utilize Equations (8)–(14) to obtain the weekly-periodic energy-consuming pattern and the residual time series of the office building.

As mentioned previously, the weekly-periodic energy-consuming pattern should include two parts, which are the weekday pattern and the weekend pattern. The obtained weekday pattern is depicted in Figure 11a, while the weekend pattern is shown in Figure 11b. We can observe that the energy consumption in weekends is quite different from that in weekdays. After removing the energy-consuming pattern, the residual time series of the office building is demonstrated in Figure 11c. This residual time series is utilized to train the MDBN in the hybrid model.

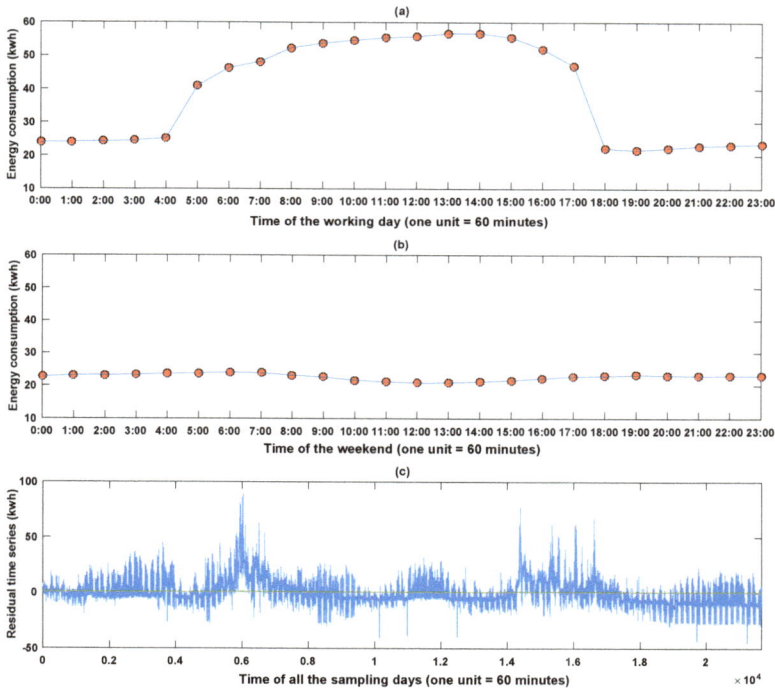

Figure 11. Periodicity knowledge and the residual time series of the office building data set: (a) the energy-consuming pattern of weekdays; (b) the energy-consuming pattern of weekends; (c) the residual time series.

4.4.2. Configurations of the Prediction Models

Similarly, we run $3^3 = 27$ trials to determine the optimal structure of the MDBN model for the office building energy consumption prediction. The experimental results are listed in Table 4. As shown in Table 4, the trail 13 obtains the best performance. Consequently, the optimal structure of the MDBN in the hybrid model for office building has three hidden layers, 100 hidden units in each layer and four input variables.

Table 4. Experimental results of the MDBN in 27 trails under the consideration of three design factors and their corresponding levels.

Trails	Factor			Residual Data			Trails	Factor			Residual Data		
	i	*ii*	*iii*	MAE (kwh)	MRE (%)	RMSE (kwh)		*i*	*ii*	*iii*	MAE (kwh)	MRE (%)	RMSE (kwh)
1	1	1	1	2.30	12.67	3.69	15	2	2	3	2.35	12.99	3.69
2	1	1	2	2.22	12.29	3.61	16	2	3	1	2.25	12.49	3.65
3	1	1	3	2.32	12.74	3.67	17	2	3	2	2.30	12.78	3.68
4	1	2	1	2.23	11.97	3.63	18	2	3	3	2.36	13.10	3.71
5	1	2	2	2.35	12.81	3.71	19	3	1	1	2.21	12.19	3.65
6	1	2	3	2.40	13.10	3.71	20	3	1	2	2.23	12.29	3.66
7	1	3	1	2.17	11.93	3.58	21	3	1	3	2.27	12.54	3.67
8	1	3	2	2.29	12.70	3.67	22	3	2	1	2.17	12.06	3.60
9	1	3	3	2.27	12.55	3.63	23	3	2	2	2.26	12.51	3.65
10	2	1	1	2.26	12.53	3.65	24	3	2	3	2.23	12.25	3.67
11	2	1	2	2.31	12.83	3.68	25	3	3	1	2.14	11.91	3.60
12	2	1	3	2.36	13.10	3.70	26	3	3	2	2.32	12.64	3.73
13	2	2	1	**2.09**	**11.62**	**3.54**	27	3	3	3	2.21	12.30	3.64
14	2	2	2	2.31	12.84	3.68							

For the other four comparative predictors, their parameter configurations for the office building energy consumption prediction are listed as follows:

- For the BPNN, there were 200 neurons in the hidden layer. Furthermore, the sigmoid function was chosen to realize the nonlinear transformation of features. Additionally, we ran the BP algorithm 1000 times to obtain the final outputs.
- For the GRBFNN, the 5-fold cross-validation was utilized to determine the optimized spread of the radial basis function. Furthermore, the spread was chosen from 0.01 to 2 with a 0.1 step length.
- For the ELM, there were 150 neurons in the hidden layer, and the hardlim function was chosen as the activation function for converting the original features into another space.
- For the SVR, the penalty coefficient was set to be 10 and the sigmoid function was chosen as the kernel function to realize the nonlinear transformation of input features.

4.4.3. Experimental Results

For the testing data of the office building, parts of the prediction results of the five predictors are illustrated in Figure 12. Again, for better visualization, the prediction error histograms of the five predictors are shown in Figure 13.

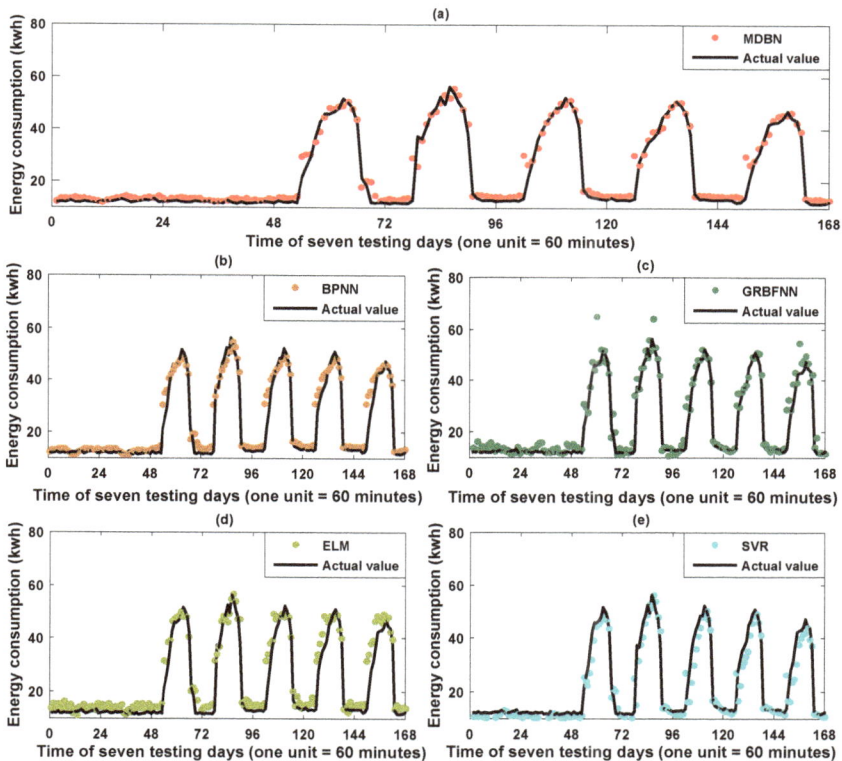

Figure 12. Parts of the prediction results of the five predictors constructed by utilizing the energy-consuming pattern: (**a**) hybrid DBN model; (**b**) BPNN; (**c**) GRBFNN; (**d**) ELM; and (**e**) SVR.

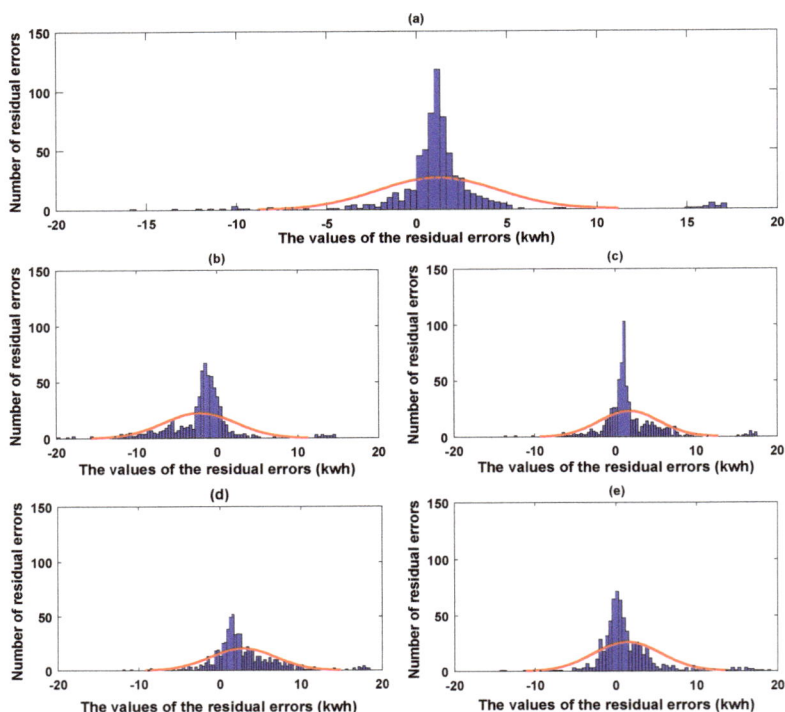

Figure 13. Prediction error histograms of the five predictors constructed by utilizing the energy-consuming pattern: (**a**) hybrid DBN model; (**b**) BPNN; (**c**) GRBFNN; (**d**) ELM; and (**e**) SVR.

Then, in order to examine the superiority of the hybrid model for the office building energy consumption prediction, the five prediction models are compared under the consideration of different data types (the original and residual data). Experimental results are demonstrated in Table 5.

Table 5. The performances of the five models with different data types for the office building energy consumption prediction.

Methods	Data Type	MAE (kwh)	MRE (%)	RMSE (kwh)	r	R^2
MDBN	Residual data	**2.09**	**11.62**	**3.54**	**0.97**	**0.93**
	Original data	2.32	11.50	4.19	0.95	0.90
BPNN	Residual data	2.57	12.64	4.04	0.96	0.93
	Original data	3.85	23.21	4.75	0.95	0.91
GRBFNN	Residual data	2.54	12.62	4.39	0.95	0.91
	Original data	4.35	21.94	5.98	0.93	0.87
ELM	Residual data	3.50	17.18	4.92	0.96	0.92
	Original data	4.61	25.52	5.92	0.90	0.82
SVR	Residual data	3.23	14.89	4.98	0.94	0.88
	Original data	6.13	34.42	7.55	0.92	0.85

4.5. Comparisons and Discussions

As discussed previously, smaller values of the MAE, RMSE and MRE represent better prediction results while lager values of r and R^2 correspond to better performance. Considering all the values of such indices as shown in Tables 3 and 5 (It is worth noting that the values of the indices in Table 3 are about the retail energy consumption while the values in Table 5 are about the office energy consumption. The retail building consumed much more energies than the office building. As a result, some values of the MAE, RMSE and MRE in Table 3 are larger than those in Table 5), the predictors constructed by utilizing the energy-consuming patterns perform better than those designed only by the original data. Taking the RMSE index for example, in the first experiment, the accuracies of the MDBN, BPNN, GRBFNN, ELM and SVR based hybrid models are promoted by 11.1%, 7.0%, 4.2%, 21.6% and 9.6%, respectively, while, in the second experiment, the accuracy improvements of such models are 15.6%, 14.8%, 26.5%, 16.9% and 34.0%, respectively. As a result, we can draw a conclusion that the periodicity knowledge is helpful to improve the accuracy for building energy consumption prediction.

From Figures 9 and 12, we can see that the hybrid DBN model can not only predict the regular testing data well for both the retail store and the office building energy consumption from the global perspective, but also give the best prediction results for the noisy irregular data, e.g., the sampling points from 25 to 50 in Figure 9 in the retail store experiment. These irregular testing data can reflect the uncertainties in the energy consumption time series. In other words, the proposed hybrid DBN model has the most powerful ability to deal with the uncertain and/or the randomness in the historical building energy consumption data.

Figures 10 and 13 demonstrated the prediction error histograms of the five models designed through using the periodicity knowledge in the two experiments. In the histograms, the horizontal direction depicts the exact values of the prediction errors, while the vertical direction indicates the number of the prediction errors in different partitioned intervals. The more the prediction errors float around zero, the better performance the predictors will achieve. From both figures, we can clearly observe that the proposed hybrid DBN model has more prediction errors floating near zero compared with the other four artificial intelligence techniques—that is to say, the approximation capability of the proposed hybrid DBN model is promising for the two experimented buildings. Furthermore, to further validate the accuracy of the MDBN based hybrid model, scatter plots of the actual and predicted values in the two experiments are demonstrated in Figure 14a,b, respectively. From Figure 14, we can observe that the predicted values from the hybrid DBN model can duplicate the actual values well.

Among all the predictors constructed by both the original and residual data, the proposed MDBN based hybrid model has the best prediction accuracy in the two experiments as shown in Tables 3 and 5. This phenomenon indicates that the proposed deep learning method has the miraculous learning and prediction abilities in time series forecasting applications. This also verifies the powerful feature extraction ability of the deep learning algorithm and the effectiveness of the modified learning strategies.

One thing to be mentioned is that the numbers of the data used in this paper are not very big (about the ten thousand scale). Even though the hybrid MDBN model is not learned by big data in both experiments, it still shows us excellent performances. This is also consistent with some other application results where the DBNs were trained without a mass of data. For example, in [49,50], the DBNs were applied to the time series prediction and the wind power prediction, which also do not have a large quantity of data. In both applications, the experimental results demonstrated that the DBN approach performs best compared with the traditional techniques. All these applications verified the learning ability of the DBN models for not very large data applications.

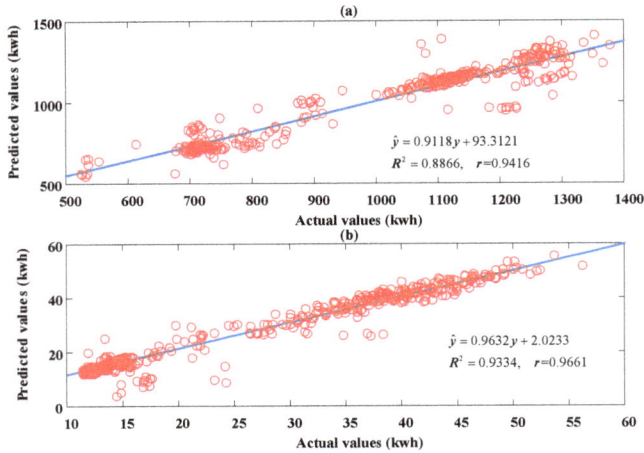

Figure 14. Scatter plots of the actual and predicted values of the energy consumptions in the retail building (**a**) and the office building (**b**).

5. Conclusions

In this paper, a hybrid model is presented to further improve the prediction accuracy for building energy consumption prediction. The proposed model combines the MDBN model with the periodicity knowledge to obtain the final prediction results. The theoretical contributions of this study consist of two aspects: (1) the periodicity knowledge was extracted and encoded into the prediction model. In addition, the prediction accuracy can be greatly improved through utilizing this kind of prior knowledge; (2) a novel learning algorithm that combines the contrastive divergence algorithm and the least squares method was proposed to optimize the parameters of the MDBN. This is the first time that the DBN is applied to the building energy consumption prediction. On the other hand, this study applied the proposed approach to the energy consumption prediction of two kinds of buildings. Experimental and comparison results verified the effectiveness and superiorities of the proposed hybrid model.

As is well known, many kinds of time series data, e.g., the traffic flow time series and the electricity consumption time series, have the periodicity property. The hybrid model can be expected to yield better performance in the predictions of such time series. In the future, we will extend our approach to these applications. On the other aspect, our study only focuses on the data science that tries to utilize the data to realize the energy-consumption prediction without considering any scientific or practical information of energy related principles. Theoretically, the energy related principles are very helpful to improve the prediction performance. We are now exploring the strategies to construct the novel hybrid prediction models through combining the energy related principles and observed data to further improve the prediction accuracy.

Acknowledgments: This work is supported by the National Natural Science Foundation of China (61473176, 61105077, 61573225), and the Natural Science Foundation of Shandong Province for Young Talents in Province Universities (ZR2015JL021).

Author Contributions: Chengdong Li, Jianqiang Yi and Yisheng Lv have contributed to developing ideas about energy consumption prediction and collecting the data. Zixiang Ding and Guiqing Zhang programmed the algorithm and tested it. All of the authors were involved in preparing the manuscript.

Conflicts of Interest: The authors declare no conflict of interest.

References

1. Ahmad, M.W.; Mourshed, M.; Rezgui, Y. Trees vs Neurons: Comparison between random forest and ANN for high-resolution prediction of building energy consumption. *Energy Build.* **2017**, *147*, 77–89.
2. Banihashemi, S.; Ding, G.; Wang, J. Developing a hybrid model of prediction and classification algorithms for building energy consumption. *Energy Procedia* **2017**, *110*, 371–376.
3. Naji, S.; Keivani, A.; Shamshirband, S.; Alengaram, U.J.; Jumaat, M.Z.; Mansor, Z.; Lee, M. Estimating building energy consumption using extreme learning machine method. *Energy* **2016**, *97*, 506–516.
4. Hsu, D. Comparison of integrated clustering methods for accurate and stable prediction of building energy consumption data. *Appl. Energy* **2015**, *160*, 153–163.
5. Dong, B.; Cao, C.; Lee, S.E. Applying support vector machines to predict building energy consumption in tropical region. *Energy Build.* **2005**, *37*, 545–553.
6. Jung, H.C.; Kim, J.S.; Heo, H. Prediction of building energy consumption using an improved real coded genetic algorithm based least squares support vector machine approach. *Energy Build.* **2015**, *90*, 76–84.
7. Hong, W.C.; Dong, Y.; Zhang, W.Y.; Chen, L.Y.; Panigrahi, B.K. Cyclic electric load forecasting by seasonal SVR with chaotic genetic algorithm. *Int. J. Electr. Power Energy Syst.* **2013**, *44*, 604–614.
8. Fan, G.F.; Peng, L.L.; Hong, W.C.; Sun, F. Electric load forecasting by the SVR model with differential empirical mode decomposition and auto regression. *Neurocomputing* **2016**, *173*, 958–970.
9. Hong, W.C. Chaotic particle swarm optimization algorithm in a support vector regression electric load forecasting model. *Energy Convers. Manag.* **2009**, *50*, 105–117.
10. Robinson, C.; Dilkina, B.; Hubbs, J.; Zhang, W.; Guhathakurta, S.; Brown, M.A.; Pendyala, R.M. Machine learning approaches for estimating commercial building energy consumption. *Appl. Energy* **2017**, *208*, 889–904.
11. Hinton, G.E.; Osindero, S.; Teh, Y.W. A fast learning algorithm for deep belief nets. *Neural Comput.* **2006**, *18*, 1527–1554.
12. Lv, Y.; Duan, Y.; Kang, W.; Li, Z.; Wang, F.Y. Traffic flow prediction with big data: A deep learning approach. *IEEE Trans. Intell. Transp. Syst.* **2015**, *16*, 865–873.
13. Yang, H.F.; Dillon, T.S.; Chen, Y.P.P. Optimized structure of the traffic flow forecasting model with a deep learning approach. *IEEE Trans. Neural Netw. Learn. Syst.* **2017**, *28*, 2371–2381.
14. Li, C.; Ding, Z.; Zhao, D.; Yi, J.; Zhang, G. Building energy consumption prediction: An extreme deep learning approach. *Energies* **2017**, *10*, 1525.
15. Xiao, Y.; Wu, J.; Lin, Z.; Zhao, X. A deep learning-based multi-model ensemble method for cancer prediction. *Comput. Methods Programs Biomed.* **2018**, *153*, 1–9.
16. Galea, C.; Farrugia, R.A. Forensic face photo-sketch recognition using a deep learning-based architecture. *IEEE Signal Process. Lett.* **2017**, *24*, 1586–1590.
17. Masoumi, M.; Hamza, A.B. Spectral shape classification: A deep learning approach. *J. Vis. Commun. Image Represent.* **2017**, *43*, 198–211.
18. Sarikaya, R.; Hinton, G.E.; Deoras, A. Application of deep belief networks for natural language understanding. *IEEE/ACM Trans. Audio Speech Lang. Process. (TASLP)* **2014**, *22*, 778–784.
19. Zhang, X.L.; Wu, J. Deep belief networks based voice activity detection. *IEEE Trans. Audio Speech Lang. Process.* **2013**, *21*, 697–710.
20. Chen, C.C.; Li, S.T. Credit rating with a monotonicity-constrained support vector machine model. *Expert Syst. Appl.* **2014**, *41*, 7235–7247.
21. Wang, L.; Xue, P.; Chan, K.L. Incorporating prior knowledge into SVM for image retrieval. In Proceedings of the International Conference on Pattern Recognition, Cambridge, UK, 23–26 August 2004; pp. 981–984.
22. Wu, X.; Srihari, R. Incorporating prior knowledge with weighted margin support vector machines. In Proceedings of the Tenth ACM SIGKDD International Conference on Knowledge Discovery and Data Mining, Seattle, WA, USA, 22–25 August 2004; ACM: New York, NY, USA, 2004; pp. 326–333.
23. Li, C.; Zhang, G.; Yi, J.; Wang, M. Uncertainty degree and modeling of interval type-2 fuzzy sets: definition, method and application *Comput. Math. Appl.* **2013**, *66*, 1822–1835.
24. Abonyi, J.; Babuska, R.; Verbruggen, H.B.; Szeifert, F. Incorporating prior knowledge in fuzzy model identification. *Int. J. Syst. Sci.* **2000**, *31*, 657–667.
25. Li, C.; Yi, J.; Zhang, G. On the monotonicity of interval type-2 fuzzy logic systems. *IEEE Trans. Fuzzy Syst.* **2014**, *22*, 1197–1212.

26. Chakraborty, S.; Chattopadhyay, P.P.; Ghosh, S.K.; Datta, S. Incorporation of prior knowledge in neural network model for continuous cooling of steel using genetic algorithm. *Appl. Soft Comput.* **2017**, *58*, 297–306.

27. Kohara, K.; Ishikawa, T.; Fukuhara, Y.; Nakamura, Y. Stock price prediction using prior knowledge and neural networks. *Intell. Syst. Account. Financ. Manag.* **1997**, *6*, 11–22.

28. Li, C.; Gao, J.; Yi, J.; Zhang, G. Analysis and design of functionally weighted single-input-rule-modules connected fuzzy inference systems. *IEEE Trans. Fuzzy Syst.* **2016**, doi:10.1109/TFUZZ.2016.2637369.

29. Lin, H.; Lin, Y.; Yu, J.; Teng, Z.; Wang, L. Weighing fusion method for truck scales based on prior knowledge and neural network ensembles. *IEEE Trans. Instrum. Meas.* **2014**, *63*, 250–259.

30. Bengio, Y.; Lamblin, P.; Popovici, D.; Larochelle, H. Greedy layer-wise training of deep networks. In Proceedings of the Twentieth Annual Conference on Neural Information Processing Systems, Vancouver, BC, Canada, 4–7 December 2006; pp. 153–160.

31. Fischer, A.; Igel, C. An introduction to restricted Boltzmann machines. *Prog. Pattern Recognit. Image Anal. Comput. Vis. Appl.* **2012**, *7441*, 14–36.

32. Bengio, Y. *Learning Deep Architectures for AI*; Now Publishers: Boston, MA, USA. 2009; pp. 1–127.

33. Hinton, G. Training products of experts by minimizing contrastive divergence. *Neural Comput.* **2002**, *14*, 1771–1800.

34. Roux, N.L.; Bengio, Y. Representational power of restricted boltzmann machines and deep belief networks. *Neural Comput.* **2008**, *20*, 1631–1649.

35. Bu, S.; Liu, Z.; Han, J.; Wu, J.; Ji, R. Learning high-level feature by deep belief networks for 3-D model retrieval and recognition. *IEEE Trans. Multimed.* **2014**, *16*, 2154–2167.

36. Huang, G.B.; Wang, D.H.; Lan, Y. Extreme learning machines: A survey. *Int. J. Mach. Learn. Cybern.* **2011**, *2*, 107–122.

37. Huang, G.B.; Zhu, Q.Y.; Siew, C.K. Extreme learning machine: Theory and applications. *Neurocomputing* **2006**, *70*, 489–501.

38. Huang, G.B.; Zhu, Q.Y.; Siew, C.K. Extreme learning machine: A new learning scheme of feedforward neural networks. In Proceedings of the 2004 IEEE International Joint Conference on Neural Networks, Budapest, Hungary, 25–29 July 2004; Volume 2, pp. 985–990.

39. Huang, G.B.; Chen, L.; Siew, C.K. Universal approximation using incremental constructive feedforward networks with random hidden nodes. *IEEE Trans. Neural Netw.* **2006**, *17*, 879–892.

40. Erb, R.J. Introduction to backpropagation neural network computation. *Pharm. Res.* **1993**, *10*, 165–170.

41. Uzlu, E.; Kankal, M.; Akpınar, A.; Dede, T. Estimates of energy consumption in Turkey using neural networks with the teaching-learning-based optimization algorithm. *Energy* **2014**, *75*, 295–303.

42. Yedra, R.M.; Diaz, F.R.; Nieto, M.D.M.C.; Arahal, M.R. A neural network model for energy consumption prediction of CIESOL bioclimatic building. *Adv. Intell. Syst. Comput.* **2014**, *239*, 51–60.

43. Lu, J.; Hu, H.; Bai, Y. Generalized radial basis function neural network based on an improved dynamic particle swarm optimization and AdaBoost algorithm. *Neurocomputing* **2015**, *152*, 305–315.

44. Friedrichs, F.; Schmitt, M. On the power of Boolean computations in generalized RBF neural networks. *Neurocomputing* **2005**, *63*, 483–498.

45. Awad, M.; Khanna, R. Support vector regression. *Neural Inf. Process. Lett. Rev.* **2007**, *11*, 203–224.

46. Wu, C.H.; Ho, J.M.; Lee, D.T. Travel-time prediction with support vector regression. *IEEE Trans. Intell. Transp. Syst.* **2004**, *5*, 276–281.

47. Buildings Datasets. Available online: https://trynthink.github.io/buildingsdatasets/ (accessed on 13 May 2017).

48. Glantz, S.A.; Slinker, B.K. *Primer of Applied Regression and Analysis of Variance*; Health Professions Division, McGraw-Hill: New York, NY, USA, 1990.

49. Hirata, T.; Kuremoto, T.; Obayashi, M.; Mabu, S.; Kobayashi, K. Time series prediction using DBN and ARIMA. In Proceedings of the International Conference on Computer Application Technologies, Atlanta, GA, USA, 10–14 June 2016; pp. 24–29.

50. Tao, Y.; Chen, H. A hybrid wind power prediction method. In Proceedings of the Power and Energy Society General Meeting, Boston, MA, USA, 17–21 July 2016; pp. 1–5.

![energies logo] *energies*

MDPI

Article

A High Precision Artificial Neural Networks Model for Short-Term Energy Load Forecasting

Ping-Huan Kuo [1] [iD] **and Chiou-Jye Huang** [2,*] [iD]

[1]   Computer and Intelligent Robot Program for Bachelor Degree, National Pingtung University, Pingtung 90004, Taiwan; phkuo@mail.nptu.edu.tw
[2]   School of Electrical Engineering and Automation, Jiangxi University of Science and Technology, Ganzhou 341000, Jiangxi, China
*   Correspondence: chioujye@163.com; Tel.: +86-137-2624-7572

Received: 14 December 2017; Accepted: 9 January 2018; Published: 16 January 2018

Abstract: One of the most important research topics in smart grid technology is load forecasting, because accuracy of load forecasting highly influences reliability of the smart grid systems. In the past, load forecasting was obtained by traditional analysis techniques such as time series analysis and linear regression. Since the load forecast focuses on aggregated electricity consumption patterns, researchers have recently integrated deep learning approaches with machine learning techniques. In this study, an accurate deep neural network algorithm for short-term load forecasting (STLF) is introduced. The forecasting performance of proposed algorithm is compared with performances of five artificial intelligence algorithms that are commonly used in load forecasting. The Mean Absolute Percentage Error (MAPE) and Cumulative Variation of Root Mean Square Error (CV-RMSE) are used as accuracy evaluation indexes. The experiment results show that MAPE and CV-RMSE of proposed algorithm are 9.77% and 11.66%, respectively, displaying very high forecasting accuracy.

Keywords: artificial intelligence; convolutional neural network; deep neural networks; short-term load forecasting

---

1. Introduction

Nowadays, there is a persistent need to accelerate development of low-carbon energy technologies in order to address the global challenges of energy security, climate change, and economic growth. The smart grids [1] are particularly important as they enable several other low-carbon energy technologies [2], including electric vehicles, variable renewable energy sources, and demand response. Due to the growing global challenges of climate, energy security, and economic growth, acceleration of low-carbon energy technology development is becoming an increasingly urgent issue [3]. Among various green technologies to be developed, smart grids are particularly important as they are key to the integration of various other low-carbon energy technologies, such as power charging for electric vehicles, on-grid connection of renewable energy sources, and demand response.

The forecast of electricity load is important for power system scheduling adopted by energy providers [4]. Namely, inefficient storage and discharge of electricity could incur unnecessary costs, while even a small improvement in electricity load forecasting could reduce production costs and increase trading advantages [4], particularly during the peak electricity consumption periods. Therefore, it is important for electricity providers to model and forecast electricity load as accurately as possible, in both short-term [5–12] (one day to one month ahead) and medium-term [13] (one month to five years ahead) periods.

With the development of big data and artificial intelligence (AI) technology, new machine learning methods have been applied to the power industry, where large electricity data need to be carefully managed. According to the Mckinsey Global Institute [14], the AI could be applied in the electricity industry for power demand and supply prediction, because a power grid load forecast affects many stakeholders. Based on the short-term forecast (1–2 days ahead), power generation systems can determine which power sources to access in the next 24 h, and transmission grids can timely assign appropriate resources to clients based on current transmission requirements. Moreover, using an appropriate demand and supply forecast, electricity retailers can calculate energy prices based on estimated demand more efficiently.

The powerful data collection and analysis technologies are becoming more available on the market, so power companies are beginning to explore a feasibility of obtaining more accurate results using AI in short-term load forecasts. For instance, in the United Kingdom (UK), the National Grid is currently working with the DeepMind [15,16], a Google-owned AI team, which is used to predict the power supply and demand peaks in the UK based on the information from smart meters and by incorporating weather-related variables. This cooperation tends to maximize the use of intermittent renewable energy and reduce the UK national energy usage by 10%. Therefore, it is expected that electricity demand and supply could be predicted and managed in real time through deep learning technologies and machines, optimizing load dispatch, and reducing operation costs.

The load forecasting can be categorized by the length of forecast interval. Although there is no official categorization in the power industry, there are four load forecasting types [17]: very short term load forecasting (VSTLF), short term load forecasting (STLF), medium term load forecasting (MTLF), and long term load forecasting (LTLF). The VSTLF typically predicts load for a period less than 24 h, STLF predicts load for a period greater than 24 h up to one week, MTLF forecasts load for a period from one week up to one year, and LTLF forecasts load performance for a period longer than one year. The load forecasting type is chosen based on application requirements. Namely, VSTLF and STLF are applied to everyday power system operation and spot price calculation, so the accuracy requirement is much higher than for a long term prediction. The MTLF and LTLF are used for prediction of power usage over a long period of time, and they are often referenced in long-term contracts when determining system capacity, costs of operation and system maintenance, and future grid expansion plans. Thus, if the smart grids are integrated with a high percentage of intermittent renewable energy, load forecasting will be more intense than that of traditional power generation sources due to the grid stability.

In addition, the load forecasting can be classified by calculation method into statistical methods and computational intelligence (CI) methods. With recent developments in computational science and smart metering, the traditional load forecasting methods have been gradually replaced by AI technology. The smart meters for residential buildings have become available on the market around 2010, and since then, various studies on STLF for residential communities have been published [18,19]. When compared with the traditional statistical forecasting methods, the ability to analyze large amounts of data in a very short time frame using AI technology has displayed obvious advantages [10].

Some of frequently used load forecast methods include linear regression [5,6,20], autoregressive methods [7,21], and artificial neural networks [9,22,23]. Furthermore, clustering methods were also proposed [24]. In [20,25] similar time sequences were matched while in [24] the focus was on customer classification. A novel approach based on the support vector machine was proposed in [26,27]. The other forecasting methods, such as exponential smoothing and Kalman filters, were also applied in few studies [28]. A careful literature review of the latest STLF method can be found in [8]. In [13], it was shown that accuracy of STLF is influenced by many factors, such as temperature, humidity, wind speed, etc. In many studies, the artificial neural network (ANN) forecasting methods [9–11,29] have been proven to be more accurate than traditional statistical methods, and accuracy of different ANN methods has been reviewed by many researchers [1,30]. In [31], a multi-model partitioning algorithm (MMPA) for short-term electricity load forecasting was proposed. According to the obtained experimental results, the MMPA method is better than autoregressive integrated moving average (ARIMA) method. In [17], authors used the ANN-based method reinforced by wavelet denoising algorithm. The wavelet method was used to factorize electricity load data into signals with different frequencies. Therefore, the wavelet denosing algorithm provides good electricity load data for neural network training and improves load forecasting accuracy.

In this study, a new load forecasting model based on a deep learning algorithm is presented. The forecasting accuracy of proposed model is within the requested range, and model has advantages of simplicity and high forecasting performance. The major contributions of this paper are: (1) introduction of a precise deep neural network model for energy load forecasting; (2) comparison of performances of several forecasting methods; and, (3) creation of a novel research direction in time sequence forecasting based on convolutional neural networks.

2. Methodology of Artificial Neural Networks

Artificial neural networks (ANNs) are computing systems inspired by the biological neural networks. The general structure of ANNs contains neurons, weights, and bias. Based on their powerful molding ability, ANNs are still very popular in the machine learning field. However, there are many ANN structures used in the machine learning problems, but the Multilayer Perceptron (MLP) [32] is the most commonly used ANN type. The MLP is a fully connected structure artificial neural network. The structure of MLP is shown in Figure 1. In general, the MLP consists of one input layer, one or more hidden layers, and one output layer. However, the MLP network presented in Figure 1 is the most common MLP structure, which has only one hidden layer. In the MLP, all the neurons of the previous layer are fully connected to the neurons of the next layer. In Figure 1, $x_1, x_2, x_3, \ldots, x_6$ are the neurons of the input layer, h_1, h_2, h_3, h_4 are the neurons of the hidden layer, and y_1, y_2, y_3, y_4 are the neurons of the output layer. In the case of energy load forecasting, the input is the past energy load, and the output is the future energy load. Although the MLP structure is very simple, it provides good results in many applications. The most commonly used algorithm for MLP training is the backpropagation algorithm.

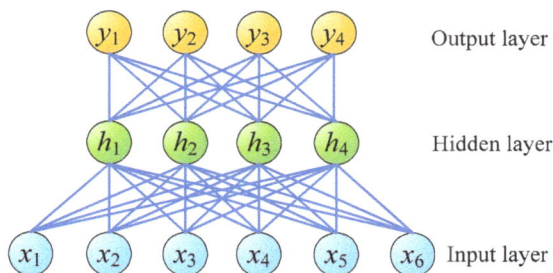

Figure 1. The Multilayer Perceptron (MLP) structure.

Although MLPs are very good in modelling and patter recognition, the convolutional neural networks (CNNs) provide better accuracy in highly non-linear problems, such as energy load forecasting. The CNN uses the concept of weight sharing. The one-dimensional convolution and pooling layer are presented in Figure 2. The lines in the same color denote the same sharing weight, and sets of the sharing weights can be treated as kernels. After the convolution process, the inputs x_1, $x_2, x_3, \ldots, x_6$ are transformed to the feature maps c_1, c_2, c_3, c_4. The next step in Figure 2 is pooling, wherein the feature map of convolution layer is sampled and its dimension is reduced. For instance, in Figure 2 dimension of the feature map is 4, and after pooling process that dimension is reduced to 2. The process of pooling is an important procedure to extract the important convolution features.

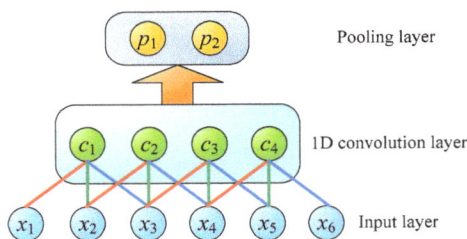

Figure 2. The one-dimensional (1D) convolution and pooling layer.

The other popular solution of the forecasting problem is Long Short Term Memory network (LSTM) [33]. The LSTM is a recurrent neural network, which has been used to solve many time sequence problems. The structure of LSTM is shown in Figure 3, and its operation is illustrated by the following equations:

$$f_t = \sigma(W_f \cdot [h_{t-1}, x_t] + b_f) \tag{1}$$

$$i_t = \sigma(W_i \cdot [h_{t-1}, x_t] + b_i) \tag{2}$$

$$\tilde{C}_t = \tanh(W_C \cdot [h_{t-1}, x_t] + b_C) \tag{3}$$

$$C_t = f_t \times C_{t-1} + i_t \times \tilde{C}_t \tag{4}$$

$$o_t = \sigma(W_o \cdot [h_{t-1}, x_t] + b_o) \tag{5}$$

$$h_t = o_t \times \tanh(C_t) \tag{6}$$

where x_t is the network input, and h_t is the output of hidden layer, σ denotes the sigmoidal function, C_t is the cell state, and $\tilde{C}_t$ denotes the candidate value of the state. Besides, there are three gates in LSTM: i_t is the input gate, o_t is the output gate, and f_t is the forget gate. The LSTM is designed for solving the long-term dependency problem. In general, the LSTM provides good forecasting results.

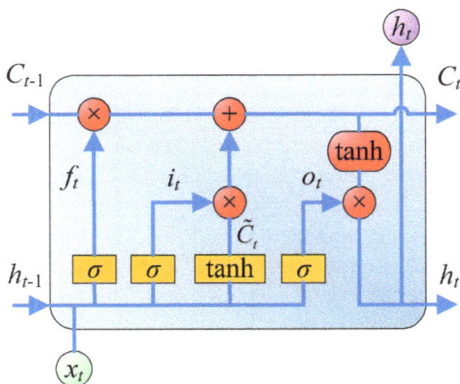

Figure 3. The Long Short Term Memory network (LSTM) structure.

3. The Proposed Deep Neural Network

The structure of the proposed deep neural network DeepEnergy is shown in Figure 4. Unlike the general forecasting method based on the LSTM, the DeepEnergy uses the CNN structure. The input layer denotes the information on past load, and the output values represent the future energy load. There are two main processes in DeepEnergy, feature extraction, and forecasting. The feature extraction in DeepEnergy is performed by three convolution layers (Conv1, Conv2, and Conv3) and three pooling layers (Pooling1, Pooling2, and Pooling3). The Conv1–Conv3 are one-dimensional (1D) convolutions, and the feature maps are all activated by the Rectified Linear Unit (ReLU) function. Besides, the kernel sizes of Conv1, Conv2, and Conv3 are 9, 5, 5, respectively, and the depths of the feature maps are 16, 32, 64, respectively. The pooling method of Pooling1 to Pooling3 is the max pooling, and the pooling size is equal to 2. Therefore, after the pooling process, the dimension of the feature map will be divided by 2 to extract the important features of the deeper layers.

In the forecasting, the first step is to flat the Pooling3 layer into one dimension and construct a fully connected structure between Flatten layer and Output layer. In order to fit the values previously normalized in the range [0, 1], the sigmoidal function is chosen as an activation function of the output layer. Furthermore, in order to overcome the overfitting problem, the dropout technology [34] is adopted in the fully connected layer. Namely, the dropout is an efficient way to prevent overfitting in artificial neural network. During the training process, neurons are randomly "dead". As shown in Figure 4, the output values of chosen neurons (the gray circles) are equal to zero in certain training iteration. The chosen neurons are randomly changed during training process.

Furthermore, the flowchart of proposed DeepEnergy is represented in Figure 5. Firstly, the raw energy load data are loaded into the memory. Then, the data preprocessing is executed and data are normalized in the range [0, 1] in order to fit the characteristic of the machine learning model. For the purpose of validation of DeepEnergy generalization performance, the data are split into training data and testing data. The training data are used for training of proposed model. After the training process, the proposed DeepEnergy network is created and initialized. Before the training, the training data are randomly shuffled to force the proposed model to learn complicated relationships between input and output data. The training data are split into several batches. According to the order of shuffled data, the model is trained on all of the batches. During the training process, if the desired Mean Square Error (MSE) is not reached in the current epoch, the training will continue until the maximal number of epochs or desired MSE is reached. On the contrary, if the maximal number of epochs is reached, then the training process will stop regardless the MSE value. Final performances are evaluated to demonstrate feasibility and practicability of the proposed method.

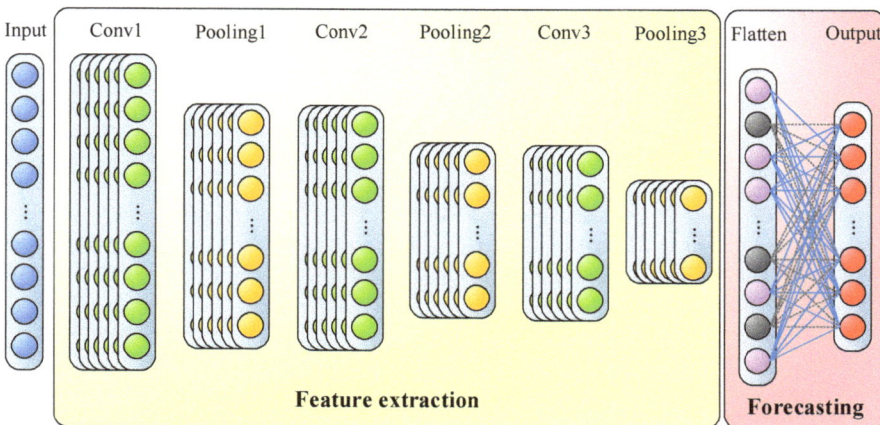

Figure 4. The DeepEnergy structure.

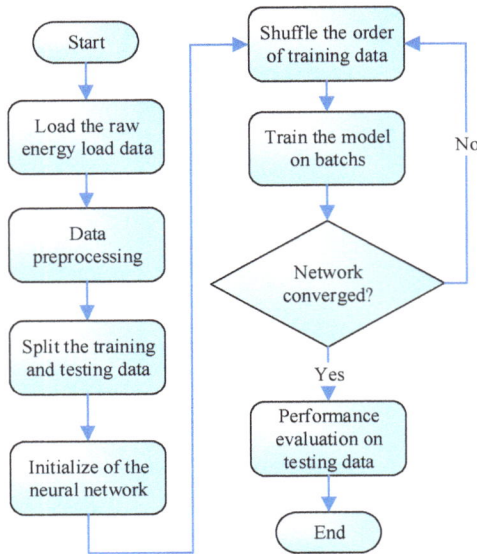

Figure 5. The DeepEnergy flowchart.

4. Experimental Results

In the experiment, the USA District public consumption dataset and electric load dataset from 2016 provided by the Electric Reliability Council of Texas were used. Since then, the support vector machine (SVM) [35] is a popular machine learning technology, in experiment; the radial basis function (RBF) kernels of SVM were chosen to demonstrate the SVM performance. Besides, the random forest (RF) [36], decision tree (DT) [37], MLP, LSTM, and proposed DeepEnergy network were also implemented and tested. The results of load forecasting by all of the methods are shown in Figures 6–11. In the experiment, the training data were two-month data, and test data were one-month data. In order to evaluate the performances of all listed methods, the dataset was divided into 10 partitions. In the first partition, training data consisted of energy load data collected in January and February 2016, and test data consisted of data collected in March 2016. In the second partition, training data were data collected in February and March 2016, and test data were data collected in April 2016. The following partitions can be deduced by the same analogy.

In Figures 6–11, red curves denote the forecasting results of the corresponding models, and blue curves represent the ground truth. The vertical axes represent the energy load (MWh), and the horizontal axes denote the time (hour). The energy load from the past (24 × 7) h was used as an input of the forecasting model, and predicted energy load in the next (24 × 3) h was an output of the forecasting model. After the models received the past (24 × 7) h data, they forecasted the next (24 × 3) h energy load, red curves in Figures 6–11. Besides, the correct information is illustrated by blue curves. The differences between red and blue curves denote the performances of the corresponding models. For the sake of comparison fairness, testing data were not used during the training process of models. According to the results presented in Figures 6–11, the proposed DeepEnergy network has the best prediction performance among all of the models.

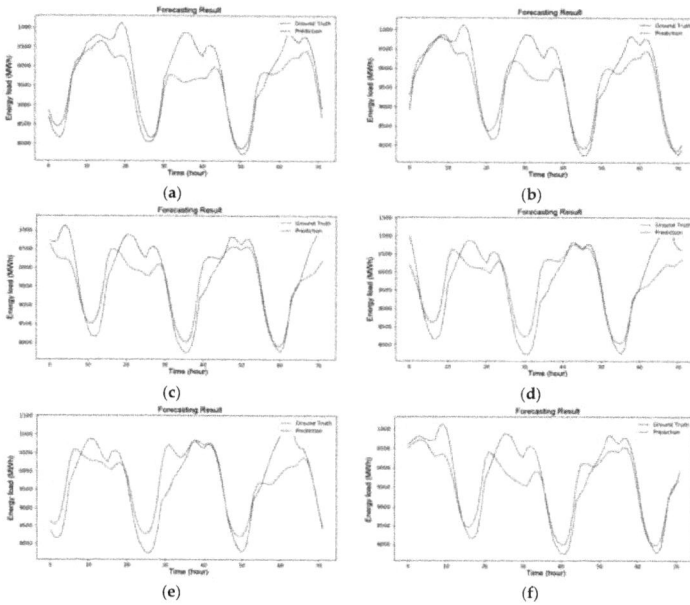

Figure 6. The forecasting results of support vector machine (SVM): (**a**) Partial results A; (**b**) Partial results B; (**c**) Partial results C; (**d**) Partial results D; (**e**) Partial results E; (**f**) Partial results F.

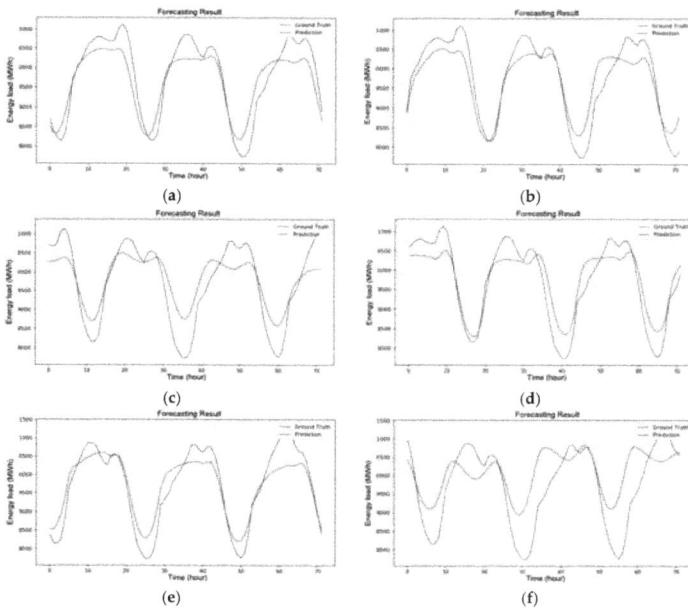

Figure 7. The forecasting results of random forest (RF): (**a**) Partial results A; (**b**) Partial results B; (**c**) Partial results C; (**d**) Partial results D; (**e**) Partial results E; (**f**) Partial results F.

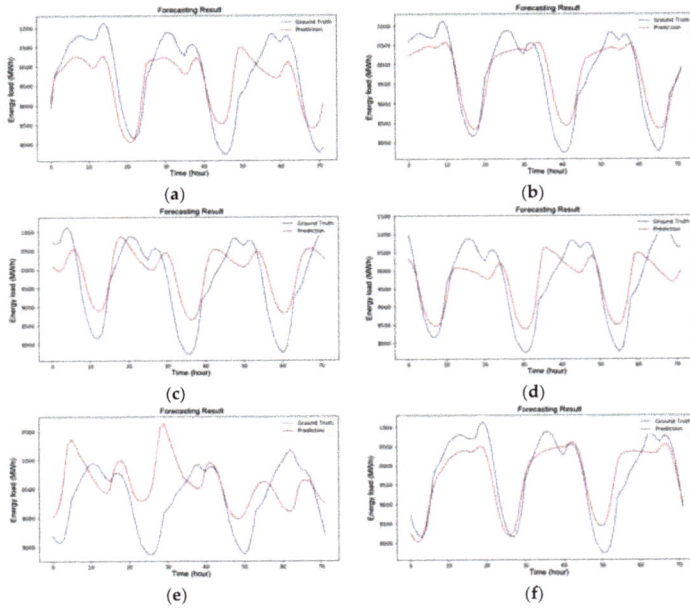

Figure 8. The forecasting results of decision tree (DT): (**a**) Partial results A; (**b**) Partial results B; (**c**) Partial results C; (**d**) Partial results D; (**e**) Partial results E; (**f**) Partial results F.

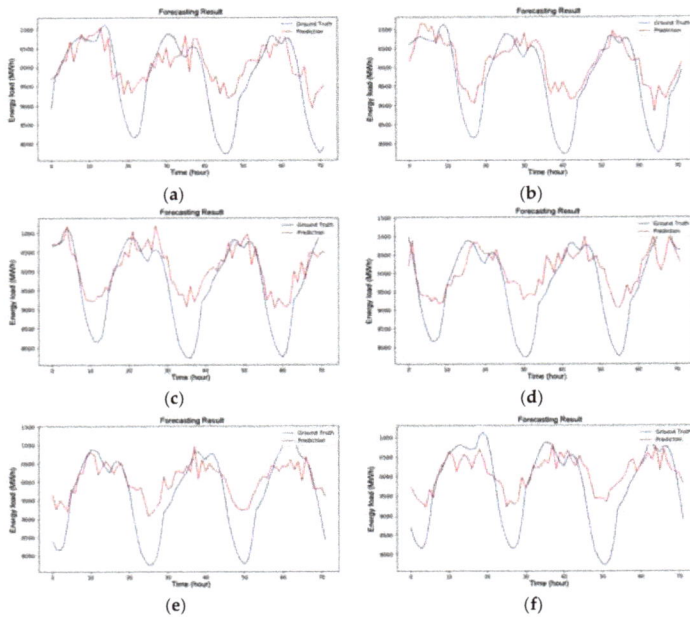

Figure 9. The forecasting results of Multilayer Perceptron (MLP): (**a**) Partial results A; (**b**) Partial results B; (**c**) Partial results C; (**d**) Partial results D; (**e**) Partial results E; (**f**) Partial results F.

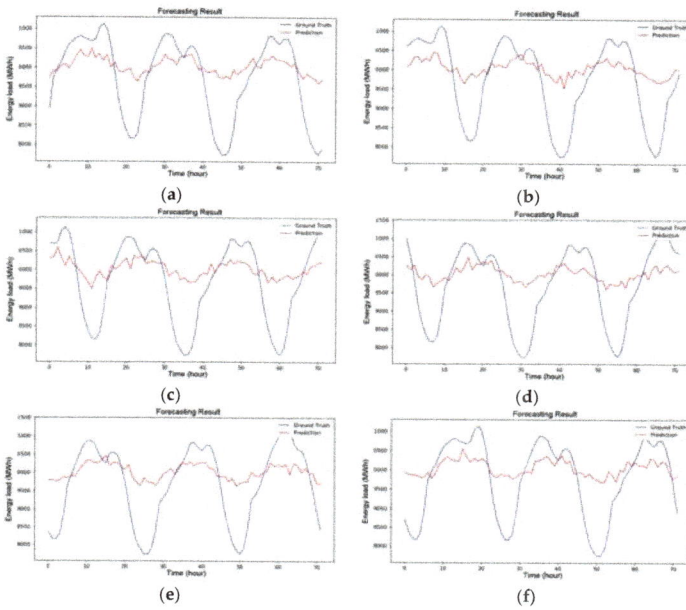

Figure 10. The forecasting results of LSTM: (**a**) Partial results A; (**b**) Partial results B; (**c**) Partial results C; (**d**) Partial results D; (**e**) Partial results E; (**f**) Partial results F.

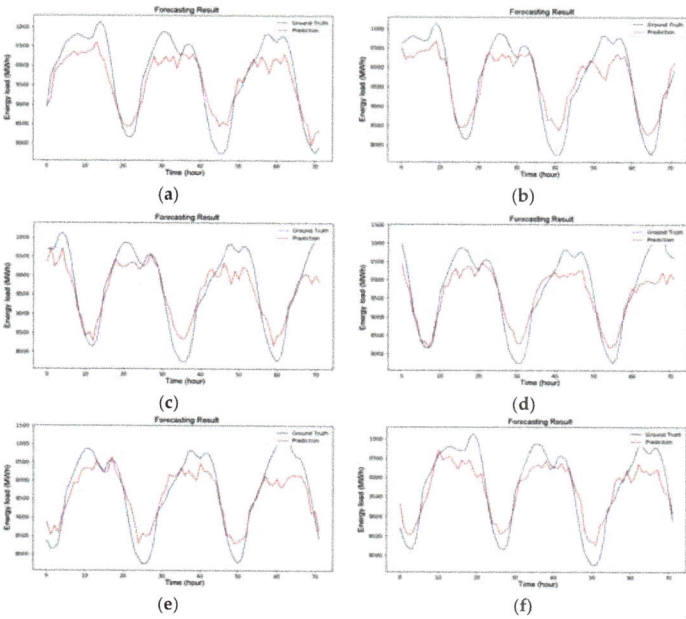

Figure 11. The forecasting results of proposed DeepEnergy: (**a**) Partial results A; (**b**) Partial results B; (**c**) Partial results C; (**d**) Partial results D; (**e**) Partial results E; (**f**) Partial results F.

In order to evaluate the performance of forecasting models more accurately, the Mean Absolute Percentage Error (MAPE) and Cumulative Variation of Root Mean Square Error (CV-RMSE) were employed. The MAPE and CV-RMSE are defined by Equations (7) and (8), respectively, where y_n denotes the measured value, $\hat{y}_n$ is the estimated value, and N represents the sample size.

$$\text{MAPE} = \frac{1}{N} \sum_{n=1}^{N} \left| \frac{y_n - \hat{y}_n}{y_n} \right| \tag{7}$$

$$\text{CV} - \text{RMSE} = \frac{\sqrt{\frac{1}{N} \sum_{n=1}^{N} \left(\frac{y_n - \hat{y}_n}{y_n} \right)^2}}{\frac{1}{N} \sum_{n=1}^{N} y_n} \tag{8}$$

The detailed experimental results are presented numerically in Tables 1 and 2. As shown in Tables 1 and 2, the MAPE and CV-RMSE of the DeepEnergy model are the smallest and the goodness of error is the best among all models, namely, average MAPE and CV-RMSE are 9.77% and 11.65%, respectively. The MAPE of MLP model is the largest among all of the models; an average error is about 15.47%. On the other hand, the CV-RMSE of SVM model is the largest among all models; an average error is about 17.47%. According to the average MAPE and CV-RMSE values, the electric load forecasting accuracy of tested models in descending order is as follows: DeepEnergy, RF, LSTM, DT, SVM, and MLP.

Table 1. The experimental results in terms of Mean Absolute Percentage Error (MAPE) given in percentages.

Test	SVM	RF	DT	MLP	LSTM	DeepEnergy
#1	7.327408	7.639133	8.46043	9.164315	10.40804813	7.226127
#2	7.550818	8.196129	10.23476	11.14954	9.970662683	8.244051
#3	13.07929	10.11102	12.14039	19.99848	14.85568499	11.00656
#4	16.15765	17.27957	19.86511	22.45493	12.83487893	12.17574
#5	5.183255	6.570061	8.50582	15.01856	5.479091542	5.41808
#6	10.33686	9.944028	11.11948	10.94331	11.7681534	9.070998
#7	8.934657	6.698508	8.634132	7.722149	7.583802292	9.275215
#8	18.5432	16.09926	17.17215	16.93843	15.6574951	13.2776
#9	49.97551	17.9049	21.29354	29.06767	16.31443679	11.18214
#10	11.20804	8.221766	10.68665	12.20551	8.390061493	10.80571
Average	14.82967	10.86644	12.81125	15.46629	11.32623153	9.768222

Table 2. The experimental results in terms of Cumulative Variation of Root Mean Square Error (CV-RMSE) given in percentages.

Test	SVM	RF	DT	MLP	LSTM	DeepEnergy
#1	9.058992	9.423908	10.57686	10.65546	12.16246177	8.948922
#2	10.14701	10.63412	12.99834	13.91199	12.19377007	10.46165
#3	17.02552	12.42314	14.58249	23.2753	16.9291218	13.30116
#4	21.22162	21.1038	24.48298	23.63544	14.13596516	14.63439
#5	6.690527	7.942747	10.10017	15.44461	6.334195125	6.653999
#6	11.88856	11.6989	13.39033	12.20149	12.96057349	10.74021
#7	10.77881	7.871596	10.35254	8.716806	8.681353107	10.85454
#8	19.49707	17.09079	18.95726	17.73124	16.55737557	14.51027
#9	54.58171	19.91185	24.84425	29.37466	17.66342548	13.01906
#10	13.80167	10.15117	13.06351	13.39278	10.20235927	13.47003
Average	17.46915	12.8252	15.33487	16.83398	12.78206008	11.65942

It is obvious that red curve in Figure 11, which denotes the DeepEnergy algorithm, is better than other curves in Figures 6–10, which further verifies that the proposed DeepEnergy algorithm has the best prediction performance. Therefore, it is proven that the DeepEnergy STLF algorithm proposed in the paper is practical and effective. Although the LSTM has good performance in time sequence problems, in this study, the reduction of training loss is still not fast enough to handle this forecasting problem because the size of input and output data is too large for the traditional LSTM neural network. Therefore, the traditional LSTM is not suitable for this kind of prediction. Finally, the experimental results show that proposed DeepEnergy network provides the best results in energy load forecasting.

5. Discussion

The traditional machine learning methods, such as SVM, random forest, and decision tree, are widely used in many applications. In this study, these methods also provide acceptable results. In aspect of SVM, the supporting vectors are mapped into a higher dimensional space by the kernel function. Therefore, the selection of kernel function is very important. In order to achieve the goal of nonlinear energy load forecasting, the RBF is chosen as a SVM kernel. When compared with the SVM, the learning concept of decision tree is much simpler. Namely, the decision tree is a flowchart structure easy to understand and interpret. However, only one decision tree does not have the ability to solve complicated problems. Therefore, the random forest, which represents the combination of numerous decision trees, provides the model ensemble solution. In this paper, the experimental results of random forest are better than those of decision tree and SVM, which proves that the model ensemble solution is effective in the energy load forecasting. In aspect of the neural networks, the MLP is the simplest ANN structure. Although the MLP can model the nonlinear energy forecasting task, its performance in this experiment is not outstanding. On the other hand, the LSTM considers data relationships in time steps during the training. According to the result, the LSTM can deal with the time sequence problems, and the forecasting trend is marginally correct. However, the proposed CNN structure, named the DeepEnergy, has the best results in the experiment. The experiments demonstrate that the most important feature can be extracted by the designed 1D convolution and pooling layers. This verification also proves the CNN structure is effective in the forecasting, and the proposed DeepEnergy gives the outstanding results. This paper not only provides the comparison of the traditional machine learning and deep learning methods, but also gives a new research direction in the energy load forecasting.

6. Conclusions

This paper proposes a powerful deep convolutional neural network model (DeepEnergy) for energy load forecasting. The proposed network is validated by experiment with the load data from the past seven days. In the experiment, the data from coast area of the USA were used and historical electricity demand from consumers was considered. According to the experimental results, the DeepEnergy can precisely predict energy load in the next three days. In addition, the proposed algorithm was compared with five AI algorithms that were commonly used in load forecasting. The comparison showed that performance of DeepEnergy was the best among all tested algorithms, namely the DeepEnergy had the lowest values of both MAPE and CV-RMSE. According to all of the obtained results, the proposed method can reduce monitoring expenses, initial cost of hardware components, and long-term maintenance costs in the future smart grids. Simultaneously, the results verify that proposed DeepEnergy STLF method has strong generalization ability and robustness, thus it can achieve very good forecasting performance.

Acknowledgments: This work was supported by the Ministry of Science and Technology, Taiwan, Republic of China, under Grants MOST 106-2218-E-153-001-MY3.

Author Contributions: Ping-Huan Kuo wrote the program and designed the DNN model. Chiou-Jye Huang planned this study and collected the energy load dataset. Ping-Huan Kuo and Chiou-Jye Huang contributed in drafted and revised manuscript.

Energies **2018**, *11*, 213

Conflicts of Interest: The authors declare no conflict of interest.

References

1. Raza, M.Q.; Khosravi, A. A review on artificial intelligence based load demand forecasting techniques for smart grid and buildings. *Renew. Sustain. Energy Rev.* **2015**, *50*, 1352–1372. [CrossRef]
2. Da Graça Carvalho, M.; Bonifacio, M.; Dechamps, P. Building a low carbon society. *Energy* **2011**, *36*, 1842–1847. [CrossRef]
3. Jiang, B.; Sun, Z.; Liu, M. China's energy development strategy under the low-carbon economy. *Energy* **2010**, *35*, 4257–4264. [CrossRef]
4. Cho, H.; Goude, Y.; Brossat, X.; Yao, Q. Modeling and forecasting daily electricity load curves: A hybrid approach. *J. Am. Stat. Assoc.* **2013**, *108*, 7–21. [CrossRef]
5. Javed, F.; Arshad, N.; Wallin, F.; Vassileva, I.; Dahlquist, E. Forecasting for demand response in smart grids: An analysis on use of anthropologic and structural data and short term multiple loads forecasting. *Appl. Energy* **2012**, *96*, 150–160. [CrossRef]
6. Iwafune, Y.; Yagita, Y.; Ikegami, T.; Ogimoto, K. Short-term forecasting of residential building load for distributed energy management. In Proceedings of the 2014 IEEE International Energy Conference, Cavtat, Croatia, 13–16 May 2014; pp. 1197–1204. [CrossRef]
7. Short Term Electricity Load Forecasting on Varying Levels of Aggregation. Available online: https://arxiv.org/abs/1404.0058v3 (accessed on 11 January 2018).
8. Gerwig, C. Short term load forecasting for residential buildings—An extensive literature review. *Smart Innov. Syst.* **2015**, *39*, 181–193.
9. Hippert, H.S.; Pedreira, C.E.; Souza, R.C. Neural networks for short-term load forecasting: A review and evaluation. *IEEE Trans. Power Syst.* **2001**, *16*, 44–55. [CrossRef]
10. Metaxiotis, K.; Kagiannas, A.; Askounis, D.; Psarras, J. Artificial intelligence in short term electric load forecasting: A state-of-the-art survey for the researcher. *Energy Convers. Manag.* **2003**, *44*, 1524–1534. [CrossRef]
11. Tzafestas, S.; Tzafestas, E. Computational intelligence techniques for short-term electric load forecasting. *J. Intell. Robot. Syst.* **2001**, *31*, 7–68. [CrossRef]
12. Ghayekhloo, M.; Menhaj, M.B.; Ghofrani, M. A hybrid short-term load forecasting with a new data preprocessing framework. *Electr. Power Syst. Res.* **2015**, *119*, 138–148. [CrossRef]
13. Xia, C.; Wang, J.; McMenemy, K. Short, medium and long term load forecasting model and virtual load forecaster based on radial basis function neural networks. *Int. J. Electr. Power Energy Syst.* **2010**, *32*, 743–750. [CrossRef]
14. Bughin, J.; Hazan, E.; Ramaswamy, S.; Chui, M. *Artificial Intelligence—The Next Digital Frontier?* Mckinsey Global Institute: New York, NY, USA, 2017; pp. 1–80.
15. Oh, C.; Lee, T.; Kim, Y.; Park, S.; Kwon, S.B.; Suh, B. Us vs. Them: Understanding Artificial Intelligence Technophobia over the Google DeepMind Challenge Match. In Proceedings of the 2017 CHI Conference on Human Factors in Computing Systems, Denver, CO, USA, 6–11 May 2017; pp. 2523–2534. [CrossRef]
16. Skilton, M.; Hovsepian, F. Example Case Studies of Impact of Artificial Intelligence on Jobs and Productivity. In *4th Industrial Revolution*; National Academies Press: Washingtom, DC, USA, 2018; pp. 269–291.
17. Ekonomou, L.; Christodoulou, C.A.; Mladenov, V. A short-term load forecasting method using artificial neural networks and wavelet analysis. *Int. J. Power Syst.* **2016**, *1*, 64–68.
18. Valgaev, O.; Kupzog, F. Low-Voltage Power Demand Forecasting Using K-Nearest Neighbors Approach. In Proceedings of the Innovative Smart Grid Technologies—Asia (ISGT-Asia), Melbourne, VIC, Australia, 28 November–1 December 2016.
19. Valgaev, O.; Kupzog, F. Building Power Demand Forecasting Using K-Nearest Neighbors Model—Initial Approach. In Proceedings of the IEEE PES Asia-Pacific Power Energy Conference, Xi'an, China, 25–28 October 2016; pp. 1055–1060.
20. Humeau, S.; Wijaya, T.K.; Vasirani, M.; Aberer, K. Electricity load forecasting for residential customers: Exploiting aggregation and correlation between households. In Proceedings of the 2013 Sustainable Internet and ICT for Sustainability, Palermo, Italy, 30–31 October 2013.

21. Veit, A.; Goebel, C.; Tidke, R.; Doblander, C.; Jacobsen, H. Household electricity demand forecasting: Benchmarking state-of-the-art method. In Proceedings of the 5th International Confrerence Future Energy Systems, Cambridge, UK, 11–13 June 2014; pp. 233–234. [CrossRef]

22. Jetcheva, J.G.; Majidpour, M.; Chen, W. Neural network model ensembles for building-level electricity load forecasts. *Energy Build.* **2014**, *84*, 214–223. [CrossRef]

23. Kardakos, E.G.; Alexiadis, M.C.; Vagropoulos, S.I.; Simoglou, C.K.; Biskas, P.N.; Bakirtzis, A.G. Application of time series and artificial neural network models in short-term forecasting of PV power generation. In Proceedings of the 2013 48th International Universities' Power Engineering Conference, Dublin, Ireland, 2–5 September 2013; pp. 1–6. [CrossRef]

24. Fujimoto, Y.; Hayashi, Y. Pattern sequence-based energy demand forecast using photovoltaic energy records. In Proceedings of the 2012 International Conference on Renewable Energy Research and Applications, Nagasaki, Japan, 11–14 November 2012.

25. Chaouch, M. Clustering-based improvement of nonparametric functional time series forecasting: Application to intra-day household-level load curves. *IEEE Trans. Smart Grid* **2014**, *5*, 411–419. [CrossRef]

26. Niu, D.; Dai, S. A short-term load forecasting model with a modified particle swarm optimization algorithm and least squares support vector machine based on the denoising method of empirical mode decomposition and grey relational analysis. *Energies* **2017**, *10*, 408. [CrossRef]

27. Deo, R.C.; Wen, X.; Qi, F. A wavelet-coupled support vector machine model for forecasting global incident solar radiation using limited meteorological dataset. *Appl. Energy* **2016**, *168*, 568–593. [CrossRef]

28. Soubdhan, T.; Ndong, J.; Ould-Baba, H.; Do, M.T. A robust forecasting framework based on the Kalman filtering approach with a twofold parameter tuning procedure: Application to solar and photovoltaic prediction. *Sol. Energy* **2016**, *131*, 246–259. [CrossRef]

29. Hahn, H.; Meyer-Nieberg, S.; Pickl, S. Electric load forecasting methods: Tools for decision making. *Eur. J. Oper. Res.* **2009**, *199*, 902–907. [CrossRef]

30. Zhang, G.; Patuwo, B.E.; Hu, M.Y. Forecasting with artificial neural networks: The state of the art. *Int. J. Forecast.* **1998**, *14*, 35–62. [CrossRef]

31. Pappas, S.S.; Ekonomou, L.; Moussas, V.C.; Karampelas, P.; Katsikas, S.K. Adaptive load forecasting of the Hellenic electric grid. *J. Zhejiang Univ. A* **2008**, *9*, 1724–1730. [CrossRef]

32. White, B.W.; Rosenblatt, F. Principles of neurodynamics: Perceptrons and the theory of brain mechanisms. *Am. J. Psychol.* **1963**, *76*, 705. [CrossRef]

33. Hochreiter, S.; Urgen Schmidhuber, J. Long short-term memory. *Neural Comput.* **1997**, *9*, 1735–1780. [CrossRef] [PubMed]

34. Srivastava, N.; Hinton, G.; Krizhevsky, A.; Sutskever, I.; Salakhutdinov, R. Dropout: A Simple Way to Prevent Neural Networks from Overfitting. *J. Mach. Learn. Res.* **2014**, *15*, 1929–1958. [CrossRef]

35. Suykens, J.A.K.; Vandewalle, J. Least squares support vector machine classifiers. *Neural Process. Lett.* **1999**, *9*, 293–300. [CrossRef]

36. Liaw, A.; Wiener, M. Classification and Regression by randomForest. *R News* **2002**, *2*, 18–22. [CrossRef]

37. Safavian, S.R.; Landgrebe, D. A Survey of Decision Tree Classifier Methodology. *IEEE Trans. Syst. Man Cybern.* **1991**, *21*, 660–674. [CrossRef]

MDPI
St. Alban-Anlage 66
4052 Basel
Switzerland
Tel. +41 61 683 77 34
Fax +41 61 302 89 18
www.mdpi.com

Energies Editorial Office
E-mail: energies@mdpi.com
www.mdpi.com/journal/energies

www.ingramcontent.com/pod-product-compliance
Lightning Source LLC
Chambersburg PA
CBHW051704210326
41597CB00032B/5365